NATO ASI Series

Advanced Science Institutes Series

A series presenting the results of activities sponsored by the NATO Science Committee, which aims at the dissemination of advanced scientific and technological knowledge, with a view to strengthening links between scientific communities.

The Series is published by an international board of publishers in conjunction with the NATO Scientific Affairs Division

A Life Sciences B Physics	Plenum Publishing Corporation London and New York
C Mathematical and Physical Sciences D Behavioural and Social Sciences E Applied Sciences	Kluwer Academic Publishers Dordrecht, Boston and London
F Computer and Systems Sciences G Ecological Sciences H Cell Biology I Global Environmental Change	Springer-Verlag Berlin Heidelberg New York London Paris Tokyo Hong Kong Barcelona Budapest

NATO-PCO DATABASE

The electronic index to the NATO ASI Series provides full bibliographical references (with keywords and/or abstracts) to more than 30000 contributions from international scientists published in all sections of the NATO ASI Series. Access to the NATO-PCO DATABASE compiled by the NATO Publication Coordination Office is possible in two ways:

- via online FILE 128 (NATO-PCO DATABASE) hosted by ESRIN, Via Galileo Galilei, I-00044 Frascati, Italy.
- via CD-ROM „NATO-PCO DATABASE“ with user-friendly retrieval software in English, French and German (© WTV GmbH and DATAWARE Technologies Inc. 1989).

The CD-ROM can be ordered through any member of the Board of Publishers or through NATO-PCO, Overijse, Belgium.

Series I: Global Environmental Change, Vol. 5

The ASI Series Books Published as a Result of Activities of the Special Programme on Global Environmental Change

This book contains the proceedings of a NATO Advanced Research Workshop/ Advanced Study Institute held within the activities of the NATO Special Programme on Global Environmental Change, which started in 1991 under the auspices of the NATO Science Committee.

The volumes published as a result of the activities of the Special Programme are:

Vol. 1: **Global Environmental Change.** Edited by R. W. Corell and P. A. Anderson. 1991.
Vol. 2: **The Last Deglaciation: Absolute and Radiocarbon Chronologies.** Edited by E. Bard and W. S. Broecker. 1992.
Vol. 3: **Start of a Glacial.** Edited by G. J. Kukla and E. Went. 1992.
Vol. 4: **Interactions of C, N, P and S Biogeochemical Cycles and Global Change.** Edited by R. Wollast, F. T. Mackenzie and L. Chou. 1993.
Vol. 5: **Energy and Water Cycles in the Climate System.** Edited by E. Raschke and D. Jacob. 1993.

Energy and Water Cycles in the Climate System

Edited by

Ehrhard Raschke
Daniela Jacob
GKSS
Forschungszentrum Geesthacht GmbH
Max-Planck-Straße
2054 Geesthacht, Germany

Springer-Verlag
Berlin Heidelberg New York London Paris Tokyo
Hong Kong Barcelona Budapest
Published in cooperation with NATO Scientific Affairs Division

Proceedings of the NATO Advanced Study Institute on Energy and Water Cycles in the Climate System held at Glücksburg, Germany, September 30–October 11, 1991

Library of Congress Cataloging-in-Publication Data
Energy and water cycles in the climate system / edited by Ehrhard Raschke, Daniela Jacob.
(NATO ASI series. Series I, Global environmental change ; vol. 5)
"Proceedings of the NATO Advanced Study Institute on Energy and Water Cycles in the Climate System held at Glücksburg, Germany, September 30-October 11, 1991" – T.p. verso. Includes index.

ISBN-13: 978-3-642-76959-7 e-ISBN-13: 978-3-642-76957-3
DOI: 10.1007/978-3-642-76957-3

1. Dynamic meteorology–Congresses. 2. Atmospheric physics–Congresses. 3. Cloud physics–Congresses. 4. Atmospheric radiation–Congresses. 5. Hydrological cycle–Congresses. I. Raschke, Ehrhard. II. Jacob, Daniela, 1961- . III. NATO Advanced Study Institute on Energy and Water Cycles in the Climate System (1991: Glücksburg, Schleswig-Holstein, Germany) IV. Series. QC880.E46 1992 551.5–dc20

Softcover reprint of the hardcover 1st edition 1993

Typesetting: Camera ready by authors
31/3145 - 5 4 3 2 1 0 – Printed on acid-free paper

Contents

FOREWORD

Water is the most effective agent in the climate system to modulate energy transfer by radiative processes, through its exchanges of latent heat and within cascades of chemical processes. It is the source of all life on earth, and once convective clouds are formed, it enables large vertical transports of momentum, heat and various atmospheric constituents up to levels above the tropical tropopause. Water triggers very complex processes at the earth's continental surfaces and within the oceans. At last, water in its gaseous phase is the most important greenhouse-gas!

Numerical modelling and measurements of the state of the present climate system needs a very thorough understanding of all these processes and their various interactions and forcings. This is a prerequisite for more substantial forecasts of future states in all scales of time, from days to centuries. Therefore, the management of the World Climate Research Programme established in 1988 the new programme GEWEX (Global Energy and Water Cycle Experiment). GEWEX is specifically defined to determine the energy and water transports in the fast components of the climate system with the presently available modelling and measurement means and to provide new capabilities for the future. Research in GEWEX must further develop methods to determine the influence of climatic anomalies on available water resources.

GEWEX needs interdisciplinary cooperation between hydrologists, atmospheric and oceanic scientists; it needs new developments in numerical modelling, data handling and acquisitions - and also in measurement capabilities, both space-borne and from the ground. Moreover, it needs also young scientists; well-trained for this cooperative work and full of new and fresh ideas.

Within the International Geosphere and Biosphere Programme (IGBP) a subprogramme has been established, the BAHC, to study the fundamental Biological Aspects of the Hydrological Cycle. This Project, unlike GEWEX, favours the interactions between water transports and the terrestrial vegetation systems, but it needs accurate informations on energy and water fluxes. Therefore the projects GEWEX and BAHC complement each other.

The NATO - Advanced Study Institute, held from 30 September to 11 October 1991 in Glücksburg, in the most northern corner of Germany, has been designed entirely to introduce a new generation of scientists into this challenging programme. NATO's support and the co-sponsorships by the WCRP directorate in Geneva and by the GKSS-Research Centre in Geesthacht, Germany (the employer of both editors), and various other sponsors enabled us to meet 15 lecturers of all involved disciplines with about 80 students from 14 countries. The lecturers did their best to transfer their knowledge to all attendees. Many of the students added to this wealth of wisdom new information on their own research. Their posters could be discussed during the entire session and further explained by their authors. This was a most encouraging entertainment.

A little bit of culture (concerts, sightseeing, and just the disco) with German and Danish flavours added to the success of this event. At the end, two poor editors even succeeded in convincing all lecturers to send in their papers within reasonable time limits.

We left the contents of all papers to the author's responsibilities. Here and there we asked for some changes or supplements; we found a few typos and small uncertainties in the figures. A small index has been prepared for the reader's convenience. It may not be complete.

"Our book" is not perfect. But its contents and references should serve all those interested scientists who want to find their way in GEWEX and related disciplines, and also in the complementary research within the framework of IGBP.

Our efforts have been made possible with the enormous amount of work, which all authors invested into their lectures and articles. Finally Manuela Boin and Luis Kornblueh translated all manuscripts from various strange formats into the unique shape of this book (formatted by LaTeX- thanks to D. Knuth and L. Lamport). Heidi Schäfer, Andreas Macke and Markus Quante helped us in various earlier steps before and during the event in Glücksburg.

Ehrhard Raschke Daniela Jacob

Participants of the NATO-ASI **Energy and Water Cycles in the Climate System** *in Glücksburg, Germany.*

Chapter 1

ATMOSPHERIC ENERGETICS AND THE WATER CYCLE

José Pinto Peixoto
Instituto Geofisico do Infante d'Luis
Universidade de Lisboa
P-1200 Lisboa
Portugal

1.1 The climate system

1.1.1 Introduction

In order to understand the mechanisms and the physical processes responsible for the climate it is first necessary to have a clear picture of the characteristic features of the structure and behaviour of the climate.

Since the atmosphere is a thermo-hydro-dynamical system it can be characterized by its composition, its thermodynamical state as specified by the thermodynamic variables and its mechanical state (motions). A complete description of the state of the atmosphere should also include other variables such as cloudiness, precipitation, heating distribution, etc. which affect the large-scale behavior of the atmosphere. Traditionally the most important elements of the climate are temperature and precipitation. It is on the basis of these elements that the climates are usually classified. Their geographical distribution shows warm and moist climates in the low latitudes, warm and much drier climates in the subtropics, temperate and moist climates in mid to high latitudes and, finally, cold and dry climates in the polar and subpolar regions.

NATO ASI Series, Vol. I 5
Energy and Water Cycles in the Climate System
Edited by E. Raschke and D. Jacob

1.1.2 The structure of the climate system

The climate system ($\mathcal{S}$) is a composite system consisting of five major interactive adjoint components: the atmosphere ($\mathcal{A}$), the hydrosphere ($\mathcal{H}$) with the oceans ($\mathcal{O}$), the cryosphere ($\mathcal{C}$), the lithosphere ($\mathcal{L}$) and the biosphere ($\mathcal{B}$), i.e.

$$\mathcal{S} \equiv \mathcal{A} \cup \mathcal{H} \cup \mathcal{C} \cup \mathcal{L} \cup \mathcal{B}\,.$$

As shown, schematically in Fig. 1.1 all the subsystems are open and non-isolated. The global climate

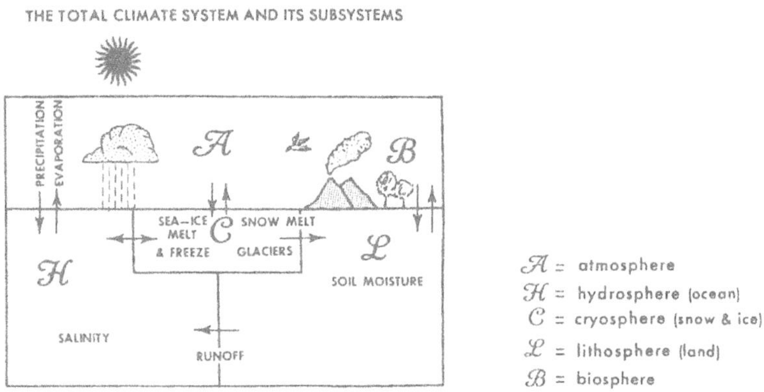

Figure 1.1: *Schematic diagram of the total climate system and its subsystems, highlighting some aspects of the hydrological cycle (from Peixoto and Oort, 1984)*

system $\mathcal{S}$, as a whole, is assumed to be a non-isolated system for the energy but a closed system for the exchange of matter with outer space. The atmosphere, hydrosphere, cryosphere and biosphere act as a cascading system linked by complex physical processes involving fluxes of energy, momentum and matter across the boundaries and generating numerous feedback mechanisms.

The components of the climatic system are heterogeneous thermo-hydro-dynamical systems, which can be characterized by their chemical composition and their thermodynamic and mechanical state. The thermodynamic state are specified, in general, by intensive variables (temperature, pressure, specific humidity, specific energy, density, salinity, etc.) whereas the mechanical state is defined by variables that characterize the motions (force, velocities, etc.).

Due to the complexity of the internal climatic systems and on the basis of the different response times, it is convenient to consider a hierarchy of internal systems taking first the systems with the shortest response times, so that all other components are considered to be part of the external system. For example, for time scales of days to weeks the atmosphere can be regarded as the sole internal component of the climatic system ($\mathcal{S} \equiv \mathcal{A}$) with the oceans, ice masses and land surfaces treated as the external forcings or boundary conditions. For time scales of months to years, the climatic internal system must include the atmosphere and the oceans ($\mathcal{S} \equiv \mathcal{A} \cup \mathcal{O}$) and, at times, snow cover and sea ice. For the study of the variability of climate during the past ages the entire cryosphere must also be included in the internal system ($\mathcal{S} \equiv \mathcal{A} \cup \mathcal{O} \cup \mathcal{C}$) considering the influences of $\mathcal{L}$ and $\mathcal{B}$ as external forcings.

1.1.3 The climate state

Even under steady external forcing the internal system is always subject to random oscillations in time and space. We may then consider a large ensemble of the states corresponding to the same external forcing. Thus the climate may be defined in terms of the ensemble of internal states and in terms of a probability distribution. For fixed external forcings we assume the uniqueness of the limiting set of statistics, and we will accept the ergodic hypothesis, so that we can replace the ensemble averages by time averages. Then we can define a climate state as a set of averages over the ensemble completed with higher moment statistics, such as variances, correlations, etc., together with a description of the state of the external system.

For a different set of the complete external conditions we may obtain a different climate state of the internal system so we can define a climate change as the difference between two climate states of the same kind, such as the difference between the climate states of two typical August months, two typical decades, etc.. This difference should include the differences between the averages and moments of higher order. A climate anomaly will then be defined as a departure of a particular climate state for a given interval of time from the ensemble of the equivalent states.

In climate studies it is necessary to consider not only the internal effects but also the complex interactions between the atmosphere and its external system. For weather prediction the atmosphere behaves almost inertially, so that slowly acting boundary conditions can be ignored.

Starting with a given initial state the solutions of the equations that govern the dynamics of a nonlinear system, such as the atmosphere, result for an infinite interval of time in a set of infinitely long-term statistics. For different initial conditions the limiting solutions may or may not be unique. If in the integration of the equations all initial states lead to the same set of statistical properties, the system is *ergodic or transitive.* If instead there are two or more different sets of statistical properties, where some initial states lead to another, the system is called *intransitive.* If there are different sets of statistics which a transitive system may assume in its evolution from different initial states through a long, but finite, period of time, the system is called *almost intransitive* (Lorenz, 1969; Saltzmann, 1978, 1983). In the *transitive case* the equilibrium climate statistics are both stable and unique, whereas in the *almost intransitive case* the system in the course of its evolution will show finite periods during which distinctly different climatic regimes prevail. This may be due to internal feedback or instabilities involving the different components of the climatic system. The glacial and interglacial periods in the earth's history may be manifestations of an *almost intransitive system.*

1.1.4 Climate variability

The terrestrial climate has varied significantly and continuously on time scales ranging from years to glacial periods and to the age of the earth.

Essentially the variability of climate can be expressed in terms of two basic modes. The *forced* variations which are the response of the climatic system to changes in the *external forcing* and the *free variations* due to internal instabilities and feedbacks, leading to nonlinear interactions among the various components of the climatic system.

The changes in the purely external factors that affect the climatic system, but are not influences by the climatic variables, constitute what may be called the *external causes* of climatic changes, whereas those changes that are related to nonlinear interactions among the various physical processes in the internal system are called *internal causes.* The distinction between the two classes of causes is not always very clear.

The *external causes* comprise variations in both *astronomical and terrestrial forcings.* The astronomical factors would include changes

(a) in the intensity of solar irradiance;

(b) in the orbital parameters of the earth (eccentricity of the orbit, axial precession and obliquity of the ecliptic: the Milankovitch parameters);

(c) in the rate of rotation of the earth.

In the terrestrial forcing we must consider:

(a) variations in atmospheric composition (mixing ratios of carbon dioxide and ozone, aerosol loading, etc.) due to volcanic eruptions and human activity;

(b) variations of the land surface due to man's land use (deforestation, desertification, etc.);

(c) long-term changes of tectonic factors such as continental drift, mountain building processes, polar wandering, etc.

Some other possible terrestrial and astronomical forcing mechanisms have been suggested, such as changes in solar output, the collision of the earth with interplanetary matter, changes in volcanic activity and changes in the geothermal flux.

The *internal causes* are associated with many positive and negative feedback mechanisms and other strong interactions between the atmosphere, oceans and cryosphere. These processes can lead to *instabilities or oscillations* of the system which can either operate independently or introduce strong modifications on the external forcings.

1.2 The water cycle

1.2.1 The nature of problem

The global hydrosphere is formed by various reservoirs interconnected by the transfers of water in any of its three phases. In decreasing order of water amount held in storage, the five reservoirs of the hydrosphere are: the world oceans, the ice masses and snow deposits, the terrestrial waters, the atmosphere and, finally, the biosphere.

Vast quantities of water are continuously on the move in the climatic system. Under the direct or indirect influence of solar energy, water evaporates from the oceans and the continents and is transpired by plants and animals into the atmosphere. Within the atmosphere the ascending moist air develops, by cooling, into clouds (condensation). Thus, in the atmosphere, water is stored and transported in the condensed phase (liquid water and ice crystals) as clouds, or in the vapor phase (water vapor). Under the action of gravity it falls on the continents and oceans in the form of rain, snow, hail, or other forms of precipitation. The precipitated water then returns partly to the atmosphere, through evaporation and evapotranspiration, infiltrates into the ground, or runs off over or under the ground to the rivers and streams, which carry it back to the oceans and seas and becomes ready to start a new journey.

This gigantic and complex system of destillation, pumping and transport of water substance in its many forms originates an unending circulation that constitutes the hydrological cycle .

The hydrological cycle has two major branches - the terrestrial and the atmospheric branches. The terrestrial branch consists of the inflow, outflow and storage of water in its various forms on and in the continents and in the oceans, while the atmospheric branch consists of the atmospheric transports of water, mainly in the vapor phase.

The loss or 'output' of water from the earth's surface, through evaporation and evatranspiration, is the 'input' of water for the atmospheric branch, whereas precipitation, the atmospheric 'output', may be regarded as a gain for the terrestrial branch of the hydrological cycle. It is clear that the atmospheric and terrestrial branches of the hydrological cycle have to be taken as a whole, bringing together the sister disciplines of meteorology and hydrology.

As we know (see 1.2.5) the precipitation and evaporation over the globe are unevenly distributed. However, on the average, there is an excess of precipitation over evaporation in the equatorial region, associated with the intertropical convergence zone (ITCZ) , and a similar excess in midlatitudes associated with the perturbations along the polar front, whereas in the subtropical and polar regions evaporation exceeds precipitation. Thus the water vapor released, mainly over the subtropical oceans, is continuously transported both equatorward and poleward to maintain the moisture supply for the observed precipitation belts. This shows how fundamental and relevant is the role played by the atmosphere and its general circulation as a forcing factor in maintaining the hydrological cycle.

Water can exist in the climatic system in various phases, and the amounts of thermodynamic energy involved in the phase transitions (latent heats) are very large. For example, condensation and freezing release considerable amounts of heat, while evaporation and melting absorb equivalent amounts of heat. Evaporation occurs mainly at the surface of the oceans and continents leading to a transfer of latent heat into the atmosphere, that is released as sensible heat when the water vapor condenses.

The water on the continents is distributed in several reservoirs, namely in glaciers ($25 \cdot 10^{15} m^3$), ground water ($8.4 \cdot 10^{15} m^3$), lakes and rivers ($0.2 \cdot 10^{15} m^3$), and in the living matter of the biosphere ($0.0006 \cdot 10^{15} m^3$). The amount of water locked in the polar ice is impressively large, totaling some

1.8% of all the water in the hydrosphere. Of the total amount of underground water, vadose water (water present in soils) accounts for only $0.066 \cdot 10^{15} m^3$. The remainder is about evenly divided between reservoirs deeper than 800 m and reservoirs shallower than that level (Peixoto and Kettani, 1973). The residence time of water in the various reservoirs can be deduced from the ratio of the

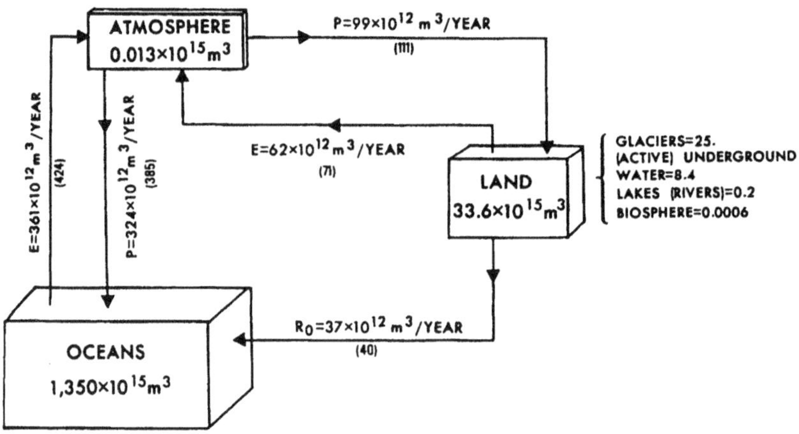

Figure 1.2: *The amounts of water stored in the oceans, land and atmosphere and the amounts exchanged annually between the different reservoirs through evaporation, precipitation and runoff (estimates are from Peixoto and Kettani, 1973, and, in parentheses, from Baumgartner and Reichel, 1975)*

amount of water in the particular reservoir and the accumulation or depletion rate as presented in Fig. 1.2. It is found to vary from about nine days for atmospheric water vapor to thousands of years for the polar ice and the oceans (Peixoto and Oort, 1983).

1.2.2 Balance requirements

Classic equation of hydrology

The water balance requirements for the terrestrial branch of the hydrological cycle leads to the so-called classic equation of hydrology. Applying the principle of continuity to a specific region, the balance equation for the terrestrial branch may be written as

$$S = P - (E + R_0 + R_U)\,, \tag{1.1}$$

where S is the rate of storage of water, P the precipitation rate (in liquid and solid phase), E the evaporation rate (which includes evapotranspiration over land and sublimation over snow and ice); R_0 is surface runoff and R_U the subterranean runoff. For a large land region, the net subterranean runoff is usually small so that the classic equation can be simplified to the form

$$\{\overline{S}\} = \{\overline{P - E}\} - \{\overline{R_0}\}\,, \tag{1.2}$$

where $(\overline{\ })$ denotes a time average and $\{\ \}$ a space average over the region of area A, $\{\overline{S}\}$ is the rate of change of total surface and subterranean water storage, $\{\overline{P-E}\}$ is the average rate of precipitation minus evaporation per unit area, and $\{\overline{R_0}\}$ is the average rate of runoff. For long periods of time and large areas $\{\overline{S}\}$ tends to be small compared to the other terms, and the equation can be reduced to

$$\{\overline{E}\} = \{\overline{P}\} - \{\overline{R_0}\}\,. \tag{1.3}$$

The quantity of major interest for the classical hydrology has been runoff which can be measured fairly accurately through gauging the extensive networks of streams. Precipitation has been measured, very extensively, since it is the primary cause for runoff.

Measurements of evaporation and evapotranspiration and of the change in water storage are very difficult to make, although some semi-empirical estimates of these quantities are available for local areas.

Balance equation for water vapor

The study of atmospheric branch of the hydrological cycle is based on the balance requirements of water vapor in the atmosphere. For a general discussion see Peixoto (1973) and Peixoto and Oort (1983).

The amount of water vapor, W, contained in a unit area column of air which extends from the earth's surface to the top of the atmosphere is given by the expression

$$W = \int_0^{p_0} q\,\frac{dp}{g}\,, \tag{1.4}$$

where q is the specific humidity, g the acceleration of gravity and p the pressure.

It is called the precipitable water in the atmosphere. It represents the amount of liquid water that would result if all the water vapor in the unit column of the atmosphere was condensed. It is usually expressed in units of $kg\,m^{-2}$ or mm.

The total horizontal transport vector, integrated along the vertical, is the "aerial runoff" $\vec{Q}$ defined as:

$$\vec{Q} = \int_0^{p_0} q\vec{v}\,\frac{dp}{g} = Q_\lambda\vec{\imath} + Q_\phi\vec{\jmath}\,, \tag{1.5}$$

where $\vec{v}$ is the wind vector, Q_λ the zonal component (parallel to the latitude circles), Q_ϕ the meridional component and $\vec{\imath}$ and $\vec{\jmath}$ the unit vectors tangent to a parallel and to a meridian, respectively.

The main sources and sinks, $s(q)$, of water vapor in the atmosphere are primarily evaporation and condensation and, to a lesser extent (except near the earth's surface), diffusion from the surroundings. Thus, $s(q) = e - c$, where e is the rate of evaporation (plus sublimation) and c the rate of condensation per unit mass (in units of grams of water per kg moist air per s). For the total atmosphere, integrating along the vertical, the previous difference becomes $E - P$ at the earth's surface, since the evaporated water within the atmosphere is very small.

From the balance requirements it is easy to show that the excess of evaporation over precipitation $(E-P)$, at the earth's surface, is balanced by the local rate of change of water vapor storage $\frac{\partial W}{\partial t}$ and by the net inflow, or the net outflow, of water vapor, given by the divergence of $\vec{Q}$, $\nabla \vec{Q}$ (Peixoto, 1973). Thus, we obtain a simplified general balance equation for the water vapor in the atmosphere:

$$\frac{\partial W}{\partial t} + \nabla \vec{Q} = E - P. \tag{1.6}$$

Averaging in time and in space a region bounded by a conceptual vertical wall (for example, a river-drainage based or an interior sea) the previous equation can be written in the form:

$$\left\{\frac{\partial \overline{W}}{\partial t}\right\} + \left\{\nabla \vec{\overline{Q}}\right\} = \left\{\overline{E} - \overline{P}\right\}. \tag{1.7}$$

Using the Gauss theorem, this equation may be written in a form

$$\left\{\frac{\partial \overline{W}}{\partial t}\right\} + \frac{1}{A}\oint \left(\vec{Q} \cdot \vec{n}\right) dl = \left\{\overline{E} - \overline{P}\right\}, \tag{1.8}$$

where $\vec{n}$ denotes the outward normal unit vector to the boundary of the region with an area A.

For short intervals of time, the rate of change of precipitable water, $\frac{\partial W}{\partial t}$, is very small compared with the other terms. For sufficiently long periods of time, this term can be disregarded, so that $\left\{\nabla \vec{\overline{Q}}\right\} = \left\{\overline{E} - \overline{P}\right\}$. Thus the divergence of water vapor is found over those regions of the globe where evaporation exceeds precipitation, whereas convergence is found where precipitation is greater than evaporation.

The term $\left\{\overline{E} - \overline{P}\right\}$ establishes the connection between the terrestrial and the atmospheric branches of the hydrological cycle. Elimination of $\overline{E} - \overline{P}$ between the two equations yields

$$\left\{\overline{R_0}\right\} + \left\{\overline{S}\right\} = -\left\{\nabla \vec{\overline{Q}}\right\} - \left\{\frac{\partial \overline{W}}{\partial t}\right\}. \tag{1.9}$$

This equation which combines both branches of the hydrological cycle can be regarded as a general equation of hydrology. If, besides the aerological terms, $\left\{\overline{R_0}\right\}$ and $\left\{\overline{P}\right\}$ are known over a certain region, one can estimate the rate of changes in ground water and the rate of evaporation.

Over long periods of time, such as a year, changes in storage in the land and in the atmosphere become small so that, for example, for a continent the surface and subsurface runoff are exactly balanced by the aerial "runoff" into that continent from the surrounding ocean areas. Furthermore, when the entire global atmosphere is considered over a long period of time, all transport and storage terms vanish, and we can conclude that the global-mean evaporation has to be equal to the global-mean precipitation, as we will see later (see section 1.2.5).

1.2.3 Atmospheric branch of the hydrological cycle

Data

The basic data necessary to evaluate the various fields, W, $\vec{Q}$ and $\nabla \vec{Q}$ are the daily values of specific humidity and wind components at various levels. This was done using the rawinsonde data for the ten year period, May 1963 through April 1973 by the Geophysical Fluid Dynamics Laboratory/NOAA, Princeton University. The network of rawinsonde stations covered the entire globe from pole to pole. In addition to the mean annual fields for the entire 10-year period, statistics of higher orders were also evaluated (Oort, 1983).

Water vapor storage in the atmosphere

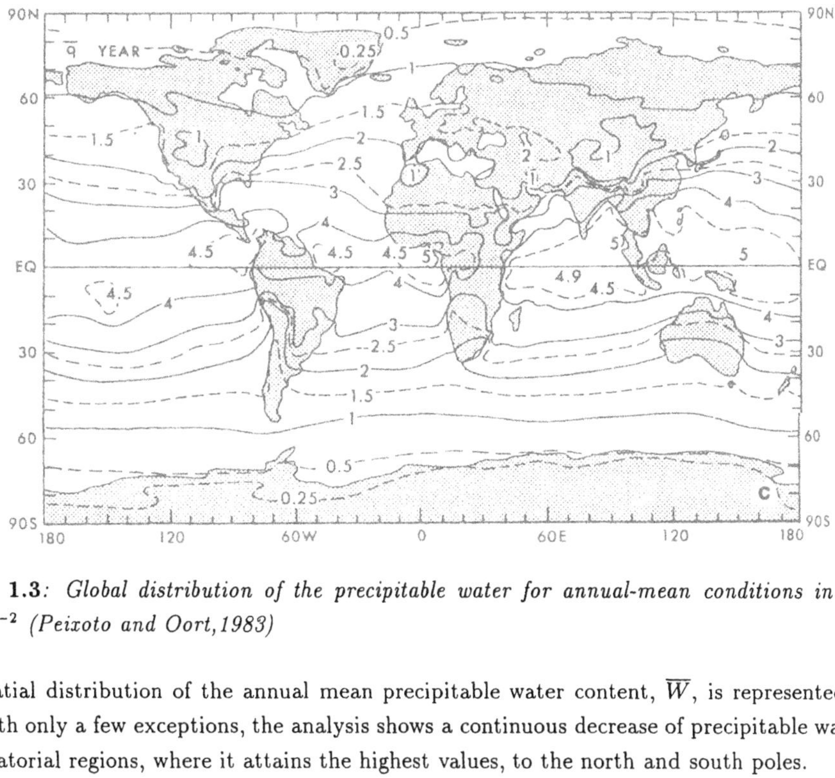

Figure 1.3: *Global distribution of the precipitable water for annual-mean conditions in units of* $10\,kg\,m^{-2}$ *(Peixoto and Oort, 1983)*

The spatial distribution of the annual mean precipitable water content, $\overline{W}$, is represented in Fig. 1.3. With only a few exceptions, the analysis shows a continuous decrease of precipitable water from the equatorial regions, where it attains the highest values, to the north and south poles.

The departures from zonal symmetry are associated with the physiography of the earth's surface in both hemispheres. As a general rule, the precipitable water is higher over the oceans than over the continents. The distribution over the Southern Hemisphere is practically zonal, since the ocean coverage exceeds by far that of the continents. As expected, the lowest values of W ($5\,kg\,m^{-2}$) occur over subpolar and polar region.

The precipitable water over the desert areas is considerably smaller than the corresponding zonal average, mainly due to strong subsidence. This effect is pronounced in the eastern portions of the large semi-permanent anticyclones of the subtropics. In addition, the effects of high terrain on the precipitable water distribution are illustrated by relatively dry areas (often $W < 10\,kg\,m^{-2}$) over the major mountain ranges, such as the Rockies, Himalayas, highlands of Ethiopia, and the Andes. The effects of topography and the land-sea contrast in the Southern Hemisphere are shown by the dipping of the $20\,kg\,m^{-2}$ isoline towards lower latitudes.

The humidity decreases rapidly with height, almost following an exponential law. It also decreases with latitude. More than 50% of the water vapor is concentrated below the 850 hPa surface (1500 m), while more than 90% is confined to the layer below 500 hPa (5.600 m). The seasonal variations are more intense in the Northern than in the Southern Hemisphere, as expected from the corresponding temperature variations.

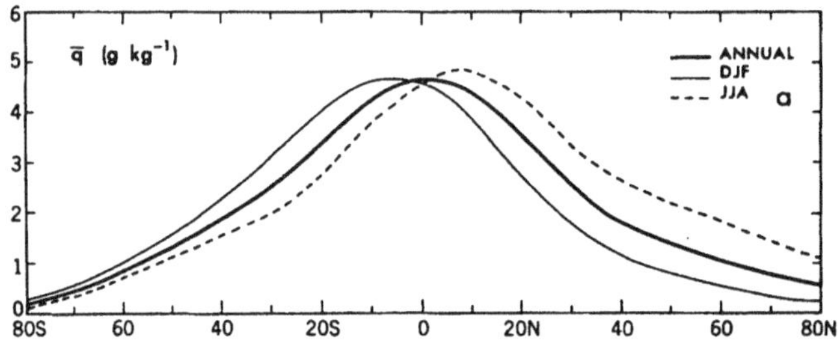

Figure 1.4: *Meridional profiles of the vertical and zonal mean values of the time-mean specific humidity (a), the day-to-day standard deviation of the specific humidity (b), and the east-west standard deviation for the time-mean specific humidity (c) in g* kg^{-1} *for annual, DJF and JJA mean conditions (Peixoto and Oort, 1983)*

Using the grid point values of the W maps, the zonally averaged storage of water in the atmosphere $\left[\overline{W}\right]$ were evaluated (Fig. 1.4). The seasonal profiles are also included. They show a maximum in the equatorial zone, with a slight seasonal migration into the summer hemisphere, and a monotoric decrease to polar latitudes with the steepest gradients in the subtropics.

The global water content in the atmosphere obtained from the $\left[\overline{W}\right]$ - profile, is on the order of $13.1 \cdot 10^{15} kg$ which is equivalent to a uniform layer of about 2.5 cm of water covering the globe. Seasonal changes in the hemisphere water content are more pronounced in the Northern than in the Southern Hemisphere.

Assuming a mean annual precipitation value over the globe of 1 m, the ratio of the amount of water in the air and the precipitation rate leads to a residence time of water in the atmosphere of (0.025/1.00) yr or about 9 days. This indicates that the water vapor in the atmosphere is replenished about forty times a year.

The mean water vapor transport in the atmosphere

The role of the general circulation in the hydrological cycle commences with an analyses of the vertically integrated atmospheric moisture flow in terms of the $\vec{Q}$ vector field, the aerial runoff, as shown in Fig. 1.5.

The $\vec{Q}$ maps provide a good indication of the prevailing movements of the main moist air masses in the atmosphere. The $\vec{Q}$ field is not uniform in intensity and direction. The intensity is largest over ocean areas (e.g. Pacific Atlantic, South Indian Ocean, etc.). There is, in general, a transfer of water vapor from the oceans into the continents. Even when the vectors are mainly zonal, their intensities

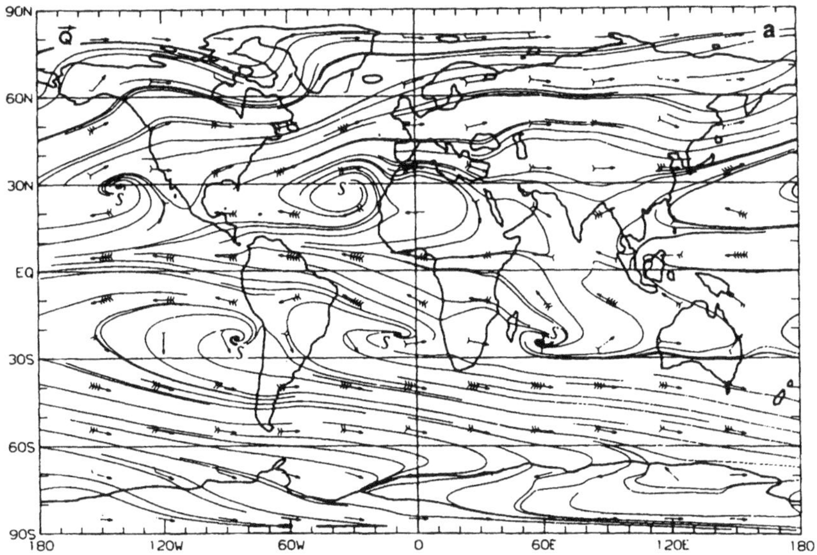

Figure 1.5: *Global distribution of the total aerial runoff $\vec{Q}$ and some corresponding streamlines for annual mean conditions (from Peixoto and Oort, 1983)*

over oceans are larger, by far, than over the continents so that a net inland flow of water results. The non uniformity of $\vec{Q}$ field distribution leads immediately to the impossibility of accepting the evaporation-precipitation *in situ* theory.

Although the zonal (east-west) component exceeds, in general, the meridional (south-north) component, the $\vec{Q}$ field is far from being truly zonal. The meridional component is far more important for the water balance of the globe than the zonal component.

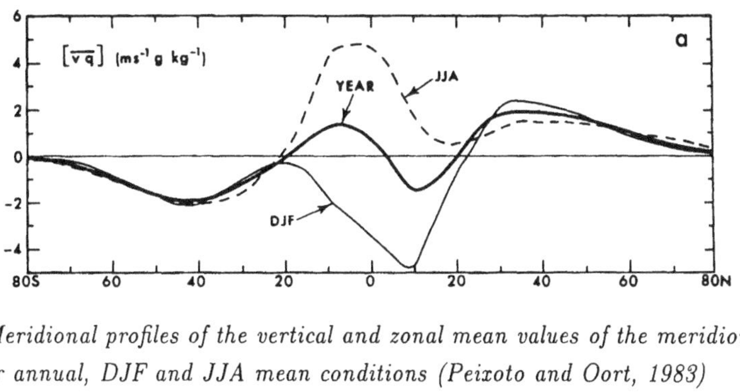

Figure 1.6: *Meridional profiles of the vertical and zonal mean values of the meridional transport of water vapor for annual, DJF and JJA mean conditions (Peixoto and Oort, 1983)*

To illustrate the main characteristics of the mean meridional flux and its importance for the water budget of the earth, meridional profiles of averaged values of the meridional flux for yearly and seasonal conditions are shown in Fig. 1.6. The meridional transports are poleward in both hemispheres, with maxima near 40° latitude and with small seasonal variations. In the tropical zone, the mean annual transports are positive south of the equator and negative to the north of it. There is evidence for a strong interaction between the two hemispheres as shown by the seasonal curves. In fact, the cross-equatorial flow in the Hadley cells changes direction with the seasons leading to a water vapor flux into the Northern Hemisphere during JJA and a flow into the Southern Hemisphere during DJF. For the year as a whole, there is a net influx into the Northern Hemisphere of $3.2 \cdot 10^8\, kg\, s^{-1}$, so that the Southern Hemisphere supplies a considerable amount of water vapor to the Northern Hemisphere. The cross-equatorial flow of water vapor implies an annual excess of precipitation over evaporation in the Northern Hemisphere of $39\, mm\, yr^{-1}$.

Although the meridional flux of moisture over midlatitudes varies considerably with the seasons, it is predominantly poleward throughout the year. This transport is mainly accomplished by baroclinic lows associated with the polar front and by stationary eddies, such as subpolar lows and subtropical anticyclones, together with their transient pulsations. The largest variability during the year is associated with the movement and changes in strength of the Hadley cells. The lower branches of these cells, in both Hemispheres, are very effective in transporting moisture into the intertropical convergence zone (ITCZ).

Vertical transport of water vapor

The vertical transport of water vapor in the atmosphere plays an essential role in the hydrological cycle, since it links the terrestrial and atmospheric branches. Convection associated with vertical instability and cumulus activity play an important role in this transport particularly in the tropics and in summer over the continent.

The belt of maximum upward transport over the equatorial region is associated with the ascending branches of the Hadley cells, whereas the upward flux in middle and high latitudes must be connected with the quasi-stationary low-pressure systems. The centers of maximum downward flux occur mainly in the eastern parts of the subtropical anticyclones over the oceans with the prevailing subsidence.

1.2.4 Mean planetary water balance and the general circulation of the atmosphere

The mean divergence of water vapor

The importance of the divergence field of water vapor in the atmosphere depends on it relationship with $\left(\overline{E} - \overline{P}\right)$, as discussed earlier. Thus, divergence maps are of great interest for the study of the planetary water balance, since regions of mean positive divergence $\left(\overline{E} - \overline{P} > 0\right)$ constitute main source regions of water vapor, whereas the regions of convergence $\left(\overline{E} - \overline{P} < 0\right)$ are sink regions of water vapor for the atmosphere. Using the grid point values of the meridional and zonal components of $\vec{Q}$ it is possible to evaluate the divergence of $\vec{Q}$ $\left(\nabla \vec{Q}\right)$. The resulting map for the mean yearly conditions is presented in Fig. 1.7.

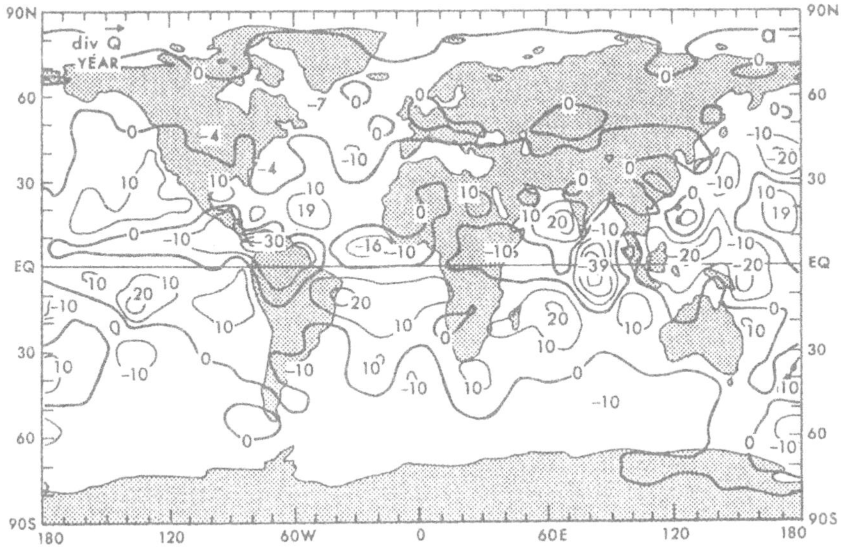

Figure 1.7: *Global distribution of the horizontal divergence of the total water vapor transport* $\nabla\vec{Q}$, *in units of* $0.1\,m\,yr^{-1}$ *for annual mean conditions. Negative values indicate areas of convergence of moisture or of excess of precipitation over evaporation (from Peixoto and Oort, 1983)*

Convergence (negative divergence) generally prevails over the equatorial and mid to high latitude zones, while divergence predominates in the subtropics. The equatorial convergence of water vapor is associated with the ITCZ. Water vapor is carried towards the region of mean rising motion by the lower branches of the Hadley cell, leading to heavy precipitation. The belt of convergence of water vapor consists of various centers located over the headwaters and drainage basins of large river systems, such as the Amazon in South America, The Ubangi, Congo, Senegal and Blue Nile in Africa and the Indus, Ganges, Mekong and Yangtze in Southeast Asia.

The subtropical belts of divergence coincide largely with the arid zones of the globe. They are associated with strong evaporation over the oceans and with subsidence that prevails over the large subtropical anticyclones. The distribution of divergence over the oceans is easy to understand. However, the situation is more difficult to explain when divergence occurs over land. In this case, surface and underground flows from less arid regions must supply the water required to counterbalance the observed excess of evaporation over precipitation (Starr and Peixoto, 1958).

The middle of high latitude convergence in both hemispheres is mainly associated with the transient cyclonic lows that accompany the polar front. Over the polar regions, there are some indications of a slight divergence, especially in the vicinity of the north pole.

The evaporation is very large over the oceanic regions with strong divergence of water vapor and consequently with high salinity values. On the other hand, in regions of convergence, such as in equatorial latitudes, the excess of fresh water from rain will dilute the ocean water, leading to lower salinity values.

Zonally averaged profiles of the $\left[\nabla\vec{Q}\left(\simeq\overline{E}-\overline{P}\right)\right]$ fields are shown in Fig. 1.8. In all cases, they indicate a strong convergence with an excess of mean precipitation over mean evaporation, $\left[\nabla\vec{Q}\right]\simeq$

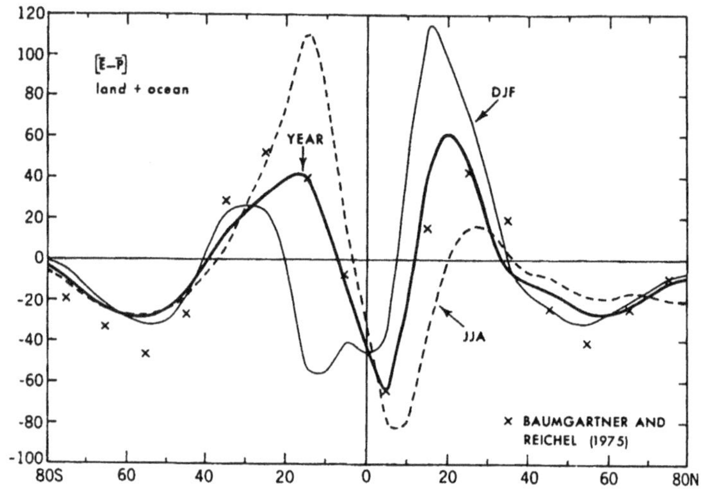

Figure 1.8: *Meridional profiles of the zonal mean divergence of the total water vapor transport,* $[\nabla\vec{Q}] \approx [E-P]$, *in* $0.01\,m\,yr^{-1}$ *for annual, DJF and JJA mean conditions. Some estimates of* $E-P$ *by Baumgartner and Reichel (1975) are added for comparison (see also Table 1.1)*

$\left[\overline{E}-\overline{P}\right] < 0$, over the equatorial region and strong divergence in the subtropics. The divergence intensities during the winter season.

It is interesting to compare the meridional transport $\overline{Q}_\phi$ and $\left[\nabla\vec{Q}\right]$ profiles. They show a large export of water vapor from the zones of divergence and a strong import of moisture into the belts of convergence.

The aerial runoff and water balance

The role of the general circulation in the hydrological cycle was shown in terms of the $\vec{Q}$ vector field, the so-called aerial runoff. From the $\vec{Q}$ field it is possible to draw the corresponding streamlines (see Fig. 1.5).

They provide a good indication of the prevailing movements of the main moist air masses in the atmosphere and the sites of their formation. They show, again, that the main sources of water vapor for the atmosphere are located over the subtropical oceans and that most of the water vapor, necessary for precipitation over the continents, comes from the oceans. In steady state conditions, this net inflow of moisture to the continents must be compensated by runoff of the rivers into the oceans. The air masses also receive some moisture over the continents due to evapotranspiration and evaporation from lakes, soil, etc. However, this last moisture supply constitutes only a small fraction of the water that falls locally, as precipitation over land.

Since the water vapor transport occurs, mainly, in the lower troposphere, the realm of weather phenomena, it is clearly affected by the earth's topography. Indeed, the absence of large mountains along the Atlantic coast in Europe favors the deep penetration of moisture from the Atlantic Ocean into the Eurasian Continent and the Mediterranean region. On the other hand, the existence of the Rocky Mountains parallel to the west coast of North America does not allow moisture from the Pacific Ocean to penetrate deeply into the American continent. Most of the moisture falling as precipitation over North America seems to be supplied by water vapor originating over the warm waters of the Gulf of Mexico, with a deep northward intrusion of water vapor in all seasons. On the other hand, most of the moisture over South America comes from the Atlantic Ocean.

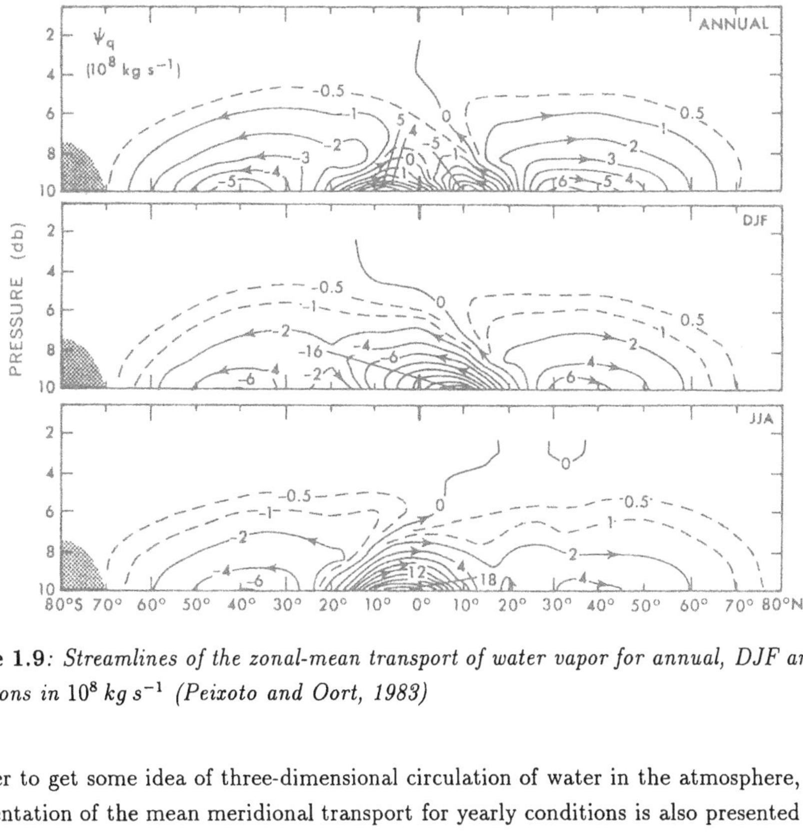

Figure 1.9: *Streamlines of the zonal-mean transport of water vapor for annual, DJF and JJA mean conditions in* 10^8 *kg* s^{-1} *(Peixoto and Oort, 1983)*

In order to get some idea of three-dimensional circulation of water in the atmosphere, a streamline representation of the mean meridional transport for yearly conditions is also presented in Fig. 1.9.

Inspection of the cross-section confirms that the water vapor circulates in the lower troposphere, and that the maximum transports occur in the planetary boundary layer. The yearly cross-section shows that there is an export of water vapor from the zones, between 12 ° N and 35 ° N, and 10 ° S and 35 ° S. In the Northern hemisphere, a considerable portion of this water vapor (to the right of the zero streamline) is exported polewards throughout the lower half of the atmosphere, while the remainder, confined to much lower levels, is directed to the equatorial regions, mainly to feed the ITCZ. In Southern hemisphere low latitudes, a large part of the water vapor is exported to the north in a shallow layer near the earth's surface, some of it crossing the equator and falling as precipitation in the ITCZ. The other part is transported southwards, at much higher levels, up to 400 hPa. Polewards of 40 ° latitude, in both hemispheres, the streamlines terminate at the surface, indicating the existence of an excess of precipitation over evaporation.

The interactions between the atmospheric branch and the terrestrial branch are essential to achieve and maintain the state of quasi-equilibrium of the climatic system. Occasionally, however, imbalances may persist for an extended period of time, leading to drought or to flood conditions.

1.2.5 Observed precipitation, evaporation and runoff

Precipitation

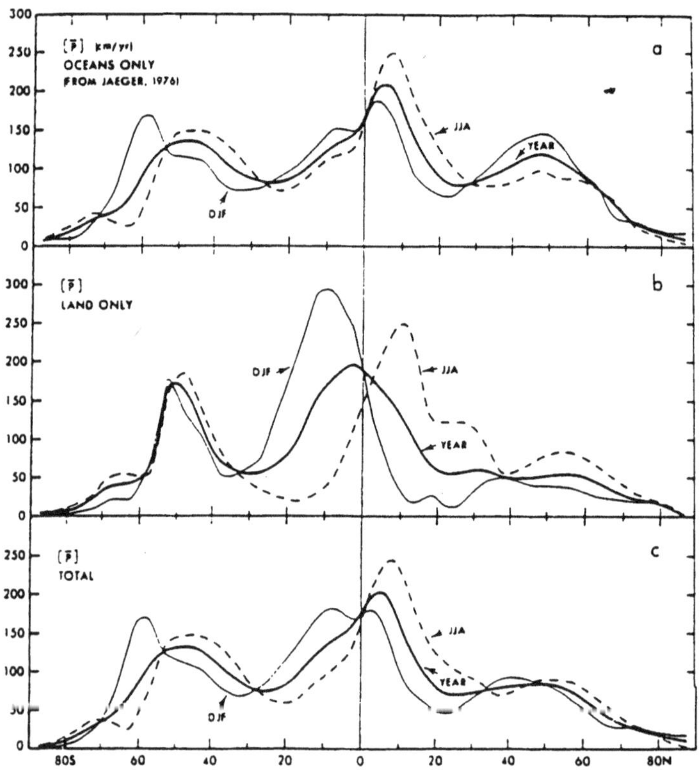

Figure 1.10: *Meridional profiles of the zonal mean precipitation rate in $cm\,yr^{-1}$ for the ocean area (a), the land area (b) and the total land plus ocean area (c) for annual, DJF and JJA mean conditions after Jäger (1976)*

Precipitation is highly variable in space and time. Nevertheless its average values are fairly stable and can be represented in map form (Figure 1.10).

The most significant feature of the distribution is the high rainfall that occurs in equatorial in the ITCZ which corresponds to the thermal equator. It is interesting to note the very high values observed over the equatorial regions in South America, Africa, and Indonesia and in the equatorial Pacific Ocean where precipitation may exceed $3\,m\,yr^{-1}$. During the annual cycle the ITCZ migrates north and south in phase with the solar insulation which explains the shift of the observed maxima.

In subtropical regions under the influence of the large semi-permanent anticyclones where subsidence predominates the precipitation is low, less than $0.2\,m\,yr^{-1}$, and large parts of the subtropical continents are covered by deserts, where the precipitation is even lower, e.g., in Africa and Australia.

Over the midlatitude belts, where the polar fronts with the associated disturbances predominate, the precipitation shows a secondary maximum. Over the polar regions the moisture content of the atmosphere is very low, and the amounts of precipitation are less than $0.2\,m\,yr^{-1}$ during all seasons.

The seasonal shifts of the ITCZ are found to be more pronounced over land than over the oceans. The latitudinal range of the shifts in the ITCZ is associated with the existence of the marginal climates bordering the arid and desert-like regions, such as occurs in the Sahel belt in Africa.

Evaporation

The evaporation rate depends on many factors. The most important ones are the temperature, vertical diffusion coefficients, the vertical gradient of humidity, wind speed and the availability of water. Evaporation is often measured with a shallow circular pan. However, such values are very much influenced by local conditions and local exposure. Thus these measurements, useful as they may be for local purposes, such as assessing evaporation from water reservoirs, small lakes, irrigated areas, etc., are of little use in computing the water budget for larger regions of the earth.

Surface runoff

After considering the precipitation and evaporation fields separately it is useful to compare the behavior of these two quantities, since they are closely related elements of climate and hydrology. Thus, we present in Table 1.1 the mean zonal averages of precipitation and evaporation for 10° latitude belts as well as the hemispheric and global averages according to Baumgartner and Reichel (1975). Similarly, annual mean values of the same quantities for individual continents and oceans are presented in Table 1.2 according to the same authors. The tables include also the $P - E$ differences, the so-called discharge or runoff, and the values of the evaporation ratio E/P and the runoff ratio $(P - E)/P$. These last two quantities are of interest since they are sometimes used as climatic indices or in hydrology studies. Table 1.2 further includes the estimated river discharge R_0 from the continents as measured by the peripheral runoff that reaches the oceans.

Analysis of results

The $P - E$ values in Tab. 1.1 show an excess of precipitation over evaporation at mid and high latitude as well as in the equatorial zone between 10° S and 10° N, whereas a deficit of precipitation is found in the subtropics of each hemisphere between about 10° and 40° latitude. In the long-term mean, the excess or deficit in each belt has to be compensated by a net meridional divergence or convergence of water in the particular belt. The runoff ratio $(P - E)/P$ gives an idea of the fraction of the precipitation that is involved in the runoff. The values of the evaporation ratio E/P in Tab. 1.1 show clearly the high aridity of the subtropics with ratios larger than one.

One must be aware that there is not always a close agreement between the values of P and E published by different authors, as demonstrated by the comparisons in Tab. 1.1 and 1.2 with Sellers (1965) values based on similar observations.

The precipitation in the two hemispheres is almost the same whereas a large difference exists for the evaporation. The higher value of evaporation in the Southern hemisphere results because this hemisphere is largely covered by oceans. The Northern hemisphere shows a positive water balance ($P - E = 73\,mm\,yr^{-1}$) whereas in the Southern hemisphere a net negative value of $-73\,mm\,yr^{-1}$ occurs, so that a flow of water in the liquid form must take place across the equator from the Northern into the Southern hemisphere. As we have seen an equal amount of water in the vapor form has to be exported in the opposite direction to maintain the balance of the water substance.

As shown from Tab. 1.2 South America shows the highest P, E and $P - E$ values in agreement with what we have seen before; on the other hand, Australia, Africa, and Antarctica show very small runoff ($P - E$) values. Overall, the precipitation and evaporation tend to be smaller over the continents than over the oceans. The mean value of $P - E$ over all continents is estimated to be $266\,mm\,yr^{-1}$. This surplus of condensed water must be transported by rivers and glaciers from the continents into the oceans where a deficit of $-110\,mm\,yr^{-1}$ is found. The table further shows that $P - E$ is positive for the Arctic and Pacific Oceans and strongly negative for the Atlantic and Indian Oceans, (the *dry* oceans), leading to the net deficit for all oceans combined. The implications of these estimates of $P - E$ taken together with the values of the observed river discharge are that a net transfer of water must occur from the Pacific and Arctic Oceans into the Atlantic and Indian Oceans. For example, water transport by the rivers from the surrounding continents into the Atlantic is estimated to be on the order of $R_0 = 197\,mm\,yr^{-1}$ so that the equivalent of $E - P - R_0 = 175\,mm\,yr^{-1}$ must come from the Pacific and Arctic Oceans. Further for the Indian Ocean which only receives $72\,mm\,yr^{-1}$ from continental runoff an inflow of $179\,mm\,yr^{-1}$ must take place from the Pacific Ocean.

The net excess of precipitation over evaporation over the continents must be maintained by a net influx of water in the vapor form from the large ocean sources.

1.3 Energetics

1.3.1 Radiation balance and Energy balance

As a synthesis of the radiation and energy processes in the atmosphere we present an integrated picture of how the global radiation and energy balances are achieved in the mean in the climatic system. As shown in Fig. 1.11 the incoming solar radiation is not all used by the climatic system. A substantial fraction (on the order of 30%) of the incident solar radiation is reflected by clouds and to a lesser extent by the earth's surface, and, therefore, does not participate in the atmospheric heat engine. Out of the remaining 70%, 20% is absorbed by the atmosphere and 50% by the oceans and land. This last amount will be used partly to maintain the hydrological cycle through evaporation (24%) and hence heat the atmosphere indirectly through condensation of water, and partly to heat the atmosphere directly through the flux of sensible heat (6%). The remaining 20% is used to heat

Latitude	surface area	P	E	P-E	E/P	(P-E)/P
80°-90°N	3.9	46 (120)	36 (42)	10 (93)	0.78	0.22
70°-80°N	11.6	200 (185)	126 (145)	74 (124)	0.63	0.37
60°-70°N	18.9	507 (415)	276 (333)	231 (224)	0.54	0.46
50°-60°N	25.6	843 (789)	447 (469)	396 (250)	0.53	0.47
40°-50°N	31.5	874 (907)	640 (641)	234 (156)	0.73	0.27
30°-40°N	36.4	761 (872)	971 (1002)	-210 (23)	1.28	-0.28
20°-30°N	40.2	675 (790)	1110 (1246)	-435 (-435)	1.64	-0.64
10°-20°N	42.8	1117 (1151)	1284 (1389)	-167 (-322)	1.15	-0.15
0°-10°N	44.1	1885 (1934)	1250 (1235)	635 (478)	0.66	0.34
0°-10°N	44.1	1435 (1445)	1371 (1304)	64 (144)	0.96	0.04
10°-20°N	42.8	1109 (1132)	1507 (1541)	-398 (-342)	1.36	-0.36
20°-30°N	40.2	777 (857)	1305 (1416)	-528 (-312)	1.68	-0.68
30°-40°N	36.4	875 (932)	1181 (1256)	-306 (-128)	1.35	-0.35
40°-50°N	31.5	1128 (1226)	862 (895)	266 (150)	0.76	0.24
50°-60°N	25.6	1003 (1046)	553 (520)	450 (278)	0.55	0.45
60°-70°N	18.9	549 (418)	229 (174)	320 (245)	0.42	0.58
70°-80°N	11.6	230 (82)	54 (45)	176 (98)	0.23	0.77
80°-90°N	3.9	73 (30)	12 (0)	61 (32)	0.16	0.84
0°-90°N	255.0	990 (1009)	897 (944)	73 (39)	0.92	0.07
0°-90°S	255.0	975 (1000)	1048 (1064)	-73 (-39)	1.07	0.07
globe	510.0	973 (1004)	973 (1004)	-	1.00	-
units	10^6 Km^2	mm yr^{-1}	mm yr^{-1}	mm yr^{-1}	-	-

Table 1.1: *Estimated mean annual values of the precipitation P, evaporation E, runoff (P-E), evaporation ratio E/P (an aridity index), and runoff ratio (P-E)/P for 10 ° latitude belts, the hemispheres and the globe from Baumgartner and Reichel (1975). For comparison, Seller's (1965) estimates for P and E, and Peixoto and Oort's (1983) independent estimates of P-E as computed from divergence are shown in parentheses.*

the underlying surface, and will be lost later as infrared radiation to the atmosphere (14%) and mainly in the spectral window ($8 - 12\,\mu m$) to outer space (6%).

The heat absorbed by the atmosphere is used to increase the internal and potential energy which will partially (less than 1%) be converted into kinetic energy to maintain the atmospheric and oceanic general circulations against friction. Finally, the atmosphere will radiate out to space around 64% as infrared radiation, thereby closing the energy cycle.

In the following sections we will discuss how the energy is used to operate the ocean-atmosphere heat engine by applying the concept of availability of energy to the various subsystems of the climate system.

Region	surface area (10^6 km^2)	P (mm yr^{-1})	E (mm yr^{-1})	P-E (mm yr^{-1})	R_0 (mm yr^{-1})	E/P	(P-E)/P
Europe	10.0	657 (600)	375 (360)	282 (240)	-	0.57	0.43
Asia	44.1	696 (610)	420 (390)	276 (220)	-	0.60	0.40
Africa	29.8	696 (670)	582 (510)	114 (160)	-	0.84	0.16
Australia	7.6	447 (470)	420 (410)	27 (60)	-	0.94	0.06
North America	24.1	645 (670)	403 (400)	242 (270)	-	0.62	0.38
South America	17.9	1564 (1350)	946 (860)	618 (490)	-	0.60	0.40
Antarctica	14.1	169 (30)	28 (0)	141 (30)	-	0.17	0.83
All Land Areas	148.9	746 (720)	480 (410)	266 (310)		0.64	0.36
Arctic Ocean	8.5	97 (240)	53 (120)	44 (120)	307	0.55	0.45
Atlantic Ocean	98.0	761 (780)	1133 (1040)	-372 (-260)	197	1.49	-0.49
Indian Ocean	77.7	1043 (1010)	1294 (1380)	-251 (-370)	72	1.24	-0.24
Pacific Ocean	176.9	1292 (1210)	1202 (1140)	90 (70)	69	0.93	0.07
All Oceans	361.1	1066 (1120)	1176 (1250)	-110 (-130)	110	1.10	-0.10
Globe	510.0	973 (1004)	973 (1004)	0 (0)	-	1.10	0

Table 1.2: *Estimated mean annual values of the precipitation P, evaporation E, runoff P-E, river runoff R_0 from continents into the oceans, evaporation ratio E/P, and runoff ratio (P-E)/P for the various continents and oceans from Baumgartner and Reichel (1975). For comparison, estimates of P, E and P-E from Sellers (1965) have been added in parentheses.*

1.3.2 Basic forms of energy

The main intrinsic forms of the energy in the atmosphere are internal energy I, potential-gravitational energy Φ , kinetic energy K, and latent energy associated with phase transitions of water, LH (evaporation and condensation of water vapor, freezing of water, and melting and sublimation of snow and ice). Expressions for these energy forms per unit mass are:

$$I = c_v T \tag{1.10}$$

$$\Phi = gz \tag{1.11}$$

$$K = \frac{1}{2}\vec{c}\cdot\vec{c} = \frac{1}{2}\left(u^2+v^2+w^2\right) = \frac{1}{2}\left(u^2+v^2\right). \tag{1.12}$$

In case of the latent energy we will use the general form

$$LH = Lq\,. \tag{1.13}$$

It depends on the the particular phase transition how much latent energy is actually realized as sensible heat.

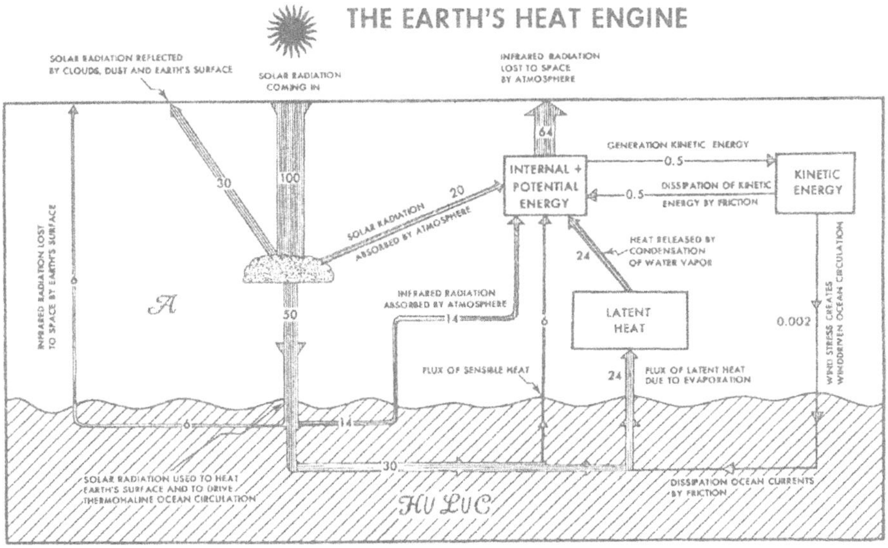

Figure 1.11: *Schematic diagram of the flow of energy in the climatic system. A value of 100 is assigned to the incoming flux of solar energy ($\approx$ 340 $W\,m^{-2}$). All values represent annual averages for the entire atmosphere. For simplicity no separate energy boxes have been drawn for the oceans ($\mathcal{H}$), land surfaces ($\mathcal{L}$) or cryosphere ($\mathcal{C}$) (Peixoto and Oort, 1984)*

Finally we have for the total energy

$$E = I + \Phi + LH + K\,. \tag{1.14}$$

The main relevant form of external energy is the solar radiation. Radiation is the only form of energy which can alter the global energy of the climate system. We may ignore geothermal, electrical, and nuclear energy since they are negligible for the energetics of the atmosphere. From the thermodynamic point of view, latent heat is a form of internal energy but we will consider it separately.

In a hydrostatic system, the internal energy for a unit area column of the atmosphere which extends from the earth's surface to the top of the atmosphere is given by

$$\int_0^\infty \rho I\,dz = \int_0^\infty \rho c_v T\,dz = \int_0^{p_0} c_v T\frac{dp}{g} \tag{1.15}$$

and the potential energy by

$$\int_0^\infty \rho\Phi\,dz = \int_0^\infty \rho g z\,dz = \int_0^{p_0} z\,dp = \int_0^\infty \rho R_d T\,dz \tag{1.16}$$

after integration by parts.

The potential and internal energy in the atmosphere are not independent forms of energy. They are proportional to each other with the ratio $R_d/c_p = 2/5$, and it is convenient to consider them together as one form of energy, the so-called total potential energy. Thus

$$\int_0^\infty \rho(\Phi + I)dz = \int_0^\infty \rho c_p T\, dz = \int_0^{p_0} c_p T \frac{dp}{g}\,, \tag{1.17}$$

where $c_p = c_v + R_d$. This sum represents the enthalpy in the atmospheric column.

Introducing the potential temperature given by

$$\theta = T\left(\frac{p}{p_0}\right)^\kappa\,, \tag{1.18}$$

where $\kappa = \frac{R_d}{c_p}$, the total potential energy becomes

$$\int_0^\infty \rho(\Phi + I)dz = \frac{c_p}{g(1+\kappa)p_0{}^\kappa}\int_0^\infty \theta\, dp^{\kappa+1}\,. \tag{1.19}$$

Integration by parts gives:

$$\int_0^\infty \rho(\Phi + I)dz = \frac{c_p}{g(1+\kappa)p_0{}^\kappa}\int_0^\infty p^{\kappa+1}\, d\theta\,. \tag{1.20}$$

1.3.3 Energy balance equations

The rate of change of the potential energy per unit mass is given by

$$\frac{d\Phi}{dt} = gw \tag{1.21}$$

and the rate of change of internal energy

$$\frac{dI}{dt} = \dot{Q} - p\,\alpha\,\nabla\vec{c}\,. \tag{1.22}$$

The diabatic heating term $\dot{Q}$ can be written explicitly in the form:

$$\dot{Q} = \dot{Q}_h + \dot{Q}_f\,, \tag{1.23}$$

$$\text{where}\quad \dot{Q}_h = -\alpha\,\nabla\vec{F}_{rad} - \alpha L(e-c) - \alpha\,\nabla\vec{J}_H^D - \alpha\vec{\vec{\tau}} : \nabla\vec{c} \tag{1.24}$$

$$\text{and}\quad \dot{Q}_f = -\alpha\vec{\vec{\tau}} : \nabla\vec{c}\,. \tag{1.25}$$

Thus we have decomposed the total heating $\dot{Q}$ into a first part $\dot{Q}_h$ containing radiational heating, latent heating and heating due to conduction, and a second part $\dot{Q}_f$ associated with frictional dissipation. The heat flux due to conduction is denoted by $\vec{J}_H^D$.
For the total potential energy we have

$$\frac{d}{dt}(\Phi + I) = gw + \dot{Q} - p\,\alpha\,\nabla\vec{c}, \tag{1.26}$$

$$\text{or}\quad \frac{d}{dt}(\Phi + I) = gw + \dot{Q} + \alpha\,\vec{c}\cdot\nabla p - \alpha\,\nabla(p\vec{c})\,. \tag{1.27}$$

The kinetic energy can be obtained immediately from equation of motion by using the common procedure of taking the scalar product with $\vec{c}$:

$$\frac{dK}{dt} = -gw - \alpha\,\vec{c}\cdot\nabla p - \alpha\,\vec{c}\cdot\nabla\vec{\vec{\tau}} \tag{1.28}$$

or in an alternative form

$$\frac{dK}{dt} = -gw + p\,\alpha\,\nabla\,\vec{c} - \alpha\,\nabla\left(p\,\vec{c} + \vec{\vec{\tau}}\cdot\vec{c}\right) + \alpha\,\vec{\vec{\tau}} : \nabla\,\vec{c}\,. \tag{1.29}$$

In hydrostatic equilibrium we have

$$gw + \alpha\vec{c}\cdot\nabla\,p = \alpha\,\vec{v}\cdot\nabla\,p \tag{1.30}$$

where $\vec{v}$ is the horizontal, two-dimensional wind vector.

Inspection of the previous energy equations shows that they are not independent of each other since they are linked by common terms. In fact, Eq. 1.21 shows that the rate of change of potential energy comes from the work done against the force of gravity, gw. This same term appears with the opposite sign in Eq. 1.28 which suggests that there is a conversion of potential into kinetic energy, or vice versa.

As shown by Eq. 1.22 the sources or sinks of internal energy are the rate of heating $\dot{Q}$ and the rate of work performed by compression against the pressure field, $-p\,\alpha\,\nabla\,\vec{c}$. This last term together with the frictional heating part $\dot{Q}_f$ occur also in Eq. 1.29 but with the opposite sign, which shows the link between the kinetic and internal energy.

The term, $-\alpha\,\nabla\left(p\vec{c} + \vec{\vec{\tau}}\cdot\vec{c}\right)$ in Eq. 1.29 indicates the work at the boundaries by pressure and friction forces. This term plays an important role in transferring energy from the atmosphere into the oceans; it generates the wind-driven ocean currents. Analogously, motions of the atmosphere with or against the pressure gradient force across isobars will also convert potential energy into or from kinetic energy. These processes are adiabatic (reversible) because they can proceed in both directions. However, frictional dissipation is an irreversible process.

For the latent heat we have

$$L\frac{dq}{dt} = L(e - c) - \alpha L\,\nabla\vec{J}_H^D\,. \tag{1.31}$$

The last term is associated with molecular diffusion of water vapor across the boundaries and can usually be neglected.

Let us consider next the total energy of the atmosphere. We must then take into account explicitly the radiation balance which results from the incoming solar radiation and the outgoing infrared radiation emitted by the earth. Thus for total energy E we obtain

$$\frac{dE}{dt} = -\alpha\,\nabla\,(p\vec{c}) - \alpha\,\vec{c}\cdot\nabla\,\vec{\vec{\tau}} + L(e - c) + \dot{Q}\,. \tag{1.32}$$

To illustrate our interpretation of the previous energy Equations 1.21 through 1.32, and to show the meaning of the various source, conversion and sink terms of energy a schematic diagram of the energy cycle is given in Fig. 1.12.

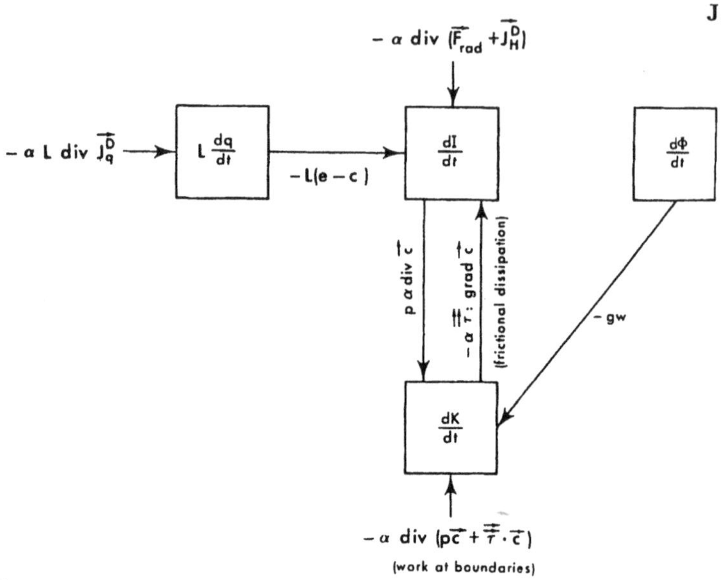

Figure 1.12: *Schematic box diagram showing the terms which connect the various forms of energy in the atmosphere (Peixoto and Oort, 1984)*

1.4 The atmospheric heat engine

1.4.1 Availability of energy in the atmosphere

We have already seen that the highest temperatures occur at low latitudes near the earth's surface and the lowest temperatures at high latitudes in the upper atmosphere. Furthermore, the atmosphere, in fact acts as a vehicle to transport heat poleward and upward. Thus the atmosphere may be regarded as a heat engine with, on the whole, heat flowing from the warm sources to the cold sinks. The work performed by the atmospheric engine is used to maintain the kinetic energy of the circulations against a continuous drain of energy by frictional dissipation.

Even for the ideal case of a Carnot machine, the efficiency η of the heat machine would be low since the difference between the temperature of the warm source T_W and that of the cold sink T_c is relatively small compared with the temperature of the warm source.

$$\eta = \frac{T_W - T_c}{T_W} \leq 10\% . \tag{1.33}$$

Later we will present other estimates of the efficiency which, as expected, lead to lower values than the limiting ideal case. Thus the amount of kinetic energy generated by the atmospheric heat engine has to be small compared with the total potential plus internal energy.

In a preliminary analysis of the generation of kinetic energy we have seen that the source for kinetic energy was the total potential energy. However, only a fraction of the potential plus internal energy in the atmosphere is available to be converted into kinetic energy, whereas most of the total potential energy is unusable. This same fact was already realized by Margules (1903).

Although the total potential energy in a barotropic atmosphere can be very high there does not exist the capacity to generate kinetic energy because the necessary mechanisms for conversion are not present. Indeed in a barotropic atmosphere the horizontal pressure gradient force is zero everywhere, and moreover if the atmosphere is in hydrostatic equilibrium there is no net vertical force since the vertical component of the pressure gradient is exactly compensated by the force of gravity. Such an atmosphere contains a large amount of potential energy but none of it is available for conversion into kinetic energy. Using the concepts of the availability theory we may say that such an atmosphere is in a *dead* state. However, the actual atmosphere is baroclinic and in quasi-hydrostatic equilibrium. Thus within the real atmosphere there exist several regions or subsystems with a non-uniform distribution of temperature on isobaric levels. Using the terminology of the availability theory we can state that there are many regional subsystems in an *alive* state, and that the atmosphere as a whole is thus in an alive state. Since the atmosphere is not a completely uniform system it will continuously evolve through many different natural states each having a different total potential energy. When the atmosphere goes from one natural state to another there is a change in total potential energy, and, thus, kinetic energy is either created or destroyed. Only the difference in total potential energy is involved in the transformation process and is thus available to be converted into kinetic energy.

The concept of available potential energy can now be defined with regard to an ideal conceptual state of the atmosphere with a minimum total potential energy. In fact, this state is founded on the conservation of mass and on its redistribution through isentropic processes. Under these conditions the sum of the internal, potential and kinetic energy is invariant. From all the possible mass-equivalent dead states we choose the one with maximum entropy and minimum total potential energy as the reference state (indicated by a subscript r). This state is characterized by a horizontal stratification with absolute stability in both pressure, potential temperature and height. Thus the availability for total potential energy is defined by

$$\Lambda = \int (\Phi + I) dm - \int (\Phi + I)_r \, dm, \tag{1.34}$$

which represents the maximum possible amount of total potential energy that can be converted into kinetic energy.

The actual operation of the general circulation leads to a continuous depletion of the available potential energy. Its maintenance requires then certain mechanisms for restoring the losses and thus ensuring the existence of alive systems in the atmosphere. To show that these mechanisms are mainly associated with the non-uniform heating of the atmosphere let us consider a dead atmosphere. If this atmosphere would be heated uniformly its total potential energy would increase but none would be available because the atmospheric structure would remain barotropic. On the other hand, if we would keep the total amount of potential energy constant but add and subtract heat differentially, alive subsystems would be generated and availability will be produced. In other words, although the atmosphere would initially be barotropic it would later become baroclinic so that circulations would develop (Bjerknes theorem). Therefore to keep the general circulation of the atmosphere alive one needs to continuously generate available potential energy through heating of warm and cooling of cold regions.

1.4.2 Available potential energy

Taking the definition of availability 1.34 and using expressions 1.17, 1.19, 1.20, we obtain the following equivalent expressions for the available potential energy

$$P = \int c_p(T - T_r)dm = \int c_p T \left(1 - \frac{T_r}{T}\right) dm \tag{1.35}$$

or

$$P = \int c_p \,\theta \frac{p^\kappa - {p_r}^\kappa}{{p_0}^\kappa} dm = \int c_p T \left\{1 - \left(\frac{p_r}{p}\right)^\kappa\right\} dm\,. \tag{1.36}$$

If we substitute the expression 1.20 for the total potential energy in the θ -coordinate system in 1.19 we obtain the *exact* expression derived by Lorenz (1955):

$$P = \frac{c_p}{g(1+\kappa){p_0}^\kappa} \int\int\int \left(p^{\kappa+1} - \tilde{p}^{\kappa+1}\right) dx\, dy\, d\theta\,. \tag{1.37}$$

If we compare expressions 1.35 and 1.36 with the corresponding expressions for $(\Phi + I)$ we see that the main difference comes from the binomial factors $\left(1 - \frac{T_r}{T}\right)$ and $\left\{1 - \left(\frac{p_r}{p}\right)^\kappa\right\}$ that, within this context, can be defined as efficiency factors.

After some simplifications of Eq. 1.37 one can derive the so-called *approximate* expression of Lorenz (1955)

$$P = \frac{c_p}{2} \int \Gamma \left(T^2 - \tilde{T}^2\right) dm\,, \tag{1.38}$$

which gives the available potential energy in terms of the two-dimensional variance of temperature on a constant pressure surface. This expression is most convenient to use in actual calculations. The quantity Γ is an inverse measure of the gross global-mean static stability

$$\Gamma = -\frac{\frac{\kappa\theta}{pT}}{\frac{\partial\tilde{\theta}}{\partial p}} = \frac{\gamma_d/\tilde{T}}{\gamma_d - \tilde{\gamma}}\,. \tag{1.39}$$

Because of the zonal character of the general circulation it is useful to partition the kinetic energy into the kinetic energy of the zonal-mean flow K_M and the kinetic energy of the perturbations K_E. We have called these components mean and eddy kinetic energy, respectively. Similarly the available potential energy may be partitioned into the components P_M and P_E. These partitions are done using the expansion

$$\left[\overline{u^2}\right] = [\overline{u}]^2 + \left[\overline{u'^2}\right] + \left[\overline{u^{*}}^2\right] \tag{1.40}$$

and similar expansions for $\left[\overline{v^2}\right]$ and $\left[\overline{T^2}\right]$. The bar operator u defined by the temporal average $\overline{(\,)} \equiv \frac{1}{\Delta t}\int(\,)\,dt$, so that $u = \overline{u} + u'$ and $\overline{u'} = 0$. The $[\,]$ operator is the mean zonal value defined by $[(\,)] \equiv \frac{1}{2\pi}\oint(\,)\,d\lambda$, so that $u = [u] + u^*$ and $[u^*] = 0$.

Therefore we have

(1) for kinetic energy

$$K_M = \frac{1}{2}\int \left([\overline{u}]^2 + [\overline{v}]^2\right) dm \tag{1.41}$$

$$K_E = K_{TE} + K_{SE} = \frac{1}{2}\int \left(\left[\overline{u'^2} + \overline{v'^2}\right] + \left[\overline{u^*}^2 + \overline{v^*}^2\right]\right) dm \tag{1.42}$$

(2) for available potential energy.

$$P_M = \frac{c_p}{2}\int \Gamma\left(\left[\tilde{T}\right]^2 - T^2\right) dm \tag{1.43}$$

$$P_E = P_{TE} + P_{SE} = \frac{c_p}{2}\int \Gamma\left(\left[\overline{T'^2}\right] + \left[\overline{T^{*2}}\right]\right) dm \tag{1.44}$$

The balance equations for these basic forms of energy permit a clear interpretation of the mechanisms involved in the generation and dissipation of energy and of the transformations between the various forms of energy.

1.4.3 Balance equations for kinetic and available potential energy

Balance equations for mean and eddy kinetic energy in the atmosphere

As an application of the previous methodology we will derive the balance equations for the mean and eddy kinetic energy, K_M and K_E. Starting with the two horizontal equations of motion (Oort and Peixoto, 1983):

$$\frac{\partial u}{\partial t} = -u\frac{\partial u}{R\cos\phi\,\partial\lambda} - v\frac{\partial u}{R\,\partial\phi} - \omega\frac{\partial u}{\partial p} + fv - g\frac{\partial z}{R\cos\phi\,\partial\lambda} + F_\lambda \tag{1.45}$$

$$\frac{\partial v}{\partial t} = -u\frac{\partial v}{R\cos\phi\,\partial\lambda} - v\frac{\partial v}{R\,\partial\phi} - \omega\frac{\partial v}{\partial p} - fu - g\frac{\partial z}{R\,\partial\phi} + F_\phi\,, \tag{1.46}$$

multiplying them by $[u]$ and $[v]$, respectively, and adding, we obtain a balance equation for K_M. Integrating over the entire mass of the atmosphere we can disregard the boundary terms so that we can write the balance equation in the following symbolic form:

$$\frac{\partial K_M}{\partial t} = C(P_M, K_M) + C(K_E, K_M) - D(K_M)\,, \tag{1.47}$$

where $C(A, B)$ indicates the rate of conversion from A to B and $D(K_M)$ the rate of dissipation of K_M. Thus

$$C(P_M, K_M) = -\int [v]\,g\frac{\partial[z]}{R\,\partial\phi}\,dm = -\int [\omega][\alpha]\,dm \tag{1.48}$$

$$C(K_E, K_M) = -\int [v^*u^*]\frac{\partial[u]}{R\,\partial\phi}\,dm + \int [\omega^*u^*]\frac{\partial[u]}{\partial p}\,dm + \int [v^{*2}]\frac{\partial[v]}{R\,\partial\phi}\,dm + \int [\omega^*v^*]\frac{\partial[v]}{\partial p}\,dm \tag{1.49}$$

and

$$D(K_M) = -\int \left([u][F_\lambda] + [v][F_\phi]\right) dm\,. \tag{1.50}$$

The term $C(P_M, K_M)$ shows that kinetic energy in the zonal mean flow can be generated by horizontal cross isobaric flow down the north-south pressure gradient or by the rising of light air and the sinking of dense air in zonally symmetric meridional overturnings. The term $C(K_E, K_M)$ will lead to a conversion from eddy into mean kinetic energy as long as the north-south and up-down transports of momentum by eddies are up the gradient of the mean zonal wind (*negative* viscosity phenomenon).

The rate of conversion between mean potential and mean kinetic energy can be written in either the $v\,\nabla z$ or the $\omega\alpha$ form.

Similarly as for K_M, the balance equation for K_E can be obtained by multiplying the first and second equations of motion by u^* and v^*, respectively, adding and finally integrating over the entire atmosphere. Thus we may write in symbolic form:

$$\frac{\partial K_E}{\partial t} = C(P_E, K_E) - C(K_E, K_M) - D(K_E)\,, \tag{1.51}$$

New terms are the eddy conversion and eddy dissipations terms

$$C(P_E, K_E) = -\int g\,[\vec{v}^* \cdot \nabla z^*]\,dm = -\int [\omega^* \alpha^*]\,dm \tag{1.52}$$

and

$$D(K_E) = -\int [u^* F_\lambda{}^*]\,dm - \int [v^* F_\phi{}^*]\,dm\,. \tag{1.53}$$

The rate of conversion between eddy potential and eddy kinetic energy, $C(P_E, K_E)$, generates eddy kinetic energy when, along a latitude circle, the wind anomalies tend to be negative correlated with the anomalies in the pressure gradient or when light (warm) air tends to rise and dense (cold) air tends to sink.

In the mixed space-time domain, transient and stationary eddy terms would appear rather than only spatial eddy terms. For example, instead of the simple term $[\omega^*\alpha^*]$ in the conversion rate form eddy potential to eddy kinetic energy in the space domain, we would have $[\overline{\omega'\alpha'}] + [\overline{\omega}^*\,\overline{\alpha}^*]$, showing explicitly the influence of both transient and stationary eddies.

Balance equations for mean and eddy available potential energy in the atmosphere

The balance equation for P_M can be derived using the first law of thermodynamics in the form:

$$\frac{\partial T}{\partial t} = -\vec{v} \cdot \nabla T - \omega \frac{T}{\theta} \frac{\partial \theta}{\partial p} + \frac{\dot{Q}}{c_p}\,, \tag{1.54}$$

by first multiplying it by $c_p \Gamma([T] - \tilde{T})$, where $([T] - \tilde{T})$ is the departure of the zonal average temperature from its global mean, and then integrating it over the entire atmosphere so that the boundary terms disappear. Thus we can write in symbolic form:

$$\frac{\partial P_M}{\partial t} = G(P_M) - C(P_M, P_E) - C(P_M, K_M)\,. \tag{1.55}$$

Here

$$G(P_M) = \int \Gamma\left([T] - \tilde{T}\right)\left([\dot{Q}] - \tilde{\dot{Q}}\right) dm \tag{1.56}$$

indicates the rate of generation of P_M mainly through heating of relatively warm air at low latitudes and cooling of relatively cold air at high latitudes, and

$$C(P_M, P_E) = -c_p \int \Gamma [v^* T^*] \frac{\partial [T]}{R\,\partial \phi}\,dm - c_p \int p^{-\kappa} [\omega^* T^*] \frac{\partial}{\partial p}\left(\Gamma p^{\kappa}([T] - \tilde{T})\right) dm \tag{1.57}$$

indicates the rate of conversion from mean to eddy available potential energy through eddy poleward or eddy upward heat transports down the mean zonal temperature gradient.

Using the same methodology the balance equation for P_E can be derived by multiplying Eq. 1.54 by $c_p \Gamma T^*$ and integrating over the entire atmosphere:

$$\frac{\partial P_E}{\partial t} = G(P_E) + C(P_M, P_E) - C(P_E, K_E). \tag{1.58}$$

The only new term is the rate of generation of P_E:

$$G(P_E) = \int \Gamma[T^* Q^*]\, dm. \tag{1.59}$$

This term operates through the heating of warm air anomalies and the cooling of cold air anomalies along each latitude circle, thereby increasing the east-west variance of temperature and P_E.

The balance equations for the four basic forms of energy are summarized below:

$$\begin{aligned}
\frac{\partial K_M}{\partial t} &= C(P_M, K_M) + C(K_E, K_M) - D(K_M) \\
\frac{\partial K_E}{\partial t} &= C(P_E, K_E) - C(K_E, K_M) - D(K_E) \\
\frac{\partial P_M}{\partial t} &= G(P_M) - C(P_M, K_M) - C(P_M, P_E) \\
\frac{\partial P_E}{\partial t} &= G(P_E) - C(P_E, K_E) + C(P_M, P_E)
\end{aligned}$$

The connecting links between the various energy forms as given by these equations are illustrated

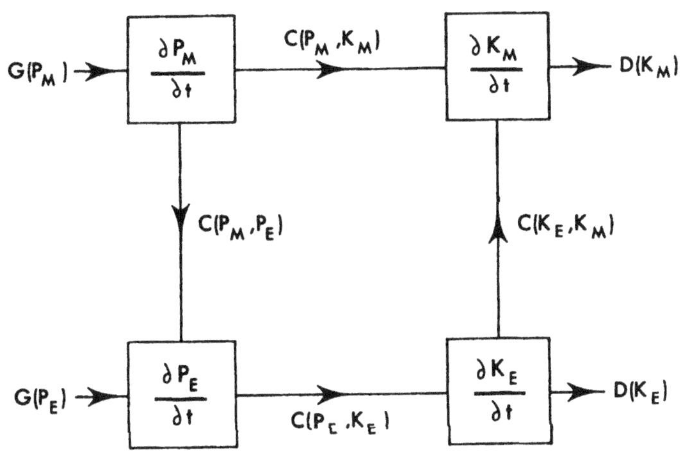

Figure 1.13: *Schematic diagram of the energy cycle in the atmosphere showing the rates of generation, conversion and dissipation of the various forms of energy. (Peixoto and Oort, 1984)*

schematically in Fig. 1.13 in the form of a box diagram. As we have seen before, the generation of available potential energy is proportional to the covariance of the diabetic heating and temperature. Therefore, when regions of high temperature are heated and regions of low temperature are cooled available potential energy will be generated. The generation of available potential energy resembles the behavior of a heat engine where fuel has to be added to heat the arm furnace and cooling has to be provided to maintain the required temperature difference. Basically, the heating of the warm regions and the cooling of the cold regions amount to decreasing the entropy of the system. Obviously, the opposite processes would lead to an increase in entropy and a destruction of available potential energy.

1.4.4 Observed energy cycle in the atmosphere

Spatial distributions of the energy and energy conversions

Meridional and vertical profiles of the mean and eddy energy components are shown in Fig. 1.14 and 1.15. The curves represent the integrands that contribute to the global integrals of the available potential and kinetic energy components. However, the integrands of P_M and P_E are not necessarily meaningful by themselves.

The meridional profiles in Fig. 1.14 show important differences between the Northern and Southern Hemispheres, namely:

(1) Poleward of 50° latitude the contributions to P_M are greater in the Southern than in the Northern Hemisphere for all seasons, because of the larger meridional temperature gradients in the Southern Hemisphere.

(2) On the other hand, the eddy activity shown in transient P_{TE} and standing eddies P_{SE} is higher in the Northern than in the Southern Hemisphere. The very large values of P_{TE} in the northern midlatitudes for the year are due primarily to the large seasonal cycle of the temperature over the northern continents.

(3) The K_M values are in general smaller in the Northern than in the Southern Hemisphere, except during winter.

(4) The K_{TE} values of the transient eddies are of comparable magnitude in both hemispheres.

The vertical profiles of the global averages of the various forms of energy (as illustrated in Fig. 1.15) show that there is a seasonal variation even when the global energy content of the atmosphere is considered. It seems, from the energetics point of view, that the troposphere is more active during the northern winter (DJF) than during the northern summer (JJA) for all energy components.

1.4.5 The energy cycle and the general circulation

The contributions to the directly measurable global integrals of the energy conversions terms, $C(P_M, P_E)$, $C(K_E, K_M)$ and $C(P_M, K_M)$, are shown in Fig. 1.16 and 1.17. The most striking differences between the two hemispheres in Fig. 1.16 are due to the differences in the amplitude of the seasonal cycle. Aside from this fact, the two hemispheres appear to behave much alike.

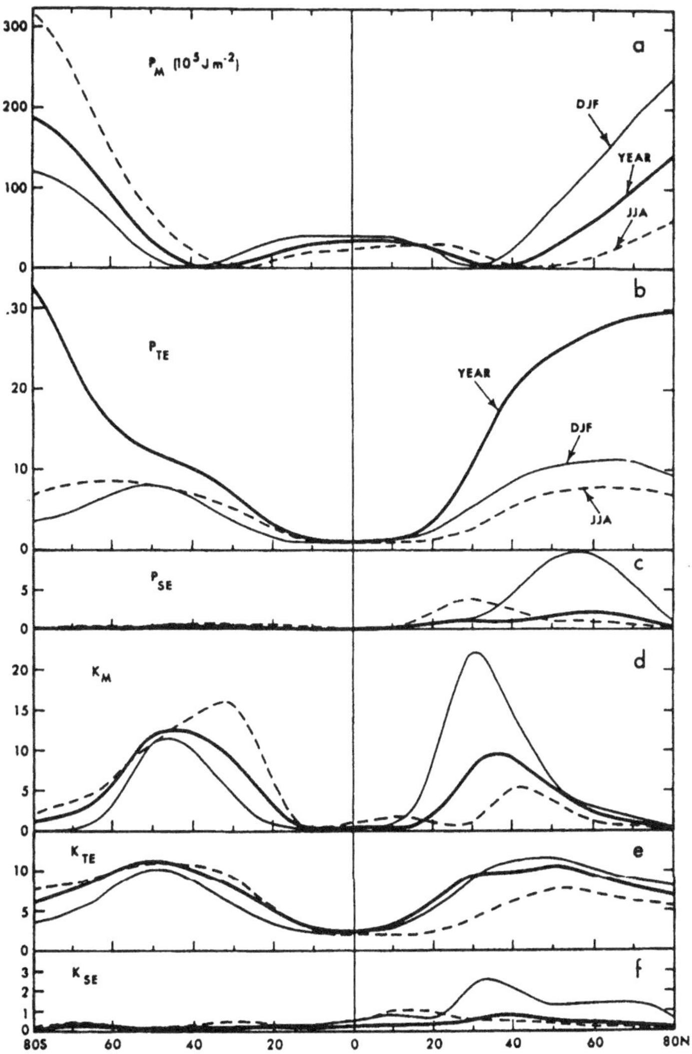

Figure 1.14: *Meridional profiles of the contributions to the global integrals of the mean available potential energy (a), transient eddy available potential energy (b), stationary eddy available potential energy (c), mean kinetic energy (d), transient eddy kinetic energy (e) and stationary eddy kinetic energy (f) in the atmosphere in units of* $10^5\, J\, m^{-2}$ *(Oort and Peixoto, 1984)*

The annual-mean energetics of the global atmosphere are summarized in Fig. 1.18 in the form of a schematic box diagram of the type shown before in Fig. 1.13.

We have now reached the point where we can begin to describe the general mechanisms by which the incoming solar radiation maintains the kinetic energy of the atmosphere and oceans against frictional dissipation. As primary forcing, the radiational heating and cooling in the atmosphere tend to generate an almost zonally symmetric distribution of temperature in both hemispheres with

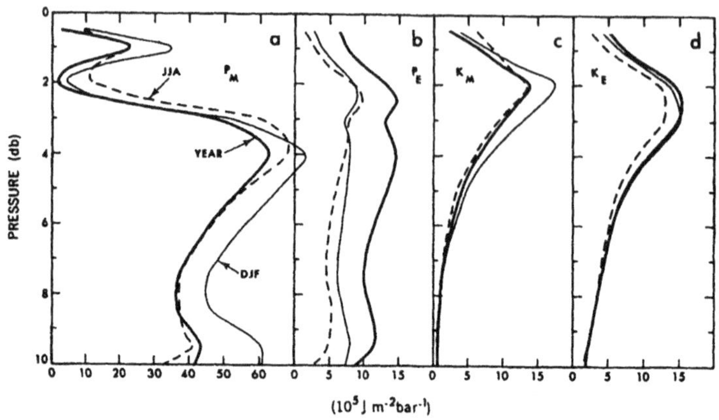

Figure 1.15: *Vertical profiles of the contributions to the global integrals of the mean available potential energy (a), eddy available potential energy (b), mean kinetic energy (c) and eddy kinetic energy (d) in the atmosphere in units of* $10^5\ J\,m^{-2}\,bar^{-1}$. *The transient and stationary eddies are combined in the eddy terms (Oort and Peixoto, 1984)*

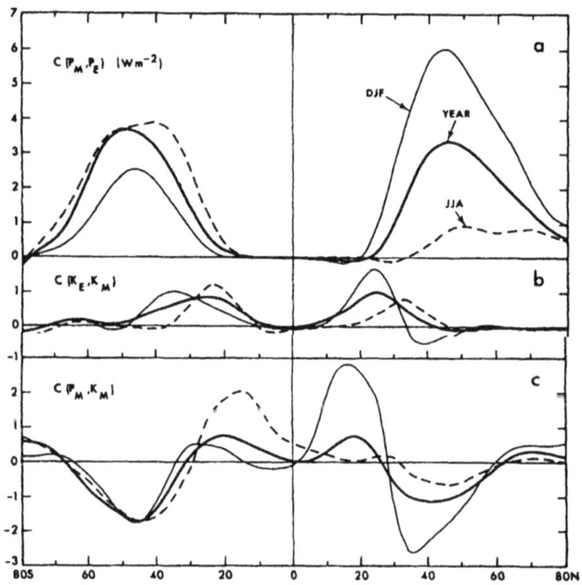

Figure 1.16: *Meridional profiles of the contributions to the global integrals of the conversion rates from mean to eddy available potential energy (a), eddy to mean kinetic energy (b) and mean available potential to mean kinetic energy (c) in the atmosphere in units of* $W\,m^{-2}$ *(Oort and Peixoto, 1983)*

a strong north-south gradient which is most intense in middle latitudes. The north-south variance of the zonal-mean temperature at each level is a measure of the mean zonal available potential energy P_M. However, the meridional transport of heat by the eddies in midlatitudes deforms the ideal zonally symmetric distribution of temperature leading to an additional variance in the east-west direction which is proportional to the eddy available potential energy P_E. Through mechanisms such

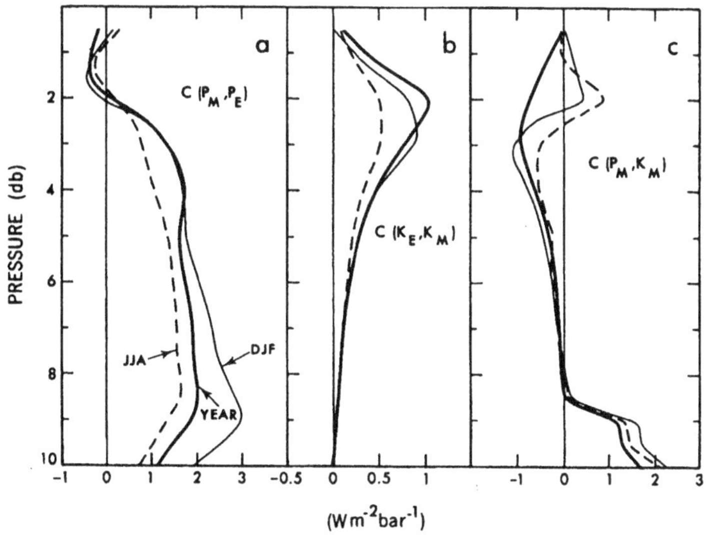

Figure 1.17: *Vertical profiles of the contributions to the global integrals of the conversion rates from mean to eddy available potential energy (a), eddy to mean kinetic energy (b) and mean available potential to mean kinetic energy (c) in the atmosphere in units of* $W\,m^{-2}\,bar^{-1}$ *(Oort and Peixoto, 1983)*

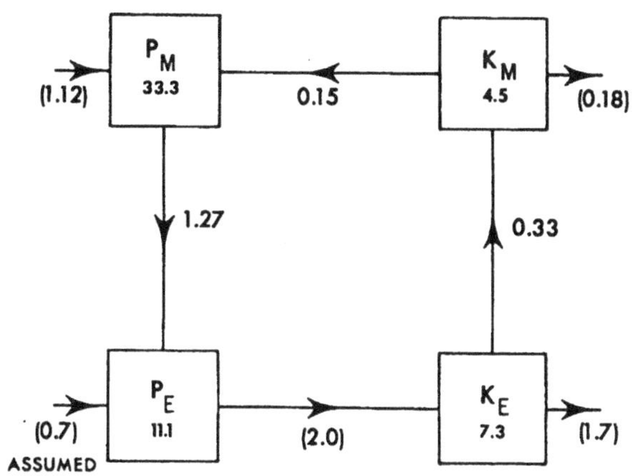

Figure 1.18: *The observed energy cycle in the global atmosphere for annual mean conditions. Since the rates of change of energy are negligible, the energy amounts are given instead inside each box in units of* $10^5\,J\,m^{-2}$. *The rates of generation, conversation and dissipation are in* $W\,m^{-2}$. *Not directly measured terms are shown in parentheses (modified after Oort and Peixoto, 1983)*

as baroclinic instability, some of the eddy available potential energy is transformed into kinetic energy of the growing disturbances K_E. These eddies also transport momentum poleward increasing the zonal motion, and thus generating zonal mean kinetic energy K_M. A large part of the eddy kinetic energy is directly dissipated by friction or cascades into smaller and smaller eddies (turbulence) until it is finally converted into heat by molecular viscosity. This is a consequence of the second law of thermodynamics. In the absence of external forcing the system tends to evolve to states of larger and larger disorder. Of course, the maximum disorder or entropy is obtained when the ordered energy is converted into random molecular motions. The dissipation seems to be most intense in the eddy kinetic energy.

The mean meridional circulations in the tropics (Hadley cells) generated zonal angular momentum in their upper branches and thereby mean zonal kinetic energy. However, the indirect circulations in midlatitudes of each hemisphere (Ferrel cells) consume mean zonal kinetic energy at a rate slightly exceedings the production in the Hadley cells. Therefore, for the atmosphere as a whole part of the mean zonal kinetic energy is converted back into mean zonal available potential energy.

Considering presently the numerical estimates from Oort and Peixoto (1983) shown in Fig. 1.18, we are led to the following scheme for the energy cycle. Both the net heating in the tropics by the absorption of solar radiation and by the release of latent heat and the net cooling in high latitudes by the loss of terrestrial radiation result in the generation of zonal mean available potential energy at the rate of about $1.1 \cdot W m^{-2}$. This energy is converted into eddy available potential energy by the growing baroclinic disturbances at a rate of about $1.3 \cdot W m^{-2}$. The fastest growing eddies are in the range of wavenumber 5 to 8 about the pole. Some of the eddy available potential energy may be dissipated by a greater loss of infrared radiation to space in warm than in cold air and by a similar heat exchange with the earth's surface.

However, latent heating is believed to surpass these effects leading to a small positive generation of P_D, assumed to be on the order of $0.7 \cdot W m^{-2}$. The eddy available potential energy is next converted into eddy kinetic energy at the rate of about $2.0 \cdot W m^{-2}$ through the sinking of relatively cold air and the rising of warm air in the eddies. As we know, there is a weak but prevalent tendency of counter gradient flow of momentum up the gradient of zonal mean angular velocity. Therefore, some of the eddy kinetic energy is converted into kinetic energy of the mean flow in a barotropic process leading to a decascade of energy from the small into the larger scales ($0.3\, W m^{-2}$), which can be interpreted as a negative viscosity phenomenon (Starr, 1968). However, the bulk of the kinetic energy of the large-scale eddies is dissipated by friction in a normal cascade regime ($1.7\, W m^{-2}$).

Finally, some of the mean zonal kinetic energy is dissipated by friction and turbulence ($0.2\, W m^{-2}$), while a small residual is converted into mean zonal available potential energy ($0.15\, W m^{-2}$) by the combined action of the direct and indirect mean meridional circulations. If one compares the total amount of kinetic energy with the rate of dissipation of kinetic energy we find that it would take about 8 days to dissipate the kinetic energy.

1.4.6 The efficiency of the energy cycle

Schematically we have found that the energy cycle in the atmosphere proceeds from P_M to K_M through the following scheme

$$P_M \rightarrow P_E \rightarrow K_E \rightarrow K_M \,. \tag{1.60}$$

Confirming that the eddies play a crucial role in regulating the general circulation of the atmosphere as suggested by Starr (1948).

For the year as a whole, the eddies (transient plus stationary) form an important fraction of the global kinetic (62%) and global available potential energy (25%). The ratio of the amount of kinetic energy to the total amount of kinetic plus available potential energy is remarkably stable with the same value of about 21% for each hemisphere independent of the season.

The intensity of the atmospheric energy cycle can be measured by estimating either the rate of conversion of available potential energy into kinetic energy $C(P,K)$, the rate of generation of available potential energy $G(P)$ or the rate of dissipation of kinetic energy $D(K)$, since in a steady state all three terms must balance. Regarded again as a heat engine, the efficiency of the atmosphere can be taken as the ratio of the kinetic energy dissipated by friction ($2.0\,W\,m^{-2}$) and the mean incoming solar radiation ($238\,W\,m^{-2}$) which leads to a value for the efficiency $\eta \approx 0.8\%$. Alternatively, we may define the efficiency as the ratio $D(K)/P$ giving a value of about 4%. These values of the efficiency are more meaningful than the estimate by Brunt (1939) who used the total energy instead of the available energy in the denominator. They are also better than those based on a Carnot cycle because they involve the real input of energy into the system. The values for the Carnot efficiency were found earlier to be on the order of 10%.

1.5 Entropy in the climate system

1.5.1 Introduction

There are numerous irreversible processes in the climate system that lead to an increase of entropy. Among these are the absorption of solar and terrestrial radiation, melting of snow and ice, condensation, evaporation, erosion by the winds and running water, and the turbulent and molecular diffusion of gases, heat and momentum.

How can we explain the high level of organization in certain atmospheric processes characterizing the weather, and the fact that these processes can evolve in an orderly way that, at first sight, would seems to lead to a decrease of entropy. All these processes seem to proceed against the law of increase of entropy, and all occur due to the high quality (i.e., low entropy) of the incoming solar radiation and its systematic variation with latitude. It is, of course, the solar energy that heats the earth's surface and the atmosphere unevenly, thereby generating the global wind systems, producing the evaporation of water as one of the vital components of the hydrological cycle, and maintaining photosynthesis, among many other processes.

Let us compare the quality of the shortwave solar radiation ($\lambda_{max} \approx 0.48\,\mu m$) with the quality of the longwave radiation ($\lambda_{max} \approx 10\,\mu m$) emitted by the earth. According to the Planck equation, the energy of a photon ε is given by $\varepsilon = h\nu$, and thus is inversely proportional to the wavelength ($\lambda = c/\nu$). Thus solar photons are richer in energy than terrestrial photons. In other words, the amount of entropy associated with the incoming solar radiation is much lower than the amount of entropy associated with the emitted terrestrial radiation, and the climate system receives high-quality *rich* energy and returns low-quality *impoverished* energy to space. Thus solar radiation revitalizes the meteorological phenomena, feeds the hydrological cycle and renovates the biosphere. If the earth were an isolated system there would be an unavoidable increase in entropy leading to a death-like uniformity of the planet earth. It is this capacity for permanent renovation that makes all natural phenomena possible in the climate system.

The second law of thermodynamics implies that energy can only change from a higher to a lower level of availability, i.e., energy can only change from a more to a less usable or a less ordered form. The second law implies the existence of a functions s, the entropy, that for an isolated system increases monotonically until it reaches its maximum value at the state of thermodynamic equilibrium

$$\frac{ds}{dt} > 0\,. \tag{1.61}$$

Entropy can be used as a measure of the unavailable energy. An increase in entropy means a decrease in available energy and evolution toward a state of greater disorder.

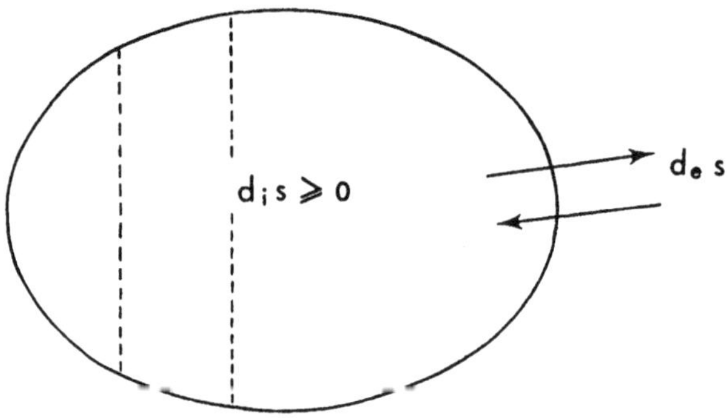

Figure 1.19: *Schematic diagram of the climate system showing the transfer of entropy across the boundaries and the production of entropy inside the system.*

The second law can also be extended to open systems that exchange energy and matter with their surroundings. We must then consider two components in the total entropy change dS. One component d_eS represents the transfer of entropy across the boundaries of the open system and the other component d_iS is the total entropy produced within the system. According to the second law, the rate of generation of entropy inside the system is always positive (see Figure 1.19) so that (Peixoto et al., 1991):

$$\frac{dS}{dt} = \frac{d_eS}{dt} + \frac{d_iS}{dt}\,, \tag{1.62}$$

and

$$\frac{d_iS}{dt} > 0\,. \tag{1.63}$$

Using this formulation the basis differences between reversible and irreversible processes become very clear. Only irreversible processes contribute to entropy production.

The rate of change of entropy per unit mass is given by

$$\frac{ds}{dt} = \frac{\dot{Q}}{T}, \tag{1.64}$$

where $\dot{Q}$ is the rate of diabatic heating per unit mass. As we have seen before (see section 1.3.2) the diabatic heating includes radiational heating, the release of latent heat, and the heating by conduction and frictional dissipation:

$$\dot{Q} = -\frac{1}{\rho}\left(\nabla \vec{F}_{rad} + \nabla \vec{F}_{LH} + \nabla \vec{J}_H^D\right) - \frac{1}{\rho}\vec{\vec{\tau}} : \nabla \vec{c}. \tag{1.65}$$

where ρ is the density of air. $\vec{F}_{LH}$ is the latent heat flux, $\vec{F}_{rad}$ is the net radiative flux (solar plus terrestrial), J_H is the sensible heat flux due to conduction, τ is the wind stress tensor, and $\vec{c}$ is the wind velocity.

1.5.2 Balance equation of entropy

General equation

A balance equation of entropy can now be obtained by substituting the expression for diabatic heating into Eq. 1.64.

$$\rho\frac{ds}{dt} = -\frac{\nabla \vec{F}_{rad}}{T} - \frac{\nabla \vec{F}_{LH}}{T} - \frac{\nabla \vec{J}_H^D}{T} - \frac{\vec{\vec{\tau}} : \nabla \vec{c}}{T} \tag{1.66}$$

or

$$\begin{aligned}\rho\frac{ds}{dt} &= -\nabla\frac{1}{T}\left(\vec{F}_{rad} + \vec{F}_{LH} + \vec{J}_H^D\right)\\ &\quad -\frac{1}{T^2}\left(\vec{F}_{rad} + \vec{F}_{LH} + \vec{J}_H^D\right)\cdot\nabla T - \frac{\vec{\vec{\tau}} : \nabla\vec{c}}{T}\end{aligned} \tag{1.67}$$

Let us expand the total derivative on the left hand side of Eq. 1.67 and let us integrate the resulting equation over the entire atmosphere. This leads to

$$\begin{aligned}\frac{\partial}{\partial t}\int_V \rho\, s\, dV &= -\int_A \rho\, s\, \vec{c}\cdot\vec{n}\, dA - \int_A \frac{1}{T}\vec{\mathcal{F}}\cdot\vec{n}\, dA\\ &\quad -\int_V \frac{1}{T^2}\left(\vec{\mathcal{F}}\cdot\nabla T\right) dV - \int_V \frac{1}{T}\left(\vec{\vec{\tau}} : \nabla\vec{c}\right) dV\end{aligned} \tag{1.68}$$

where

$$\vec{\mathcal{F}} = \vec{F}_{rad} + \vec{F}_{LH} + \vec{J}_H^D. \tag{1.69}$$

A is the area of the globe, and $\vec{n}$ is the unit vector direct outward at right angles to the boundaries of the atmosphere.

The first term on the right hand side of Eq. 1.68 is associated with the net mass transport across the boundaries which we can assume to be zero for the entire atmosphere. The second term represents the flux of entropy across the boundaries, and the third and fourth terms the rate of generation of entropy associated with the various irreversible processes inside the atmosphere.

Simplified equation

It is convenient to rewrite the various boundary terms in Eq. 1.68 introducing an appropriate reference or equivalent temperature T^* for each component so that

$$\frac{1}{T^*} \equiv \frac{\int_A \frac{1}{T} \vec{F} \cdot \vec{n}\, dA}{\int_A \vec{F} \cdot \vec{n}\, dA} . \tag{1.70}$$

For example, for the radiation we find

$$-\int_A \frac{\vec{F}_{rad}}{T} \cdot \vec{n}\, dA = -\frac{1}{T^*_{rad}} \int_A \vec{F}_{rad} \cdot \vec{n}\, dA = \frac{G_{rad}}{T^*_{rad}} , \tag{1.71}$$

where G_{rad} is the radiative flux across the boundaries into the atmosphere and T^*_{rad} the reference radiative temperature. For the other boundary terms we can obtain similar expressions so that Eq. 1.68 can be rewritten symbolically:

$$\frac{\partial S}{\partial t} = \frac{G_{rad}}{T^*_{rad}} + \frac{G_{LH}}{T^*_{LH}} + \frac{G_{SH}}{T^*_{SH}} + \sigma_{rad} + \sigma_{LH} + \sigma_{SH} + \sigma_{dis} . \tag{1.72}$$

In this equation S is the total entropy of the atmosphere and σ_{rad}, σ_{LH}, σ_{SH} and σ_{dis} denote the rate of generation of entropy by radiation, latent heat release, conduction of heat and kinetic energy dissipation, respectively, which are all associated with irreversible processes in the atmosphere.

The latent heat flux $G_{LH} = -\int \vec{F}_{LH} \cdot \vec{n}\, dA$, when integrated over a unit volume, will lead to a term $-L \int (e - c)\, dV$ where e is the rate of evaporation plus sublimation and c the rate of condensation and freezing. For the atmosphere as a whole we can assume that most of the evaporation takes place at the earth's surface and that the condensation occurs mainly in clouds away from the surface, leading to a net upward transport of heat from the surface into the atmosphere. Because the evaporation takes place at lower levels and at higher temperatures than the condensation, a sizeable generation of entropy must be connected with these phase transitions. The mixing of tropical and polar air masses (with different temperatures and humidities) also contributes to the generation of entropy.

For a sufficiently long interval of time we may assume that, on the average, the atmosphere is in a steady state. Therefore, averaging of the global entropy balance equation (1.72) over time leads to

$$\overline{\left(\frac{G_{rad}}{T^*_{rad}}\right)} + \overline{\left(\frac{G_{LH}}{T^*_{LH}}\right)} + \overline{\left(\frac{G_{SH}}{T^*_{SH}}\right)} + \overline{\sigma}_{rad} + \overline{\sigma}_{LH} + \overline{\sigma}_{SH} + \overline{\sigma}_{dis} = 0 . \tag{1.73}$$

Under steady state conditions the corresponding energy terms must also obey the energy balance equation:

$$\overline{G}_{rad} + \overline{G}_{LH} + \overline{G}_{SH} = 0 . \tag{1.74}$$

Let us discuss now further the reference temperature. If we assume that the fluctuations of the reference temperatures are small, i.e., $T'/\overline{T^*} \ll 1$, then

$$\overline{\left(\frac{G}{T^*}\right)} = \frac{\overline{G}}{\overline{T^*}} - \frac{\overline{G'T'}}{\overline{T^*}^2} \tag{1.75}$$

where $T' \equiv T^* - \overline{T^*}$ and $\overline{T^*}$ is the time average of the instantaneous reference temperature and not the reference temperature of the mean state. As preliminary calculations show, we can use the globally-averaged mean temperature, $\tilde{T}$, instead of $\overline{T^*}$. Thus, Eq. 1.73 can be written as

$$\frac{\overline{G}_{rad}}{\tilde{T}^*_{rad}} + \frac{\overline{G}_{LH}}{\tilde{T}^*_{LH}} + \frac{\overline{G}_{SH}}{\tilde{T}^*_{SH}} + \overline{\sigma}_{rad} + \overline{\sigma}_{LH} + \overline{\sigma}_{SH} + \overline{\sigma}_{dis} = 0\,, \tag{1.76}$$

which is used in the following calculations.

1.5.3 Observed entropy budget of the atmosphere

To evaluate the entropy budget in the atmosphere we begin by estimating the first four terms in Eq. 1.76, that represent the fluxes of entropy across the top and bottom boundaries of the atmosphere. This requires computing both the mean energy fluxes G across the boundaries and the corresponding (energy-flux weighted) reference temperatures T^* for these boundaries.

The annual-mean energy flux values at the top and bottom of the atmosphere given before in Fig. 1.11 of section 1.3.1 (as percentages of the total incoming solar radiation) can be used to estimate the values of the entropy flux terms, assuming that the mean incident solar radiation is $340\,W\,m^{-2}$ (a solar constant of $1360\,W\,m^{-2}$ is considered). The results are presented in Fig. 1.20. The figure shows the chosen values of the mean temperatures, the energy fluxes and the estimated entropy fluxes at the top and bottom of the atmosphere. The mean temperatures for the downward solar radiation flux and the upward fluxes at the earth's surface were assumed to be 5760 K (the effective temperature of the sun) and 288 K, respectively. We will use a positive sign for the fluxes of energy and entropy when they represent a gain of energy or entropy for the atmosphere and a negative sign in the case of a loss of energy or entropy for the atmosphere.

The three terms in the upward flux of entropy at the top of the atmosphere are associated with the fluxes of infrared radiation emitted by the earth's surface (through the spectral window $8.5-11.0\,\mu m$), the atmosphere and the clouds. For the emission of infrared radiation by the atmosphere a mean temperature of 252 K was used and for the emission by clouds a value of 259 K.

At the bottom of the atmosphere we estimated the values of the fluxes of entropy associated with the downward flux of solar radiation ($-29.5 mWm^{-2}K^{-1}$), the upward sensible heat flux ($71 mWm^{-2}K^{-1}$), the upward latent heat flux ($283\,mW\,m^{-2}\,K^{-1}$) and the emitted longwave radiation ($236 mWm^{-2}K^{-1}$). As expected from the high quality of the incoming solar radiation the fluxes of entropy in the solar radiation are small both at the top and bottom of the atmosphere with values of 41.3 and $-29.5\,mWm^{-2}K^{-1}$, respectively.

When we add all boundary fluxes together, the total flux of entropy at the top of the atmosphere is $-884\,mWm^{-2}K^{-1}$ and the total flux at the bottom of the atmosphere $278\,mWm^{-2}K^{-1}$. Thus according to Eq. 1.73 the mean rate of generation of entropy by internal processes in the atmosphere under steady conditions must be $\sigma = (884 - 278)\,mWm^{-2}K^{-1} = 606\,mWm^{-2}K^{-1}$.
As we see from the values given in Fig. 1.20 the total amount of entropy exported by the climate system to space is $(-71 - 513 - 341)\,mWm^{-2}K^{-1} = -925\,mWm^{-2}K^{-1}$. This value is twenty two times the amount of entropy imported by the incoming solar radiation at the top of the atmosphere

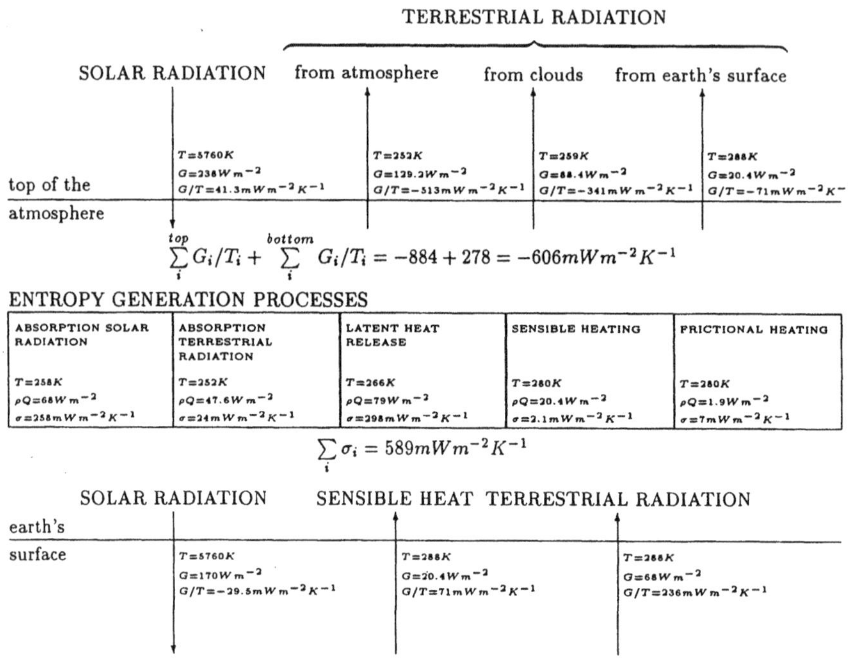

Figure 1.20: *Entropy budget of the global atmosphere estimated for annual mean conditions. Shown are the equivalent temperature T^* in K, the energy flux G in $W\,m^{-2}$ and the entropy flux G/T^* in units of $mW\,m^{-2}\,K^{-1}$ for each energy component at the top of the atmosphere and at the earth's surface. A positive (negative) sign is used when the atmosphere gains (loses) energy or entropy. The boxes in the middle of the figure contain for each component estimates of the atmospheric temperature where the absorption takes place in K, the rate at which energy is absorbed $\rho\dot{Q}$ in $W\,m^{-2}$ and the rate of entropy production σ in units of $mW\,m^{-2}\,K^{-1}$.*

($41.3\,mWm^{-2}K^{-1}$). The largest fluxes of entropy are associated with the longwave radiation fluxes ($236\,mWm^{-2}K^{-1}$ at the earth's surface and $-925\,mWm^{-2}K^{-1}$ at the top).

Another approach is to compute directly the rate of production of entropy within the atmosphere by evaluation the individual contributions by the irreversible processes that occur inside the atmosphere. We will again make use of the energy flux values given in Fig. 1.11.

Let us consider as an example the evaluation of the generation of entropy due to the absorption occurs near the $500\,hPa$-level where the temperature is about $252\ K$, so that

$$\sigma_{rad}^{sw} = \left(\frac{68}{252} - \frac{68}{5760}\right) W\,m^{-2}\,K^{-1} = 258\,mW\,m^{-2}\,K^{-1}$$

For the latent heat release we assume a net precipitation rate of $1\,m\,yr^{-1}$ and a condensation temperature of 266 K, so that

$$\sigma_{LH} = \frac{\rho_w LP}{T} = \frac{79}{266} W\,m^{-2}\,K^{-1} = 298\,mW\,m^{-2}\,K^{-1},$$

For the sensible heat exchange processes we assumed an average temperature of 280 K as representative of the bottom layer of the atmosphere. The estimated temperatures, the rate of diabatic heating and the rate of entropy generation for the various individual processes are shown in the rectangular boxes in the middle of Fig. 1.20.

Summing the entropy generation rates due to the various processes we find a value of $589\, mW\, m^{-2}\, K^{-1}$ for the total mean rate of generation. This value can be compared with the previous value of $606\, mW\, m^{-2}\, K^{-1}$ obtained from the global budget of the atmospheric energy. The difference is only $17\, mW\, m^{-2}\, K^{-1}$ and can be attributed to (apart from errors) the approximations made and some neglected effects (e.g. mixing of air masses).

In summary, the values presented in Fig. 1.20 show that the absorption of solar radiation and the release of latent heat are, by far, the largest source of entropy in the atmosphere. The rate of production of entropy associated with the absorption of longwave radiation is an order of magnitude smaller since the temperatures of infrared emission and absorption are not very different. Among the nonradiative processes, the water phase transitions dominate the entropy generation at low latitudes.

1.6 References

Baumgartner A, Reichel E (1975) The World Water Balance. *Elsevier*, Amsterdam, 179 pp.

Houghton JT (ed.) (1984) The Global Climate. *Cambridge Univ Press*, 233 pp.

Houghton HG (1985) Physical Meteorology. *The MIT Press*, Cambridge, Massachusetts, 442 pp.

Jäger L (1976) Monatskarten des Niederschlags für die ganze Erde. *Ber. Dt. Wetterdienstes* 18: 38 pp.

Lorenz EN (1955) Available potential energy and the maintenance of the general circulation. *Tellus* 7: 157-167.

Lorenz EN (1967) The nature and theory of the general circulation of the atmosphere. *WMO Publ* 218: World Meteorological Organization, Geneva, Switzerland, 161 pp.

Lorenz EN (1969) The predictability of a flow which processes many scales of motion. *Tellus* 21: 289-307.

Margules M (1903) Über die Energie der Stürme. *Jahrb Zentralanst Meteor* 1–26.

Oort AH (1983) Global atmospheric circulation statistics, 1958-1973. *NOAA Professional Paper* No. 14, U.S. Gov't Printing Office, Washington D.C., 180 pp. + 47 microfiches.

Oort AH, Peixoto JP (1983) Global angular momentum and energy balance requirements from observations. *Adv Geophys* 25: 355-490.

Peixoto JP (1973) Atmospheric vapor flux computations for hydrological purposes. *WMO Publ* 357: Geneva, Switzerland, 83 pp.

Peixoto JP, Kettani MA (1973) The control of the water cycle. *Sci Am* 228: 46-61.

Peixoto JP, Oort AH (1983) The atmospheric branch of the hydrological cycle and climate. *In: Variations of the Global Water Budget.* Reidel, London, England, 5-65.

Peixoto JP, Oort AH (1984) Physics of climate. *Rev Modern Phys* 56: 365-429.

Peixoto JP, Oort AH, de Almeida N, Tome A (1991) Entropy budget of the atmosphere. *J Geophys Res* 95: D6, 10981-10988.

Saltzmann B (1978) A survey of statistical-dynamical models of the terrestrial climate. *Adv Geophys* 20: 183-304.

Saltzmann B (1983) Climate systems analysis. *Adv Geophys* 25: 173-233.

Sellers WD (1965) Physical Climatology. *The University of Chicago Press*, Chicago, 272 pp.

Starr VP (1968) Physics of negative viscosity phenomena. *McGraw-Hill*, New York, 256 pp.

Starr VP, Peixoto JP (1958) On the global balance of water vapor and the hydrology of deserts. *Tellus* 10: 189-194.

Chapter 2

ASPECTS OF LARGE SCALE MODELLING

Ernst Klinker
European Centre for Medium-Range Weather Forecasts
Shinfield Park
Reading, Berkshire R62 9AX
United Kingdom

2.1 Introduction

Large scale modelling plays in important part in several areas, which will be covered by GEWEX activities. One objective, the determination of the hydrological cycle and the energy fluxes from observations, requires an analysis system that can process satellite data and in situ data in an optimum way. As the available observational data is insufficient for a comprehensive description of the state of the atmosphere, the information from current observations has to be complemented by short range forecast fields. In this way the numerical model acts as a tool to interpolate the fields in space and extrapolate them in time. The quality of numerical weather prediction models (NWP-models) has been improved to such an extent, that short range forecast errors in data dense areas may be more accurate than observations. Even in data sparse regions the forecast fields can provide useful information for the quality control of observations.

The quality of the large scale analyses has been improved in line with more accurate NWP-models. However, using those models in data assimilation systems some analysis errors can be traced back to model deficiencies. The handling of subgrid scale processes in large scale models is one area where model errors are most likely.

The finite difference form of the hydrodynamic equations describes only motions for scales down to twice the grid length. However, small scale processes can transport a significant amount of heat, moisture and momentum. In large scale models only the statistical behaviour of these processes can be described by expressing the subgrid scale transports in terms of the large scale variables. The fluxes obtained in this parameterization approach depend very much on the required simplifying

NATO ASI Series, Vol. I 5
Energy and Water Cycles in the Climate System
Edited by E. Raschke and D. Jacob

assumptions. Especially in the tropics different types of parametrization for convection can lead to a significantly different three dimensional circulation in the forecast, which also enters the analysis as the same forecast model is normally used for data assimilation. Therefore analyses from NWP centres should be used with caution. The dependence of the analysis on a frequently changing assimilation system poses a special problem for using long time series for climate studies.

The modelling of the global hydrological cycle, which is also part of a GEWEX objective, can be investigated with NWP-models in the deterministic range or with climate models for long time integration. Climate simulations are normally run with a rather low resolution compared to NWP-models. Though the parametrization schemes are of similar standard as those of operational models, the simulation of the hydrological cycle suffers from errors in the simulation of the flow pattern in these climate runs. This poses a difficult problem to identify deficiencies in the parametrization of subgrid-scale processes.

Using the deterministic forecast range to describe the simulation of the hydrological cycle as the advantage that a direct comparison with observations is more meaningful. Especially for the short time range the predicted large scale flow pattern is still close enough to the analyzed state of the atmosphere that more direct conclusions on the simulation of the hydrological cycle are possible.

At the NWP-centres the development of prediction models is an ongoing process. At ECMWF the vertical and horizontal resolution have been changed in several steps. In 1985 from a T63L16 (spectral resolution T63 = 63 resolved waves on a latitude circle with half wavelength of 317 km, 16 levels in the vertical) to T106L19. The next step followed with an increase of the horizontal resolution to T213 and the vertical resolution to 31 levels in September 1991. Changes in the parametrization part of the ECMWF model involved processes like radiation (Morcrette, 1991), cumulus convection (Tiedtke, 1989) and surface flux of latent heat (Miller et al., 1992, Palmer et al., 1992). These processes are all of high importance for a realistic simulation of the hydrological cycle. The frequent change of the model formulation does not allow a static description of the representation of the hydrological cycle in the ECMWF forecasting system. In effect, the way in which components of the energy budget and the hydrological cycle respond to model changes, provides an excellent help to understand the importance of some physical mechanisms involved.

The next section contains some aspects of the analysis, which are relevant for the hydrological cycle. This will be followed by a discussion of the zonal mean and the geographical distribution of single components of the hydrological cycle and the radiation budget in sections 2.3 to 2.5. The global mean of the energy budget and the hydrological cycle in section 2.6 provides an overall view of the performance of a large scale model.

2.2 Aspects of large scale analyses

Before discussing details of the hydrological cycle it is worth spending some thoughts on the quality of the analysis. Especially the humidity analysis has an important influence on a realistic simulation of the surface latent heat flux and precipitation. Illari (1989) has shown that the satellite derived precipitable water content (PWC) has useful information, which contributes to an improvement of the humidity analysis. However, the statistical retrieval method to derive PWC from radiance measurements of the TIROS-N Vertical Sounder (TOVS) produces rather smooth spatial and temporal structures. This can have a detrimental effect in areas with large horizontal or vertical gradients or during times of rapid changes. Tjemkes and Stephens (1990) have shown that the precipitable water derived from the Special Sensor Microwave/Imager (SSM/I) produces a much higher temporal variance and does not have the zonalization effect seen in the TOVS derived precipitable water.

In Fig. 2.1 average PWC value, produced in the Remote Sensing System, Sausalito, California are compared with the PWC of the ECMWF forecast and analysis for July 1990 (Klinker, 1990). The difference between the SSM/I PWC for the three July months and the ECMWF analysis for July 1990 is shown in Fig. 2.2.

In the tropical convergence zone the ECMWF analysis underestimates PWC. The microwave measurements generally exceed 50 $kg\,m^{-2}$ in this area and 55 $kg\,m^{-2}$ in the Gulf of Bengal, the East Pacific and the extreme West Pacific. Along the tropical convergence zone the ECMWF analysis is drier by around 5 $kg\,m^{-2}$. The differences are particularly large in the Indian Ocean just south of the Equator where the ECMWF analysis is up to 10 $kg\,m^{-2}$ drier.

In the sub-tropics the opposite picture emerges. The SSM/I analysis is drier in the oceanic trade wind areas. The largest differences are found on the eastern side of the continents where the sea surface temperature is relatively low. Here the ECMWF analysis contains around 10 $kg\,m^{-2}$ more precipitable water. From data assimilation experiments with and without satellite data it is known that the use of TOVS-derived PWC in the humidity analysis is mainly responsible for the moistening of the sub-tropics.

There are remarkable changes of the PWC in the first 5 days of the forecast. The difference map between the day-5 forecast and the analysis for July 1990 (Fig. 2.2) suggests that the model has a general tendency of drying the sub-tropics and moistening large areas of the tropics where the deep convection occurs. The similarity of the PWC forecast change to the PWC difference between the SSM/I analysis and ECMWF analysis (Fig. 2.2) suggests that the model approaches a state that agrees more with the SSM/I analysis than with the ECMWF analysis. The forecast changes in the Indian Ocean along the equator are a particularly good example for this. The underestimate of PWC in the ECMWF analysis compared to SSM/I of up to 10 $kg\,m^{-2}$ is almost identical with the increase during the operational forecast by day-5.

The dependence of the analysis on the model formulation used in the assimilation system shows in a time series of humidity for the ECMWF analysis and forecast (Fig. 2.3). The model change in May 1985 (higher resolution and introduction of shallow convection) resulted in a reduction of humidity in the forecast and in the analysis. The tendency to dry the layer around 700 hPa was only corrected with the introduction of the mass-flux scheme for convection in May 1989. This model change affected the analysis as well by increasing the analyzed humidity at that level.

Large scale analysis fields are often used to derive the diabatic heating as a residual of the adiabatic tendency equation for temperature. At ECMWF this has been done in a simple way by averaging over a large number adiabatic initial model tendencies. The negative value of this tendency, corrected by the long term local change, represents the required diabatic forcing to close the budget equation

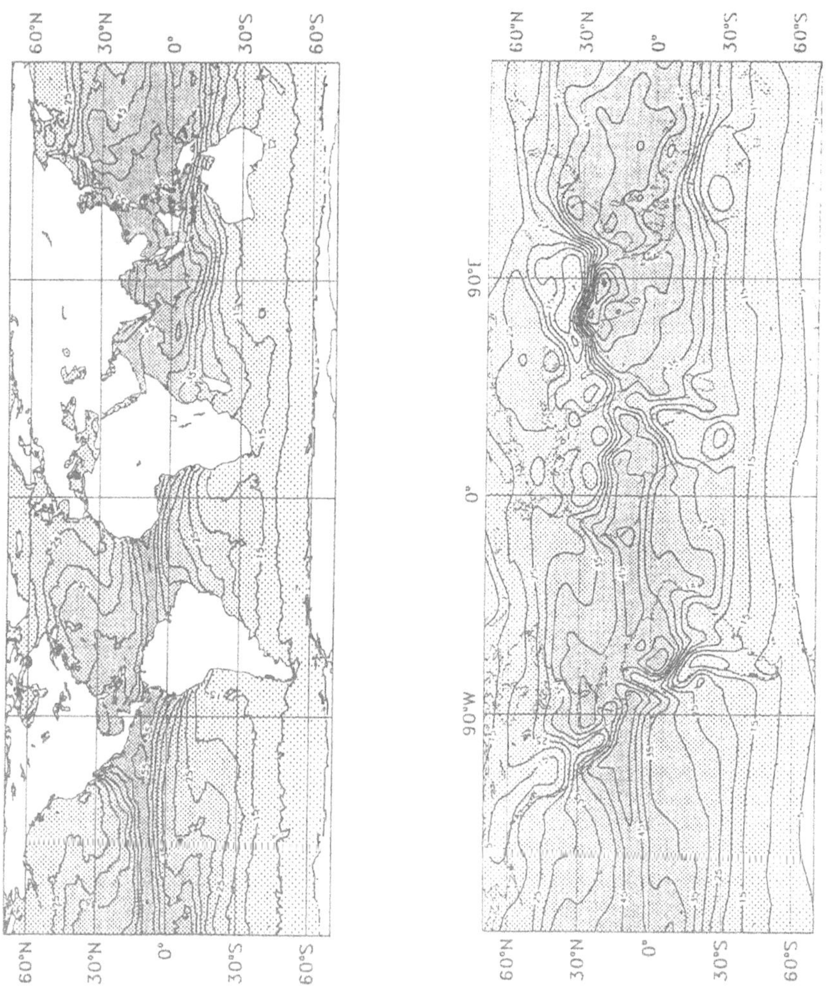

Figure 2.1: *PWC derived from SSM/I observations as a mean over July 1987, July 1988 and July 1989 (upper figure), and mean PWC of the ECMWF analysis for July 1990. (lower figure) Units:* $kg\,m^{-2}$

temperature. Errors in the parametrization of subgrid-scale heating can be found by comparing the required heating with the output of the model's parametrization package (Klinker and Sardeshmukh, 1987). The adiabatic model tendencies calculated from analyses of successive summers shows that the introduction of the new radiation scheme and the mass-flux scheme in May 1989 produced a remarkable increase of the inferred diabatic heating (Fig. 2.4). A comparison with OLR (Outgoing

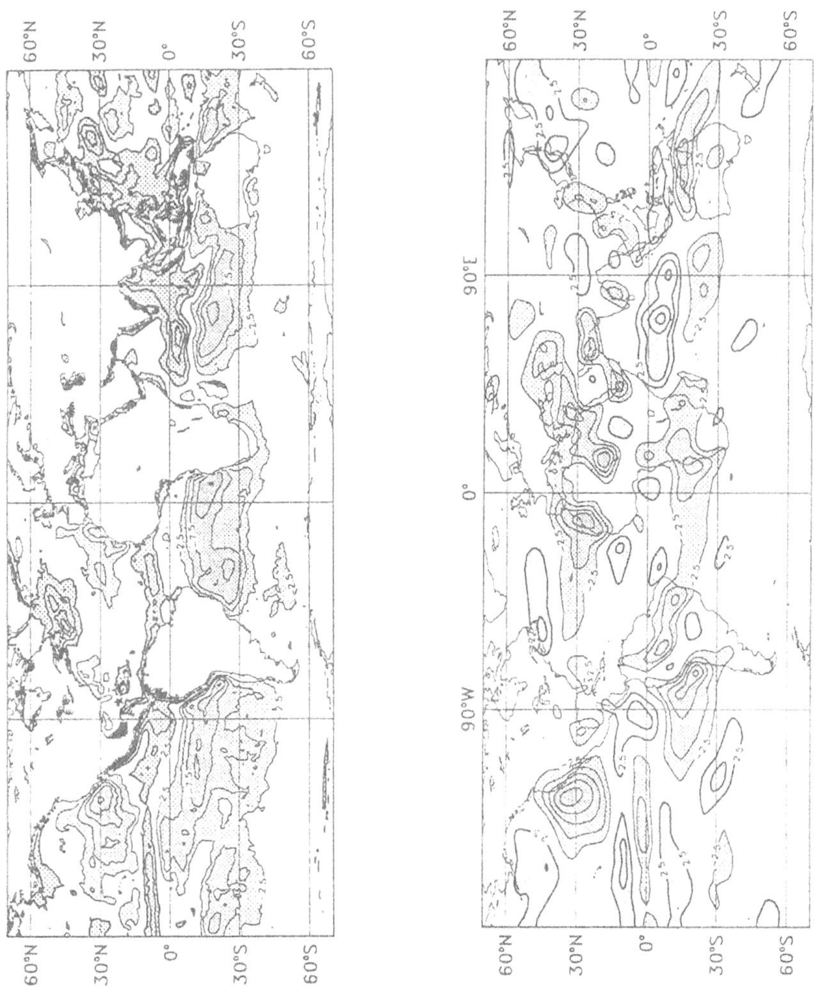

Figure 2.2: *Difference between the mean SSM/I derived PWC and the mean ECMWF analysis PWC for July 1990 (upper figure), and difference between the mean PWC of the 5-day forecast and the analysis for July 1990. (lower figure) Units:* $kg\,m^{-2}$

Longwave **R**adiation) derived rainfall rates (not shown) suggests that the heating rates over the tropical ocean areas are probably more realistic in July 1989 than before. The values over land seem to represent an overestimate.

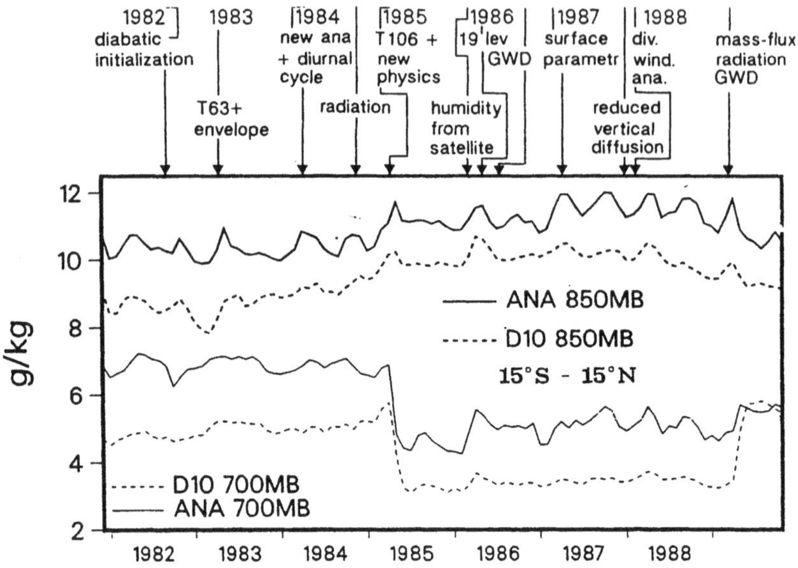

Figure 2.3: *(From Arpe, 1989). Zonal mean mixing ratio in the tropics (15 °S – 15 °N) in the analysis (solid lines) and the day 10 forecasts (dashed lines) at 850 hPa (thick lines) and at 700 hPa (thin lines).*

2.3 Precipitation

For the verification of model precipitation only a limited amount of observational data is available. Over land weather station reports are rather inhomogeneous in their geographical distribution. However, in some areas like Europe there is sufficient coverage in space and time to provide useful information for model verification. In the tropics the use of precipitation data derived from satellite OLR measurements has become an important part to verify convection schemes. Climatological estimates are often used to evaluate in long time integrations.

2.3.1 Zonal mean precipitation

Arpe (1990) has compared the zonal mean precipitation of the short and medium range forecast with Jägers climatology (Jäger, 1976) and rainfall derived from IR measurements (Janowiak and Arkin, 1990). The "spin up" or better the "spin down" of convective precipitation in the tropics is apparent from the much larger values of the first 24 hours compared to the 9-10 day values (Fig. 2.5). There is a fairly good agreement between model values and satellite estimates. Especially the secondary maximum of model produced precipitation in the Southern Hemisphere tropics does not show up in the climatological data. This comparison suggests that Jäger's climatology underestimates the tropical precipitation. In the extra-tropical storm track a "spin up" of large scale precipitation is evident, especially over the Southern Hemisphere oceans. However, even the higher values for the day 9-10 forecast in this latitude band look like an underestimate compared to climatology, though

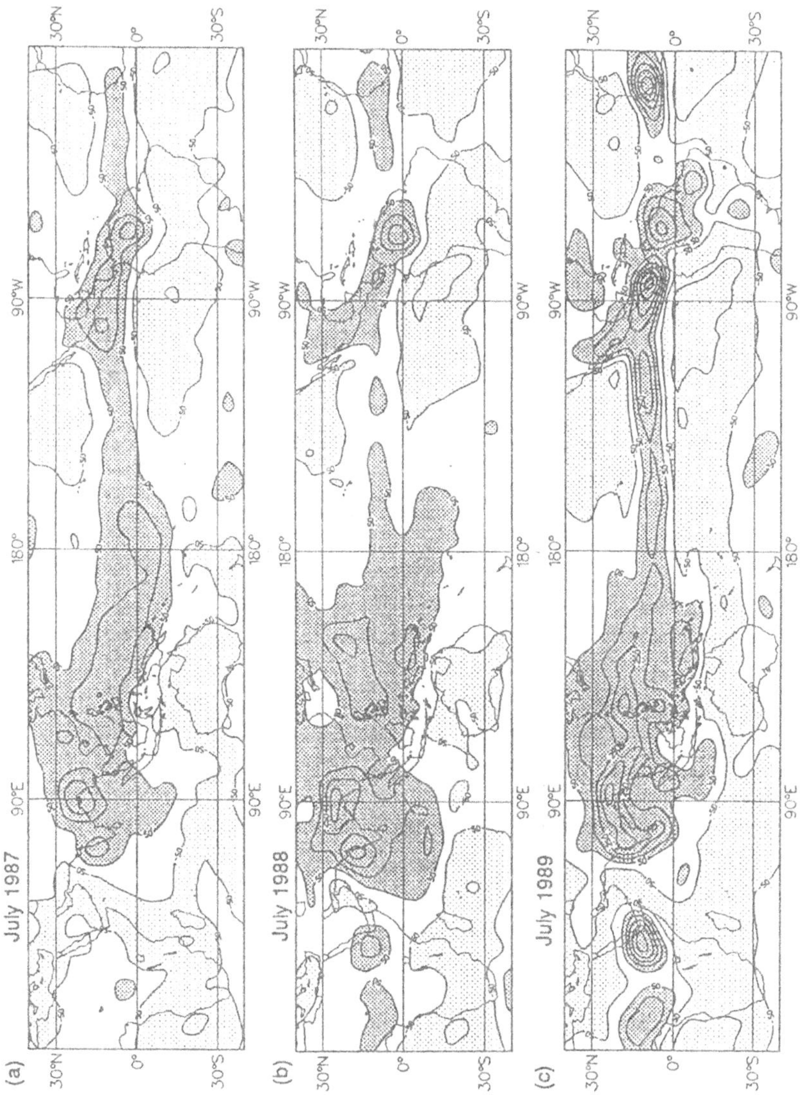

Figure 2.4: *Monthly mean of the negative of the adiabatic temperature tendencies (= required diabatic forcing), vertically integrated. Units:* Wm^{-2}

the confidence in the climatology of rainfall is certainly not very high over the Southern Hemisphere oceans. The increase of evaporation in June 1990 has lead to an increase of precipitation for the Year 1990/91 so that the difference between climatology and model simulation is smaller than in the previous year.

The uncertainty in the climate simulation of the hydrological cycle can be seen by comparing the zonal mean precipitation from different climate models shown in Fig. 2.6 for the summer and winter season. At least in the summer sample different models agree in the location of the major precipitation zone,

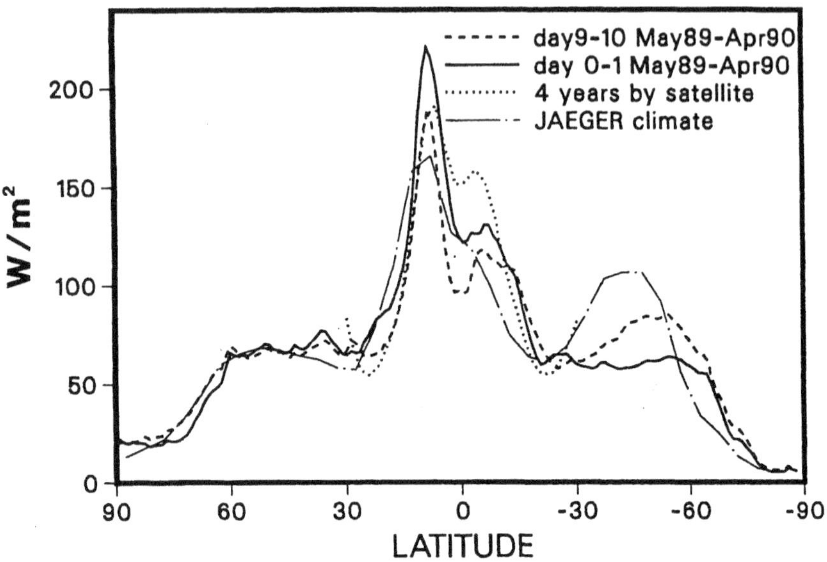

Figure 2.5: *(From Arpe, 1990). Zonal and annual means of precipitation. Climatological estimates by Jäger, 1976 are compared with estimates from satellite measurements over 4 years (Janowiak and Arkin, 1990) and with ECMWF day 0-1 and day 9-10 forecasts during May 1989 to April 1990.*

however, the total amount differs by up to 2 $mm\,d^{-1}ay$ in the extra-tropics and 4 $mm\,d^{-1}$ in the tropics. The differences between OLR based estimates of rainfall and Jäger's climatology are small against the differences between these climate simulations of rainfall.

The impact of model changes at ECMWF is shown with time-series of zonal mean precipitation. Figs. 2.7(a) and (b) shows a latitude-time diagram of precipitation averaged separately over land and ocean. After the introduction of the mass-flux scheme for convection the large scale rain over sea has been reduced. This arises from a changed partition between large scale rain and convective rain as the convective rain shows an increase over the mid- latitude regions.

A marked change in the convective rain over the ocean is noticeable with the same model change in May 1989. The precipitation has become more concentrated around the ITCZ and is larger in intensity. The enhancement of the moisture flux over the oceans in June 1990 seems to have had the effect of enhancing the convective precipitation over sea especially in the Northern Hemisphere autumn. Over tropical land areas the model change to higher surface radiation fluxes and consequently a more active convection scheme in May 1989 is apparent from the increased convective rainfall during the hemispheric summer season. The increase is also noticeable in the Northern Hemisphere at middle latitudes.

2.3.2 Geographical distribution of precipitation

The geographical distribution of annual rainfall shows a fairly good agreement between the Jäger climatology and the ECMWF forecast in the Northern Hemisphere (Fig. 2.8). Over some land

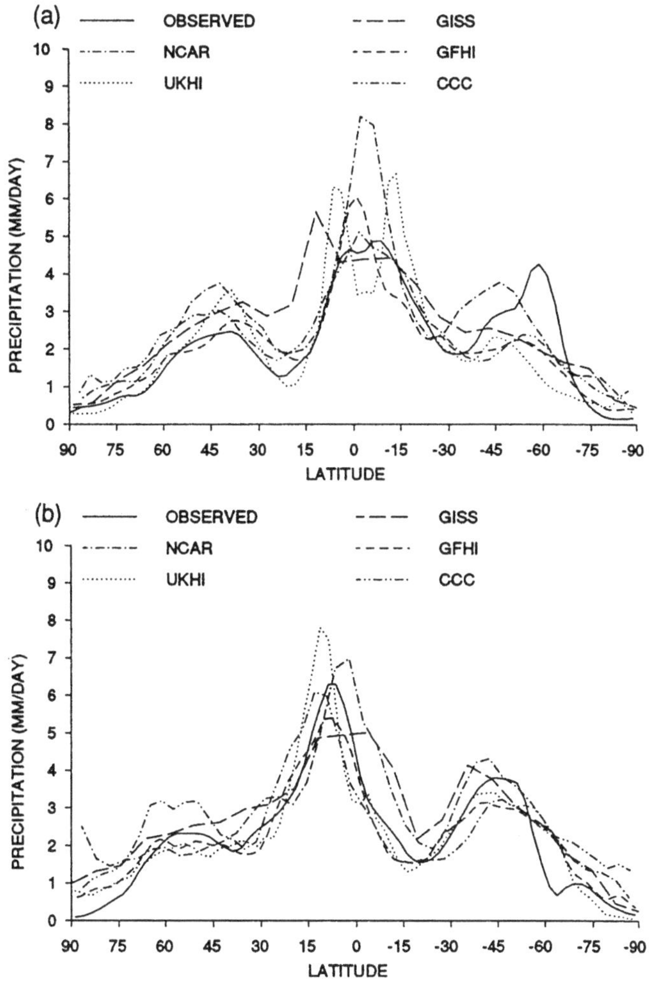

Figure 2.6: *(From Gates et al., 1990). Zonally averaged precipitation ($mm\,d^{-1}$) for various models and as observed (Jäger, 1976) for: (a) December - January - February, (b) June - July - August.*

areas the average forecast amounts are larger than climatology. From comparisons of forecast values against synoptic reports it is known that the ECMWF model produces too large rainfall rates in summer, especially in the short range.

In the tropical convergence zone the model precipitation over the oceans seems closer to satellite estimates than climatology. The comparison between the 1-day and 9-10 day forecasts shows that the model "spin down" is most pronounced over land. Over mountainous areas the short range forecast produces an especially large bias. The general reduction of tropical rainfall with forecast time leads to rather low values towards the end of the forecast. The rainfall over tropical land areas in the day 9-10 forecast is well below the climatological estimate and the satellite derived values.

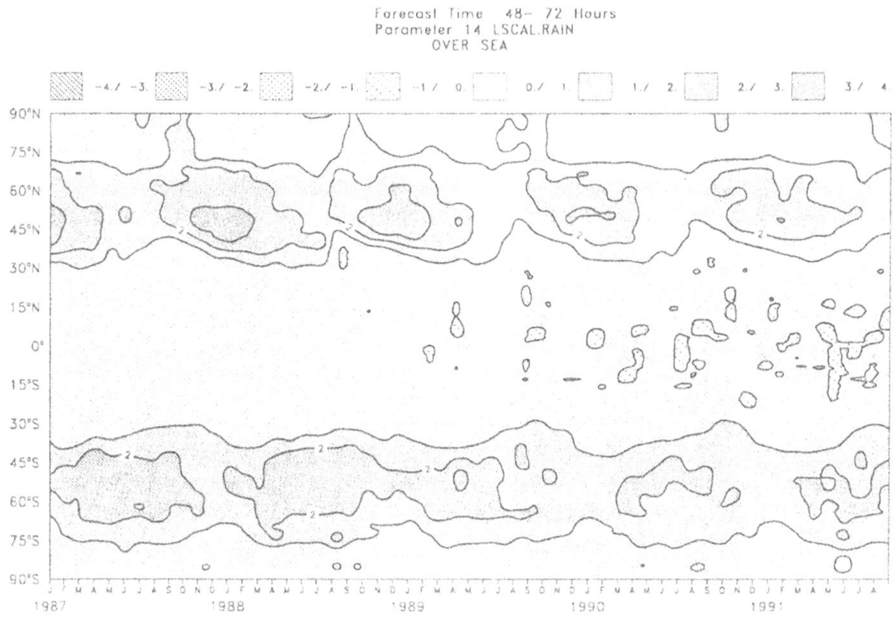

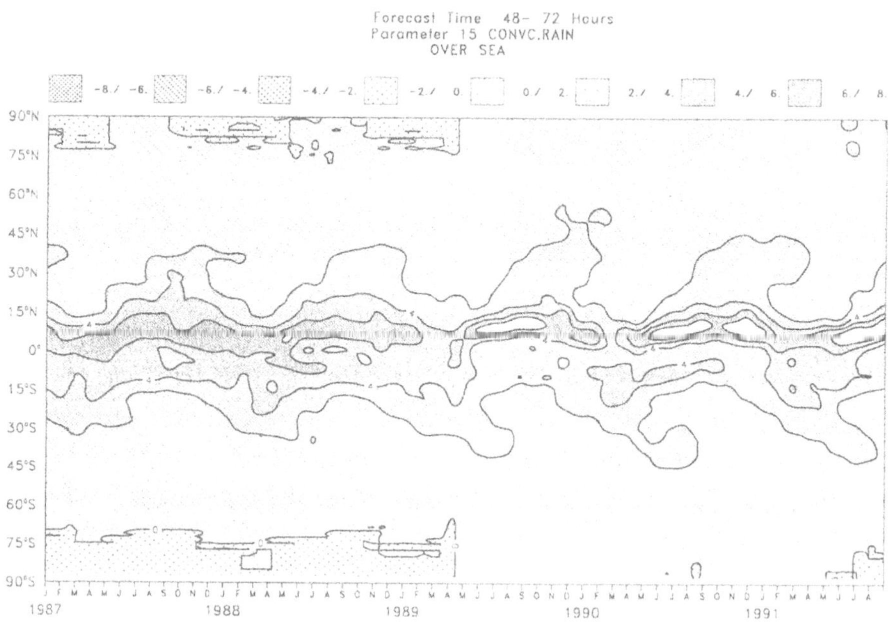

Figure 2.7: *(a) Time evolution of precipitation (48-72 hour forecast) averaged over sea points.*

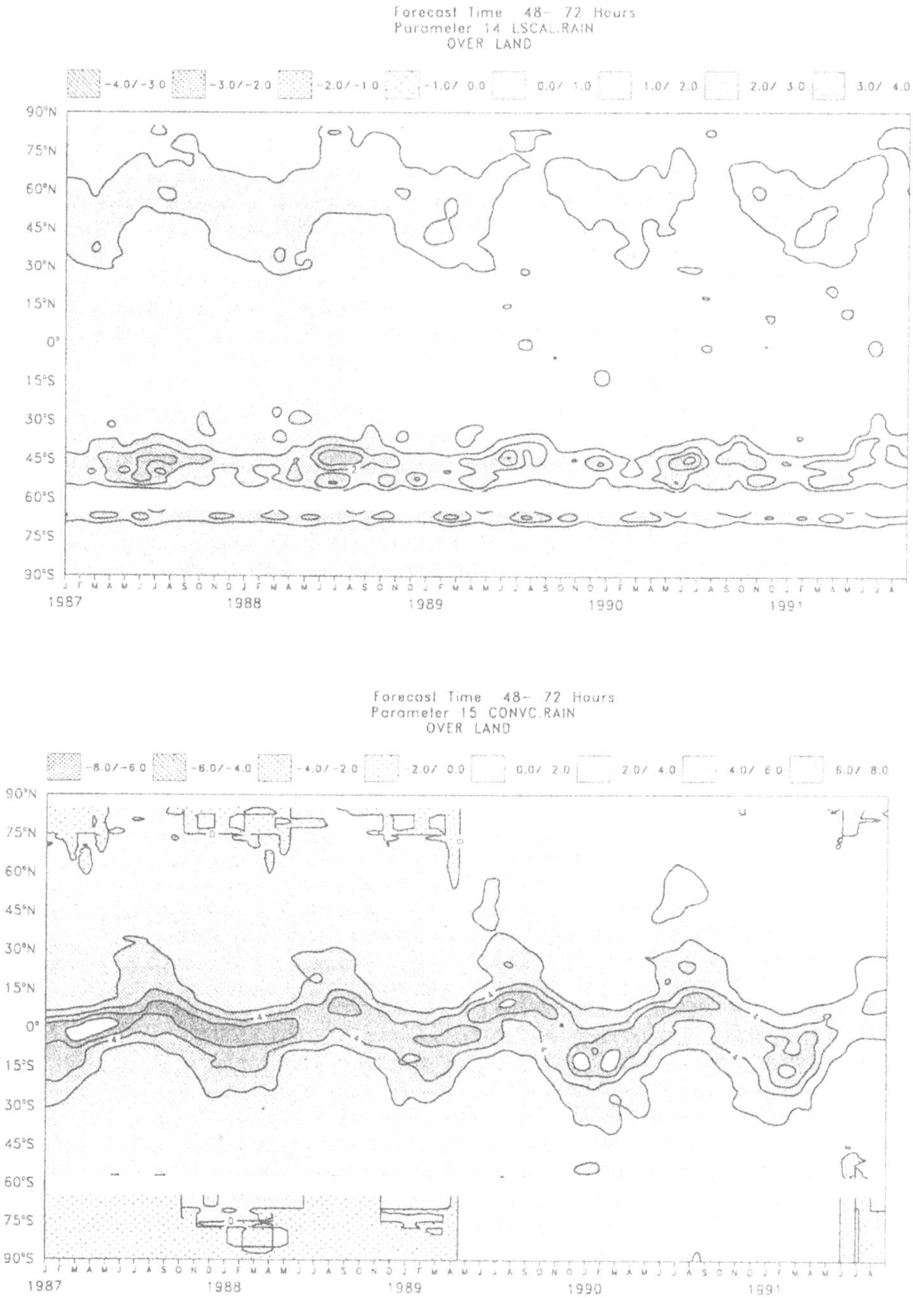

Figure 2.7: *(b) Time evolution of precipitation (48-72 hour forecast) averaged over land points.*

Figure 2.8: *Annual mean precipitation, (a) estimate form OLR measurements (Janowiak and Arkin, 1990), (b) Jäger, 1976 climatology.*

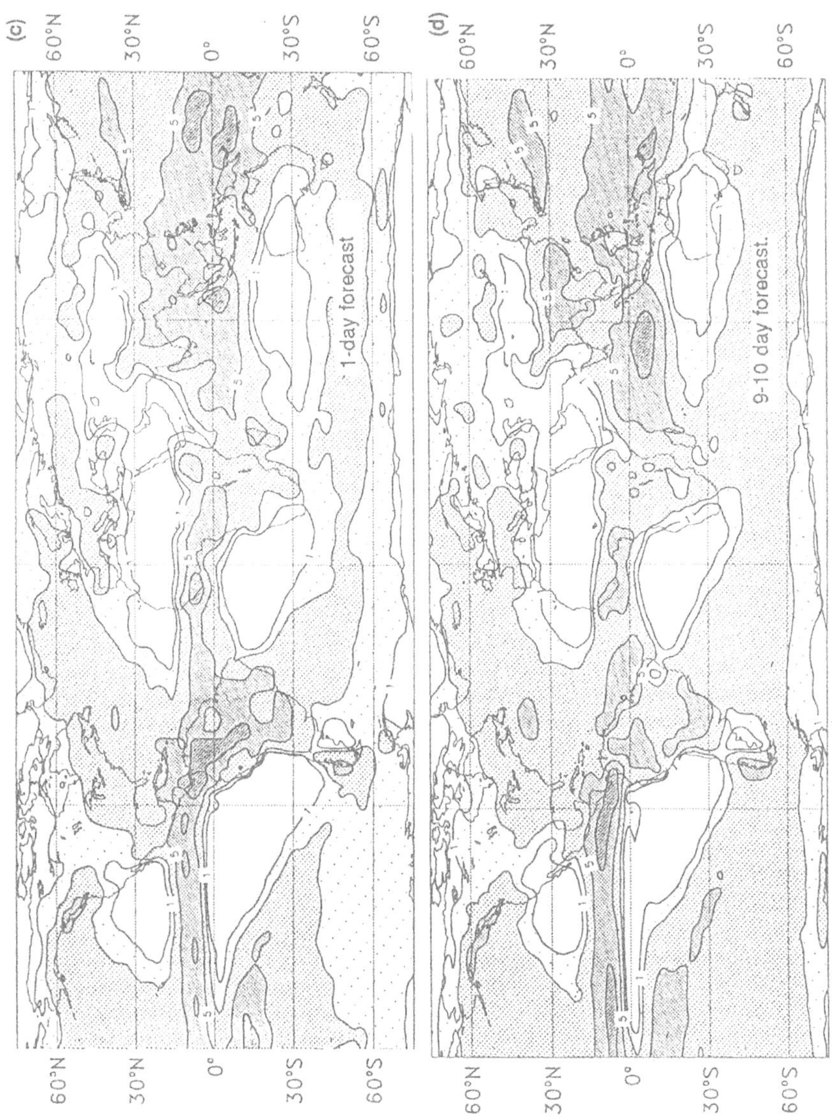

Figure 2.8: *Annual mean precipitation, ECMWF forecast values from May 1990 to April 1991, (c) 1-day forecast, (d) 9-10 day forecast. Units: mm* d^{-1}

Over the Indian Ocean the model has the tendency to enhance a precipitation maxima south of the equator. The lack of support for large precipitation amounts in this area from the two climatological data sets suggests a model problem in this area. The "spin up" of precipitation in the extra-tropical oceans shows up from the differences between the 1-day forecast and 9-10 day forecast. Over large areas in the middle latitudes of the Southern Hemisphere the precipitation increases from under 2 $mm\,d^{-1}$ to over 2 $mm\,d^{-1}$.

2.4 Latent heat flux

The evaluation of model produced latent heat flux is very much hindered by a general lack of observational data. The main source for a comparison is climatological data over oceans. These data sets are normally constructed from ship observation by using a bulk formulation for the calculation of the fluxes. Uncertainties arise from errors in wind measurements and from problems in the definition of the exchange coefficient in the bulk formula.

2.4.1 Zonal mean latent heat flux

From experiments with different sea surface temperatures (SST's) it was found that the ECMWF model was rather sensitive to changes in sea surface temperature in warm pool areas of the tropical oceans (Miller et al., 1992). Further research led to a modification of the parametrization of latent heat fluxes by increasing them in low wind speed situations over sea. Only a revised version of the flux parametrization produced a realistic response of the tropical circulation to SST anomalies.

In a latitude-time graph (Fig. 2.9) the increase of the surface heat flux is noticeable in the maxima that appear in the subtropical highs during the hemispheric winter season. However, also along the equator, where the latent heat flux has a minimum, the increase is in the order of 20 Wm^{-2}. Compared against the Oberhuber climatology (not shown) the change in the latent heat flux in June 1990 seems to have gone mostly into the right direction. A general underestimate north of the equator and in middle to low latitudes of the Southern Hemisphere has been noticeably reduced.

2.4.2 Geographical distribution of latent heat flux

In 1990/91 the magnitude of the latent heat flux in the medium range is closer to climatology than in previous years (Fig. 2.10(a)). However, the problem of an underestimate in the 1-day forecast (Fig. 2.10(b)) has not been solved with the increase of the flux in low wind speed situations. Over large parts of the Southern Hemisphere subtropical oceans the latent heat flux is noticeably smaller in the short range forecast compared to the medium range forecast. Arpe (1990) has also pointed to the underestimate of the latent heat flux in the short range forecast. Investigations at ECMWF (Illari, internal ECMWF report) suggest that the satellite soundings introduced a large positive temperature bias in the analysis of low level temperatures. Since temperature data is analyzed separately from moisture data constrained by constant relative humidity, the warm temperature bias leads to a too

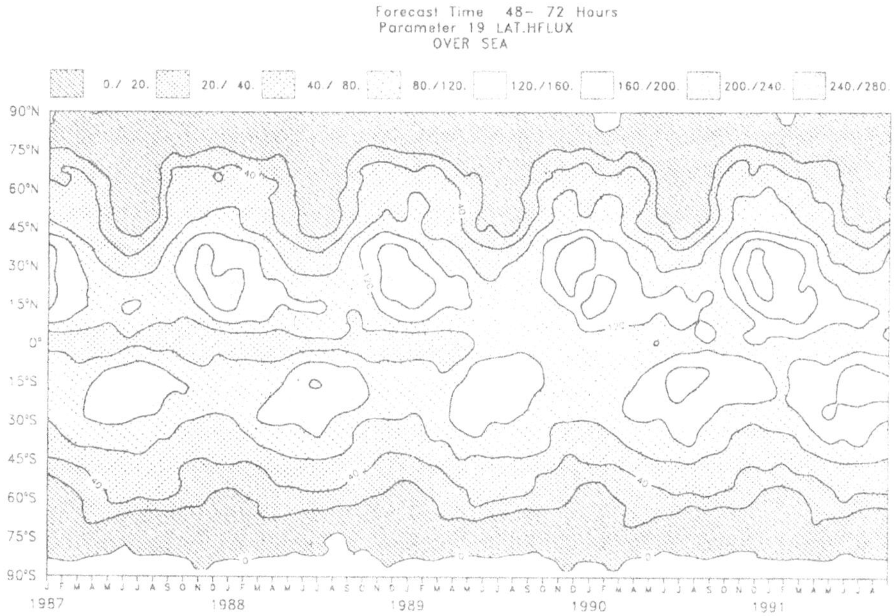

Figure 2.9: *Time evolution of the surface latent heat flux (48-72 hour forecast values) zonally averaged over sea points. Units:* Wm^{-2}

moist lower boundary layer in the analysis, which explains the reduced level of evaporation in the early stages of the forecast.

2.5 Radiation

After the measurements of the first year of the Earth Radiation Budget Experiment (ERBE) have become available they are widely used for model verification. Though these measurements are often applied to a non- simultaneous comparison they still provide useful information for the evaluation of model radiation codes.

2.5.1 Greenhouse factor

As first general estimate of the quality of the long-wave radiationfluxes we compare the model simulated greenhouse effect with ERBE observation. The atmospheric, greenhouse effect which is defined as the amount of long-wave radiation trapped in the atmosphere, is dependant on the precipitable water content the three dimensional distribution of clouds and their optical properties. The evaluation of the model greenhouse effect can provide useful information about possible errors in the hydrological cycle of the model and their influence on a climate drift in long integrations.

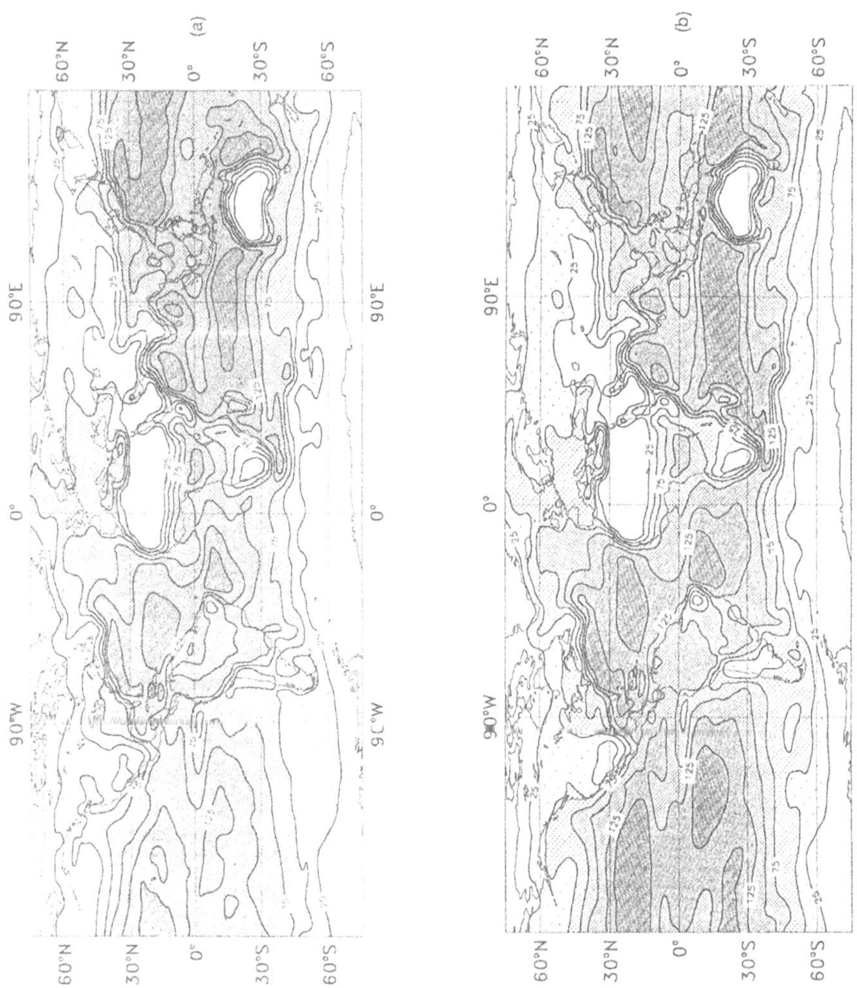

Figure 2.10: *Annual mean surface latent heat flux, forecast values for May 1990 to April 1991, (a) 1-day forecast, (b) 9-10 day forecast. Units:* Wm^{-2}

The observed greenhouse effect can be calculated from satellite measurements of OLR and analyzed sea surface temperatures (T_s). As the emitted radiation at the surface increases as the fourth power of T_s one would expect the greenhouse effect to increase with increasing surface temperature as well. Here the greenhouse factor

$$G = \frac{I}{OLR}$$

is defined as the ratio between the long-wave radiation emitted from the surface ($I = \sigma T_s^4$) and the outgoing long-wave radiation at the top of the atmosphere (OLR). This definition removes the T_s^4 dependence of G in a similar way as the normalized greenhouse effect $G^* = (I - OLR)/I$ used by Raval and Ramanathan (1989).

The calculation of the observed greenhouse factor is based on the sea surface temperature (SST) analysis of July 1985. For OLR at the top of the atmosphere ERBE measurements have been used. The scatter-diagram (Fig. 2.11(a)) shows the observed greenhouse factor $G = I/OLR$ as a function of T_s. Only sea points with a surface temperature of more than 270 K have been included. Over a wide range of SST's the greenhouse factor increases slightly with T_s due to the H_2O feedback.

For SST values higher than 299 K the greenhouse factor has a strong nonlinear increase with T_s, which arises from the additional greenhouse effect of convective clouds in the tropical convergence zone. The optical thickness of penetrative convective clouds is very high, therefore the emission of long-wave radiation from the cloud tops is relatively low.

For the comparison of the greenhouse factor in the ECMWF model with observations two sets of model OLR are used. One is extracted from 6 hour forecasts, which are run four times per day to produce a first guess for the analysis (Fig. 2.11(b)). The second set is calculated from 5-day forecasts and represents the average outgoing long-wave radiation for the first five days of the forecast (Fig. 2.11(c)). In the comparison one has to consider that the OLR from ERBE and the appropriate SST are monthly means for July 1985 whereas the model OLR and the corresponding SST are for July 1990. Despite the different periods a reasonable general agreement between the observed and model simulated greenhouse factor for the 6 hour forecast emerges. Especially the representation of the cloud forcing at high SST's represents a marked improvement compared to July 1988 (not shown). The new radiation scheme introduced in May 1989 produces a more realistic OLR in the tropical areas of deep convection. There are, however, noticeable differences, which are an indication of model problems.

For temperatures typical for the tropical convergence zone (296 -302 K), the nonlinear increase of the greenhouse factor with T_s is steeper than in the observations. A positive bias of the greenhouse effect for these high SST values suggests that the cloud-forcing in the tropics is too large for the 6-hour forecast. For a forecast range of 5 days the greenhouse factor distribution against the SST shows some changes compared to the 6-hour forecast distribution. The nonlinear increase in the 5-day forecast is as steep as in the 6-hour forecast, but the positive bias of the greenhouse factor compared to observation is even larger than for the 6-hour forecast. This indicates a further increase in cloud forcing.

(a)

(b)

(c)

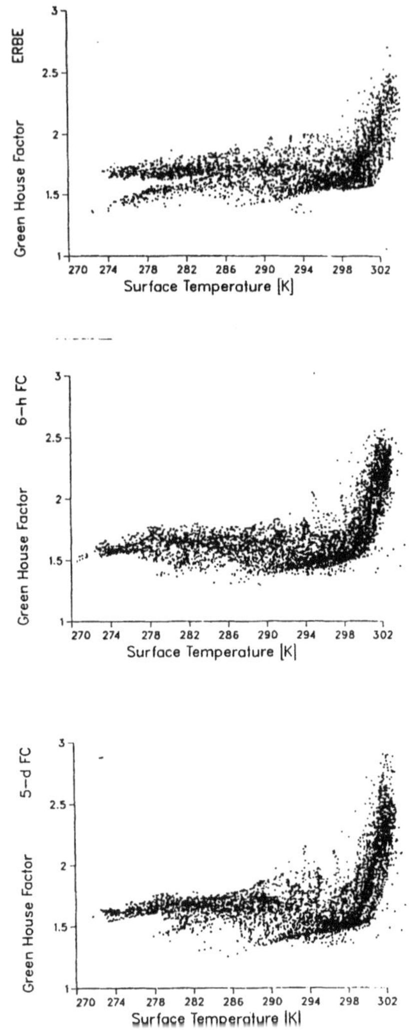

Figure 2.11: *(a) Scatter diagramm for the greenhouse factor (definition see text) versus the surface temperature using ERBE data of July 1985 for the OLR. (b) using the 6-hour forecast radiation fluxes of July 1985 for the OLR. (c) using the 5-day forecast radiation fluxes of July 1985 for the OLR.*

2.5.2 Zonal mean radiation

The disadvantage of comparing model radiation fluxes with non-simultaneous ERBE measurements can be reduced by using zonal means. Fig. 2.12 shows differences between 4 years of model produced fluxes and one year of ERBE fluxes. The fluxes are split into long-wave radiation and short wave radiation, and the averaging is done separately for sea and land points.

The difference between model OLR and ERBE observation shows the marked improvement of the radiation fluxes at the top of the atmosphere with the introduction of the new radiation scheme in

(a)

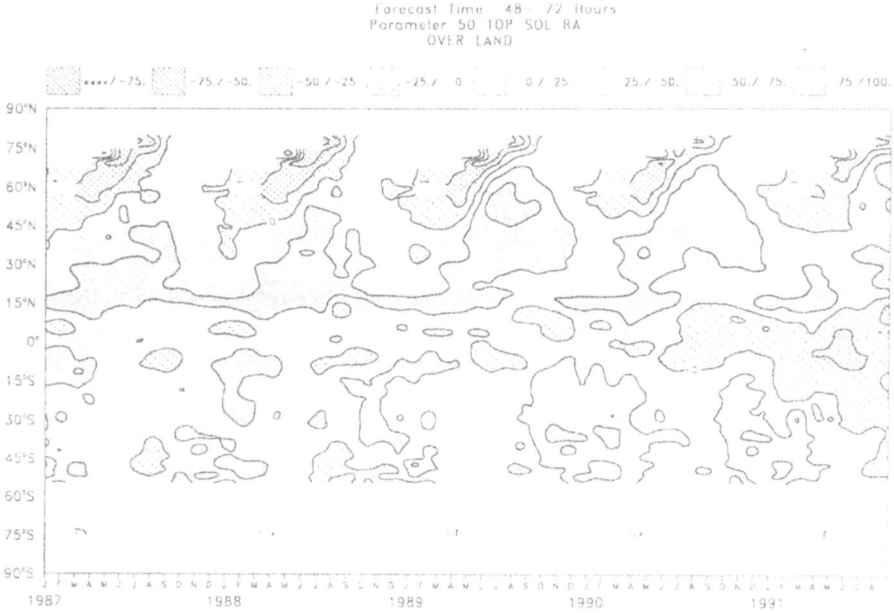

(b)

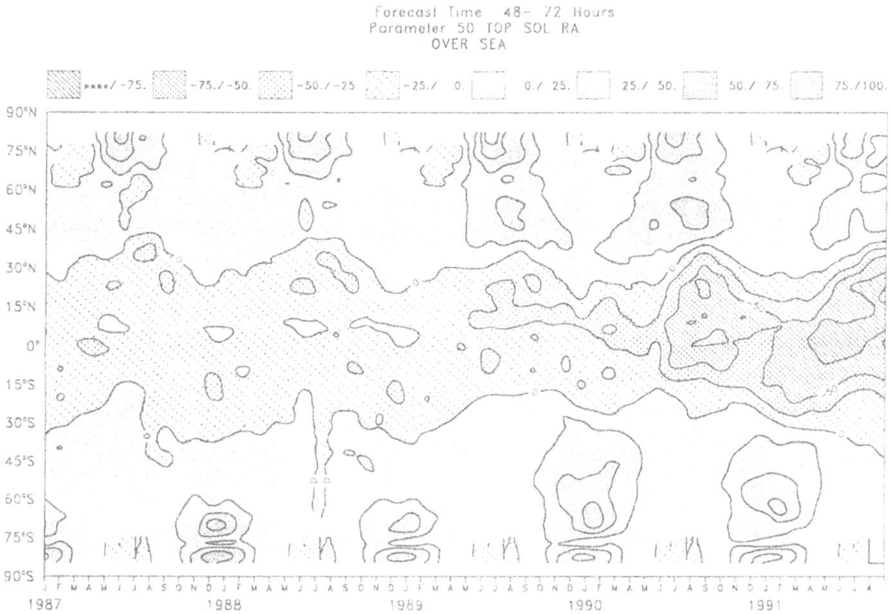

Figure 2.12: *Time evolution of the difference with ERBE observations at the net solar radiation at the top of the atmosphere zonally averaged, (a) over land points, (b) over sea points. Units:* Wm^{-2}

May 1989. The old radiation scheme, which was not sensitive enough to convective clouds had a fairly flat distribution in the tropics, which shows a positive model bias of up to 60 Wm^{-2}. The model change led to an increased geographical contrast in the OLR field at top of the atmosphere with marked minima in convective areas and maxima over clear sky regions.

The enhancement of the evaporation in low wind speed situations and the diagnosis of shallow convective clouds from the condensation rate certainly improved several features in the ECMWF model (Miller et al., 1990). However, with respect to clouds and radiation at the top of the atmosphere this change had some drawbacks. The increased moisture content had the effect of enhanced deep convection with larger fractional cloud cover and a slightly increased liquid water content. Over ocean these clouds are too reflective in the short wave spectrum and too cold in the long-wave range, which shows in Fig. 2.12 after June 1990. Over tropical land areas a similar negative impact can be seen on the solar radiation. Over Northern Hemisphere land areas the May 1989 model change produced a positive bias of the solar radiation that was due to an increased absorption of solar radiation at the surface under a too dry boundary layer. A reduced run-off of rainfall helped to increase the moisture in the boundary layer and to increase the cloud amount, which subsequently reduced the positive bias of solar radiation over the Northern Hemisphere land areas.

2.5.3 Geographical distribution of long-wave radiation

From the OLR it is possible to identify the intensity and area coverage of deep convection systems in the tropics. However, when we evaluate the model OLR by using ERBE observations we have to be aware of the non- simultaneous nature of this comparison. Some local convective systems tend to have a noticeable inter-annual variation. Despite this shortcoming the comparison can still give useful information about specific model problems in certain areas.

In Figure 2.13 observed OLR from July 1985 is compared with model OLR from july 1990. Fig. 2.13(a) shows the observed ERBE OLR for July 1985, Fig. 2.13(b) the model OLR for the first 6 hours of the forecast averaged over 4 forecast per day. In the observed and model produced radiation fluxes for July the ITCZ is well defined as a minimum of OLR north of the equator for most regions. Along the ITCZ the OLR values in the 6-hour forecast are generally lower than in the ERBE observations. The observed OLR suggests that the maximum convective activity over the West and East Pacific is separated by a region around 140 degrees west with reduced convection. In contrast to observation the 6 hour forecast produces a fairly uniform band of OLR over the Pacific ITCZ. Similarly the 6-hour forecast OLR represents an underestimation in the ITCZ of the western Atlantic.

Over the tropical land areas too low predicted OLR values are found at higher terrain, such as the mountains of Ethiopia and Mexico, and the Atlas Mountains in North Africa. These differences point to excessive convective rainfall over mountains. The operational verification of model rainfall against station data over the Alpine region in summer suggests the same. Also over the Amazon basin in central parts of Brazil the OLR difference suggests excessive convective rainfall.

During the first 5 days of the forecast there is a further decrease of the model OLR in the tropics compared to the first 6 hours as shown from the difference between the model OLR and ERBE observation (Fig. 2.13(c)). This process shifts the model OLR even more away from the observation than the short range forecast. The decrease of the OLR in the model is particular evident in the West Pacific where values go down to 180 Wm^{-2} compared to 220 or 200 Wm^{-2} in the ERBE

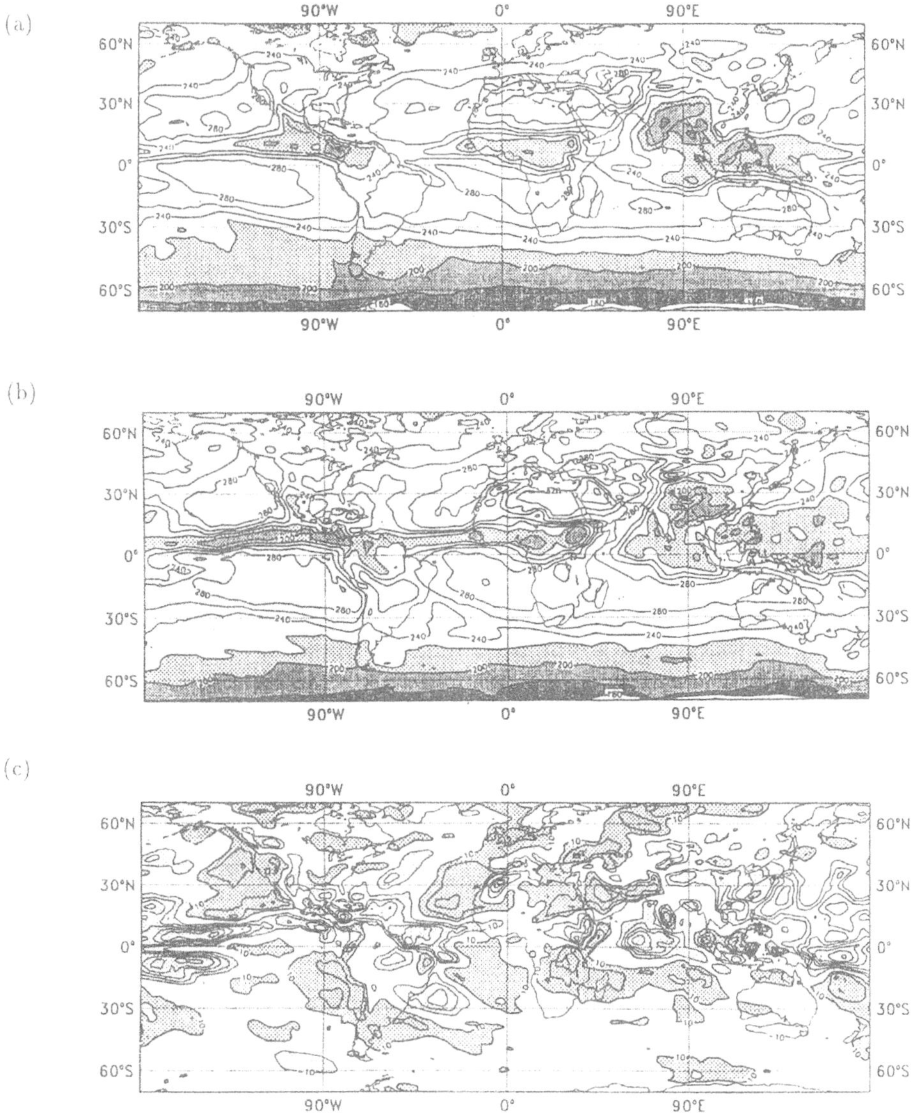

Figure 2.13: *(a) OLR from ERBE measurements for July 1985, (b) OLR of the 6-hour forecast for July 1990, (c) difference between (b) and (a). Units:* Wm^{-2}

observations. Also west of New Guinea the model OLR seems to suggest a further intensification of convection.

Too high values of OLR are found over large parts of the oceans indicating a general lack of cloud cover. This is particularly the case in the high pressure areas on the west coast of the continents. Over the Northern Hemisphere land mass there are some areas with a positive bias. This is, however, a noticeable improvement on the differences in July 1989 when the major parts of the Northern Hemisphere continents showed an overestimate of OLR in the model. This large positive bias in July

1989 was related to insufficient cloud cover and rather low soil moisture values. Both problems were alleviated by model changes in 1990.

2.6 Global energy and hydrological cycle

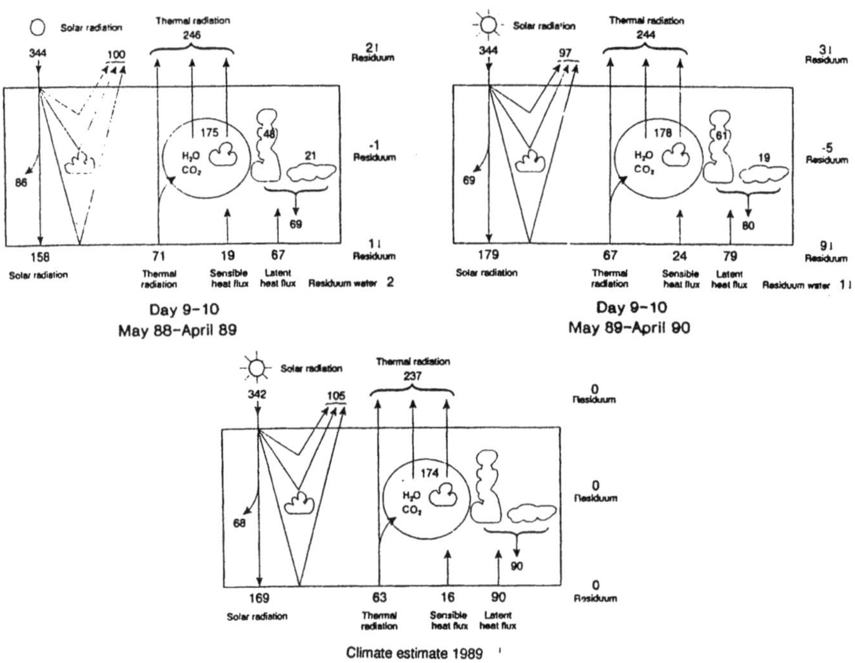

Figure 2.14: *(From Arpe, 1990) Global annual mean energy and water cycle derived from day 9-10 forecasts and climatological estimates of these cycles by Ramanathan et al., 1989. Model data from two years with different parametrization schemes are compared. Units:* Wm^{-2}

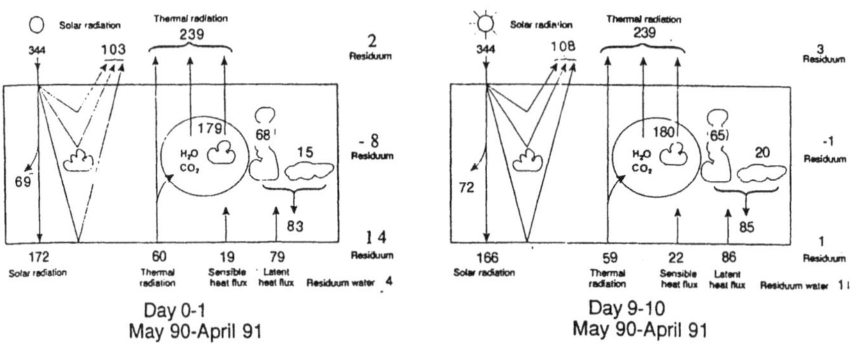

Figure 2.15: *As Figure before, but for 1990/91.*

The performance of a large scale model regarding the energy and hydrological cycle can best be summarized by presenting global and annual means in which we expect a near balance between some components of the budget. These budgets can be compared with climatological estimates that are constructed on the basis of such balances. Fig. 2.14 shows the two years of day 9-10 forecasts values from Arpe (1990) and Fig. 2.15 updated values of the year from May 1990 to April 1991 and additionally the values for the first 24 hours of the forecast in 1990/91. In general the ECMWF model agrees reasonably well with climate estimates. However, model changes had a noticeable influence on the simulation of the energy cycle.

The change in the parametrization of radiation and convection in 1989 increased the solar radiation at the surface by an amount that exceeded the climatological estimate. The warmer surface led to an increase of the surface fluxes of sensible and latent heat. Consequently, the surface flux of sensible heat, which obtains its maximum contribution from land areas seems too large in the year 1989/90 compared to climatology. The effect could be seen in the positive forecast errors of temperature at 2 m as well and in the lower troposphere over the Northern Hemisphere land areas in summer. The higher amount of absorption of solar radiation at the ground implies increased net solar radiation at the top of the atmosphere, giving a too low earth-atmosphere albedo.

The outgoing long-wave radiation has been reduced with the new radiation scheme due to a better representation of the effects of clouds. The enhanced convection increased the rainfall amounts, though some of the climatological estimates are still higher (Ramanathan et al., 1989: 90 Wm^{-2}, Jäger, 1976: 82 Wm^{-2}, Hoyt, 1976: 75-84 Wm^{-2}).

The model change in June 1990 contributed to an increase of the annual latent heat flux to 86 Wm^{-2}, a value that is closer to the climatological estimate. The increased moisture flux with an improved diagnosis of shallow convective clouds changed several features in the ECMWF model. The enhanced evaporation increased the moisture content in the tropical convergence zone, which had the consequence of enhanced deep convection and larger fractional cover of high clouds of slightly increased liquid water content. However, these clouds appear as too reflective in the short-wave range as the global budget for 1990/91 shows. At the surface the increased cloud amount had the beneficial effect in reducing the net solar radiation. Consequently the long-wave radiation flux the surface and the surface sensible heat flux were reduced as well.

One important constraint, which has been used for the construction of climatological estimates of the hydrological cycle, is the balance between the evaporation and precipitation. This balance has been nearly achieved in the day 9-10 forecast period. However, in the short range forecast (day 0-1) of 1990/91 there is a noticeable imbalance of 4 Wm^{-2} between the two. A *spin up* of evaporation and a *spin down* of precipitation in the first few days of the forecast are the main reasons for this discrepancy. The model change of increased evaporation in June 1990 improved the initial discrepancy between evaporation and precipitation, which was 30 Wm^{-2} in 1989/90. However, the typical features of a "spin up" in the hydrological cycle are still present.

2.7 Conclusions

During the last years the development of large scale models has led to a noticeable improvement of the simulation of the hydrological cycle. Taking the ECMWF model as an example the gain in forecast quality was partly due to an increase in horizontal resolution and partly due to further developments of physical parametrization schemes.

The impact of model changes on the hydrological cycle has proved that changes in the radiation scheme and in the parametrization of cumulus convection had a particularly large impact on the energy budget and the hydrological cycle. The verification of forecast parameters like temperature and wind fields does not pose a special problem because they are included in the analysis. There is, however, a large uncertainty in verifying some important quantities like cloud cover and precipitation for which only a limited amount of verification data is available. Therefore not only direct measurements but also quantities related to physical processes in the hydrological cycle are used for validation. For example, ERBE radiation measurements play an important part as a direct verification of the model's radiation scheme and as an indirect way of verifying the spatial and temporal distribution of convective activity in the tropics.

The disadvantage of using some satellite measurements is that they have to go through a time consuming post-processing procedure, which makes it impossible to use these observations for a simultaneous verification. For testing new parametrization schemes, which have an important impact on the simulation of the hydrological cycle, it seems therefore appropriate to rerun analyses for special observation periods with updated versions of the analysis and the model.

At ECMWF July 1987 and August 1987 have been reanalysed with an analysis system and forecast model that was used in operation in July 1991. Probably most importantly the model formulation of that time contained a new radiation and convection scheme, which gave a more realistic tropical circulation than the model formulation used in July 1987 operationally. First results indicate that the ERBE radiation measurements and SSM/I measurements of total precipitable water give a good indication where model problems in the simulation of the hydrological cycle occur.

2.8 References

Arpe K (1989) Impacts of changes in the ECMWF analysis-forecasting scheme on the systematic error of the model. *ECMWF seminar on 10 years of medium-range weather forecasting*, 4-9 September 1989, 69-114. Available at ECMWF. Shinfield Park, Reading RG2 9AX, U.K.

Arpe K (1990) The hydrological cycle in the ECMWF short range forecasts. *Workshop proceedings for the ECMWF/WCRP workshop on clouds, radiative transfer and the hydrological cycle.* 151-179. Available at ECMWF. Shinfield Park, Reading RG2 9AX, U.K.

Gates WL, Rowntree PR, Zeng Q-C (1990) Validation of climate models. *Climate change.* The IPCC scientific assessment.

Hoyt DV (1976) The radiation and energy budgets of the earth using both ground-based and satellite derived values of total cloud cover. *NOAA Tech. Report ERL 362-ARL4*. U.S. Dept. of Commerce, Washington D.C., 124 pp.

Illari L (1989) The quality of satellite precipitable water content data and their impact on analyzed moisture fields. *Tellus* 41A: 319-337

Jäger L (1976) Monatskarten des Niederschlags für die ganze Erde. *Berichte des Deutschen Wetterdienstes* 139 (Band 18).

Klinker E (1990) Use of satellite data to validate the hydrological cycle of the ECMWF model. *Workshop proceedings for the ECMWF/WCRP workshop on clouds, radiative transfer and the hydrological cycle.* Available at ECMWF. Shinfield Park, Reading RG2 9AX, U.K.

Klinker E, Sardeshmukh PD (1987) The diagnosis of systematic errors in numerical prediction models. *Proceedings of a workshop on diabatic forcing held at ECMWF*. 209-244. Available at ECMWF. Shinfield Park, Reading RG2 9AX, U.K.

Miller MJ, Beljaars ACM, Palmer TN (1992) The sensitivity of the ECMWF model to parametrization of evaporation from tropical oceans. Accepted for publication. *Journal of Climate.*

Morcrette J-J (1990) Impact of changes to the radiation transfer parameterizations plus cloud optical properties in the ECMWF model. *Mon Wea Rev* 118: 848-873.

Palmer TN, Brankovič Č, Viterbo P, Miller MJ (1992) Modelling interannual variations of summer monsoons. Accepted for publication. *J Climate.*

Ramanathan V, Barkstrom BR, Harrison EF (1989) Climate and earth's radiation budget. *Physics Today*, American Institute of Physics, 22-32.

Raval A, Ramanathan V (1989) Observational determination of the greenhouse effect. *Nature* 342: 758-761.

Tiedtke M (1989) A comprehensive massflux scheme for cumulus parametrization in large-scale models. *Mon Wea Rev* 117: 1779-1800.

Tjemkes SA, Stephens GL (1990) Intercomparison between microwave and infrared observations of precipitable water. *Fifth conference on satellite meteorology and oceanography.* September 3-7, 1990.

Wentz FJ, Mattox LA, Peteherych S (1986) New algorithms for microwave measurements of ocean winds: Applications to seasat and special sensor microwave imager. *J Geophys Res* 91: No C2, 2289-2307.

Chapter 3

RADIATION-CLOUD-CLIMATE INTERACTION

Ehrhard Raschke
GKSS Research Centre
Institute for Physics
D-2054 Geesthacht
Germany

3.1 Introduction

Our climate at any place of the world is result of a large manifold of complex processes and interactions within the climate system. Amongst them we shall discuss in this paper preferably those which are based on the complex interplay between clouds and the ambient fields of electromagnetic radiation in the short- and longwave regions of the spectrum. They obtain high attention in present climate research, since several numerical sensitivity studies have demonstrated an enormous sensitivity of the climate to such processes (e.g. Houghton et al., 1990). They are still focus of intensive research within the scopes of the World Climate Research Programme and its various subprogrammes, in particular within the GEWEX, the Global Energy and Water Cycle Experiment.

Clouds can considerably shield the ground surfaces from heating by solar radiation, but simultaneously they prevent their cooling by reemitting heat radiation back to the ground. They absorb appreciable amounts of solar radiation and reflect very effectively large fractions of the incident radiation back to space, thus effecting the solar heating of atmospheric layers above their tops. High and cold clouds emit much less thermal radiation than low clouds and the ground, thus they may even "heat" the climate system (Hansen et al., 1981). Many feedback processes occur in the climate system as shown for instance by Schneider (1972), Liou (1986) and Weatherald and Manabe (1988).

In turn, also the cloud microphysical properties are influenced by interaction with the ambient radiation fields. Radiation absorbed or emitted by cloud particles affects considerably their growth or decay; it also causes turbulent exchanges. Studies have shown that these processes can determine

NATO ASI Series, Vol. I 5
Energy and Water Cycles in the Climate System
Edited by E. Raschke and D. Jacob

the structures and as a consequence also the radiative properties of the whole cloud system itself (e.g. Jonas, 1989; Zhang et al., 1989).

Thus we shall expect interactions with the existing climate, when the radiative transfer properties of clouds are changing. Natural causes might be changing supplies of water vapour and of condensation nuclei (e.g. Charlson et al., 1987) and indeed variations in the tropospheric circulation systems. But also anthropogenic changes of the atmospheric composition could contribute to variations of cloud properties. Several sensitivity studies in numerical models have demonstrated these effects and underlined that the basic mechanisms must be understood quite accurately. Mitchell et al. (1989) showed for instance the effects of more detailed parametrizations of cloud physical processes and properties on calculated climate states. Kristjánsson (1991) demonstrated also, that the horizontal resolution in circulation models needs to be increased to simulate cloud fields and their precipitation more realistically.

In this report we shall discuss the planetary radiation budget, the basic physics and occurrence of cloud systems, radiation and its interaction within the atmosphere and at the ground, the interaction between clouds and the radiative environment, and some results derived from recent radiation budget missions in satellites. This paper cannot provide a complete review on all cloud radiation interactions. It is complemented by other papers in this book. We further refer to the discussion of most recent research activities in this field by Harshvardhan (1991).

3.2 The planetary radiation budget - effects of clouds

The earth exchanges radiative energy with space, where the Sun's shortwave irradiance, S_0, amounting to 1370 Wm^{-2} ($\pm 2Wm^{-2}$) at the mean distance of the earth from Sun, is the only energy source for all processes within the climate system. The latter consists of the atmosphere, oceans and cryosphere, and of all continental surfaces. Of this amount, one quarter reaches in an annual average the earth's disc, where about 29 to 30 % of it is reflected back to space. The remaining portion, as shown in the paper by Peixoto (this volume, and Peixoto and Oort, 1984) is absorbed by atmospheric gases, in clouds, and to a main fraction at ground. Peixoto et al. (1991) could demonstrate, that clouds participate substantially in the generation (by absorption of solar radiation, and by release of latent heat) and destruction (by emission of heat radiation to space) of entropy.

Assuming radiative equilibrium between earth and space this budget can be described with a simple relation equating the incident and emitted radiation fluxes (Eq. 3.1).

$$\frac{S_0}{4}(1-\rho) = \sigma T_e^4 \tag{3.1}$$

ρ = mean planetary albedo, T_e = effective radiation temperature of our planet.

The imbalance, or the global net radiation budget, may reach values up to 8 Wm^{-2} and shows an annual cycle for monthly averages (e.g. Raschke, 1972; Harrison et al., 1990). Eq. 3.1 was the basis of first simple climate sensitivity studies (e.g. Budyko, 1969). In turn, the same amount of radiative energy is emitted back to space by clouds, atmospheric gases and the ground. It would correspond to an equivalent black-body temperature of 251 to 252 K, while the Sun's emission corresponds to 6000 K. Both spectra overlap only weakly in the spectral range near $3 - 4\mu m$. A more thorough discussion of the interaction between radiation and matter follows in section 3.4.

The global averages and moreover the spatial distributions of the planetary radiation budget components are dominantly modulated by clouds. These are responsible for about one half of the total reflection back to space. Thus without clouds the earth's "clear sky" albedo would range between 15 and 16 percent (Raschke, 1972; Hartmann et al., 1986; Hartmann and Doelling, 1991; Stephens and Greenwald, 1991). As thorough analyses of recent satellite measurements have shown, the effect of clouds is in particular large over the dark oceans (reflectance: 6 to 8 percent) in the tropical and extratropical cloud belts. Rossow (in this volume) will show more details on the cloud climatology as derived from imaging data of operational satellites.

In the longwave part of the spectrum, terrestrial heat radiation is mainly emitted to space from cloud tops which are colder than the surface. Cloud bottoms and the lowest layers of the troposphere emit appreciably high amounts of radiation to the surface. Thus, in summary, without clouds the earth's surface would loose more radiative energy to space than with clouds .

Clouds participate in the atmosphere's greenhouse effect by reducing it considerably. While without clouds the earth's surface temperature could assume a mean value of more than 300 K (assuming the present state of the climate system unchanged) its present value is around 289 K. In the shortwave region they "cool" and in the longwave "heat" the earth where generally the cooling dominates; their net effect is shown in Fig. 3.1 (from Roeckner et al., 1991).

In the present state of our climate clouds cause a general "net cooling" over most parts of the globe, which amounts to about 60 Wm^{-2} (or about 25 percent of the total emission to space of 240 Wm^{-2}). However over some areas they even may enlarge slightly the radiative energy gain of the planet. These latter regions are known to be covered mostly by high level cirrus only, which also in earlier model studies (Hansen et al., 1981) were found to enhance the greenhouse effect of the atmosphere. Seasonal variations of these cloud effects where analyzed by Harrison et al. (1990) with data of the ERBE (Earth Radiation Budget Experiment; Barkstrom et al. 1989).

Recent climate model studies tend to forecast more high level cloudiness for future climate scenaries with increased CO_2 and CH_4 concentrations (Roeckner, 1987). Would this mean an increased greenhouse effect of our planet, or an reduction of it if such high-altitude ice clouds are optically thick enough to be less transparent for the incident solar radiation, as it has been very recently suggested by Ramanathan and Collins (1991)?

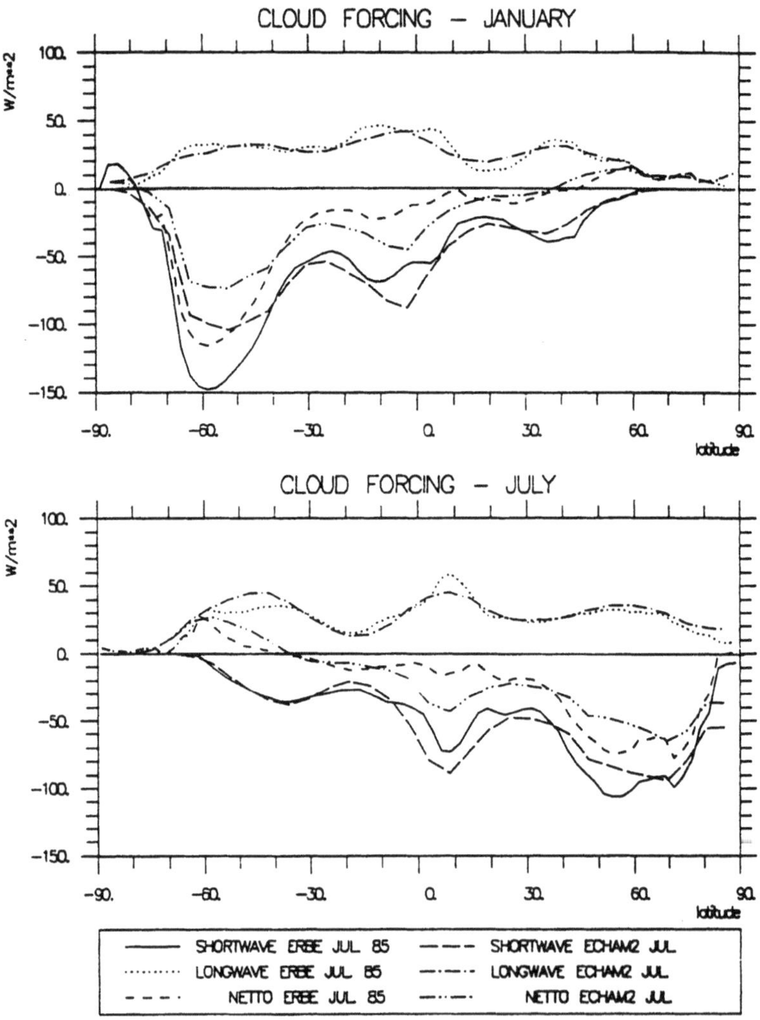

Figure 3.1: *Zonally averaged cloud radiative forcing as obtained from observations of the Earth Radiation Budget Experiment for January 1986 and July 1985 and as simulated by the climate model ECHAM 2 in a 20-year climate experiment (from Roeckner et al., 1991).*

3.3 Cloud Physics and Phenomenology

About 45 to 55 % of the earth's surface is always covered with cloud fields.Their occurrence can approximately be classified into four major classes, as summarized in Table 3.1.

cloud class	frequency of occurence
1. low stratiform cloud layers, near the top of the planetary boundary layer (1 – 2 *km*), consisting mostly of water droplets	30 - 60 %
2. medium level cloud fields in altitudes between 3 and 7 *km*, consisting often of mixture of ice and water particles	40 - 50 %
3. high-level clouds (cirrus), mostly consisting of ice particles	25 - 40 %
4. deep convective clouds, e.g. thunderstorms, penetrating often from the boundary layer up close to the tropopause	10 - 15 %

Table 3.1: *Four major cloud classes and the ranges of their frequency of occurrence.*

Fig. 3.2 (from Hahn et al., 1982) shows the relative abundance of these clouds in all latitudes as derived solely from surface observations. To describe quantitatively the radiative transfer properties of such cloud fields, we need to know their following properties.

Microphysics: particle size distributions, shapes, number densities and thermodynamic phases; index of refraction; aerosols and their properties in the cloud air.

Macrophysics: vertical and horizontal structures and extent of cloud fields; fractional coverage within a defined grid area, altitudinal ranges, temperatures.

Note: We always should keep in mind, that modelling and diagnoses of the atmosphere is mostly done in Eulerian reference frames, thus these properties must properly be defined for entire grid elements (e.g. Lee, 1989). Presently used parametrization schemes for cloud fields in General Circulation and Climate models provide only very simple information on such subgrid scale details and lead to large differences in calculated cloud fields (see e.g. Cess et al., 1990). Therefore several regional experiments are made within the scope of the ISCCP to study directly the physical properties of cloud fields. One of them is the International Cirrus Experiment (ICE; Raschke, 1988), a joint effort of European groups; also in the US a similar experiment (FIRE, e.g. Starr, 1987) is carried through.

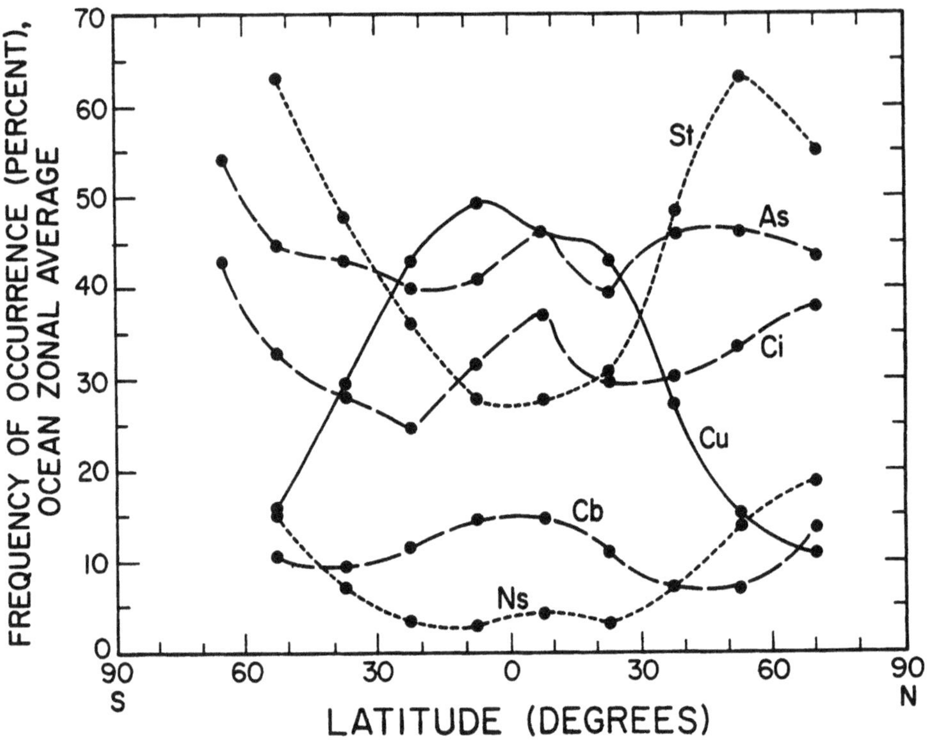

Figure 3.2: *Frequency of occurrence of each cloud type as a function of latitude, average of all four seasons (from Hahn et al., 1982).*

3.3.1 Microphysical properties of cloud fields

Clouds contain moist (saturated) air and particles (aerosols, water droplets and ice crystals).

a) Aerosols: These are smallest particles with sizes between 0.01 and 10 μm consisting mostly of soluble and insoluble materials originating from the surface - with appreciable amounts of anthropogenic origin. Most aerosols grow with increasing amounts of available water - thus already altering optical properties of aerosol layers - before smaller droplets form (Twomey, 1977 and 1991). Most of the aerosols act as cloud condensation and ice nuclei. A very complete review of all interaction between aerosols and clouds has recently been summarized in a report on an experts meeting on this subject (Vali, 1991).

b) Water droplets form when the supersaturation ratio reaches a critical value around 1. Their further growth depends on the availability of water in the ambient air, atmospheric turbulence and radiation. Of particular importance for further growth are collisions with other droplets and changes between the water and ice phases. The size spectra range from 0.1 to about 50 μm, where the latter droplets are already large enough to cause "drizzle" at the surface. One

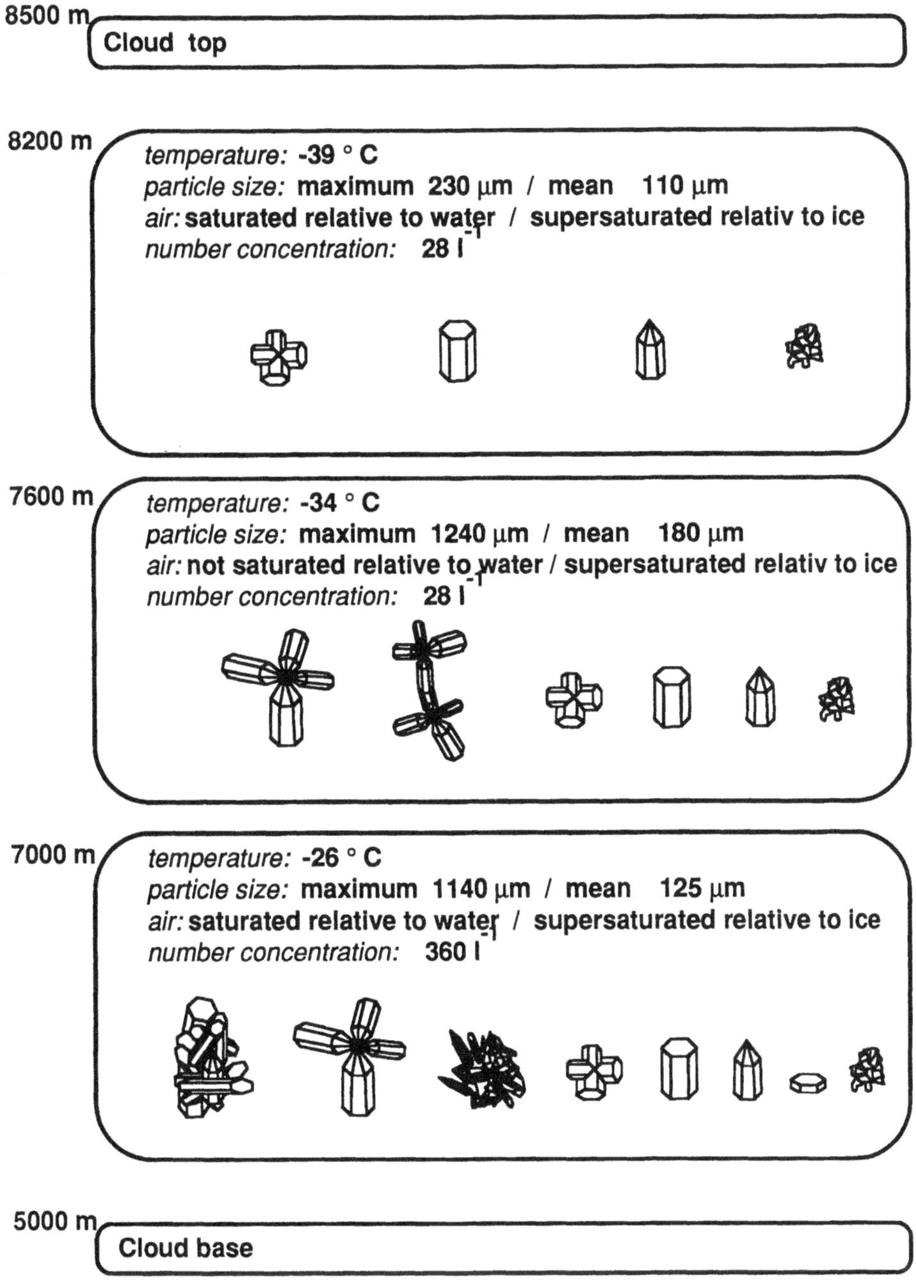

Figure 3.3: *Ice particle sizes, shapes and number concentrations derived from holographic data taken with the British aircraft C130 during the field measurements of the International Cirrus Experiment in 1989 (from Krupp, 1991).*

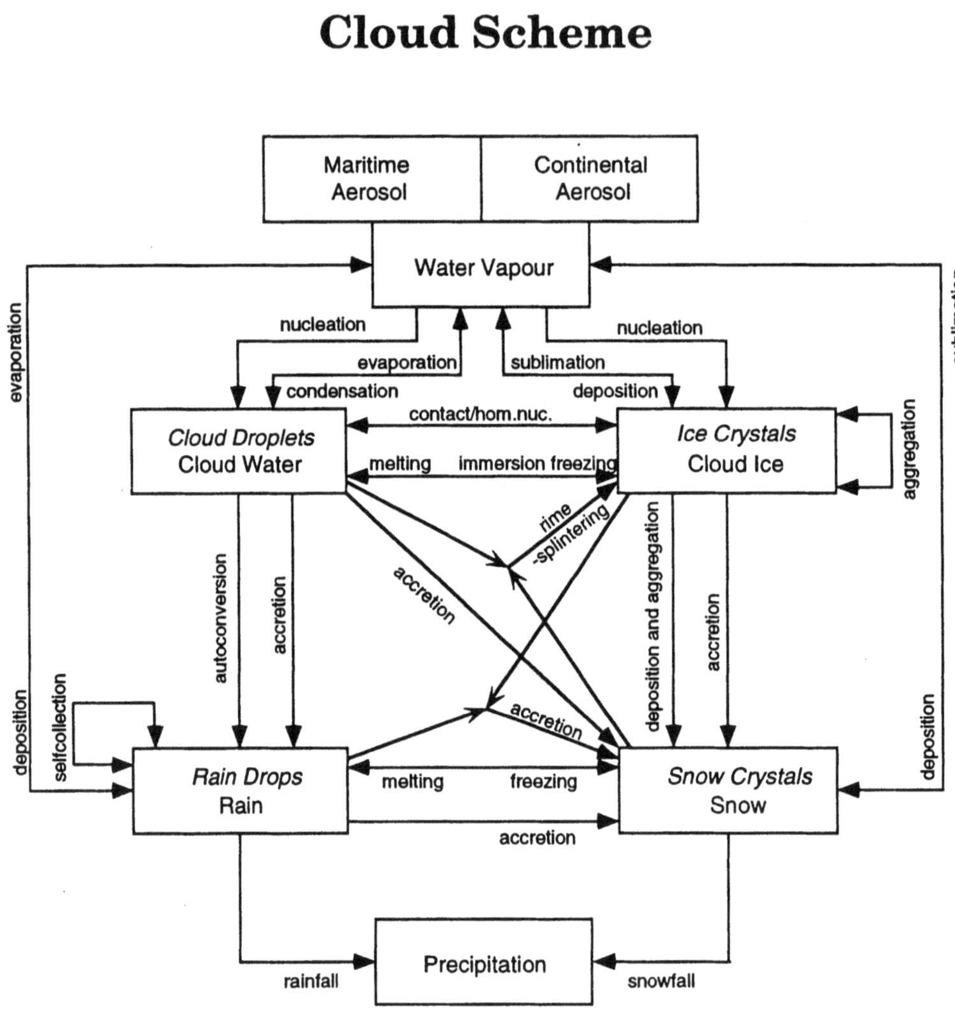

Figure 3.4: *Processes occurring in a cloud with liquid drops, ice and snow particles, which should be considered in a numerical model (summarized by Levkov and Rockel, 1991 - private communication).*

often finds in lower level stratiform clouds larger drops near the top, and smallest above of them (e.g. Nicholls and Leighton, 1986).

Multiple collisions and collections, and in particular transitions from the ice-phase (snow process of Bergeron-Findeisen) produce rain drops. The largest may reach diameters up to $5 - 8\,mm$. In a first approximation all water droplets can be considered to be nearly spherical, which allows an easier treatment in radiative transfer calculations.

In convective cloud layers above the freezing level snow and hailstones may form, depending on the water supply and internal turbulences, and on the length of vertical paths through the cloud as well. These, in turn, affect the rain fields. Browning (1990) described extensively the occurrence of cloud and rain fields.

c) Water in its solid phase (snow and ice) often shows crystal forms of extremely complicated structures. Fig. 3.3 summarizes recent analyses of holograms made from aircraft in "dry" cirrus (C. Krupp, 1991). All of them show a hexagonal basic structure, which even can be seen in cuts of hailstones. In cirrus the ice needles can reach lengths up to $800\,\mu m$. Various processes are responsible for the formation of observed ice size spectra in cirrus. Obviously the splindering is the most important one to increase the particle number over that determined by the number density of available ice nuclei (Mossop, 1985). These very small particles of sizes less than about $20\,\mu m$ cannot be observed in the cirrus level with either holographic or other means. Several numerical studies have shown that they might be responsible for high reflectances of cirrus (e.g. Stackhouse and Stephens, 1991). Thus, future field measurements should pay more attention for their direct observations. Several analyses of satellite data and ground-based lidar and radiometer measurements lead to speculations on their occurrence in cirrus fields (e.g. Platt et al., 1989).

d) A model of microphysical processes in a cloud may need to consider all of these processes or at least a limited number of them, depending on the cloud type and external forcings. These processes (Levkov, Rockel, 1991: private communication) are often still incompletely known and need further investigations in the laboratories, by numerical simulations and also "in the field". Most of them are summarized in the scheme shown in Fig. 3.4. Even rather sophisticated numerical studies of cloud growth and decays under different external forcings make use of simplifications and parametrized processes (e.g. Zhang et al., 1989).

3.3.2 Macrophysical forms of cloud fields

Areal photographs now obtained routinely from many satellites are best suited to show the horizontal structures of cloud fields.

We can observe complex, but often exiting structures in the convective cloud fields near the equator and also in frontal zones (extratropical belts). The lower level stratiform cloud fields above subtropical oceans often show a cellular structure, similar to cells forming in cold air outbreaks from polar sea ice over the warmer oceans. High-level cirriform cloud fields often show a "fuzzy" structure. They stretch along the wind field.

Roles and streets and many other complex arrangements can be found in low and medium level clouds, indicating intensive interactions with the ground below by exchanges of momentum and heat. Most complex structures can be observed in all cloud fields occurring in highly unstable air.

Usually traditional statistical measures, such as areal fractional coverage, size distributions of cloud elements, mutual distances between them, and areal distribution functions are used to describe these cloud field, in particular to simulate their areal radiative transfer properties. These "measures" can mostly be related to momentum and heat fluxes in the ambient atmosphere. More recently also analyses of the fractal structures and the intermittency parameter have been made in analogy to the generalization of boundary layer properties (e.g. Lee, 1989; Welch and Wielicki, 1989).

Vertical structures in cloud fields (layers, cloud free areas inside, etc.) are result of many interactions between the cloud itself and the ambient atmosphere with internal (e.g. turbulence) and external forcings (radiation, advection of momentum or water vapor, etc.). Even the lower level stratiform cloud layers, when visible for the eye to be extremely homogeneous, show strong internal structures. These provide difficulties to characterize microphysical properties on the basis of aircraft measurements.

Further the vertical appearance of cloud fields is also result of the atmospheric stability and of large scale motion fields. These structures are phenomenologically understood, however, not yet very accurately described in models to provide a solid basis for a quantitative description of areal cloud field parameters. Therefore, one often restricts such descriptions to very few parameters: areal coverage vs. altitude, water and ice content, typical size distributions of particles, and the thermodynamic phase.

3.4 Radiative energy transport in the atmosphere

In studies of radiation-cloud-climate interactions we must consider the integral radiative energy in the spectra of solar (shortwave) and terrestrial (longwave) radiation . The spectral components of them are important to understand the fractional absorption within atmospheric layers and inside of clouds. Radiative energy in narrow spectral regions is also of vital importance for chemical processes (e.g.: aerosol formation, ozone in lower tropospheric layers) and for biological activities at ground and in near surface layers of all water bodies.

In this chapter, we will briefly review these interactions to derive quantities which could be used to describe quantitatively the radiative characteristics of cloud fields. Our qualitative arguments cannot replace any of the now many basic treatments of radiative energy transfer within absorbing and scattering media (e.g. Goody and Yung, 1989; Schanda, 1986).

3.4.1 Basics of radiative transfer in the molecular atmosphere

Interactions between electromagnetic radiation and the molecular atmosphere, are primarily determined by elastic scattering, by absorption and emission processes. At ground surfaces reflections and emissions occur, respectively. Scattering processes follow strictly the Rayleigh-law with a λ^{-4}-dependence (λ = wavelength). We should not consider here the polarization and depolarization due to multiple scattering, although these properties could be of vital interest for the cloud detection in remote sensing data.

Complex line and band models have been developed to describe accurately the emission/absorption along a ray. Absorption occurs in numerous bands of O_3, O_2, H_2O, CH_4 and other minor constituents. At wavelengths larger than about $4\,\mu m$ these gases (except O_2) also emit large amounts of thermal

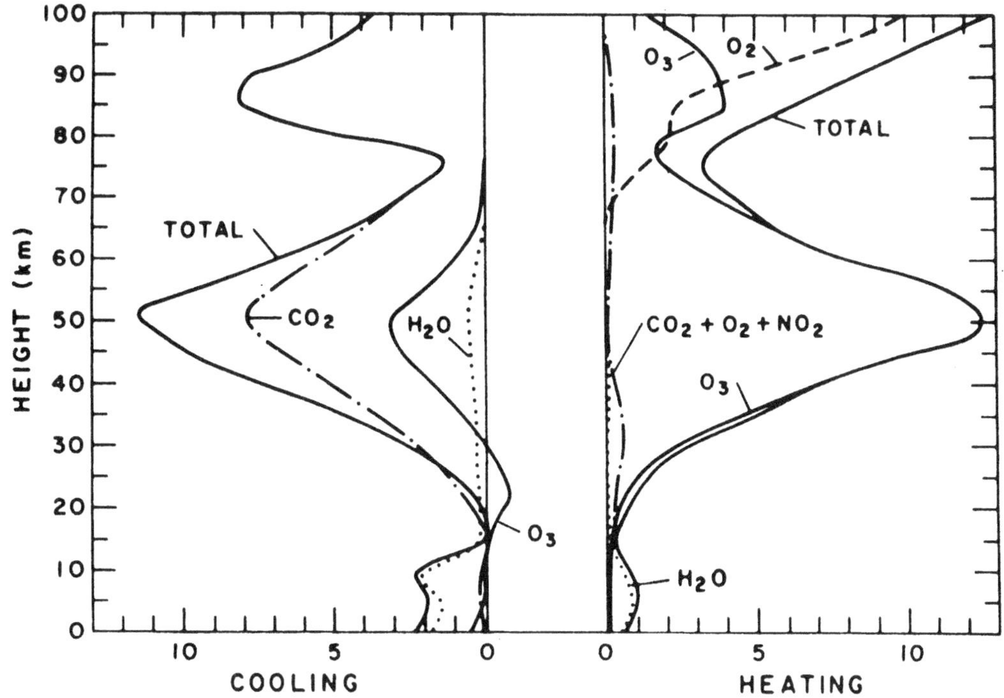

Figure 3.5: *Vertical distribution of heating due to absorption of solar radiation (right) and cooling due to emission of infrared radiation (left). (Source: D.G. Andrews et al., 1987).*

radiation . Fig. 3.5 shows an estimate of atmospheric heating and cooling which is due to absorption and emission by atmospheric gases only, at mean atmospheric conditions (for further information; e.g. Goody and Yung, 1989). Accordingly water vapour is the dominating Greenhouse gas in the troposphere (Shine and Sinha, 1991; Starr and Melfi, 1991), while ozone and carbon dioxide are responsible for the heat budget in the layers above.

There are now many radiative transfer codes available treating these processes with various degrees of complexity - where in altitudes above about 40 km also the deviations from local thermodynamic equilibrium must be considered. An attempt starts to "calibrate" successfully in particular those models which are used in climate and circulation models (ICRCCM: International Comparison of Radiation Codes for Climate Models). First results have recently been published in a series of papers (Ellingson and Fouquart, 1991).

3.4.2 Radiative energy transfer in the turbid atmosphere

Particles (i.e.: aerosols and cloudparticles) which are many orders of magnitude larger than air molecules cause the "turbidity" in the atmosphere. Since their sizes are similar or larger than wavelength ranges of ambient radiation fields, the simple Rayleigh law does not apply anymore. Either the more complex "Mie-calculations" or the laws of geometrical optics (i.e.: "ray tracing") should be used. The absorption and scattering effects are calculated on the basis of values of the complex index of refraction, whose imaginary part describes the absorptive properties. Usually spherical shapes are assumed in most calculations to allow for the simpler spherical symmetry in the radiative transfer codes. But more complex particle structures, even hexagonal cirrus crystals, have been studied in more recent work.

Single-scattering functions for various size distributions of spherical particles show that in particular the forward peak ($\theta = 0^\circ$) is heavily dependent on the particle size. Fig. 3.6 (Rockel et al., 1991) presents a display of the wavelength and size dependence of the single scattering albedo for ice spheres. Fig. 3.7 (A. Macke, private communication, 1992) shows a very recent calculation of the scattering function for a bullet-shaped ice crystal, in comparison with that of a sphere of the same equivalent area. Such scattering functions can be calculated for various particle sizes and shapes must be "mixed" according to the relative presence of particles in each atmospheric layer. It has been speculated by Stackhouse and Stephens (1991) that backward scattering from very small nonspherical ice crystals might be responsible for high albedos measured above cirrus fields.

There is now a variety of radiative transfer codes available to simulate the transfer in one- or even two- (and three-) dimensions with inclusion of multiple scattering in each layer. They are enormous time consuming and, therefore cannot be coupled with models for the circulation of the global atmosphere.

3.4.3 Various approximation

In reality it is not possible to simulate in all details and in addition with computational economy the transfer of radiative energy through realistic cloud ensembles. Therefore, one-dimensional approximations are made, where clouds are described by their vertical column of liquid or ice water content, by mean particle size distributions (profile, if known) and their fractional cover in each layer. Random or maximum overlap of cloud elements is assumed in dependence on the atmospheric stability; e.g. in a highly unstable troposphere cumulus towers occur often in all layers up to the troposphere (Roeckner, 1987). Kristjánsson (1991) demonstrated the effects of spatial resolutions of clouds and orography on computed precipitation and vertical motion fields.

Thus, in general for calculations of the radiative transfer properties clouds are mostly taken as horizontally infinite and homogeneous sheets, with given values of water or ice content and size-distributions. Aerosols and water vapor need to be added. The radiative transfer properties are assumed as inherent to the cloud itself, since measurements may often contain additional effects (e.g. high reflectance of the ground, etc.).

Fields of broken clouds, which occur more often than the above-described homogeneous layers, show quite different behaviour. Some studies are made on the basis of aircraft measurements, laboratory work and numerical simulations. Their reflectance for solar radiation is smaller than of plane-parallel layers, due to shadows in the "holes". These holes act in the downward direction also as "pipelines", due to multiple forward reflections (Lee, 1989).

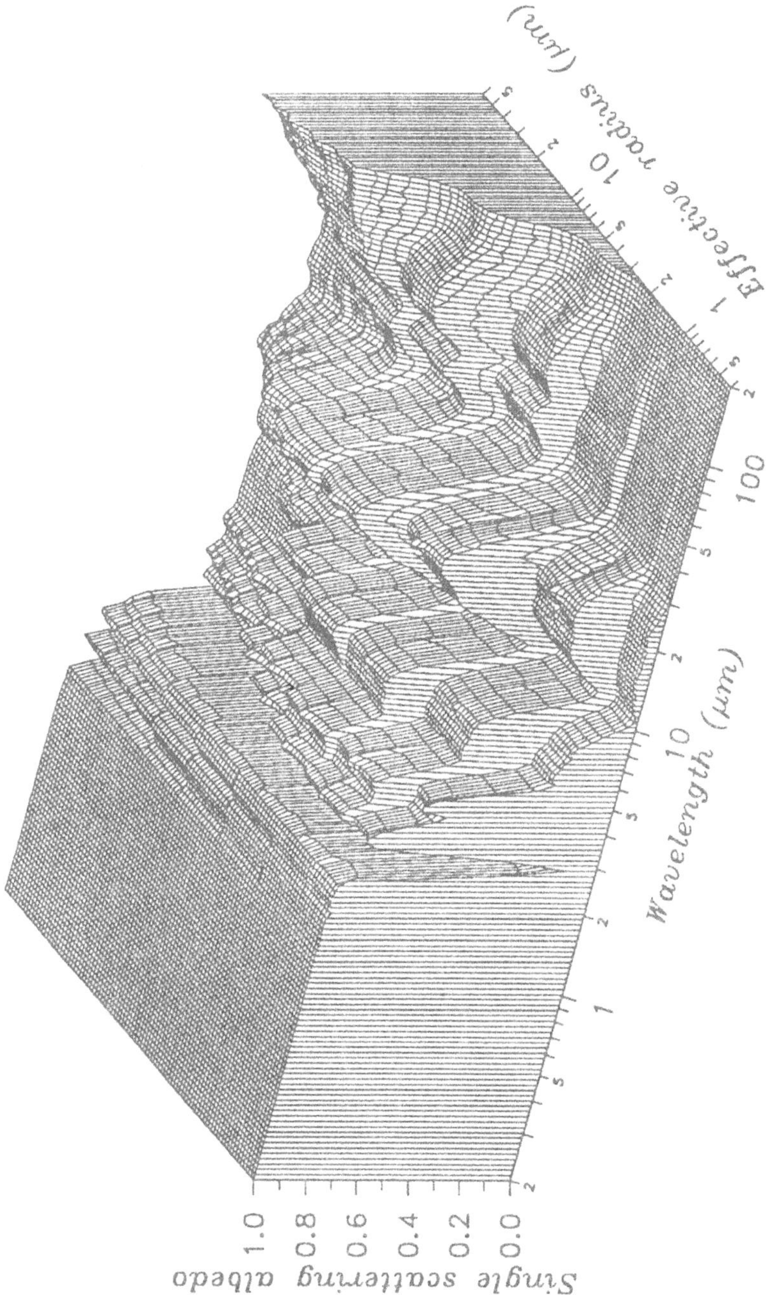

Figure 3.6: *Single scattering albedo of ice spheres from Mie calculations based on the theory by Hansen and Travis (1974). Calculations are made by Rockel et al. (1991).*

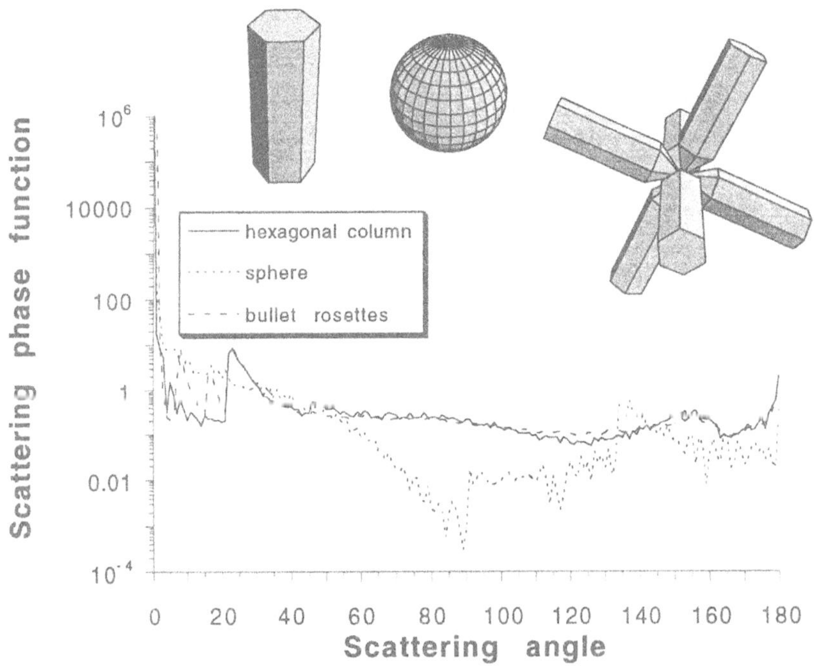

Figure 3.7: *Single scattering functions of large ice crystals, which are randomly orientated in 3 directions and illuminated with radiation of the wavelength* $\lambda = 0.55\,\mu m$. *For comparison also the scattering function of an equivalent sphere is shown. (A. Macke, 1992 - private communication)*

3.5 Empirical studies of radiative cloud effects on climate

Eq. 3.1, balancing the incoming solar with the outgoing terrestrial (heat) radiation has often been origin of simples climate sensitivity studies. One early study has been made by Budyko (1969). Later many authors modified it to treat more effectively in various ways the cloud effects. In this treatment we simply should follow a very recent summary by Ridgway (1991). Other studies are made by Stephens and Greenwald (1991); Hartmann and Doelling (1991) and many others.

3.5.1 Cloud-radiation sensitivity parameters

If we assume a mean cloud albedo ρ_c and ground albedo ρ_0, both given with fractions A_c and A_o, then the incident solar radiation Q_s can be calculated from Eq. 3.2.

$$Q_s = \frac{S_0}{4}\left(1 - \rho_c A_c - \rho_0(1 - A_c)\right) \tag{3.2}$$

Its sensitivity to changes of the fractional cloud cover are given by Eq. 3.3.

$$\frac{\partial Q_s}{\partial A_c} = -\frac{S_0}{4}(\rho_c - \rho_0) \tag{3.3}$$

Similarity we can define the outgoing longwave radiation Q_t as

$$Q_t = \sigma T_c^4 = Q_{tc}A_c + (1 - A_c)Q_{t0} \tag{3.4}$$

Q_{tc} and Q_{t0} would be mean emissions to space from cloud tops and the ground, respectively.

Since we consider here global averages of all parameters no distinction is made between clouds of different altitudes.

The net sensitivity to cloud amount can then be expressed by Eq. 3.5:

$$\sigma = -\frac{S_0}{4}(\rho_c - \rho_0) \cdot \left[1 - \frac{\partial Q_t/\partial A_c}{\partial Q_s/\partial A_c}\right] \tag{3.5}$$

where over most surfaces - except above polar snow fields - $\rho_c > \rho_0$.

We observe large variations in these parameters which are basically due to choices of different values for the various parameters involved. But we can see from Eq. 3.5 the general tendency, that the reflectance of clouds for solar radiation tends to "cool" the planet more than it is "heated" by heat radiation which is trapped on earth by the same clouds.

There are a number of earlier empirical studies made with climate data and later ones on the basis of most recent satellite measurements of the radiation budget at the top of the atmosphere. We try to summarize them here in Table 3.2.

Authors	$\partial Q_s/\partial A_c$	$\partial Q_t/\partial A_c$	net	with $\rho_c(\rho_0)$
Schneider(1972)	-130 Wm^{-2}	75 Wm^{-2}	-55 Wm^{-2}	0.5(0.12)
Cess(1976)	-85.5 Wm^{-2}	86 Wm^{-2}	$\sim$ 0 Wm^{-2}	0.42(0.17)
Ohring(1981) (satellite data)	-104 Wm^{-2}	35 Wm^{-2}	-71 Wm^{-2}	

Table 3.2: *Sensitivity of radiation budget to cloud amount A_c (from Arking, 1990).*

3.5.2 "Cloud forcing" - or cloud effects on the radiation budget

In a slightly modified consideration, which allows also for regional studies of the cloud effects on radiation fields, several authors consider differences of Q_s and Q_t over the same area, but at mean (average) conditions and at clear skies. These differences are often named "cloud forcing".

$$\begin{aligned} C_t &= Q_{t0} - Q_t \\ C_s &= Q_s - Q_{s0} \end{aligned} \tag{3.6}$$

Analyses of the more recent measurements of the Earth Radiation Budget Experiment (ERBE) yield values for sums of C_t and C_s, which range from about -60 Wm^{-2} to +40 Wm^{-2}, where then in the net positive values of the sums of both can be found over the Sahara (where upper level thin cirrus clouds often occur - and no others) and the polar zones while larger negative values occur over regions with frequent lower level cloud fields. These values are subject to errors, in particular when clear sky scenes cannot be observed accurately. They also depend on the temperature and albedo contrast between the cloud and clear scene. Li and Leighton (1991) made a thorough error analyses of observations made above the Arctic Sea ice with respect to their sensitivity to errors in the scene identification. Hartmann and Doelling (1991) provide in their paper a mnemotechnical explanation of the heating and cooling effects of clouds in relation to the magnitude of their albedo and emission of heat radiation. Their Fig. 2 is reproduced here in Fig. 3.8.

We should not overlook the effect of clouds on the surface energy budget over polar regions and responses in the sea ice properties.

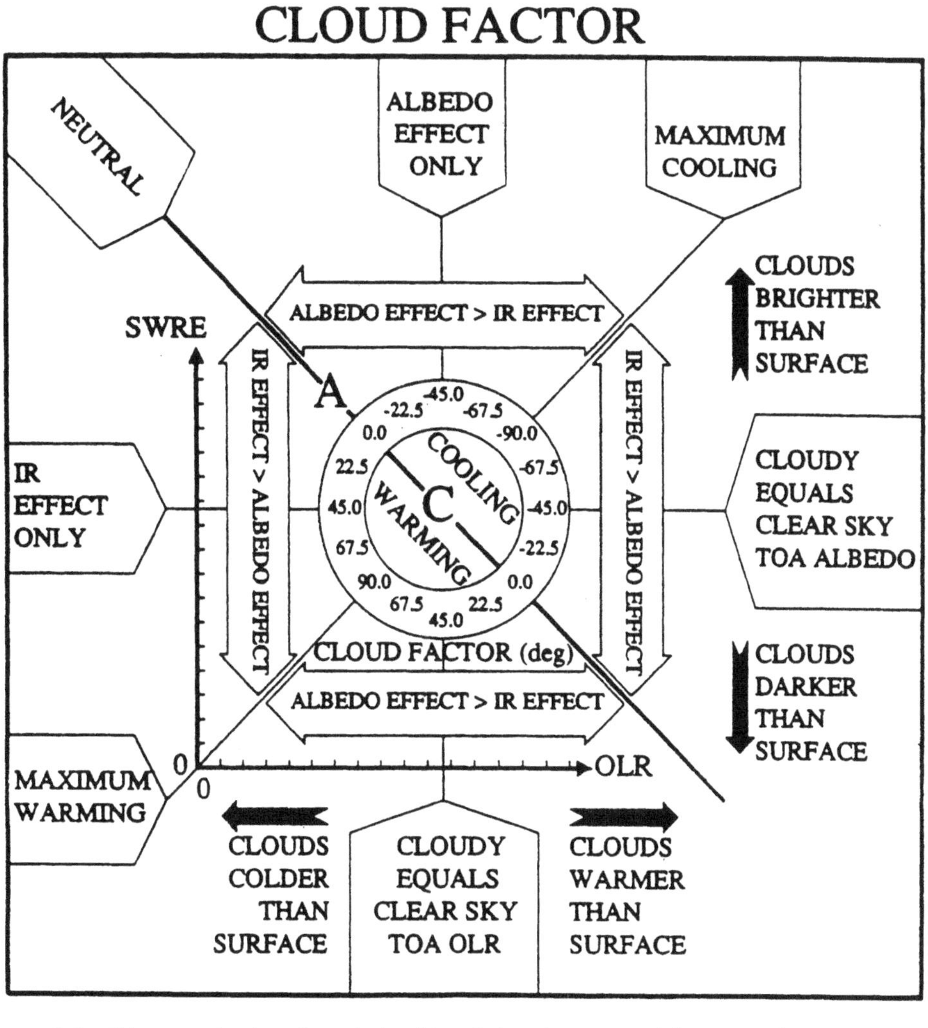

Figure 3.8: *Diagram showing the relationship of the alignment of data in reflected solar versus outgoing longwave coordinates to the net effect of cloudiness on the radiation balance and their relationship to the cloud factor angle. The diagonal with slope -1 represents radiatively neutral clouds, while a diagonal with slope +1 represents those cases where the shortwave and longwave effects of cloud reinforce each other to produce cooling (first quadrant) or warming (third quadrant). Figure and legend from Hartmann and Doelling, 1991.*

3.6 Numerical studies

The analyses of the radiation budget, as discussed above, can analogously be applied to the radiation budget results of numerical studies with atmospheric circulation models. In fact they provide also a convenient tool to "measure" the performance of such models in calculating the present climate state.

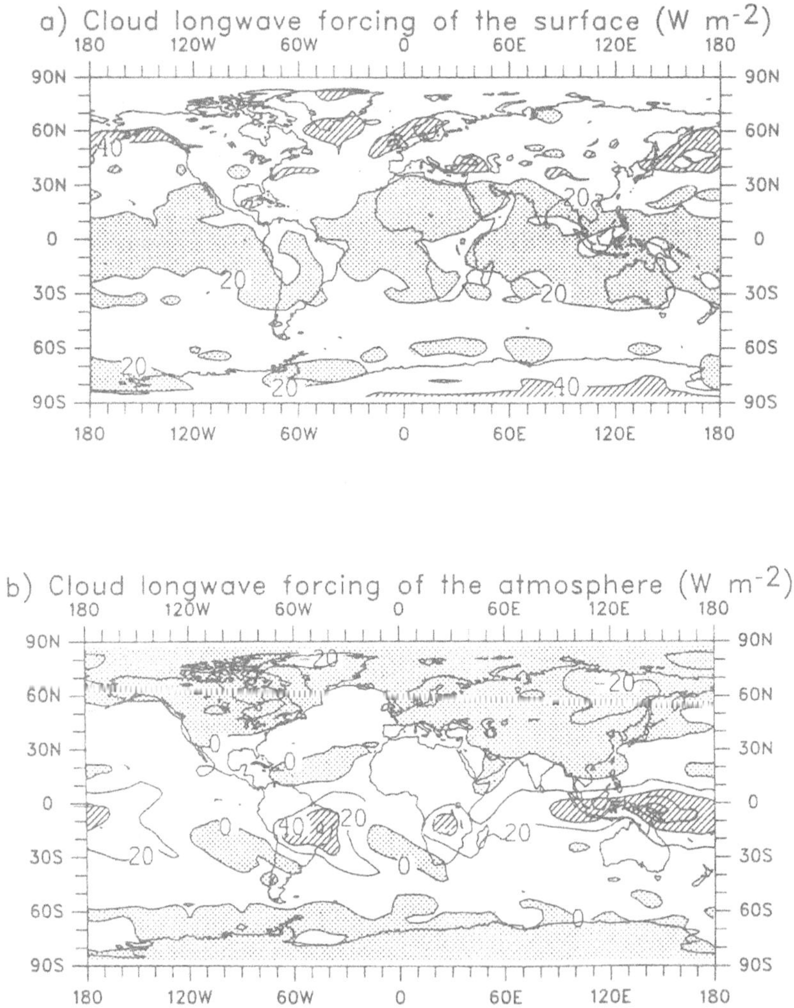

Figure 3.9: *Cloud longwave forcing of (a) the surface and (b) the atmosphere, computed with a GCM. The contour interval is 20 Wm^{-2}. Values greater than 40 Wm^{-2} are hatched on both plots. Values less than 20 Wm^{-2} are stippled on (a) and negative values are stippled on (b) (from Slingo and Slingo, 1988).*

Amongst the many published studies we should quote here only very few. Slingo and Slingo (1988 and 1991) calculated the cloud forcing for both the radiation budgets at the top and bottom of the atmosphere, respectively. The difference of both even allow to determine the cloud influences on the absorption (or loss) of radiative energy within the atmosphere as a net results. One of their results is reproduced in Fig. 3.9.

Similar studies with observations are not yet possible since we are lacking in relevant observations of the radiation budget parameters at the earth's surface. However attempts are already underway to compute the surface radiation budget parameters from satellite and other measurements. Here it has been proven, that the solar budget can be obtained with accuracies of better than 10% (e.g. Raschke and Rieland, 1989), while much larger errors are still to be expected for the longwave radiation budget at ground (Schmetz et al., 1986). Stuhlmann et al. (1990) made very careful ground truth comparisons with values of the downward solar radiation as derived from measurements of the geostationary satellite Meteosat 2 over Africa and Europe. Their individual results are published in a detailed atlas (Raschke, et al. 1991).

During the last years the modellers recognized that their results may not be comparable in all cases, therefore they began with studies using quite similar or even the same initial conditions. It is the purpose of such studies to understand the performance of models better, but also to identify more clearly their various deficiencies. One of such intercomparisons has been made by Cess et al. (1990) with particular concentration on climate feedback processes. In another, Boer et al. (1991), considered in particular components of the water cycle.

Cess et al. (1990) compiled the results of 19 atmospheric circulation models. They find for another climate sensitivity factor

$$\lambda = \left(\frac{\Delta Q_t}{\Delta T_0} - \frac{\Delta Q_s}{\Delta T_0}\right)^{-1} \tag{3.7}$$

which relates changes of the radiation budget components at the top of the atmosphere to induced changes of the sea surface temperature T_0 a large variation of the model responses as shown in Fig. 3.10. A closer look into the individual results showed that these variations are due to the fact that each model treats clouds different from the other.

For comparison we will simply assume no interactive feedback mechanisms. Then

$$\frac{\Delta Q_s}{\Delta T_s} = 0 \quad \text{and} \quad \frac{\Delta Q_t}{\Delta T_s} = 4 \quad ; \quad \frac{Q_t}{T_s} \equiv 3.3 W m^{-2} K^{-1} \tag{3.8}$$

for typical conditions: $Q_t \sim 240\, W m^{-2}$, $T_s \sim 288\, K$. One obtains in this case

$$\lambda = 0.3 K m^2 W^{-1}. \tag{3.9}$$

If one includes water vapor, the main greenhouse gas, into such considerations, than one arrives at values of

$$\lambda_0 \sim 0.5 K m^2 W^{-1}. \tag{3.10}$$

which is close to the "clear sky" values shown in Fig. 3.9.

Cess et al. looked also in the net cloud forcing of these 19 models and found values between -2 and -45 $W m^{-2}$ with a mean of -21 $W m^{-2}$ (stand. dev.: 9 $W m^{-2}$).

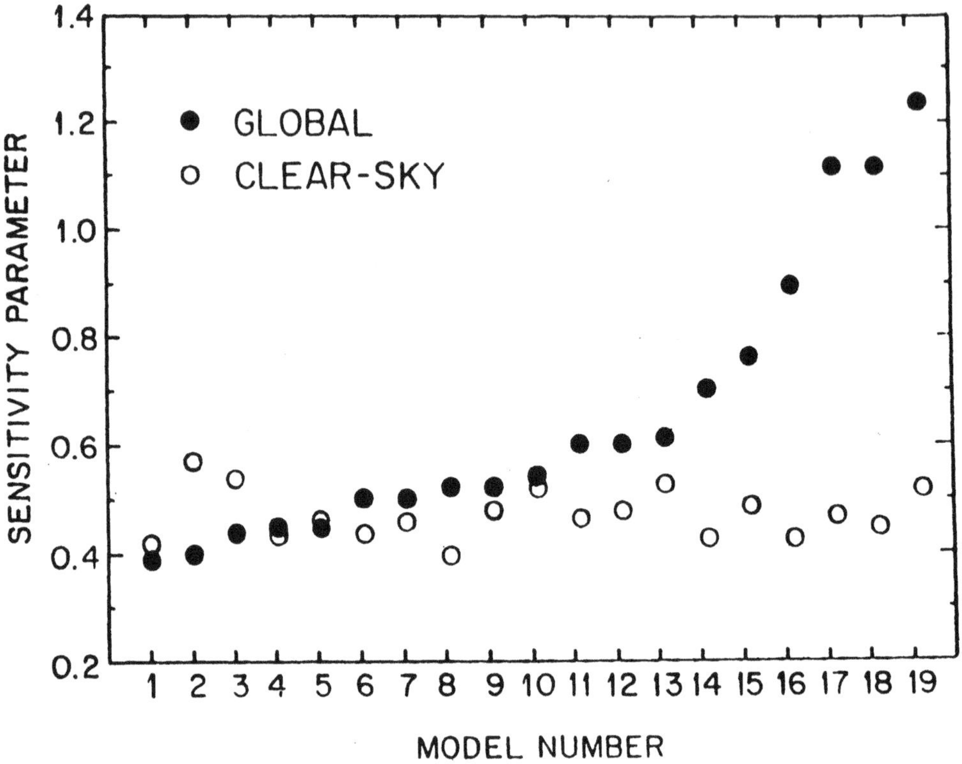

Figure 3.10: *Clear-sky and global sensitivity parameters ($K\,m^2\,W^{-1}$) for the 19 GCMs. The model numbers correspond to the ordering in Table 9 (from Cess et al., 1990).*

Some of these numerical simulations of future climate regimes with increased carbon dioxide and other Greenhouse gases forcings indicate, that the near tropopause cloudiness might increase at all latitudes and the low level clouds over both polar regions, while there is an indication of reduced cloudiness in the midtroposphere regions. This development would result in an overall enhancement of the Greenhouse effect by clouds, except over apolar regions and a stabilization of the troposphere, since in the latter case more radiative energy is absorbed near the tropopause than now (e.g. Weatherald and Manabe, 1988).

3.7 Some concluding remarks

It has been the purpose of this paper to highlight our present capabilities to observe and model the effects of clouds on radiative energy transports and of related climate feedbacks. The records of observations contain now multi-year measurements of the planetary radiation budget components, of various cloud parameters derived within the ISCCP (Rossow and Schiffer, 1991) from the operational satellites' images data, and indeed many other meteorological and ground state parameters. These can indeed be used to derive some sensitivity parameters, such as the quantities λ and δ relating the planetary radiation budget to changes in the cloud cover. The ISCCP data sets can now be used to distinguish even between the effects of low middle and high clouds. Once also reliable data on the surface radiation budget over the entire globe are available - as they are now under consideration within a project of the WCRP - one might get a clearer picture on the effects of clouds on the radiative energy absorption and loss within the atmosphere.

Can we relate these investigations to possible climate interaction? Possibly not on longer time scales. However, an investigation of shorter time scales might be already of greater benefit, in particular in regions of high "sensitivity" to climate variations. Amongst them might be the Arctic sea ice regions, where small variations in the radiation budget at ground - as induced by variations in the cloudiness - could cause considerable changes in the sea ice thickness and, thus, its stability against mechanical stress. In the tropics and subtropics the radiative and thermodynamic effects of convective cloud systems might already act as some kind of a greenhouse regulator compensating for increases of sea surface temperatures above some critical point of about 302 K (Ramanathan and Collins, 1991). Flohn and Kapala (1989) show, that in the tropical regions the tropospheric water vapour might already have increased leading to more convection.

Numerical circulation models would in principle be best suited to study radiation - cloud - climate interactions, since they allow for large varieties of sensitivity studies. But their performance is still too controversial in this respect. We have still to learn in both disciplines: data analyses and numerical modelling and the assimilation of real time observations into numerical models for dynamical processes. None of these models is yet complete enough to answer all open questions. One of them is related to the atmospheric water vapour and its changes when climate might warm due to the increased Greenhouse forcing. Shine and Sinha (1991) suggest, that indeed the planetary radiation budget would change most sensitive to changes of the lower tropospheric water vapour content, which is indeed higher than in layers above, due to the temperature effects. Some controversy arose around Lindzen's (1990) remarks on a possible drying-out of the upper tropical troposphere due to increased convection in the future climate. Betts (1990) used some qualitative thermodynamic arguments to indicate that it is still unclear how the cloud water content and optical thickness might change in future. On the other side, Roeckner et al. (1987) found in their model an increase of both and thus a larger negative feedback.

There is still considerable research (field measurements and numerical simulations as well) required to fully understand the physical processes which maintain cloud fields and their properties. Further we are still lacking in adequate measuring capabilities to map globally and with sufficient accuracy the upper tropospheric water vapour.

In this paper we have not discussed in any detail the expected effects of human activities, such as the often observed shiptrails and aircraft contrails or some altered clouds above or downwind of larger urban areas.

3.8 References

Andrews DG, Holton JR, Leovy CB (1987) Middle Atmosphere Dynamics. *Academic Press* Toronto, 489 pp.

Arking A (1990) The radiative effects of clouds and their impact on climate. *Bull Am Meteor Soc* 71: 795-813.

Barkstrom B, Harrison E, Smith G, Green R, Kibler J, Cess R, ERBE Science Team (1989) The Earth Radiation Budget Experiment (ERBE) archival and April 1985 results. *Bull Am Meteorol Soc* 70: 1254-1262.

Betts AK (1990) Greenhouse warming and the tropical water budget. *Bull Am Meteor Soc* 71: 1464-1465.

Boer GJ, Bauthin I (1991) An Intercomparison of the climate simulated by 14 atmospheric General Circulation Models. *WCRP-58* WMO-TD 425, available through World Meteorological Organization, Geneva.

Browning KA (1990) Rain, rainclouds and climate. *Quart Journ Roy Met Soc* 116: 1025-1051.

Budyko MI (1969) The effect of solar radiation variations on the climate of the earth. *Tellus* 21: 611-619.

Charlson RJ, Lovelock JE, Andreae MO, Warren SG (1987) Oceanic phytoplankton, atmospheric sulfur, cloud albedo and climate. *J Geophys Res* 89: 9668-9672.

Cess RD, Potter GL, and 30 authors (1990) Intercomparison and interpretation of climate feedback processes in 19 atmospheric circulation models. *J Geophys Res* 95: 16601 - 16615.

Ellingson R, Fouquart Y (1991) The intercomparison of radiation codes in climate models. *J Geophys Res* 96: D5 - 16 papers.

Flohn H, Kapala A (1989) Changes of tropical air-sea interaction processes over a 30-year period. *Nature* 338: 244-246.

Goody RM, Yung YL (1989) Atmospheric Radiation (Second Edition). *Oxford University Press* ISBN 0-19-505 134-3.

Hahn CJ, Warren SG, London J, Chervin RM, Jenne R (1982) Atlas of simultaneous occurrence of different cloud types over the ocean. *NCAR Tech Note* TN-201+STR.

Hansen J, Johnson D, Lacis A, Lebedeff S, Lee P, Rind D, Russel G (1981) Climate impact of atmospheric carbon dioxide. *Science* 213: 957-966.

Harrison EF, Minnis P, Barkstrom BR, Ramanathan V, Cess RD, Gibson GG (1990) Seasonal variation of cloud radiative forcing derived from the Earth Radiation Budget Experiment. *J Geophys Res* 95: 18681-18703.

Harshvardhan (1991) Atmospheric radiation. *Rev in Geophys* Suppl 56-68.

Hartmann DL, Doelling D (1991) On the net radiative effectiveness of clouds. *J Geophys Res* 96: D1, 869-891.

Hartmann DL, Ramanathan V, Berroir A, Hunt GE (1986) Earth radiation budget data and climate research. *Rev of Geophys* 24: 439-468.

Houghton JT (ed.) (1990) Climate change - The IPCC scientific assessment. *World Met Org* 358 pp.

Jonas, P (1989) Effects of clouds on radiation. *Atmosph Res* 23: 259-286.

Kristjánsson JE (1991) Cloud parameterization at different horizontal resolutions. *Quart J Roy Met Soc* 117: 1255-1280.

Krupp C (1991) Holographische Messungen an Eiskristallen in Cirruswolken während des Internationalen Cirrus Experimentes ICE. Masters Thesis, Inst.for Geoph. and Meteorol. *University of Cologne* 131 pp.

Lee FY (1989) On airborne measurements of radiation from broken cloud fields. Doctoral Thesis. *University of Cologne* 105 pp.

Li Z, Leighton HG (1991) Scene identification and its effects on cloud radiative forcing in the Arctic. *J Geophys Res* 96: D5, 9175-9188.

Lindzen RS (1990) Some coolness concerning global warming. *Bull Amer Meterol Soc* 71: 288-299.

Liou K-N (1986) Review - Influence of cirrus clouds in weather and climate processes: A global perspective. *Mon Wea Rev* 114: 1167-1199.

Mitchell JFB, Senior CA, Ingram WJ (1989) CO_2 and climate: A missing feedback? *Nature* 341: 132-134.

Mossop SC (1985) The origin and concentration of ice crystals in clouds. *Bull Am Meteorol Soc* 66: 264-273.

Nicholls S, Leighton J (1986) An observational study of the structure of stratiform cloud sheets. Part I: Structure. *Quart J Roy Met Soc* 112: 431-460.

Ohring G, Clapp PF, Heddinghaus TR, Krueger A (1981) The quasi-global distribution of the sensitivity of the earth-atmosphere radiation budget to clouds. *J Atmos Sci* 38: 2539-2541.

Peixoto JP, Oort AA (1984) Physics of climate. *Reviews of Modern Physics* 56: 365-429.

Peixoto JP, Oort AH, de Almeida M, Tom A (1991) Entropy budget of the atmosphere. *J Geophys Soc* 96: D6, 10,981-10,988.

Platt CMR, Spirnhirne JD, Hart DW (1989) Optical and microphysical properties of cirrus clouds: Evidence for regions of small ice particles. *J Geophys Res* 94: 11.151-11.164.

Ramanathan V, Cess RD, Harrison EF, Minnis P, Barkstrom BR (1989) Cloud radiative forcing and climate: Results from the Earth Radiation Budget Experiment. *Science* 243: 57-63.

Ramanathan V, Collins W (1991) Thermodynamic regulation of ocean warming by cirrus clouds deduced from observations of the 1987 El-Nino. Nature, 351, 27-32.

Raschke E (1972) The planetary radiation budget from satellites. *Ztschr. f. Geophys* 38: 967-1000.

Raschke E (1988) The International Satellite Cloud Climatology Project, ISCCP, and its European regional experiment ICE, *Atmos Res* 21: 191-201.

Raschke E, Rieland M (1989) Downward solar radiation from geostationary satellite data: A first global view. *Weather* 44: 311-314.

Raschke E, Stuhlmann R, Palz W, Steemers TC (1991) Solar Radiation Atlas of Africa- Global and Diffuse Radiation Fluxes at Ground Level Derived from Imaging Data of the Geostationary Satellite METEOSAT 2. Comm. of the European Communities, *A.A. Balkema Publisher* Rotterdam, 155 pp.

Ridgway WL, Harshvardhan, Arking A (1991) Computation of atmospheric cooling rates by exact and approximate methods. *J Geophys Res* 96: No. D5, 20

Rockel B, Raschke E, Weyres B (1991) A parameterization of broad band radiative transfer properties of water, ice and mixed clouds. *Contr Atmos Sci* 64: 1-12

Roeckner E (1987) Cloud-radiation feedbacks in a climate model. Meteorol. Inst of the University Hamburg, *Report No. 2* pp 19.

Roeckner E, Schlese U, Biercamp J, Loewe P (1987) Cloud optical depth feedbacks and climate modelling. *Nature* 79: 138-140.

Roeckner E, Rieland M, Keup E (1991) Modelling of Cloud and Radiation in the ECHAM-Model. *ECMWF/WCRP Workshop "Clouds, Radiative Transfer and the Hydrological Cycle"* S. 199-222, Reading.

Rossow WB, Schiffer RA (1991) ISCCP cloud data products. *Bull Am Meteor Soc* 72: 1-20.

Schanda E (1986) Physical Fundamentals of Remote Sensing. *Springer Verlag* Berlin, pp 180.

Schmetz J, Schmetz P, Raschke E (1986) Estimation of daytime downward longwave radiation at the surface from satellite and gridpoint data. Theory and Appl. *Climatol* 37: 136-149.

Schneider, S 1972: Cloudiness as a global feedback mechanism. The effects of the radiation balance and surface temperature on variations in cloudiness. *J Atmos Sci* 29: 1413-1422.

Slingo A, Slingo JM (1988) The response of a general circulation model to cloud longwave radiative forcing. I: Introduction and initial experiments. *Meteor Soc* 114: 1027-1062.

Slingo A, Slingo JM (1991) Response of the National Center of Atmospheric Research Community climate model to improvements in the representation of clouds. *J Geoph Res* 96: D8, 15341-15357.

Shine KP, Sinha A (1991) Sensitivity of the Earth's climate to height-dependent changes in the water vapour mixing ratio. *Nature* 345: 382-384.

Stackhouse PW, Stephens G (1991) A theoretical and observational study of the radiative properties of cirrus: Results from FIRE 1986. *J Atmos Sci* 48: 2044-2059.

Starr DO'C (1987) A cirrus-cloud experiment: Intensive field observations planned for FIRE. *Bull Am Meteor Soc* 68: 119-124.

Starr DO'C, Melfi SH (1991) The role of water vapor in climate. *NASA Conference Publ* 3120, available through GEWEX Office, 50 pp.

Stephens GL, Greenwald TJ (1991) The Earth's radiation budget and its radiation to atmospheric hydrology. Parts I and II. *J Geophys Res* 96: D8, 15311-15340.

Stuhlmann R, Rieland M, Raschke E (1990) An improvement of the IGMK model to derive total and diffuse solar radiation at the surface from satellite data. *J Appl Meteorol* 29: 568-603.

Twomey S (1977) The influence of pollution on the shortwave albedo of clouds. *J Atmos Sci* 34: 1149-1152.

Twomey S (1991) Aerosols, clouds, radiation. *Atmos Environ* 25 A: 2435-2442.

Vali G (ed.) (1991) Report of the experts meeting on interaction between aerosols and clouds. WCRP-59, *WMO/TD-No.423* available from World Meteorol. Organiz., Geneva.

Warren SG, Hahn CJ, London J, Chervin RM, Jenne RL (1986) Global distribution of total cloud cover and cloud type amounts over land.*NCAR Tech Note* TN-273+STR.

Warren SG, Hahn CJ, London J, Chervin RM, Jenne RL (1989) Global distribution of total cloud cover and cloud type amounts over the ocean. *NCAR Tech Note* TN-317+STR.

Weatherald RT, Manabe S (1988) Cloud feedback processes in a General Circulation Model. *J Atmos Sci* 45: 1397-1415.

Welch R, Wielicki B (1989) Reflected fluxes for broken clouds over a Lambertian surface. *J Atmos Sci* 46: 1384-1395.

Zhang Y, Laube M, Raschke E (1989) Numerical studies of the time behaviour of cirrostratus in still air. *Contr Atm Phys* 62: 307-320.

Chapter 4

CONVECTIVE AND LARGE-SCALE CLOUD PROCESSES IN GCMS

Anthony D. Del Genio
NASA
Goddard Institute for Space Studies
New York, NY 10025
USA

4.1 Introduction

The wide range of global climate model (GCM) estimates of climate sensitivity to increasing trace gas concentrations is a result of differing treatments of poorly understood feedback processes associated with the hydrologic cycle. Within the atmosphere, these feedbacks reflect the effects of phase changes of water on radiative fluxes. For this reason, one of the highest priorities of the World Climate Research Programme is an understanding of the impact on climate of changes in cloud properties. Complementing this is the Global Energy and Water Cycle Experiment (GEWEX), whose prime objective is a description and understanding of the transport of water and energy within the atmosphere and across the atmosphere-surface interface (WCRP, 1990). Both of these goals in turn require a research emphasis on the representation of cloud radiative, dynamic, and microphysical processes in climate models.

The difficulty in modeling moist processes in climate models stems from the fact that many of the crucial interactions occur on small scales, necessitating that their effects be parameterized in global models. The problems are exacerbated by inadequate observations and poor understanding of certain cloud processes. In Section 4.2 of this paper we review the current understanding of feedbacks associated with atmospheric water. Sections 4.3, 4.4 and 4.5 discuss particular aspects of parameterized convective and stratiform cloud processes needing improvement in climate model. Validation of the feedbacks produced by GCMs is discussed in Section 4.6.

NATO ASI Series, Vol. I 5
Energy and Water Cycles in the Climate System
Edited by E. Raschke and D. Jacob

4.2 Feedbacks due to atmospheric water

The direct radiative impact of doubling CO_2 is a 1.2 K equilibrium global warming. GCMs uniformly produce much greater warming than this, though, because of the various moist feedbacks which operate. The complexity of moist processes is evidenced by the fact that the range of uncertainty in estimates of climate sensitivity to doubled CO_2 (1.5 - 5.5 K) has not narrowed over the past decade, despite substantial efforts to improve models and the appearance of new data sets from field studies and satellites. We discuss each of the feedbacks in increasing order of uncertainty in this section. For brevity, we omit feedbacks associated with changes in surface properties (snow, sea ice, vegetation, etc.), although these too are affected by changes in atmospheric moisture.

4.2.1 Water vapor/lapse rate feedback

An increase in greenhouse gas concentrations results initially in an increase in downward longwave radiation at the Earth's surface. The surface energy balance is re-established at least in part via an increase in surface evaporation as climate warms. While this is usually taken for granted, it should be remembered that evaporation depends on such things as surface wind speed and boundary layer stability, for which changes are not straightforward to predict.

Given an increase in surface evaporation, column atmospheric water vapor content rises until increased precipitation balances the increased evaporative flux. Since water vapor is itself a greenhouse gas, this causes additional warming of the climate system, a positive feedback. However, the radiative impact of increasing humidity depends on the vertical distribution of the change. Arking (1990) shows that despite the sharply decreasing concentration of water vapor with height, a given percentage change in humidity has as much radiative impact in the cold upper troposphere as in the lower troposphere. Thus, one might expect that water vapor feedback is sensitive to the modeling of dynamic and microphysical processes in GCMs.

Ellsaesser (1984) and Lindzen (1990) contend that subsidence in the environment surrounding deep cumulus clouds will dry the upper troposphere preferentially as climate warms, eliminating or reversing water vapor feedback. In equilibrium, though, the atmospheric moisture budget must be balanced, and this requires that we consider changes in all the processes which affect moisture, not only convection. GCMs with mass flux cumulus parametrizations (cf. Arakawa and Schubert, 1974; Hansen et al., 1983) include cumulus subsidence drying and therefore provide a useful test of the hypothesis. Regardless of whether the climate change is a doubling of CO_2 (cf. Hansen et al., 1984) or a prescribed sea surface temperature (SST) change (cf. Cess et al., 1990), all GCMs without exception produce increases in specific humidity of tens of percent at all altitudes in the troposphere, with the greatest increases actually occurring in the upper troposphere.

The reasons for this strong positive water vapor feedback can be understood by examining changes in the moisture budget as climate warms. Figure 4.1 illustrates this for two different versions of the GISS GCM which were subjected to global SST increases and decreases of 2 K (Del Genio et al., 1991). The two versions differ only in their convection and stratiform cloud parametrizations, but

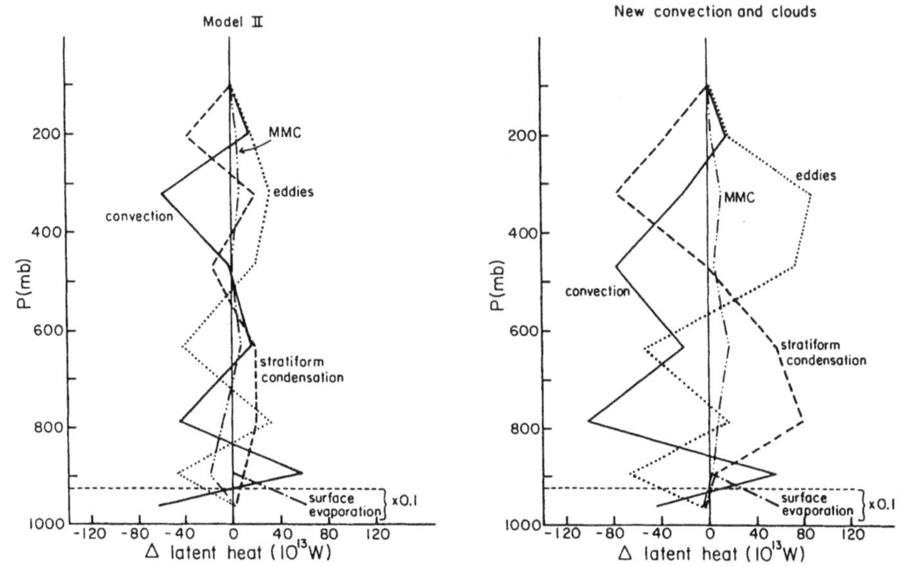

Figure 4.1: *Globally averaged changes (warmer minus cooler) in the individual components of the moisture budget for a perpetual July SST perturbation experiment with GISS Model II (left) and a newer version of the GCM (right), (Del Genio et al., 1991).*

both include penetrative convection and thus environmental subsidence. Notable differences between the two models are the presence of a quasi-equilibrium mass flux closure, convective downdraft parametrization, and prognostic liquid/ice water budget in the newer version.

In both models, cumulus subsidence drying in the upper troposphere does in fact increase in the warmer climate. On the other hand, there is increased moistening by convective detrainment at the tropopause and by the general circulation (including the Hadley and Walker circulations, midlatitude transient and stationary eddies, and grid-scale versions of mesoscale cumulus updrafts, especially in the version with the cloud water budget) at upper levels. The net result is a strong positive water vapor feedback which is almost identical in the two models despite the differences in physics. The various small- and large-scale circulation components change little as the climate changes. The increased vertical transports of water vapor are thus due to increased humidity gradients in the warmer climate arising from the nonlinear temperature dependence of the Clausius-Clapeyron equation.

The net effect of water vapor feedback on surface temperature is less than it appears, though, due to a compensating negative lapse rate feedback. From a radiative standpoint, an upward shift of the vertical profile of water vapor causes an upward shift in warming at the expense of surface warming, i.e., a decrease of lapse rate. From a dynamical perspective, at least in the tropics, convection acts as a *thermostat*, removing surface heating to higher altitudes. Since the mean tropical lapse rate is close to moist adiabatic, and the moist adiabatic lapse rate decreases with temperature, we expect the surface to warm less than the upper troposphere. This partial cancellation between feedbacks due to changes in lapse rate and the vertical humidity profile may help explain why the net water vapor feedback is so insensitive to model uncertainties (Hansen et al., 1984).

4.2.2 Cloud height feedback

GCMs generally agree that the average height of clouds will increase as climate warms. Since clouds absorb thermal infrared radiation, the higher the cloud, the colder is its radiating temperature. Therefore, increasing cloud height implies weaker infrared emission to space and a positive feedback in a warming climate. The feedback is complicated by the fact that the temperature profile changes as well.

Cloud height feedback in GCMs is caused primarily by an upward shift in upper level cirrus clouds, although changes in low and middle clouds contribute to some extent. In the tropics, and perhaps to a degree in midlatitudes, cirrus form near the detrainment level of deep convective updrafts. Thus, cloud height feedback is tied to the question of how convection depth changes in a warming climate. To examine this, we define the moist static energy h as

$$h = c_p T + gz + Lq \tag{4.1}$$

where c_p is the specific heat at constant pressure, T the temperature, g the acceleration due to gravity, z the altitude, L the latent heat of condensation, and q the specific humidity . If we ignore virtual temperature effects for simplicity, it can be shown that the detrainment or neutral buoyancy level, i.e., the level at which cloud and environment temperatures are equal, is determined by the condition $h_c = h^*$, where the asterisk denotes the saturation value. Letting U represent the surface relative humidity, and subscripts s and d the surface and detrainment levels, respectively, we can write the neutral buoyancy condition for parcels lifted from the surface as

$$c_p T_s + LUq^*(T_s) = c_p T_d + gz_d + Lq^*(T_d) \approx c_p T_d + gz_d \tag{4.2}$$

To see the effect of a climate change, we differentiate with respect to surface temperature, giving

$$c_p + Lq^*(T_s)\frac{dU}{dT_s} + LU\frac{dq^*}{dT_s} = c_p\left(1 - \Gamma_m \frac{dz_d}{dT_s} - z_d\frac{d\Gamma_m}{dT_s}\right) + g\frac{dz_d}{dT_s} \tag{4.3}$$

where Γ_m is the moist adiabatic lapse rate averaged over the depth of the convection. Rearranging, we get

$$c_p(\Gamma_d - \Gamma_m)\frac{dz_d}{dT_s} = LU\frac{dq^*}{dT_s} + Lq^*(T_s)\frac{dU}{dT_s} + c_p z_d \frac{d\Gamma_m}{dT_s} \tag{4.4}$$

where Γ_d is the dry adiabatic lapse rate. Thus, increasing boundary layer humidity associated with warming, either via increasing saturation humidity at constant relative humidity or increasing relative humidity, increases the moist static energy of rising parcels and leads to a higher cloud top. On the other hand, since the moist adiabatic lapse rate decreases with temperature, the upper troposphere warms to a greater extent than the surface in convective regions, depressing the height of convective cloud tops. At 300 K and constant relative humidity, the former effect dominates,

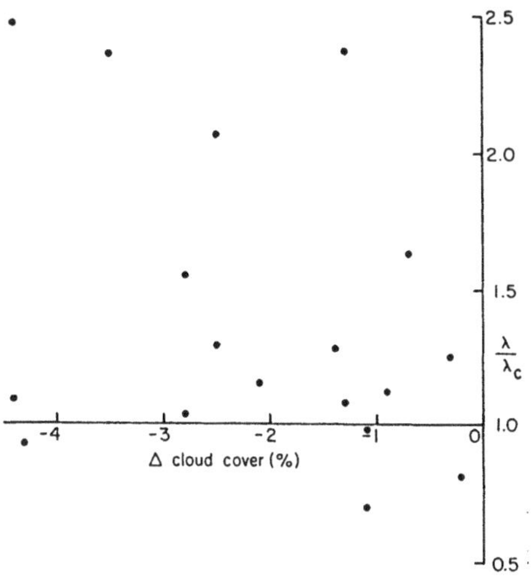

Figure 4.2: *Cloud feedback parameter vs. change in global mean total cloud cover (warmer minus cooler) for 19 GCMs performing the same perpetual July SST perturbation experiment (Cess et al., 1990).*

and the detrainment level of a 12 *km* high cumulus cloud rises about 0.35 *km* per degree surface temperature increase. To offset this, relative humidity would have to decrease by 3-4 % per degree of warming, much greater than that which occurs in GCMs or observations (Rind et al., 1991).

4.2.3 Cloud cover feedback

Cloud cover changes in a warming climate are difficult to predict from basic principles, because simultaneous warming and moistening combine to render relative humidity changes small compared to specific humidity changes. Nonetheless, almost all GCMs to date predict that global mean cloud cover should decrease as climate warms (Cess et al., 1990). The decrease occurs primarily in low and middle clouds at all latitudes outside the polar regions. From Fig. 4.1, the relevant physics tends to be increased drying by cumulus subsidence and midlatitude eddy moisture flux divergence offsetting increased moistening by evaporation, from both the surface and falling precipitation. This basic result may be sensitive to neglected aspects of the dynamics of cumulus clouds and the microphysics of stratiform clouds, as I will discuss later. At high latitudes where sea ice melts, evaporation from the added open ocean water causes a local increase in low clouds.

It is now well established that clouds on average have a net cooling effect on the current climate , i.e., net cloud forcing is negative (cf. Arking, 1991). Cloud feedback, however, depends on how cloud forcing changes as climate changes. Since the shortwave and longwave components of cloud forcing have in general different sensitivity to a change in cloud cover, the sign of the resulting cloud cover feedback is difficult to predict. For the low and middle clouds which tend to decrease preferentially in global warming simulations, the albedo effect tends to dominate, so that decreasing cloud cover is usually a positive feedback in GCMs.

Unfortunately, the situation is more complicated, because cloud cover feedback can occur even in the absence of a change in global mean cloud cover, if the spatial or temporal distribution of clouds changes. If clouds systematically shift, for example, to higher solar zenith angle (day to night, low latitudes to high latitudes, or summer to winter) as climate warms, the increase in absorbed sunlight produces a positive cloud cover feedback. The same is true if clouds shift from dark to bright overlying surfaces (ocean to desert, forest to snow). One immediate conclusion is that models without a seasonal or diurnal cycle or realistic surface albedos cannot hope to predict cloud feedback with confidence.

These caveats aside, it is perhaps not surprising that most GCMs predict positive cloud feedback since almost all produce low and middle level cloud decreases as climate warms. However, Figure 4.2 illustrates that total cloud feedback is actually poorly correlated with the change in cloud cover for a variety of GCMs performing the same experiment. The differences are due primarily to vastly different optical thickness feedbacks in different models. This feedback can be either positive or negative, and it accounts for most of the range of disagreement in GCM estimates of climate sensitivity.

4.2.4 Cloud optical thickness feedback

Climatic change can alter both the visible reflectance and infrared emissivity of clouds. Depending on which effect dominates, and whether the optical thickness τ increases or decreases, the resulting feedback can be positive or negative. For visible wavelengths and particle sizes of interest, the extinction efficiency is almost independent of the size parameter, so τ takes the simple form

$$\tau \approx \frac{3\mu\Delta z}{2\rho_w r} \tag{4.5}$$

where μ is the cloud water density, Δz the physical thickness of the cloud, ρ_w the mass density of the condensed phase, and r the effective radius of the particle size distribution.

Optical thickness feedback occurs in three different ways in climate models.

1. If τ is prescribed to be constant with time but a function of height, an increase in cloud height as climate warms indirectly induces a cloud optical thickness feedback. The feedback will be positive if $d\tau/dz < 0$, because the albedo effect dominates in most models.

2. A temperature-dependent τ can be estimated in a diagnostic cloud model based on either the instantaneous condensed water or an assumption about how the cloud water content and particle size should vary with temperature. Since more water condenses in a lifting cloud at higher temperature (Betts and Harshvardhan, 1987), $d\tau/dT > 0$, resulting typically in a negative optical thickness feedback. Sinks of cloud water do not affect the optical properties in such models.

3. A prognostic budget equation for cloud liquid/ice water can be used to calculate interactive cloud properties directly. In theory, this is the preferable approach. In practice, it introduces

many complications due to the need to parameterize sources and sinks of cloud water due to complex microphysical processes (autoconversion, collection, diffusional growth, entrainment, etc.). It also requires an assumption about particle size. The resulting feedback may be positive or negative and is difficult to predict a priori.

	GISS Model II	CCC	GISS new clouds and convection
Δ cloud cover (%)	-1.3	-4.3	-0.4
$\Delta F/\Delta T_s$ $[Wm^{-2}K^{-1}]$			
Clear	1.93	2.53	1.88
Overcast	1.92	1.22	0.56
Global	2.04	2.50	1.54
$\Delta Q/\Delta T_s$ $[Wm^{-2}K^{-1}]$			
Clear	0.00	0.12	0.22
Overcast	2.19	-2.92	0.01
Global	1.22	-0.10	0.24
λ $[Km^2W^{-1}]$			
Clear	0.52	0.42	0.60
Global	1.23	0.39	0.77
λ/λ_c	2.37	0.93	1.28

Table 4.1: *Feedback parameters derived from perpetual July $\pm 2K$ SST perturbation experiments with three GCMs utilizing different approaches to optical thickness parametrization.*

The dramatic effect of these various τ assumptions is illustrated by comparing in Table 4.1 the responses of three GCMs to prescribed globally uniform $\pm 2K$ SST increases (cf. Cess et al., 1990). The two versions of the GISS GCM correspond to those in Fig. 4.1. Model II (Hansen et al., 1983) has prescribed fixed values of τ which decrease with height. The other version incorporates an early version of a prognostic cloud water budget (Del Genio and Yao, 1990). The Canadian Climate Centre (CCC) GCM was modified to include the Betts and Harshvardhan (1987) formula for the temperature dependence of cloud water content (and τ) in a diagnostic cloud scheme.

Climate sensitivity is defined by Cess et al. (1990) as

$$\lambda = \left(\frac{\Delta F}{\Delta T_s} - \frac{\Delta Q}{\Delta T_s}\right)^{-1} \tag{4.6}$$

where F is the outgoing longwave flux, Q the absorbed solar flux, and Δ the difference in warmer minus cooler climates. By evaluating λ separately for global and clear-sky (denoted by λ_c) regions, a measure of cloud feedback can be obtained as the ratio λ/λ_c; a value > 1 denotes positive cloud feedback. Note first that the model with the largest decrease in cloud cover (CCC) has the lowest sensitivity and a negative cloud feedback because of optical thickness effects. To understand the model differences, consider the individual thermal and solar feedback derivatives. The infrared flux derivative should be smaller in overcast regions than clear regions because of the enhanced greenhouse effect of higher clouds in the warmer climate. In GISS Model II, though, this is completely offset by the associated indirect decrease in τ, which allows more thermal radiation to escape to space. The dramatic difference among the three models occurs in the overcast solar flux derivative, which is strongly positive for the higher (and thus thinner) warm climate clouds of Model II, strongly negative for the thicker clouds of the CCC GCM, and near zero for the prognostic GISS model, in which both effects seen in the other two models occur and nearly cancel.

As a result, the overall cloud feedback is either negative (CCC), slightly positive (GISS prognostic), or strongly positive (GISS Model II), depending on the sign of the optical thickness feedback. In fact, the CCC GCM would be even less sensitive were it not for the large cloud cover decrease which partly offsets its negative overcast solar flux derivative.

4.3 Issues in cumulus parametrization

Early cumulus parametrizations (Manabe et al., 1965; Kuo, 1974) either bypassed the physics of convection by simply mixing heat and moisture vertically to adjust the large-scale environment to conditions typical only of air inside small-scale cumulus clouds (moist convective adjustment), or else they assumed that convective effects were determined by direct lateral mixing of cloud temperature and moisture anomalies into the environment (Kuo schemes). It is now well-established, though, that the thermodynamic structure of the large-scale environment is very different from that inside the cloud, and that the buoyancy of a rising parcel serves only to maintain the updraft and does not directly warm the environment. Large-scale warming and drying due to convection occur instead primarily as a result of compensating subsidence of air outside the cloud (Arakawa and Chen, 1987), with secondary contributions due to detrainment of saturated cloud air and melting and evaporation of falling precipitation. This is the basis for the various penetrative mass flux cumulus schemes used in climate GCMs (cf. Arakawa and Schubert, 1974; Yao and Del Genio, 1989; Gregory and Rowntree, 1990).

Regardless of the type of scheme used, all GCMs give comparable water vapor feedbacks when perturbed (cf. Cess et al., 1990). The real justification for mass flux schemes in climate models lies in their potential (largely not yet realized) to represent the details of heat and moisture transport which control small climatic changes in relative humidity and, therefore, cloud feedback. GATE and other field experiments have revealed a variety of important processes that are missing from most GCM moist convection parametrizations (Houze and Betts, 1981). In this section we describe some of the areas in which improvement is needed.

4.3.1 Mass flux closure

Arakawa and Schubert (1974) first proposed the concept of a quasi-equilibrium between the large-scale destabilization of an atmospheric column and stabilization by moist convection. Arakawa and Schubert defined a cloud work function A, a vertically integrated measure of cloud buoyancy, and noted that in quasi-equilibrium $dA/dt \approx 0$. Given empirical estimates of A for clouds which detrain at various altitudes, they developed a spectral cumulus mass flux closure which restores the thermodynamic structure to the observed values of A at each time step after large-scale processes induce variations in A.

For the deepest cumulus clouds, which have no entrainment, the cloud work function can be written

$$\begin{aligned} A(0) &= \int_{z_b}^{z_d} \beta(z) \left(h_s - h_s^* - \int_{z_s}^{z} \frac{\partial h^*(z')}{\partial z'} dz' \right) dz \\ &= -(1-U)Lq_s^* \int_{z_b}^{z_d} \beta(z) dz + \int_{z_b}^{z_d} \beta(z) \int_{z_s}^{z} c_p(1+\gamma(z'))(\Gamma - \Gamma_m) dz' dz \end{aligned} \qquad (4.7)$$

where z_b is the cloud base height, Γ the lapse rate, p the pressure, $\gamma = (L/c_p)(\partial q^*/\partial T)_p$, and $\beta = g/c_pT(1+\gamma)$. Thus, if quasi-equilibrium holds, variations in U and $\Gamma - \Gamma_m$ should be negatively correlated in convective regions. Arakawa and Chen (1987) demonstrate that this is indeed the case in a variety of continental and marine data sets.

Despite the validity of the quasi-equilibrium assumption, Arakawa-Schubert is not well-suited for decadal climate simulations for two reasons:

1. The difficulty in solving for the ensemble of mass fluxes and entrainment rates required to maintain constant A makes this scheme computationally intensive;

2. Since the scheme relaxes the atmosphere to a predetermined set of values of A for the current climate, it does not allow for small but potentially important climatic changes in A.

For climate models, a better strategy is to demand that a mass flux closure produce quasi-equilibrium as an output rather than specifying it as an input. Since mass flux schemes are commonly triggered when a lifted cloud parcel is judged to be buoyant with respect to environmental air in the layer above, one approach is to transport enough mass to just neutralize the instability in a physics timestep. A closure of this type used in the GISS GCM (Yao and Del Genio, 1989) reproduces the observed quasi-equilibrium behavior of the tropical atmosphere (Fig. 4.3). If a longer adjustment time to neutrality is assumed, tradeoffs between a smaller instantaneous mass flux and more frequent occurrence of convection produce an almost identical monthly mean climate. Gregory and Rowntree (1990) take a similar but empirical approach to quasi-equilibrium.

Recently, several promising simplifications of the Arakawa-Schubert scheme with a similar philosophy have begun to appear. Moorthi and Suarez (1992) relax the atmosphere toward quasi-equilibrium whenever convection occurs but do not demand exact balance at each time step. Pan and Randall (1991) do not assume quasi-equilibrium, using a prognostic kinetic energy budget instead to calculate cumulus mass flux; the goal here, as in Yao and Del Genio (1989), is to achieve quasi-equilibrium as a result rather than prescribing it ab initio.

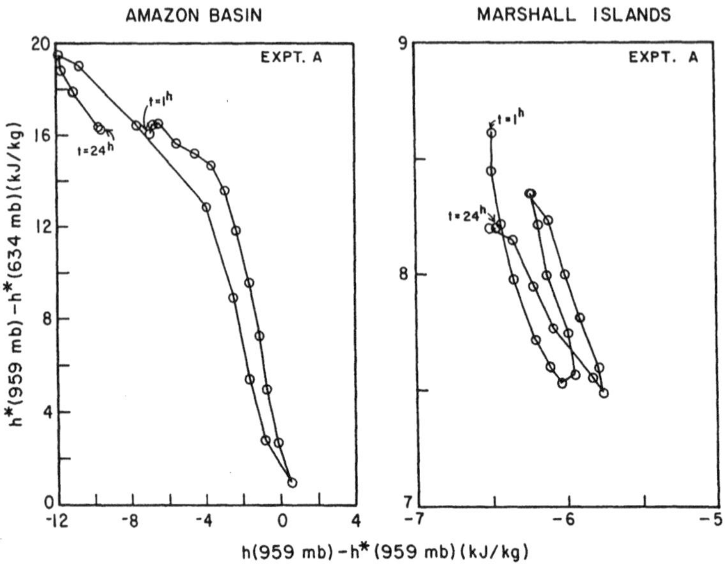

Figure 4.3: *Quasi-equilibrium behavior produced by a neutral buoyancy cumulus mass flux closure in the GISS GCM (Yao and Del Genio, 1989). The ordinate is proportional to the conditional instability between the PBL and mid-troposphere. The abscissa is a function of the relative humidity of the PBL.*

4.3.2 Convective downdrafts

A common problem in GCMs with mass flux convection schemes is excessive cumulus heating and drying in the lower troposphere. One possible reason is that the motions which compensate the cumulus updraft consist of more than just environmental subsidence. In the past decade it has become clear that convective-scale downdrafts driven by precipitation loading and evaporation are important to the heat and moisture budgets of tropical cloud clusters (Knupp and Cotton, 1985). Downdrafts cool the planetary boundary layer (PBL), and though they are often drier than the PBL itself, they are wetter than the air that is assumed to subside on the large scale in GCMs. The air carried to the surface by downdrafts typically originates in the lower or middle troposphere, with characteristics similar to air near the moist static energy minimum. Downdrafts therefore fundamentally alter the convective stability of the PBL.

The first operational GCM downdraft parametrization was implemented by Del Genio and Yao (1988). The scheme simply tests as the plume rises for the first level (if any) at which an evaporatively cooled equal mixture of cloud and environmental air is negatively buoyant. A downdraft then forms with the properties of the mixture and penetrates to cloud base, evaporating liquid water as necessary to remain as close to saturation as possible. The downdraft mass flux is specified as a fixed percentage of the updraft mass flux. The effect of downdrafts, relative to the same parametrization employing only environmental subsidence, is an increase in tropical relative humidity, especially in the lower troposphere (Fig. 4.4). Emanuel (1990) has proposed a scheme which includes a similar unsaturated downdraft driven by evaporation of precipitation, but with dynamical entrainment and

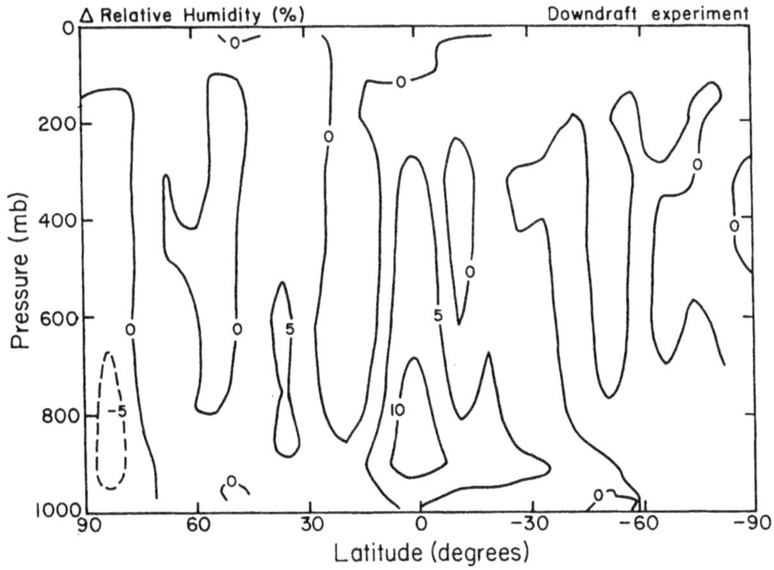

Figure 4.4: *Zonal mean distribution of the change in relative humidity caused by the addition of a convective downdraft parametrization to the GISS GCM (Del Genio and Yao, 1988).*

an ad hoc assumption about the fraction of precipitation which falls through the cloud rather than the environment.

Cheng and Arakawa (1991) have coupled a more complex and computationally intensive downdraft parametrization to the Arakawa-Schubert scheme. The scheme includes a rainwater budget, assuming that the rain falls partly inside and partly outside the updraft. The mean tilting angle of the updraft, which determines this partition and the properties of the associated downdraft, is assumed to be that required for a statistically steady updraft subensemble to maintain its buoyancy against rainwater loading. The effects on relative humidity when the scheme is used in the UCLA GCM are similar to those found by Del Genio and Yao.

From the standpoint of climate change, downdrafts may be important to climate sensitivity. Since downdrafts reduce the tendency of convection to lower the relative humidity of the PBL, they may offset the tendency of most GCMs to produce a positive component of cloud feedback due to low cloud cover decreases in a warmer climate. The Del Genio-Yao downdraft parametrization was included, for example, in the new version of the GISS GCM used for one of the SST perturbation experiments in Table 4.1. The resulting cloud cover decrease in response to a 4 K warming (0.4 %) is less than one-third the decrease experienced by GISS Model II, which utilizes only environmental subsidence to compensate cumulus updrafts.

While it is clear that downdrafts are necessary for the reliable simulation of GCM heat and moisture budgets, uncertainties abound in how best to simulate their effects. To what extent does dynamical entrainment play a role? What closure should be used for the downdraft mass flux? How much condensate is advected by the updraft, at what level does it release precipitation, and how fast does rain evaporate in the downdraft? The challenge for GCMs is to deal with these issues without resorting, e.g., to complex microphysical modeling. Another problem is how to represent the effects that different parts of a convective cluster have on the PBL. Locally, downdrafts stabilize the PBL, but adjacent to them, new updrafts form from unmodified PBL air. One possible approach, rather than mixing downdraft air with PBL air, is to assume that enhanced surface fluxes immediately restore the PBL to its pre-convective state (Cheng and Arakawa, 1991).

4.3.3 Shallow convection

Traditionally, cumulus parametrization efforts have focused on deep convection. It has long been known, though, that shallow non-precipitating trade cumulus which occupy a small fractional area of the subtropics and tropics nonetheless significantly affect the structure of the lower troposphere. Shallow convection vents moisture from the PBL through the trade inversion to the free troposphere. As such, it keeps the PBL from saturating and becoming overcast, with implications for both water vapor and cloud feedback in climate GCMs. The vapor advected above the inversion converges in the tropics, influencing deep convection as well.

Ideally, a spectral mass flux scheme should represent shallow as well as deep convection simultaneously. In practice, though, many such models underpredict shallow convection because of the absence of subgrid-scale variability in PBL properties (caused in part by the effects of deep convection on the mesoscale). The GISS GCM successfully simulates the observed bimodal distribution of tropical convective cloud top heights with a two-plume cloud model in which the cloud base mass flux is divided into undilute and entraining parts (Yao and Del Genio, 1989). The model is limited, though, by arbitrary specifications of the entrainment rate and the partitioning of mass flux between the two plumes.

The recent trend, therefore, has been to treat shallow convection separately from deep. Several approaches have been used to date in GCMs. Since the effect of shallow convection is to vertically redistribute heat and moisture, it can in principle be represented by the diffusion of dry static energy, specific humidity, and liquid water. The problem then reduces to the specification of the eddy diffusion coefficient and the gridbox-mean liquid water content. Tiedtke et al. (1988) neglect liquid water and prescribe the eddy diffusion coefficient within and below the inversion layer in the ECMWF GCM. Alternatively, a parametrization of eddy diffusion as a function of PBL turbulence, or replacement of diffusion by a higher-order turbulence closure, are possible. Betts (1986) uses an adjustment approach to shallow convection, in which the lower troposphere relaxes to an observed mixing line structure.

Tiedtke et al. find that the inclusion of explicit shallow convection is needed to prevent the trade inversion from strengthening too much and the PBL from saturating too easily. Slingo et al. (1989), by allowing detrainment to partially occur at the first level of negative buoyancy above the predicted cloud top level in the UKMO mass flux cumulus scheme, improve both the low cloud amount in the subtropics and high cloud amount in the ITCZ. A similar modification to the GISS scheme does not produce similar effects, however, so the need for separate shallow convection may be model-dependent, e.g., a function of GCM vertical resolution near the trade inversion.

4.3.4 Other issues

Betts (1986) has pointed out that while the tropical troposphere is usually judged to be conditionally unstable, it is almost neutrally stable below 600 hPa instead if convective parcels are assumed to undergo reversible moist adiabatic ascent, i.e., all condensate is retained and reduces the virtual temperature. Entrainment and precipitation, however, reduce and/or vertically redistribute condensate, so judgment of buoyancy is not a straightforward matter. For GCMs, all that matters is that a convection scheme produce a realistic vertical structure. The GISS GCM's mean tropical temperature profile, for example, is close to a 350 K virtual moist adiabat in the lower troposphere regardless of whether parcel ascent in the cumulus parametrization scheme is assumed to be pseudoadiabatic or reversible moist adiabatic. On the other hand, vertical transport of condensate is important for the parametrization of downdrafts as well as mesoscale anvil clouds, which are discussed in Section 4.5.

Existing cumulus schemes use an entraining plume model based on lateral entrainment theory, with entrainment rates specified arbitrarily or calculated to produce vanishing buoyancy. Unfortunately, a fundamental theory to predict the entrainment rate does not exist, nor can one reliably predict the thermodynamic properties of entrained air in a global model which does not resolve smaller scale structure. Emanuel (1990) proposes a scheme based instead on buoyancy sorting by small cloud-top penetrative downdrafts, which mix only once with environmental air and then sink to and detrain at their level of vanishing buoyancy. The existence of such downdrafts was first deduced by Paluch (1979), but lateral entrainment occurs as well, complicating the picture (cf. Blyth et al., 1988). One problem with implementing such a scheme is that observed downdrafts penetrate only about 1-2 km before detraining. The vertical resolution of current climate GCMs may be marginally adequate at best to resolve such features.

A major uncertainty associated with cumulus clouds is their transport of momentum. Some GCMs completely ignore cumulus momentum transport, while others adopt a *momentum mixing* or *cumulus friction* approach of downgradient momentum transport (cf. Schneider and Lindzen, 1976). This is apparently valid for isolated cumulus, but only partly so for organized clusters or lines. In particular, Le Mone et al. (1984) show that the transport normal to propagating two-dimensional bands can be upgradient. The uncertainty lies in predicting the existence and orientation of such lines in a large-scale model which does not resolve the mesoscale. At first glance, cumulus momentum transport may seem irrelevant to climate change. In GISS Model II, though, the removal of cumulus momentum mixing drastically weakens the Hadley cell (Hansen et al., 1983), suggesting that convective momentum transport may indirectly regulate tropical cloud cover.

4.4 Issues in stratiform cloud parametrization

Early climate GCMs avoided issues of cloud feedback by prescribing cloud cover. In the 1980's diagnostic cloud schemes, in which cloud cover was predicted as a function of the resolved variables of the model (Hansen et al., 1983; Washington and Meehl, 1984), began to appear. In such models clouds are generated and dissipated in the same timestep, so life cycles cannot be simulated. In addition, the optical properties must either be prescribed or based on simple assumptions (see Section

4.2). To overcome these limitations, prognostic parametrizations, in which cloud water is carried as a predicted variable, have begun to appear (Sundqvist, 1978; Roeckner, 1988; Smith, 1990; Del Genio and Yao, 1990; Le Treut and Li, 1991). Although cloud water budget schemes represent the future of cloud modeling, they are still in their infancy and present numerous challenges. I describe some of these in this section.

4.4.1 Cloud formation and cover

On small scales, clouds form close to 100 percent relative humidity, usually in the presence of updrafts, and occupy 100 percent of the volume when they occur. Saturation rarely occurs, though, on the scales resolved by climate GCMs. GCMs thus typically assume a threshold relative humidity below which cloud formation does not occur. Unfortunately, convincing observational constraints do not yet exist (cf. Slingo, 1980). The problem lies partly in the absence of an accurate measure of relative humidity on large scales and partly in the probable dependence of cloud formation on other variables.

Diagnostic parametrizations typically assume that, above the threshold, excess water vapor condenses and immediately precipitates, with or without evaporation as it falls through unsaturated air. Cloud cover is typically a function of the departure of relative humidity from the threshold value, although the threshold can be a function of cloud type, and dependences on vertical velocity, lapse rate, or convection intensity are sometimes invoked (cf. Slingo, 1980).

Prognostic schemes must parameterize cloud cover and cloud formation thresholds in the same fashion, but they differ as to the disposition of water substance in time and within the gridbox. In a prognostic scheme, the tendency of cloud water content $m = \mu/\rho$ can be written

$$\frac{\partial m}{\partial t} = D(m) + bC_c - (1-b)E_c - P \pm C_{LI} \tag{4.8}$$

where $D(m)$ represents the rate of change of cloud water due to non-stratiform sources and sinks (advection, convection, etc.), b the cloud fraction, C_c and E_c the cloud condensation and evaporation rates, P the rate of conversion of cloud water to precipitation, and C_{LI} the rate of conversion between the liquid and ice phase (zero if separate budgets for liquid and ice are not included).

Sundqvist's (1978) liquid water budget parametrization relates the stratiform latent heating to the relative humidity tendency, and estimates the latter by partitioning moisture convergence into a part which condenses into the already cloudy part of a gridbox (increasing m) and another part (including evaporation of cloud and precipitation) which increases the relative humidity of the clear fraction and the cloud cover. In his original scheme, the relative humidity of the clear fraction is fixed, but this assumption is relaxed in a later version (Sundqvist et al., 1989). Variations on this approach exist, such as assuming a subgrid-scale distribution of conserved variables (Smith, 1990).

Advantages of the prognostic approach are:

1. It allows cloud *debris* to remain long after dynamical sources of water vapor disappear, preventing the *blinking* of clouds that occurs in diagnostic models;

2. It allows in principle a direct calculation of optical properties. Advection of cloud water is difficult to treat accurately, though, and many schemes in fact ignore this aspect. Furthermore, the sink terms in the budget equation are poorly constrained yet have a major influence on cloud feedback (Mitchell et al., 1989).

For these reasons, the value of prognostic schemes at the present time is the framework they provide for future improvements in parametrization rather than the more accurate estimate of cloud feedback they yield.

4.4.2 Phase transitions

Sundqvist's original scheme ignored differences between liquid and ice, and his later version included them only implicitly. Likewise, Roeckner (1988), in the first climate change GCM simulation with a cloud water budget, ignored liquid-ice differences. The result in the Hamburg GCM was a slightly positive cloud optical thickness feedback because increases in the greenhouse effect of high cirrus outweighed increases in the albedo effect of low and middle clouds. This was the opposite of the prediction of Somerville and Remer (1984), who studied optical thickness feedback due to low clouds only in a one-dimensional model based on Feigelson's (1978) cloud liquid water content data over the USSR.

Smith (1990) was the first to parameterize different microphysical properties for liquid and ice. The key feature in this model is the faster autoconversion rate of ice, as a result of the smaller number of ice nucleating particles which leads to larger ice crystals (relative to liquid droplets) which precipitate more rapidly at smaller cloud water content. Mitchell et al. (1989) used Smith's scheme in a doubled carbon dioxide experiment with the UKMO GCM and found that optical thickness feedback was negative. Even when optical thickness feedback was neglected, the model's sensitivity was 2.5 K lower than the same model with a diagnostic cloud scheme.

The difference in optical thickness feedback between the Hamburg and UKMO models traces to the thinner ice clouds in the latter which minimize changes in cirrus greenhouse warming. The larger drop in sensitivity in the UKMO GCM with fixed cloud optics occurs because, at the altitudes at which liquid gives way to ice, the frequency of occurrence of ice relative to liquid decreases slightly as the climate warms. Since ice clouds are assumed to precipitate more quickly (i.e., have a shorter lifetime) than liquid clouds, a shift from ice to liquid increases time mean cloud cover as climate warms.

Whether this is a positive or negative feedback depends on the temperatures of transition from liquid to ice. The UKMO GCM assumes that cloud is liquid for temperatures warmer than 0 °C, ice when colder than -15 °C, and a mixture in between. At the altitudes at which these temperatures occur, the albedo effect dominates, and increasing cloud cover is a negative feedback. Observations show, however, that supercooled liquid persists down to -40 °C over land (Feigelson, 1978; Sassen et al., 1989) and almost -30 °C over ocean (Curry et al., 1990), and that ice formation does not begin until at least -4 °C over ocean and -10 °C over land (Hobbs and Rangno, 1985).

Another problem with the UKMO approach is that different physics is used for ice and liquid precipitation, the assumption being made that ice always begins to precipitate immediately. The feedback is thus quite sensitive to fall speed (Mitchell et al., 1989), and there is no provision for a small particle size component of the population. Furthermore, the scheme neglects the Bergeron-Findeisen process of diffusional growth of ice in a mixed cloud, which is important to precipitation formation at cold temperatures. As a result, excessive ice builds in the UKMO GCM in middle and high latitudes in summer, and cloud liquid water content increases with temperature at a rate twice that of Feigelson's (1978) continental observations (Smith, 1990).

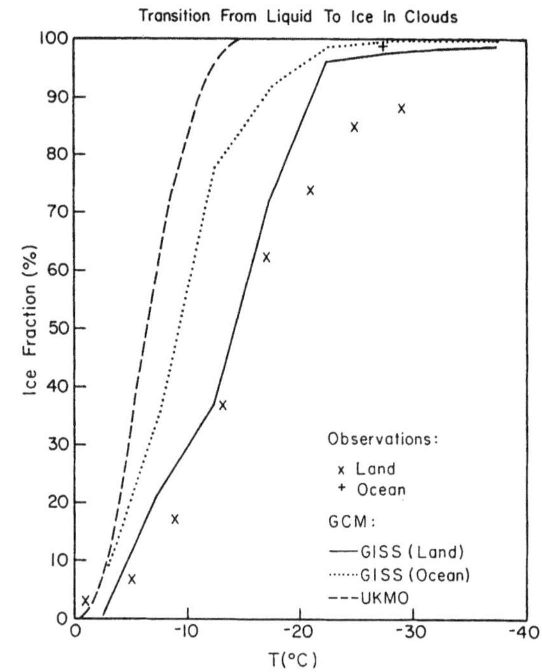

Figure 4.5: *Probability, that condensate occurs in the ice phase as a function of temperature in observations over land (Feigelson, 1978) and ocean (Curry et al., 1990), and as parameterized in the GISS (Del Genio and Yao, 1990) and UKMO (Smith, 1990) GCMs.*

By comparison, the GISS prognostic scheme (Del Genio and Yao, 1990) assumes that the transition to ice occurs between -4 and -40 °C (Figure 4.5), includes a simple representation of the Bergeron-Findeisen process for ice falling into a supercooled liquid layer, and treats ice and liquid parametrization, albeit with more efficient autoconversion rates for ice. We therefore expect it to have a higher climate sensitivity than the prognostic UKMO GCM. For prescribed SST changes, at least, this is indeed the case (see Table 4.1). Le Treut and Li (1991) obtain an even larger sensitivity in the LMD GCM by having all water in both phases precipitate immediately above a given threshold. At this stage, though, uncertainties in cloud microphysics at temperatures below freezing, especially for maritime clouds, preclude confidence in any parametrization.

4.4.3 Particle size and shape

Somerville and Remer's (1984) 1-D assessment of cloud optical thickness feedback assumed that the temperature dependence of τ was identical to that of liquid water content. Platt (1989) and others have noted, though, that cloud particle size changes can offset the radiative effects of cloud water variations. Although effective radius and mean radius are not identical, I neglect the difference for simplicity in the present discussion. Cloud water density is then related to the number density N of cloud droplets by

$$\mu = N\rho_w \frac{3}{4}\pi r^3 \tag{4.9}$$

One extreme assumption is Somerville and Remer's (r = constant), so that increasing cloud water occurs via an increasing number of droplets. In this case, τ is proportional to μ. The other extreme is to assume that N is constant and increasing cloud water implies bigger droplets, so that r goes as $\mu^{1/3}$ and τ is proportional to $\mu^{2/3}$.

In situ observations (cf. Slingo et al., 1982) and a preliminary survey of cloud particle size from satellite (Han et al., 1991) for low clouds suggest that constant N is a good assumption for small and moderate optical thicknesses, whereas constant r is a better fit to optically thick ($\tau > 10$) clouds. The aircraft remote sensing retrieval of Nakajima and King (1991), in fact, implies decreasing r as τ increases for optically thick, precipitating cases. For cirrus, on the other hand, the data of Platt (1989) indicate that crystal size increases with ice water content more rapidly than the constant N assumption would predict. Del Genio and Yao (1990) use a cloud water budget and assume N = constant to parameterize the effect of cloud particle size variations on τ. Fouquart et al. (1990) argue that cloud-to-cloud variations in N dominate instead in a global model; they suggest an empirical linear dependence of r on μ. Despite the uncertainties, the advantage of modeling particle size directly rather than implicitly is that it provides a convenient framework for incorporating improvements in understanding as well as lending itself to simulating aerosol effects, which I discuss below.

For cirrus, selection of the proper particle size is especially problematical. Aircraft measurements of ice crystal equivalent radius typically yield values several times larger than those retrieved remotely (Wielicki et al., 1990). The problem may be the inability of microphysical probes to sample the small particle end of the size distribution, although uncertainty in the near-infrared imaginary refractive index, shape of the size distribution, and appropriate phase function(s) for ice crystals with a variety of habits may be a problem for the remote retrievals. None of these complexities has yet been included in GCM cirrus cloud parametrizations.

Attention has recently focused on the possibility that increasing aerosol concentrations will provide a negative cloud optical thickness feedback. The effects are twofold: More cloud condensation nuclei (CCN) spread available liquid water over more but smaller particles, increasing τ by decreasing r. At the same time, smaller droplets rain out less efficiently, increasing τ via an increase in μ . Two sources of increasing CCN are relevant: Anthropogenic sulfate aerosols over and downwind of land, especially in industrialized areas, and marine aerosols produced by dimethylsulfide (DMS) emissions by phytoplankton (Charlson et al., 1987; Schwartz, 1988).

There are several reasons to question the global impact of changing aerosols on cloud optical properties. Increasing anthropogenic aerosol is plausible (though not well documented), but the relationship between sulfate aerosol mass concentration and CCN concentration (and between CCN concentration and N) is unknown. DMS production is a function of surface insolation rather than temperature (Bates et al., 1987), so its role in a greenhouse-gas induced climate change is indirect, via other cloud feedbacks. Foley et al. (1991) estimate from a simple climate model that the response to imposed insolation perturbations is reduced by only 7% by DMS-cloud albedo feedback.

4.4.4 Other issues

Broken cloudiness has received great attention in the climate community (cf. Fouquart et al., 1990), because radiation escaping or intercepted by the sides of a cloud can alter the plane-parallel estimate of radiation budget quantities. The effective (i.e., equivalent plane-parallel) cloud cover is less than the actual cloud cover in the longwave region, but changes from less than to greater than the actual cloud cover in the shortwave region as solar zenith angle increases. More important dynamically may be vertically broken cloudiness. Cirrus clouds especially appear to occur in multiple thin (hundreds of meters) layers; the resulting radiative flux divergence profile may destabilize the layer, drive convection, and thus alter the optical properties of the cloud (Starr and Cox, 1985). Direct simulation of these effects requires a vertical resolution about 3 times that of current climate GCMs.

A related problem, even for plane-parallel clouds, is how to overlap fractional cloud cover occurring simultaneously at different model levels. Tian and Curry (1989) find that an assumption of maximum overlap most accurately reproduces observed total cloud cover for clouds in adjacent layers, while a random overlap assumption performs better for clouds separated by clear layers. Even random overlap, though, produces systematic biases comparable to the effects of non-plane-parallel clouds.

A potentially important yet neglected influence on regional climate change is the difference between the microphysical properties of marine and continental clouds. CCN concentrations are larger over land, and the resulting clouds have smaller droplets, greater mass, and a lower probability of ice formation and precipitation than their marine counterparts (cf. Hobbs and Rangno, 1985). These differences could conceivably impact the regional response to climate forcing, e.g., via cloud cover trends of opposite sign over land and ocean or via perturbations of the regional balance between precipitation and evaporation.

4.5 Coupling between convective and stratiform clouds

Cumulus and stratiform cloud parametrizations are usually treated in separate subroutines in GCMs, and often by different people. There are at least two examples, though, both crucial to climate, of coupling between the two cloud types.

4.5.1 Mesoscale cirrus anvils

Deep cumulus clouds occupy a very small fraction of the area of the tropics and therefore do not have much of a direct radiative impact. Convective clouds are organized into mesoscale clusters, however, capped by upper troposphere anvils which are evident in satellite images. Mass flux cumulus schemes include a detrainment term which injects saturated vapor into the environment at cloud top and thereby triggers the large-scale cloud parametrization. As a result, tropical cirrus in GCMs are often associated with the occurrence of deep convection. The clouds are often too thin, however, and their frequency underpredicted. The potential climatic importance of these clouds is highlighted by Ramanathan and Collins (1991), who claim that solar reflection by cirrus shields limits the potential of the tropical oceans to warm as climate changes.

Part of the problem is that cirrus anvils are dynamic rather than passive entities. Mesoscale updrafts enhance condensation in convective clusters and produce significant stratiform precipitation (cf. Houze and Betts, 1981). This effect is parameterized by Donner (1990), who prescribes a constant *mesoscale* vertical velocity in convective gridboxes and calculates ice content as a balance between lifting condensation and sedimentation. The other problem is that some of the condensate in anvils is detrained from the accompanying cumulus clouds. Del Genio and Yao (1990) address this in the new version of the GISS GCM by including convective condensate produced in the upper troposphere in their cloud water budget rather than precipitating it immediately. The resulting cirrus is thicker than when convective condensate is rained out; the associated radiative heating drives a grid-scale version of a mesoscale updraft. Its effect on water vapor feedback can be seen in Figure 4.1 as an enhanced eddy moistening of the upper troposphere relative to that in GISS Model II.

Whether such efforts are an adequate representation of the true mesoscale dynamics for climate purposes remains to be seen. One issue yet to be addressed is the parametrization of vertical transport of convective condensate, requiring more detail than currently exists in GCM renderings of cumulus dynamics and microphysics.

4.5.2 Marine stratocumulus breakup

Low level stratus and stratocumulus over oceans may play a dominant role in climate change because of their horizontal extent and impact on planetary albedo. Unfortunately, these are among the most difficult clouds to predict in GCMs because of their small vertical scale and because they are controlled by interactions among boundary layer turbulence, radiation, and phase changes.

A key problem in cloud physics today is that of the conditions under which a nearly totally cloud covered stratocumulus deck breaks up into a field of scattered trade cumulus. Deardorff (1980) and Randall (1980) explored the conditions under which turbulent entrainment of unsaturated air above cloud top and subsequent evaporative cooling in the cloudy-clear mixture would be unstable and lead to the entire cloud being consumed by the mixing. Their criterion for this cloud-top entrainment instability (CTEI) is that the virtual temperature jump across the cloud top interface be less than a critical value.

In most GCMs marine stratus/stratocumulus are produced by the stratiform cloud parametrization, while trade cumulus are represented by the moist convection (or separate shallow convection) scheme. The transition between the two cloud types is therefore somewhat artificial. Randall et al. (1985) use an explicit mixed layer PBL parametrization in the CSU GCM and dissipate PBL cloudiness whenever the CTEI criterion is satisfied. While this approach is successful in the CSU model, it almost completely eliminates low cloud cover in the tropics and subtropics when applied in the GISS GCM (which does not use a mixed layer model).

Of more concern is observational evidence which conflicts with the traditional CTEI criterion. Kuo and Schubert (1988) show that extensive marine stratocumulus persist well beyond the Deardorff-Randall limit for CTEI. They suggest that under marginally unstable conditions, mixing proceeds slowly enough to extend the lifetime of the cloud deck by as much as several hours. Alternatively, by considering more explicitly the dynamics of mixing at the interface and the depth over which it occurs, more restictive instability criteria have been proposed (MacVean and Mason, 1990; Siems et al., 1990), but no consensus has been reached.

Another concern is whether the vertical resolution of climate models is adequate to represent PBL cloudiness processes at all. Of particular interest here is the effect of the radiative heating profile. Observations suggest that solar heating of the cloud base region can stabilize the subcloud layer and decouple the stratocumulus from its surface moisture source (cf. Nicholls, 1984), creating optical thickness variations over the diurnal cycle. The vertical scale of the relevant flux divergences is 10-20 mb, smaller than the layer thickness of most climate GCMs and inconsistent with the mixed layer approach to PBL modeling.

4.6 Validation strategies

Verification that the cloud parametrizations in a climate model are sufficiently accurate must consist of two components. First, cloud processes must be represented faithfully. Field experiments (e.g., GATE, AMEX, FIRE, ICE) already provide a wealth of data for testing convective and stratiform schemes, and future programs (e.g., TOGA/COARE, ASTEX, ARM) will only enhance this capability. Crucial gaps in information still exist, however. There is as yet no way to retrieve cloud ice water content, to detect small ice particles, nor a global sense of when the transition from liquid to ice occurs. Vertical transport of condensate in convective clouds, and its role in establishing downdrafts and anvils, is poorly understood. Large-scale relationships between cloud cover and relative humidity, vertical velocity, and other variables have not yet been established.

A potentially powerful tool for addressing some problems not amenable to data gathering is the mesoscale model. In principle, a model with a domain large enough to approach GCM grid size is a laboratory in which general relationships between variables (e.g., updraft and downdraft mass flux, updraft mass flux and anvil ice content, cloud cover and relative humidity) can be derived by conducting an ensemble of experiments. Most mesoscale model studies to date have focused on reproduction of particular case studies, but several promising attempts to link the small and large scales have recently appeared (Tao et al., 1987; Xu and Krueger, 1991).

The second aspect of validation is the more daunting challenge of verifying the sense and magnitude of cloud feedback, and thereby reducing the uncertainty in GCM estimates of climate sensitivity. Recently available satellite data sets offer hope of progress, but most analyses and model-data comparisons to date have been too simplistic to yield much useful information. Three weaknesses are immediately evident:

1. To validate a GCM's climate sensitivity, its variability, not its mean current climate, must be correct. In particular, comparisons of model- and satellite-based maps of cloud/radiation parameters are of limited use. The much employed cloud forcing diagnostic is a case in point. Consider the results of the Cess et al. SST anomaly experiment for 4 GCMs.

	CCC	GISS	ECHAM	CCM0	ERBE	N-7
Cloud forcing (Wm^{-2}):						
Shortwave	-41	-48	-60	-49	-48	-51
Longwave	15	19	15	47	31	24
Net	-26	-30	-45	-2	-17	-27
$\lambda(Km^2W^{-1})$	0.39	1.23	1.11	1.11	-	-
λ/λ_c	0.93	2.37	2.36	2.47	-	-

Table 4.2: *July cloud forcing, climate sensitivity and cloud feedback for 4 GCMs (Cess et al., 1990) and retrieved from ERBE and Nimbus-7 data (Arking, 1991).*

The first pair of GCMs have similar cloud forcing components, and both are in good agreement with the radiation budget data sets given their $\sim 10Wm^{-2}$ uncertainty, yet the GISS model is 3 times as sensitive as the CCC GCM, and their cloud feedbacks are of opposite sign. The second pair of GCMs have radically different cloud forcing components, yet almost identical sensitivity and cloud feedback. To diagnose cloud feedback, a model's variability on diurnal, seasonal, and interannual time scales is what counts.

2. Existing radiation budget data sets are not accurate enough by themselves to reveal cloud feedback, nor can they distinguish the contributions of individual cloud parameters. Consider, for example, the *greenhouse effect* diagnostic used by Ramanathan and Collins (1991), representing the difference between outgoing longwave radiation and surface emitted radiation. The temperature dependence of this diagnostic in principle contains information on the longwave component of water vapor and cloud feedback. Figure 4.6 shows the greenhouse effect from ERBE data compared with that from the two versions of the GISS GCM in Table 4.1. The two models agree with ERBE to within the scatter, yet Table 4.1 shows that they have very different longwave feedback components. In particular, the different cloud optics feedbacks are not evident.

   The problem is that the models differ in infrared feedback derivative by 0.5 $Wm^{-2}K^{-1}$. While this causes a difference of 1 K or more in sensitivity, it amounts to a 5 Wm^{-2} discrepancy over a 10 K interval, almost imperceptible amid the scatter in Figure 4.6. Thus, while ERBE data can reveal gross discrepancies of tens of Wm^{-2} in GCMs (and examples of this do exist)

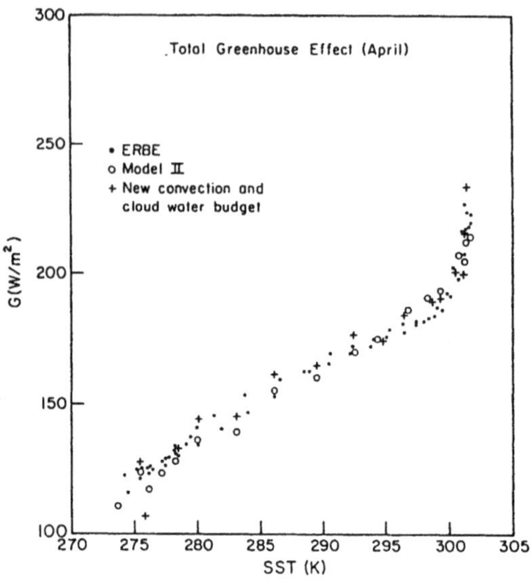

Figure 4.6: *April zonal mean total greenhouse effect* (Wm^{-2}) *for ocean areas vs. sea surface temperature from ERBE data and from two different versions of the GISS GCM.*

and regional variations, radiation budget alone is probably incapable of discriminating between positive and negative cloud feedback on a global basis.

3. Correlations of cloud/radiation parameters with temperature may actually be diagnostic of the effects of dynamics and not indicative of feedbacks. For example, the impression of a *super greenhouse effect* at high tropical temperatures cited by Ramanathan and Collins (1992) is caused in part by the lower relative humidities at 295-300 K associated with the subsiding branches of the Hadley and Walker circulations. Likewise, the positive correlation between anvil cloud height and reflectance variations deduced by Ramanathan and Collins applies only locally; averaged over the rising and/or sinking branches of the Hadley and Walker cells, compensation occurs and the correlation disappears (Fu et al., 1992). For water vapor feedback, dynamics is probably secondary to the direct thermodynamic effect (Del Genio et al., 1991; Rind et al., 1991). Dynamics also seems to be a secondary consideration for cloud optical thickness feedback, at least for low clouds, although microphysics effects may be first order (Tselioudis et al., 1991). Cloud cover and height variations, though, may be controlled as much by dynamics as by temperature, particularly in the upper troposphere.

Given that cloud feedback represents a very small signal with opposing contributions in different parts of the spectrum and for different cloud and surface types, a careful strategy might proceed as follows. Globally averaged cloud diagnostics should be the end result of a regional synthesis, with the feedback first being evaluated for individual regions (at a minimum, land vs. ocean, and tropics vs. midlatitudes vs. polar regions). Since flux variations are caused by a number of cloud, atmospheric, and surface properties, temporal variability in cloud parameters can be combined with a radiative model (cf. Rossow and Lacis, 1990) to yield flux variations that can be compared with those from radiation budget experiments. The final, and most difficult, step is to separate temperature effects from dynamics effects. Three- dimensional analyses are not yet accurate enough to provide useful vertical velocity information outside of northern midlatitudes, but regional insights may be gained by forcing single-column versions of GCM parametrizations with synoptic-scale dynamics fields from programs such as TOGA/COARE and ARM.

Most of all, though, adaptation of GCMs to knowledge gained from existing data sets must proceed at an accelerated pace. For example, the revolution in thinking about convection spawned by the GATE experiment of 1974 is only now beginning to make its way into GCM cumulus parametrizations. Data from the first FIRE experiments have proven how complex cloud behavior can be, but have not yet been distilled into insights useful to climate modelers. If the same rate of transfer from field to GCM persists, then the policy decisions of the 1990's will be forced to rely on scientific models long since known to be obsolete.

Acknowledgements

This work was supported by the NASA Climate Program and by the DOE Quantitative Links and Atmospheric Radiation Measurement Programs.

4.7 References

Arakawa A, Chen J-M (1987) Closure assumptions in the cumulus parameterization problem. Short- and Medium-Range Numerical Weather Prediction, *Suppl to J Meteor Soc Japan*: 107-131

Arakawa A, Schubert WH (1974) Interaction of a cumulus cloud ensemble with the large-scale environment, Part I. *J Atmos Sci* 31: 674-701

Arking A (1990) Feedback processes and climate response. Climate impact of solar variability. *NASA CP-3086* (Schatten KH, Arking A, eds): 219-226

Arking A (1991) The radiative effects of clouds and their impact on climate. *Bull Amer Meteor Soc* 71: 795-813

Bates TS, Charlson RJ, Gammon RH (1987) Evidence for the climatic role of marine biogenic sulfur. *Nature* 329: 319- 321

Betts AK (1986) A new convective adjustment scheme. Part I: Observational and theoretical basis. *Quart J Roy Meteor Soc* 112: 677-691

Betts AK, Harshvardhan (1987) Thermodynamic constraint on the cloud liquid water feedback in climate models. *J Geophys Res* 92: 8483-8485

Blyth AM, Cooper WA, Jensen JB (1988) A study of the source of entrained air in Montana cumuli. *J Atmos Sci* 45: 3944-3964

Cess RD, Potter GL, Blanchet JP, Boer GJ, Del Genio AD, Deque M, Dymnikov V, Galin V, Gates WL, Ghan SJ, Kiehl JT, Lacis AA, Le Treut H, Li Z-X, Liang X-Z, McAvaney BJ, Meleshko VP, Mitchell JFB, Morcrette J-J, Randall DA, Rikus L, Roeckner

E, Royer JF, Schlese U, Sheinin DA, Slingo A, Sokolov AP, Taylor KE, Washington WM, Wetherald RT, Yagai I, Zhang M-H (1990) Intercomparison and interpretation of climate feedback processes in 19 atmospheric general circulation models. *J Geophys Res* 95: 16601-16615

Charlson RJ, Lovelock JE, Andreae MO, Warren SG (1987) Oceanic phytoplankton, atmospheric sulfur, cloud albedo and climate. *Nature* 326: 655-661

Cheng M-D, Arakawa A (1991) Inclusion of convective downdrafts in the Arakawa-Schubert cumulus parameterization. Preprints, *19th Conf on Hurricanes and Tropical Meteor*, Amer Meteor Soc, Boston: 295-300

Curry JA, Ardeel CD, Tian L (1990) Liquid water content and precipitation characteristics of stratiform clouds as inferred from satellite microwave measurements. *J Geophys Res* 95: 16659-16671

Deardorff JW (1980) Cloud-top entrainment instability. *J Atmos Sci* 37: 131-147

Del Genio AD, Yao M-S (1988) Sensitivity of a global climate model to the specification of convective updraft and downdraft mass fluxes. *J Atmos Sci* 45: 2641-2668

Del Genio AD, Yao M-S (1990) Predicting cloud water variations in the GISS GCM. Preprints, *Conf on Cloud Phys*, Amer Meteor Soc, Boston: 497-504

Del Genio AD, Lacis AA, Ruedy RA (1991) Simulations of the effect of a warmer climate on atmospheric humidity. *Nature* 351: 382-385

Donner LJ (1990) Parameterizing mesoscale and large-scale ice clouds in general circulation models. Preprints, *Conf on Cloud Phys*, Amer Meteor Soc, Boston: J104-J109

Ellsaesser HW (1984) The climatic effect of CO_2: A different view. *Atmos Environ* 18: 431-434

Emanuel KA (1990) A scheme for representing cumulus convection in large-scale models. *MIT Center for Global Change*, Science Report No. 3, Cambridge, MA

Feigelson EM (1978) Preliminary radiation model of a cloudy atmosphere. Part I - Structure of clouds and solar radiation. *Beitr Phys Atmos* 51: 203-229

Foley JA, Taylor KE, Ghan SJ (1991) Planktonic dimethylsulfide and cloud albedo: An estimate of the feedback response. *Clim Change* 18: 1-15

Fouquart Y, Buriez JC, Herman M, Kandel RS (1990) The influence of clouds on radiation: A climate-modeling perspective. *Rev Geophys* 28: 145-166

Fu R, Del Genio AD, Rossow WB, Liu WT (1992) Cirrus cloud thermostat for tropical sea surface temperatures tested using satellite data. *Nature* 358: 394-397

Gregory D, Rowntree PR (1990) A mass flux convection scheme with representation of cloud ensemble characteristics and stability dependent closure. *Mon Wea Rev* 118: 1483-1506

Han Q, Rossow WB, Lacis AA (1991) Initial survey of cloud particle sizes using ISCCP data. Preprints, *5th Conf on Climate Variations*, Amer Meteor Soc, Boston: 372-375

Hansen J, Russell G, Rind D, Stone P, Lacis A, Lebedeff S, Ruedy R, Travis L (1983) Efficient three-dimensional global models for climate studies: Models I and II. *Mon Wea Rev* 111: 609-662

Hansen J, Lacis A, Rind D, Russell G, Stone P, Fung I, Ruedy R, Lerner J (1984) Climate sensitivity: Analysis of feedback mechanisms. Climate Processes and Climate Sensitivity (Hansen JE, Takahashi T, Eds), *Amer Geophys Union*, Washington, DC: 130-163

Hobbs PV, Rangno AL (1985) Ice particle concentrations in clouds. *J Atmos Sci* 42: 2523-2549

Houze RA Jr, Betts AK (1981) Convection in GATE. *Rev Geophys Space Phys* 19: 541-576

Knupp KR, Cotton WR (1985) Convective cloud downdraft structure: An interpretive survey. *Rev Geophys* 23: 183-215

Kuo H-L (1974) Further studies of the parameterization of the influence of cumulus convection on large-scale flow. *J Atmos Sci* 31: 1232-1240

Kuo H, Schubert WH (1988) Stability of cloud-topped boundary layers. *Quart J Roy Meteor Soc* 114: 887-916

Le Mone MA, Barnes GM, Zipser EJ (1984) Momentum flux by lines of cumulonimbus over the tropical oceans. *J Atmos Sci* 41: 1914-1932

Le Treut H, Li Z-X (1991) Sensitivity of an atmospheric general circulation model to prescribed SST changes: feedback effects associated with the simulation of cloud optical properties. *Climate Dyn* 5: 175-187

Lindzen RS (1990) Some coolness concerning global warming. *Bull Amer Meteor Soc* 71: 288-299

MacVean MK, Mason PJ (1990) Cloud-top entrainment instability through small-scale mixing and its parametrization in numerical models. *J Atmos Sci* 47: 1012-1030

Manabe S, Smagorinsky J, Strickler RF (1965) Simulated climatology of a general circulation model with a hydrological cycle. *Mon Wea Rev* 93: 769-798

Mitchell JFB, Senior CA, Ingram WJ (1989) CO_2 and climate: A missing feedback? *Nature* 341: 132-134

Moorthi S, Suarez MJ (1992) Relaxed Arakawa-Schubert: A parameterization of moist convection for general circulation models. *Mon Wea Rev* 20: 978-1002

Nakajima T, King MD, Spinhirne JD, Radke LF (1991) Determination of the optical thickness and effective particle radius of clouds from reflected solar radiation measurements. Part II: Marine stratocumulus observations. *J Atmos Sci* 48: 728-750

Nicholls S (1984) The dynamics of stratocumulus: Aircraft observations and comparisons with a mixed layer model. *Quart J Roy Meteor Soc* 110: 783-870

Paluch IR (1979) The entrainment mechanism in Colorado cumuli. *J Atmos Sci* 36: 2462-2478

Pan V, Randall DA (1991) Implementation of the Arakawa-Schubert cumulus parameterization with a prognostic cumulus kinetic energy. Preprints, *19th Conf on Hurricanes and Tropical Meteor*, Amer Meteor Soc, Boston: 301-304

Platt CMR (1989) The role of cloud microphysics in high-cloud feedback effects on climate change. *Nature* 341: 428-429

Ramanathan V, Collins W (1991) Thermodynamic regulation of ocean warming by cirrus clouds deduced from observations of the 1987 El Niño. *Nature* 351: 27-32

Randall DA (1980) Conditional instability of the first kind upside-down. *J Atmos Sci* 37: 125-130

Randall DA, Abeles JA, Corsetti TG (1985) Seasonal simulations of the planetary boundary layer and boundary-layer stratocumulus clouds with a general circulation model. *J Atmos Sci* 42: 641-676

Rind D, Chiou E-W, Chu W, Larsen J, Oltmans S, Lerner J, McCormick P, McMaster L (1991) Positive water vapor feedback in climate models confirmed by satellite data. *Nature* 349: 500-503

Roeckner E (1988) Cloud-radiation feedbacks in a climate model. *Atmos Res* 21: 293-303

Rossow WB, Lacis AA (1990) Global, seasonal cloud variations from satellite radiance measurements. Part II: Cloud properties and radiative effects. *J Climate* 3: 1204-1253

Sassen K, Starr DO'C, Uttal T (1989) Mesoscale and microscale structure of cirrus clouds: Three case studies. *J Atmos Sci* 46: 371-396

Schneider EK, Lindzen RS (1976) A discussion of the parameterization of momentum exchange by cumulus convection. *J Geophys Res* 81: 3158-3160

Schwartz SE (1988) Are global cloud albedo and climate controlled by marine phytoplankton? *Nature* 336: 441-445

Siems ST, Bretherton CS, Baker MB, Shy S, Briedenthal RE (1990) Buoyancy reversal and cloud-top entrainment instability. *Quart J Roy Meteor Soc* 116: 705-739

Slingo A, Nicholls S, Schmetz J (1982) Aircraft observations of marine stratocumulus during JASIN. *Quart J Roy Meteor Soc* 108: 833-856

Slingo A, Wilderspin RC, Smith RNB (1989) The effect of improved parameterizations on simulations of cloudiness and the earth's radiation budget in the tropics. *J Geophys Res* 94: 2281-2302

Slingo JM (1980) A cloud parameterization scheme derived from GATE data for use with a numerical model. *Quart J Roy Meteor Soc* 106: 747-770

Smith RNB (1990) A scheme for predicting layer clouds and their water content in a general circulation model. *Quart J Roy Meteor Soc* 116: 435-460

Somerville RCJ, Remer LA (1984) Cloud optical thickness feedbacks in the CO_2 climate problem. *J Geophys Res* 89: 9668-9672

Starr DO'C, Cox SK (1985) Cirrus clouds. Part II: Numerical experiments on the formation and maintenance of cirrus. *J Atmos Sci* 42: 2682-2694

Sundqvist H (1978) A parameterization scheme for non-convective condensation including prediction of cloud water content. *Quart J Roy Meteor Soc* 104: 677-690

Sundqvist H, Berge E, Kristjánsson JE (1989) Condensation and cloud parameterization studies with a mesoscale numerical weather prediction model. *Mon Wea Rev* 117: 1641-1657

Tao W-K, Simpson J, Soong S-T (1987) Statistical properties of a cloud ensemble: A numerical study. *J Atmos Sci* 44: 3175-3187

Tian L, Curry JA (1989) Cloud overlap statistics. *J Geophys Res* 94: 9925-9935

Tiedtke M, Heckley WA, Slingo J (1988) Tropical forecasting at ECMWF: The influence of physical parametrization on the mean structure of forecasts and analyses. *Quart J Roy Meteor Soc* 114: 639-664

Tselioudis G, Rossow WB, Rind D (1992) Global patterns of cloud optical thickness variation with temperature. *J Climate*: in press

Washington WM, Meehl GA (1984) Seasonal cycle experiment on the climate sensitivity due to a doubling of CO_2 with an atmospheric general circulation model coupled to a simple mixed-layer ocean model. *J Geophys Res* 89: 9475-9503

Wielicki BA, Suttles JT, Heymsfield AJ, Welch RM, Spinhirne JD, Wu M-LC, Starr DO'C, Parker L, Arduini RF (1990) The 27-28 October 1986 FIRE IFO cirrus case study: Comparison of radiative transfer theory with observations by satellite and aircraft. *Mon Wea Rev* 118: 2356-2376

WCRP (1990) Scientific Plan for the Global Energy and Water Cycle Experiment. *World Climate Research Programme Report* No. 40 (WMO/TD - No. 376), Geneva, Switzerland

Xu K-M, Krueger SK (1991) Evaluation of cloudiness parameterizations using a cumulus ensemble model. *Mon Wea Rev* 119: 342-367

Yao M-S, Del Genio AD (1989) Effects of cumulus entrainment and multiple cloud types on a January global climate model simulation. *J Climate* 2: 850-863

Chapter 5

SATELLITE OBSERVATIONS OF RADIATION AND CLOUDS TO DIAGNOSE ENERGY EXCHANGES IN THE CLIMATE: PART I

William B. Rossow
NASA
Goddard Institute for Space Studies
New York, NY 10025
USA

5.1 Introduction: Energy cycles of the climate

The basic energy cycle that establishes the climate of our planet is more complicated than a simple balance between the solar radiation (wavelengths from $0.4 - 4\,\mu m$) absorbed by Earth and the terrestrial radiation (wavelengths from $4 - 100\,\mu m$) emitted when the temperature of Earth reaches equilibrium. The first complication is that the spherical shape of Earth leads to an equator-to-pole difference in solar heating; the resulting temperature contrast forces motions in the atmosphere and ocean that transport some of the energy from low to high latitudes. Thus, the radiation balance at the top of the atmosphere exhibits a net heating (absorbed solar flux exceeds emitted terrestrial flux) at low latitudes and a net cooling (emitted terrestrial flux exceeds absorbed solar flux) at high latitudes which is balanced by the dynamic transports. However, the global average radiation is essentially in balance when averaged over a whole year. Thus, the solar and terrestrial radiative fluxes are the drivers for all of the phenomena we call climate.

The second complication is that Earth's atmosphere is nearly transparent to solar radiation and nearly opaque to terrestrial radiation. Thus, most of the absorbed solar flux is deposited at the surface, but the terrestrial radiation emitted by the surface cannot escape easily to space. Consequently, the surface temperature must be larger to attain equilibrium – this is the "greenhouse effect". Another

NATO ASI Series, Vol. I 5
Energy and Water Cycles in the Climate System
Edited by E. Raschke and D. Jacob

consequence of this fact is that the radiation balance at the surface is net heating, while that of the atmosphere is net cooling, setting up a vertical exchange of energy by other means.

The major cause of the opacity to terrestrial radiation is water vapor in the atmosphere, but water also creates the third complication in the energy cycle. Evaporation of water from the surface cools it; when some of this water vapor condenses to form precipitation, the atmosphere is heated. The actual process of cloud and precipitation formation also involves significant vertical re-distribution of energy as some water condenses at higher altitudes, falls and evaporates at lower altitudes and some water reaches the surface returning some energy to it. Since the water vapor can be transported by atmospheric motions, the surface cooling and atmospheric heating may not occur in the same places, implying an additional net transport of energy from warmer oceans to drier land and to colder latitudes. Thus, the energy balance at the surface is net radiative heating almost everywhere balanced by evaporative cooling and cooling by atmospheric and oceanic motions.

The water cycle also produces the fourth complication – clouds. Clouds change both the total amount of absorbed solar radiation and the total amount of emitted terrestrial radiation. In so doing, they also change the horizontal and vertical distributions of solar heating and terrestrial cooling, thereby altering the forcing for the atmospheric motions that transport heat and water. What makes this problem so challenging is that the clouds, too, are controlled by the atmospheric motions.

This basic understanding of the climate's energy cycle is accurate enough to describe the primary processes that control our environment, but not accurate enough to predict how the climate will respond to small perturbations, either natural or human-made. Predicting small transient responses or shifts of the climate equilibrium requires that we account for the changes in all of the processes that affect the radiation and water exchanges: at the center of all these exchange processes are clouds. The focus of this lecture is on the role of clouds in the radiation balance of Earth. Most of the results shown come from two major projects designed to develop the needed datasets: the International Satellite Cloud Climatology Project (ISCCP) and the Earth Radiation Budget Experiment (ERBE).

5.2 Remote sensing of clouds and radiation

Improved modeling of cloud effects on radiation is the desired outcome of the analysis of observations of clouds and radiation; however, the only practical way to obtain comprehensive measurements of cloud properties is to infer them from the radiation measurements, themselves. That is, the essence of the analysis of remote sensing data, whether from satellites, aircraft or the surface, is to model the detailed effects of clouds on radiation to link physical variables to the observed radiation (Rossow, 1989). Thus, development and improvement of radiative models including cloud effects must proceed in parallel with collection and analysis of measurements. This lecture will concern satellite observations, but the approach can be used for other types of data with some modifications.

Determination of specific cloud properties is best done with radiation measured at specific wavelengths that are selected to be more sensitive to the desired quantity and less sensitive to other quantities. However, since satellites are expensive to build and have been designed to measure other

Satellite Name	Instrument Name	Time Period	Area of Coverage	Comments
NIMBUS-4	IRIS	1970	Globe	The only IR spectra that are available
NIMBUS-7	TOMS	1978-1991	Globe	UV imager used to measure cloud amount and reflectivity
NIMBUS-7	THIR	1978-1985	Globe	IR imager used to measure cloud amount and height
NIMBUS-7	SMMR	1978-1985	Globe	Microwave scanner used to measure water vapor, cloud liquid water amount and precipitation rate
NOAA-6–12	AVHRR	1979-1991	Globe	Solar/IR imager used to measure cloud amount, optical depth, particle size, and height
	HIRS-2	1979-1991	Globe	IR sounder used to measure atmospheric temperature and water vapor profiles and cloud amount, height and optical depth
DMSP	SSM/I	1987-1991	Globe	Microwave scanner used to measure water vapor, cloud liquid water amount and precipitation rate
SMS-1,2	VISSR	1974-1982	Regional	Solar/IR imager can be used to measure cloud amount, optical depth and height
GOES-1–7	VISSR	1981-1991	Lower latitudes Americas	Solar/IR imager used to measure cloud amount, optical depth and height
	VAS		Lower latitudes Americas	IR sounder used to measure atmospheric temperature and water vapor profiles and cloud amount, height and optical depth
GMS-1–4	VISSR	1983-1991	Lower latitudes Western Pacific	Solar/IR imager used to measure cloud amount, optical depth and height
METEOSAT-1–5	VISSR	1979-1991	Lower latitudes Africa/Europe	Solar/IR imager used to measure cloud amount, optical depth and height

Table 5.1: *Some satellite datasets that can be or are being used to study clouds.*

aspects of the atmosphere and surface as well, the best choices of wavelengths are not always available. Currently available datasets that provide some information about cloud properties and their variations are listed in Table 5.1 by satellite and instrument name. Also given are the time periods covered and a simple indication of the portion of the globe covered. This list is not historically complete, rather it mentions datasets that are in good condition and readily available.

Although many cloud properties affect the radiation at any one wavelength, these properties can be arranged in order of decreasing magnitude of their effect. With a limited number of wavelengths measured, we can only measure a few of the most important properties by specifying the remaining quantities. The extent to which the specified properties vary determines the uncertainty in the measured properties.

The four most important cloud properties that can be measured from available satellite datasets are top temperature, optical thickness, particle size and liquid water content (I will discuss cloud "amount" later). Cloud top temperature can be thought of as the actual physical temperature of the topmost part of a cloud; when the cloud is opaque to radiation, then the satellite "sees" an emitted radiation equivalent to a black body at this temperature. However, when the cloud is not opaque, the radiation emitted by the cloud may originate within it and may be added to by radiation passing through the cloud from below. In other words, an optically thin cloud would appear to emit radiation equivalent to a temperature higher than that of its top (since lower atmospheric levels and the surface are usually warmer than clouds). Measurements at infrared wavelengths (usually near $10-12\,\mu m$) are most commonly used to measure cloud top temperature (e.g., Minnis and Harrison, 1984; Stowe et al., 1988; Rossow et al., 1989 and references therein); however, some microwave wavelengths measure the temperature at the top of the layer containing ice or rainfall (Mugnai and Smith, 1988).

During daytime, the optical thickness of the cloud can be measured by the amount of sunlight reflected by the cloud (e.g., Rossow et al., 1989). This method is not very sensitive to lower values, but covers the whole range of values that occur. Day or night, lower values of cloud optical thickness can be estimated from the shape of the spectrum of infrared radiation emitted by the combination of cloud, atmosphere and surface, but this method does not work for larger values. (Infrared spectrometer measurements of Earth have been obtained only in 1970 even though such measurements are "routine" for spacecraft missions to other planets.) When the optical thickness of the cloud can be measured, then its effect on the measurement of cloud top temperature can also be removed (Rossow et al., 1989; Minnis et al., 1990); however, for ice clouds the relation between the optical thickness at solar and infrared wavelengths is more uncertain because it depends on particle shape (Wielicki et al., 1990). Once cloud top temperature is known, the location of cloud top in the atmosphere can be determined by comparing this value to the profile of atmospheric temperature. IR sounder measurements, on the other hand, can be used to infer the cloud top pressure more directly; when the temperature at that pressure differs from the apparent radiometric temperature, the infrared optical thickness can also be inferred (Wylie and Menzel 1989).

The average distribution of cloud top pressures for January and July is illustrated in Fig. 5.1 using results from ISCCP; Fig. 5.2 shows the annual mean latitudinal distribution of cloud optical thicknesses and top temperature from ISCCP. The cloudiest latitudes constitute the tropical and

(a)

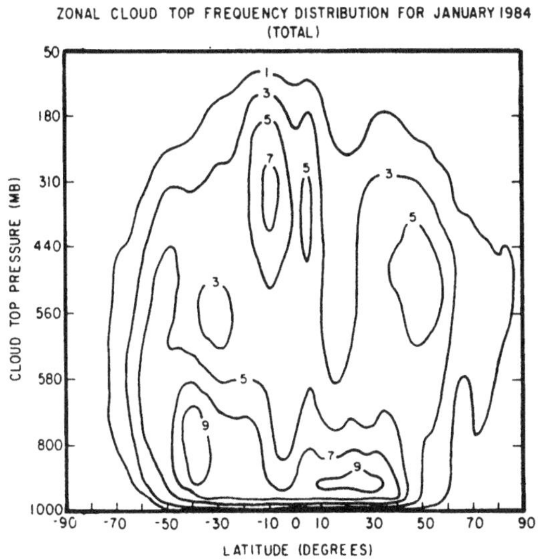

(b)

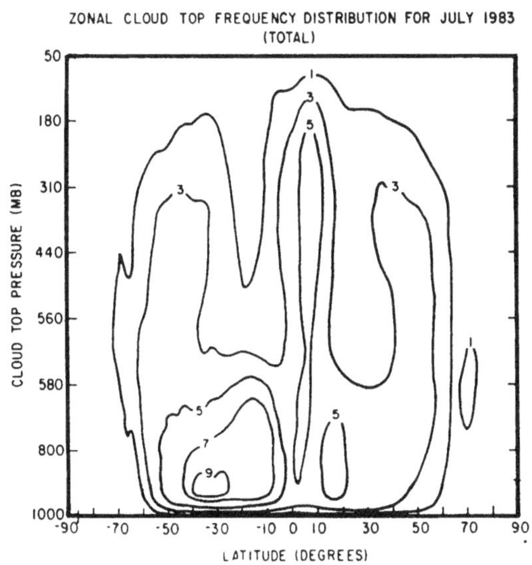

Figure 5.1: *(a) Zonal mean frequency of cloud top pressure for January 1984 from ISCCP, (b) Zonal mean frequency of cloud top pressure for July 1983 from ISCCP.*

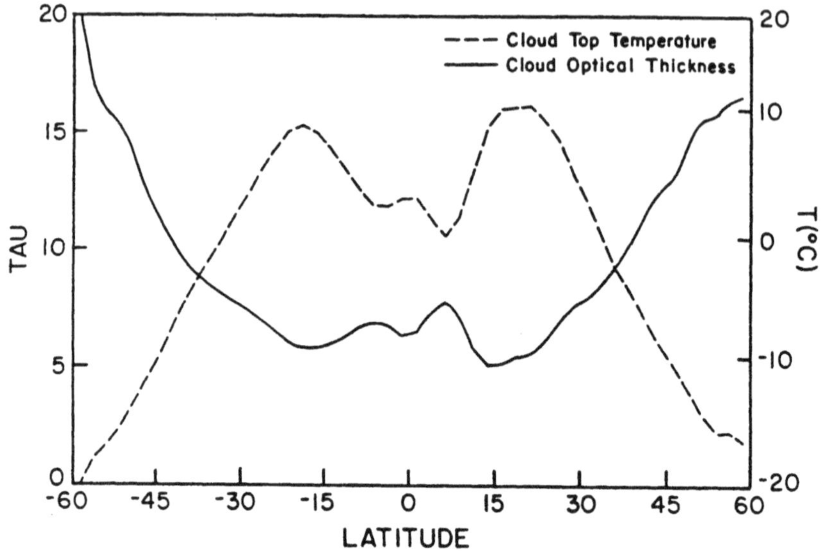

Figure 5.2: *Zonal, annual mean cloud optical thickness and top temperature* (K) *from ISCCP.*

midlatitude storm zones. The highest clouds occur in the summer hemisphere part of the tropics, whereas there are more high clouds in the winter midlatitudes (Fig. 5.1). Despite larger cloud top pressures in midlatitudes than in the tropics, cloud top temperatures are much colder at higher latitudes because the atmospheric temperatures are lower at a particular pressure level (Fig. 5.2). The northern hemisphere exhibits larger seasonal variations than the southern hemisphere. Cloud optical thicknesses are larger in the summer tropics and winter midlatitudes than in the opposite seasons; the largest optical thicknesses occur at higher latitudes (Fig. 5.2).

Cloud particle size affects both the amount of sunlight reflected by a cloud and the shape of the spectrum of infrared radiation it emits. Thus, additional measurements at these wavelengths can be used to infer the cloud particle size, together with the optical thickness and top temperature (e.g., Arking and Childs, 1985; Nakajima and King, 1990; Ackerman et al., 1990; Wielicki et al., 1990; Wetzel and Vonder Haar, 1991). Some preliminary results for low level, liquid water clouds are illustrated in Fig. 5.3, showing a systematic difference in average particle size between clouds over land and oceans: average particle radius is about $12\,\mu m$ over oceans and about $10\,\mu m$ over land.

Measurements of cloud optical thickness and particle size can be used to estimate the amount of water in a cloud. Microwave radiation at certain wavelengths can be used to measure more directly the total amount of liquid water over oceans by measuring the amount of radiation that is absorbed by the cloud (eg., Njoku, 1982; Alishouse et al., 1990; Curry et al., 1990). Figure 5.4 shows a comparison of these two measurements; the discrepancy in these results depends on the fact that the two measurements have different sensitivities to lower values. If the optical thickness values from visible wavelength measurements are truncated at about 3, that is, if the thinner clouds are removed from the ISCCP results, then the agreement between the two results is as shown; without this truncation the ISCCP results are about a factor of $2-5$ lower.

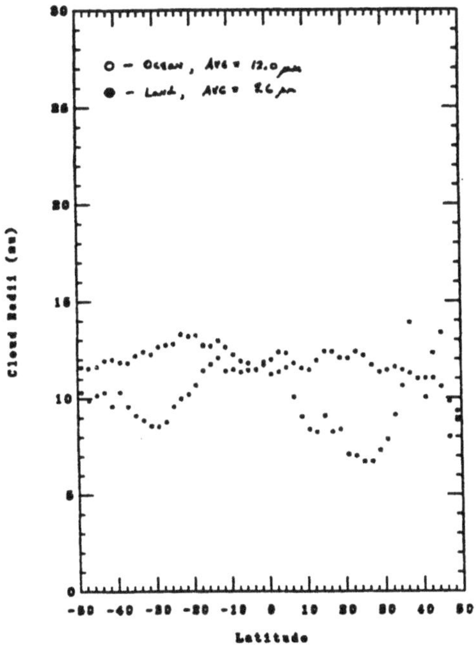

Figure 5.3: *Zonal mean cloud particle radius (μm) for warm clouds (thesis work of Q.-Y. Han).*

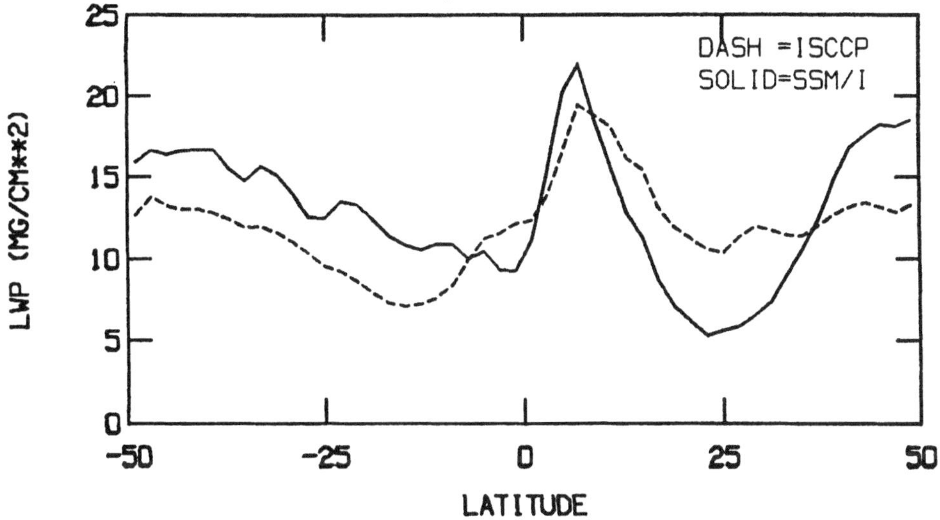

Figure 5.4: *Comparison of zonal, monthly mean liquid water path ($mg\,cm^{-2}$) inferred from SSM/I measurements and ISCCP optical thickness values (from thesis work of B. Lin).*

The accuracy with which these cloud properties can be measured depends both on the fidelity of the radiative transfer model used in the analysis and on the amount of information provided in the satellite dataset. Measurement of radiation at only one or two wavelengths limits the analysis to only one or two quantities; the dependence of the measurements on other quantities must be assumed. Measurement of more wavelengths eliminates these assumptions and improves the accuracy of the results. Future multi-wavelength datasets should produce more accurate results; however, as will be illustrated, these results already have enough accuracy to begin study of the key climate problems.

There are also some problems in performing the analyses described above that need more study. The most significant problem concerns how to account for the effects of horizontal inhomogeneities in clouds at "smaller" scales. The question that needs answering is whether there are scales at which the radiative effects of the cloud can be represented by spatially averaged properties and the effects of spatial variations at smaller scales neglected or treated in some "statistical" fashion. There are three considerations:

1. cloud properties vary on scales much smaller than those on which the radiation interacts with the cloud,
2. cloud properties exhibit a spectrum of variation that mirrors that of the atmospheric motions, with larger amplitudes at larger scales, and
3. for each spatial scale there is a characteristic time scale at which the variations occur.

The first consideration means that cloud variations at really small scales do not actually alter the radiation and the second consideration suggests that the magnitude of the cloud variations, and therefore, their importance to the radiation, is much smaller at these smaller scales. The third consideration enters when one considers that the radiative response time of the atmosphere and ocean is much larger than the time for changes of clouds at smaller scales; hence, the importance of the smaller spatial scale variations may be lessened.

This problem is usually cast in terms of determining the cloud "amount", which is thought of as the fraction of the area covered by cloud. The difficulty is that, while this concept seems straightforward, it is actually difficult in practice to define what is meant by "cloud" and where, precisely, the boundary between "cloud" and "non-cloud" is (Wielicki and Parker, 1992). Earth's atmosphere contains some particles everywhere at all times, so that a division of the atmosphere into these two parts requires some "threshold" optical thickness, as in: if the particle optical thickness is larger than 0.1, then it is cloudy. Cloud "cover" defined this way is essentially a radiative quantity that depends directly on the value of this threshold and remains, in a sense, arbitrary. The presence of a spectrum of scales of cloud variability has led to studies representing cloud cover as a fractal (e.g., Lovejoy and Schertzer 1990). Although this mathematical technique provides a valuable new way to represent this complex spectrum of variations, it does not necessarily solve any problems when used in this form. Instead, it may be better to think of "non-cloud" simply as "cloud" with small or zero optical thickness. In this form, the problem becomes the problem of treating the effects of the smaller scales of variability on the larger scale radiation fluxes. The above arguments imply that if the threshold used to divide "cloud" from "non-cloud" is well defined and small enough, then the accuracy of radiative effects

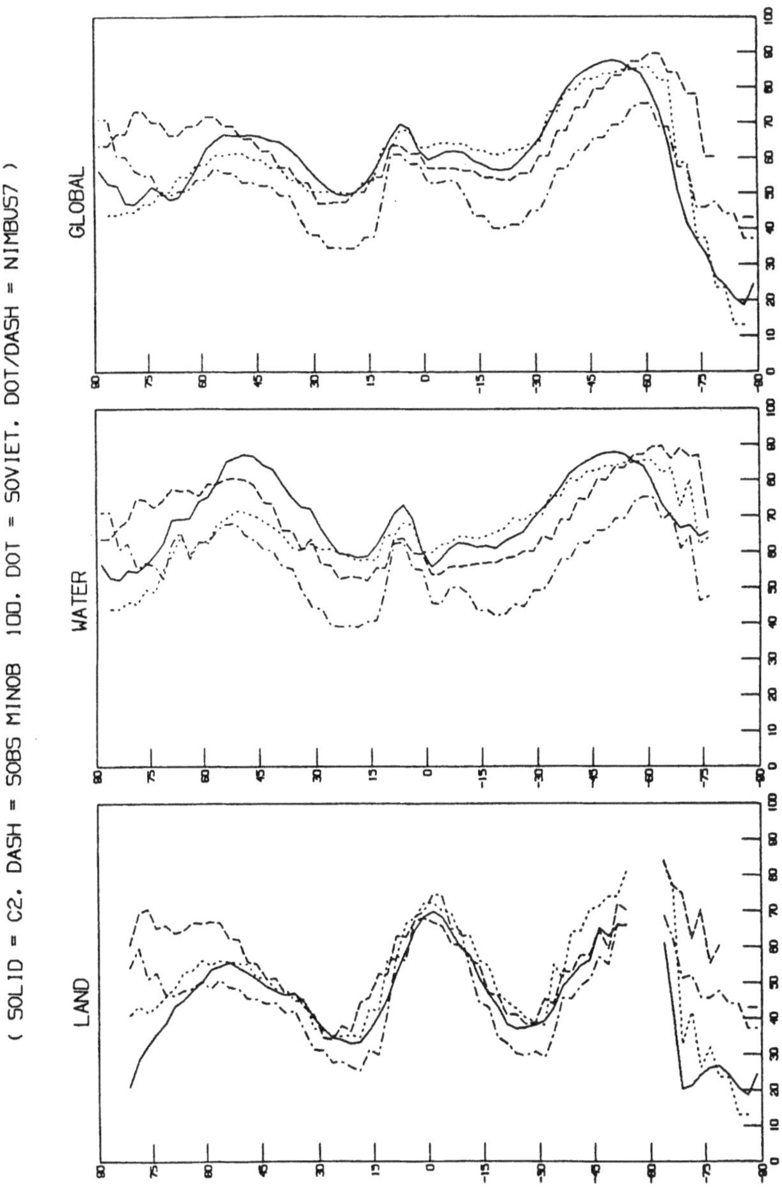

Figure 5.5: *Comparisons of zonal, annual mean cloud amounts (%) from ISCCP, surface observations, METEOR satellite and NIMBUS-7.*

calculated with this value will be high. However, such a "cloud amount" is most meaningful when the corresponding cloud optical properties are reported.

Figure 5.5 compares the average cloud amounts inferred from three different satellite systems (Rossow and Schiffer, 1991; Stowe et al., 1989, Mokhov, 1992) with those observed by surface weather observers (Warren et al., 1986, 1988). The good correspondence of these results, based on different definitions of the detection threshold, encourages one to accept that this quantity is being measured with useful accuracy. In particular, the agreement between ISCCP and surface observers is quite good and, when the ISCCP cloud detection sensitivity is reduced, cloud amounts similar to the METEOR and NIMBUS-7 results are obtained. The satellite datasets have better coverage and sampling over oceans than the surface dataset.

A more detailed comparison between individual ISCCP results and surface observations reveals that the usual distribution of differences (Fig. 5.6(a)) is actually more complex when displayed like Fig. 5.6(b). The more complex pattern is actually to be expected when comparing a "large-area-like" result like that of ISCCP with the "point-like" observation of the ground observer. When the former shows either completely clear or completely overcast conditions, which occurs about half the time, the latter shows good agreement. The disagreement at zero cloud amount and maximum difference (lower left corner of Fig. 5.6(b)) is associated with a small number of clouds that are missed by the satellite detection. When the satellite dataset indicates partial coverage, there are two different kinds of situations: one where a relatively uniform scattering of small clouds covers the whole area and one where some portion of the area is overcast. The first case explains the distribution of values near zero difference for all cloud amounts in Fig. 5.6(b), indicating good agreement for scattered cloudiness. The second case explains the values representing the maximum disagreement in Fig. 5.6(b):

1. when the fraction of the area covered by clouds is small, the probability is high that a single surface station will be under clear conditions, so that the cloud amount difference is in the upper branch and
2. when the fraction of the area that is overcast is large, the probability is high that a single surface station will be under overcast conditions, so that the cloud amount difference is in the lower branch.

At 50% coverage, both branches have equal probability. The results in Fig. 5.6 exhibit this behavior with a small shift that indicates that the ISCCP results underestimate cloud amounts over land by about 5%.

Another specific and difficult problem in the design of good radiative transfer models for radiation in cloudy atmospheres is how to treat the effects of non-spherical cloud particle shapes. The most sophisticated theories and models for radiation in a cloud are based on the assumption of spherical particle shape (Hansen and Travis 1974), which is an excellent assumption for liquid water clouds, is less accurate for rainfall, and is incorrect for most ice clouds. The shape of the cloud particles affects the detailed angular distribution of radiation that they scatter (e.g., Asano and Sato, 1980; Takano and Liou, 1989). In optically thicker clouds, where each photon is scattered many times, the shape effects are less important; however, we currently lack a general theory for treating this problem.

5.3 Calculating radiative fluxes

The most direct way to measure the total radiative fluxes (summed over all wavelengths) is with instruments that measure radiation at all wavelengths: one type makes a single measurement of

(a)

(b)

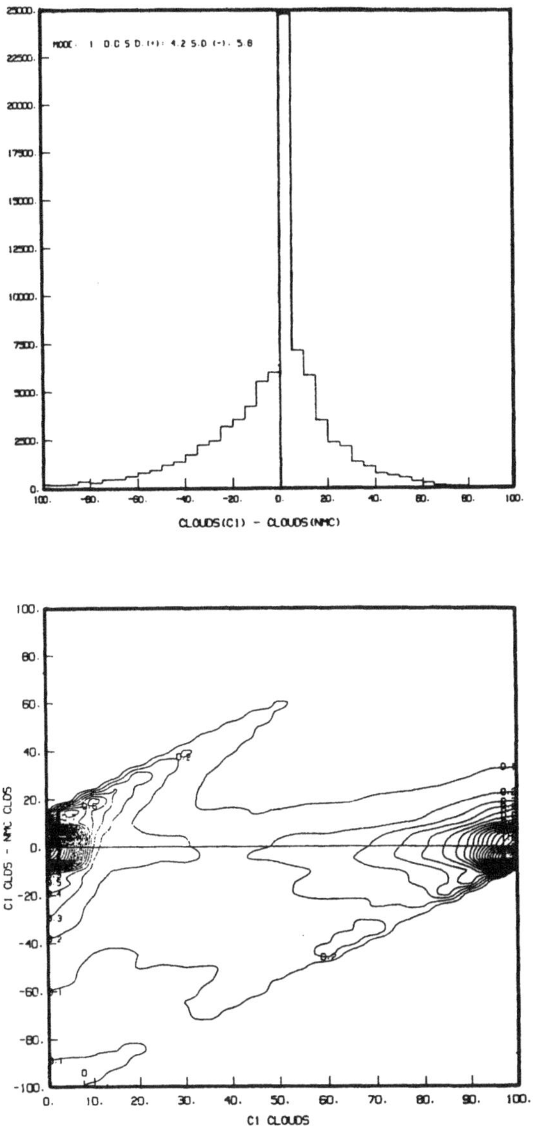

Figure 5.6: *(a) Frequency distribution of the differences between ISCCP and surface observed cloud amounts from the global distribution of weather stations for July 1985, (b) Same data shown in the upper figure as a function of ISCCP cloud amount (%). Contours indicate frequency as a fraction of the peak value.*

the radiation integrated over all wavelengths (called a "broadband" measurement) and another type makes measurements at many discrete wavelengths that can be added together to get the total. The former type of measurement has been favored (Jacobowitz et al., 1984; Barkstrom and Smith, 1986), generally because it was somewhat more difficult to calibrate instruments that make multi-wavelength measurements; however, certain types of spectrometers that make measurements that are continuous in wavelength, such as the IRIS for infrared wavelengths, may actually be easier to calibrate; but almost no spectroscopic observations of Earth have been made from space. Estimates of total fluxes have also been made based on measurements of a smaller number of wavelengths within the spectral range (Raschke et al., 1973; Gruber, 1977). Table 5.2 summarizes the datasets that have (or could be) used for this purpose.

Satellite Name	Instrument Name	Time Period	Area of Coverage	Comments
NIMBUS-7	Scanner	1978-1981	Globe	Broadband
	MFOV	1978-1990	Globe	Broadband
	WFOV	1978-1990	Globe	Broadband
NOAA-6–12	AVHRR	1979-1991	Globe	Narrowband measurements used to estimate radiation for total spectrum
	HIRS-2	1979-1991	Globe	Multi-wavelength IR measurements used to estimate radiation for total IR spectrum
ERBS (ERBE)	Scanner	1984-1990	Lat < 60°	Broadband
	WFOV	1984-1991	Lat < 60°	Broadband
NOAA-9 (ERBE)	Scanner	1985-1987	Globe	Broadband
	WFOV	1985-1991	Globe	Broadband
NOAA-10 (ERBE)	Scanner	1986-1989	Globe	Broadband
	WFOV	1986-1991	Globe	Broadband

Table 5.2: *Some satellite datasets that can be or are being used to study radiative fluxes at the top of the atmosphere and at the surface of Earth.*

In all cases, these satellite measurements are *radiances* not *radiative fluxes*, the latter is obtained by integrating the former over all angles. Since our atmosphere is a very thin shell around the planet, we usually divide the radiation going in all directions into "upward" and "downward" parts, where each part represents an integration of the radiation over a hemisphere. Since the atmosphere and surface are not homogeneous in their properties and since scattering processes, in particular, do not distribute radiation uniformly in all directions, the radiation intensity necessarily varies with the direction in which the satellite sensor points. The nature of this variation also varies strongly with wavelength. The vertical inhomogeneity of the atmosphere naturally divides the problem of inferring fluxes into two horizontal scales: one much larger than the depth of the atmosphere and one smaller than or similar to the depth of the atmosphere. At smaller scales the problem involves a three-

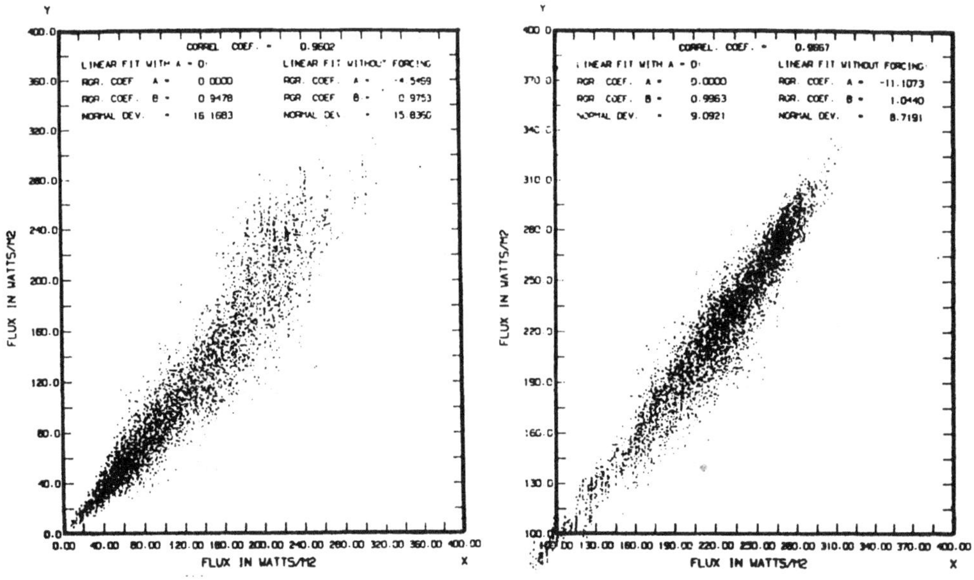

Figure 5.7: *Scatter-plot of daily values of ERBE against ISCCP against solar (left) and terrestrial (right) fluxes over the globe.*

dimensional radiation field, whereas the problem is two-dimensional at larger scales. For example, determining the "upward" flux for a region smaller than the vertical extent of the atmosphere is much more difficult because contributions from adjacent regions are just as important. Improving the treatment of the angular distribution of radiation to infer fluxes from radiance measurements is still a key problem.

Accurate determinations of total solar fluxes are made particularly difficult by the rapid daily variation of solar illumination: the fluxes change by about 1% to $>$ 10% in about 30 minutes. Most satellite datasets provide very sparse samples of this diurnal variation; in fact, only the ERBE dataset contains measurements from more than one satellite, which still only gives a statistical sample over the diurnal cycle in one month (Brooks et al., 1986).

It is also possible to calculate the radiative fluxes, if all of the necessary physical variables are measured and a radiative model of sufficient accuracy is used (Rossow and Lacis, 1990). The variables that are needed are: surface spectral albedo and emissivity, surface temperature, profiles of atmospheric temperature, humidity and ozone, abundances of other radiatively important atmospheric gases that are uniformly mixed, aerosol optical thickness and optical constants, cloud optical thickness and its vertical distribution, as well as cloud particle size, shape and phase. These variables also must be available with sufficient space and time sampling to capture the important variations of the radiation fluxes. Attempts to infer the radiation fluxes in this way are being made using datasets from the International Satellite Cloud Climatology Project (ISCCP) and from operational weather observations. Although this approach requires much more data, including special datasets needed to validate the radiative model, it allows for an examination of the detailed effects of clouds and other

Quantity	ERBE[1]	NIMBUS-7[2]	ISCCP
Downward Solar	341.3	343.0	341.4
Reflected Solar	102.1	111.1 (102.5)	113.4
Cloud effect	-48.9	-50.9	-52.2
Emitted Terrestrial	234.5	233.7 (234.9)	232.4
Cloud effect	31.0	24.1	22.2
Net Flux	4.7	-1.9 (5.6)	-4.4
Cloud effect	-17.9	-26.8	-30.0

Table 5.3: *Comparison of annual mean radiative fluxes (in Wm^{-2}) at the top of the atmosphere measured by ERBE and NIMBUS-7 with those calculated using datasets from ISCCP. The "annual" mean values for ERBE and ISCCP are the average over April, July and October 1985 and January 1986; for NIMBUS-7 these are averages over summer (June, July and August) 1979 and winter (December 1979, January and February) 1980. The cloud effects are represented by differences between the total fluxes including clouds and clear fluxes excluding clouds. The net flux is the downward solar flux minus the reflected solar minus the emitted terrestrial flux.*

parts of the climate system on the radiation balance. Tables 5.3 and 5.4 and Fig. 5.7 compare the results calculated from ISCCP data with coincident ERBE measurements.

The calculated fluxes from ISCCP datasets agree with the direct measurements about as well as they agree with each other; Table 5.4 shows the statistics from scatterplots, like those shown in Fig. 5.7, for both daily mean and monthly mean values.

	Correlation	Regression Slope (ERBE/ISCCP)	Intercept [Wm^{-2}]	RMS Difference [Wm^{-2}]
DAILY STATISTICS				
Reflected Solar	0.06	0.95	0.0	16.6
Reflected Solar - Clr	0.95	0.99	0.0	14.7
Emitted Terrestrial	0.96	1.04	-7.5	9.0
Emitted Terrestrial - Clr	0.99	1.08	-12.1	5.5
MONTHLY STATISTICS				
Reflected Solar	0.99	0.95	0.0	9.2
Reflected Solar - Clr	0.94	0.93	0.0	14.3
Emitted Terrestrial	0.99	1.07	-15.6	4.0
Emitted Terrestrial - Clr	0.99	1.11	-20.1	3.7

Table 5.4: *Comparison of daily and monthly mean fluxes measured by ERBE with values calculated daily and monthly from ISCCP results for July 1985.*

[1]Values taken from ERBE data tapes.

[2]Values taken from Ardanuy et al. (1991); values in parentheses are from Kyle et al. (1990) but for the same months as ERBE data.

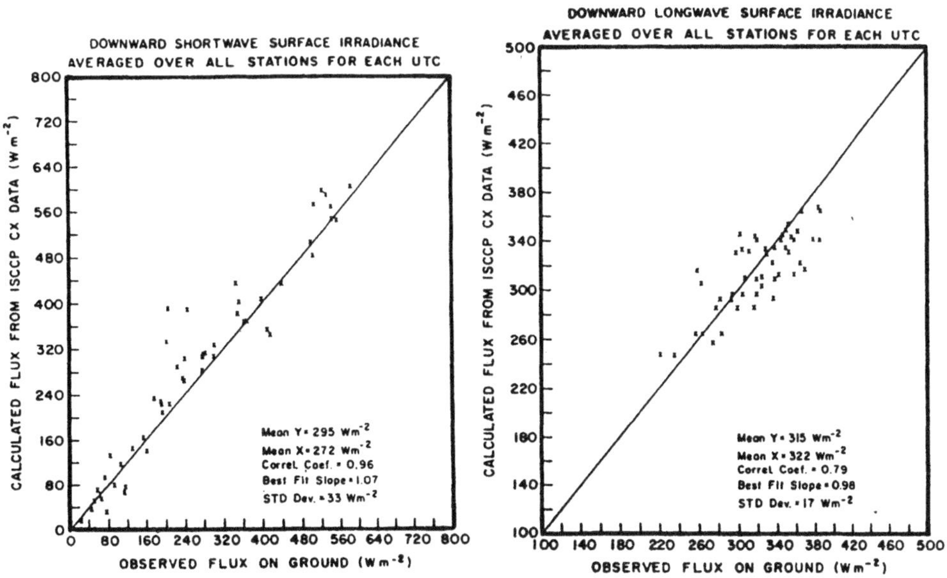

Figure 5.8: *Scatter-plot of ISCCP against surface measured downwelling solar (left) and terrestrial (right) fluxes for the Wisconsin SRB experiment.*

There is not as much data available to check the calculations of the fluxes at the surface using the same ISCCP datasets and radiative transfer model; however, Fig. 5.8 shows a comparison based on an intensive surface observation campaign conducted in support of an International Surface Radiation Budget project. The agreement is better for daily mean values:

1. for shortwave, the regression slope is 1.01 with an RMS difference of $10\,W\,m^{-2}$ and a bias of $4\,W\,m^{-2}$ and

2. for longwave, the regression slope is 0.98 with an RMS difference of $17\,W\,m^{-2}$ and a bias of $6\,W\,m^{-2}$.

5.4 Summary

Two kinds of radiation measurements have not been exploited for Earth observations as yet. One kind is measurement of continuous spectra over large portions of the electromagnetic spectrum. Although such measurements of other planets at infrared wavelengths have become common, only one spectrometer has ever been flown on an Earth- observing satellite, IRIS on NIMBUS-3 and 4 covering part of the terrestrial radiation spectrum from about $6-25\mu m$. Although such measurements are usually used to determine composition, they contain a rich amount of information about clouds, including particle properties and vertical structure (e.g., Carlson et al., 1992). Of particular interest for cloud studies would be infrared and microwave spectral measurements; infrared spectra could also be directly integrated to get the total terrestrial radiation fluxes.

The second kind of un-exploited measurement is of the polarization of reflected sunlight, which has been used to study the atmospheres of other planets (e.g., Hansen and Hovenier, 1974) but not Earth (but see Hansen, 1971). Polarization of microwave radiation is being used to examine the structure of precipitating cloud systems, but more theoretical work is required (Mugnai and Smith, 1988). The polarization of scattered solar radiation is particularly sensitive to the micro-scale properties of the scattering medium; hence it can be used to determine the cloud particle size distribution, shape and phase. It can also be used to obtain another estimate of cloud top pressure. Moreover, this type of measurement is much more sensitive to scattering at low optical thicknesses than measurements of the intensity of the reflected radiation, so that the determination of the properties of optically thin aerosols in Earth's atmosphere would be more accurate. Current plans for new instruments may produce these types of datasets within the next 5-10 years.

5.5 References

Ackerman SA, Smith WL, Spinhirne JD, Revercomb HE (1990) The 27-28 October 1986 FIRE IFO cirrus case study: Spectral properties of cirrus clouds in the $8-12\mu$ window. *Mon Wea Rev* 118:2377-2388

Alishouse JC, Snider JB, Westwater ER, Swift CT, Ruf CS, Snyder SA, Vonjsathorn J, Ferraro RR (1990) Determination of cloud liquid water content using the SSM/I. *IEEE Trans Geosci Remote Sensing* 28:817-822

Ardanuy PE, Stowe LL, Gruber A, Weiss M (1991) Shortwave, longwave and net- cloud-radiative forcing as determined from NIMBUS-7 observations. *J Geophys Res* 96:(in press)

Arking A, Childs JD (1985) Retrieval of cloud cover parameters from multispectral satellite measurements. *J Clim Appl Meteor* 24:322-333

Asano S, Sato M (1980) Light scattering by randomly oriented speroidal particles. *Appl Opt* 19:962-974

Barkstrom BR, Smith GL (1986) The Earth Radiation Budget Experiment: Science and implementation. *Rev Geophys* 24:379-390

Brooks DR, Harrison EF, Minnis P, Suttles JT, Kandel RS (1986) Development of algorithms for understanding the temporal and spatial variability of the Earth's radiation balance. *Rev Geophys* 24:422-438

Carlson BE, Lacis AA, Rossow WB (1992) Tropospheric gas composition and cloud structure of the Jovian North Equatorial Belt. *J Geophys Res* :(in press)

Curry JA, Ardeel CD, Tian L (1990) Liquid water content and precipitation characteristics of stratiform clouds as inferred from satellite microwave measurements. *J Geophys Res* 95:16, 659-16, 671

Gruber A (1977) Determination of the earth-atmosphere radiation budget from NOAA satellite data. *NOAA Tech Rep*, NESS 76. 28 pp.

Hansen JE (1971) Multiple scattering of polarized light in planetary atmospheres. II. Sunlight reflected by terrestrial water clouds. *J Atmos Sci* 28:1400-1426

Hansen JE, Hovenier JW (1974) Interpretation of the polarization of Venus. *J Atmos Sci* 31:1137-1160

Hansen JE, Travis LD (1974) Light scattering in planetary atmospheres. *Space Sci Rev* 16:527-610

Kyle HL, Mecherikunnel A, Ardanuy P, Penn L, Groveman B, Campbell GG, Vonder Haar TH (1990) A comparison of two major Earth radiation budget datasets. *J Geophys Res* 95:9951-9970

Lovejoy S, Schertzer D (1990) Multifractals, universality classes and satellite and radar measurements of cloud and rain fields. *J Geophys Res* 95:2021-2034

Minnis P, Harrison EF (1984) Diurnal variability of regional cloud and clear sky radiative parameters derived from GOES data. Part I: Analysis Method. *J Clim Appl Meteor* 23:993-1011

Minnis P, Young DF, Sassen K, Alvarez JM, Grund CJ (1990) The 27-28 October 1986 FIRE IFO cirrus case study: Cirrus parameter relationships derived from satellite and lidar data. *Mon Wea Rev* 118:2402-2425

Mugnai A, Smith EA (1988) Radiative transfer to space through a precipitating cloud at multiple microwave frequencies, Part I: Model Description. *J Appl Meteor* 27:1055-1073

Mokhov II, Shlesinger ME (1992) Analysis of global cloudiness 1. Comparison of Meteor, Nimbus-7 and ISCCP satellite data. Submitted

Nakajima T, King MK (1990) Determination of the optical thickness and effective particle radius of clouds from reflected solar radiation measurements. Part I: Theory. *J Atmos Sci* 47:1878-1893

Njoku EG (1982) Passive microwave remote sensing of the earth from space - A review. *IEEE Proc* 70:728-750

Raschke E, Vonder Haar TH, Bandeen WR, Pasternak M (1973) The annual radiation balance of the Earth-atmosphere system during 1967-70 from NIMBUS-3 measurements. *J Atmos Sci* 30:341-364

Rossow WB (1989) Measuring cloud properties from space: A review. *J Climate* 2:201-213

Rossow WB, Garder LC, Lacis LC (1989) Global, seasonal cloud variations from satellite radiance measurements. Part I: Sensitivity of analysis. *J Climate* 2:423-462

Rossow WB, Lacis AA (1990) Global, seasonal cloud variations from satellite radiance measurements. Part II: Cloud properties and radiative effects. *J Climate* 3:1204-1253

Rossow WB, Schiffer RA (1991) ISCCP cloud data products. *Bull Amer Meteor Soc* 72:2-20

Stowe LL, Wellemeyer CG, Eck TF, Yeh HYM, The NIMBUS-7 cloud data processing team (1988) NIMBUS-7 global cloud climatology. Part I: Algorithms and validation. *J Climate* 1:445-470

Stowe LL, Yeh HTM, Eck TF, Wellemeyer CG, Kyle HL, the NIMBUS-7 cloud data processing team (1989) NIMBUS-7 global cloud climatology. Part II: First year results. *J Climate* 2:671-709

Takano Y, Liou K-N (1989) Solar radiation transfer in cirrus clouds, I. Single-scattering and optical properties of hexagonal ice crystals. *J Atmos Sci* 46:3-19

Warren SG, Hahn CJ, London J, Chervin RM, Jenne RL (1986) Global distribution of total cloud and cloud type amounts over land. *NCAR Tech Note* TN-273+STR/DOE Tech. Rep. ER/60085-H1. 29 pp. + 200 maps

Warren SG, Hahn CJ, London J, Chervin RM, Jenne RL (1988) Global distribution of total cloud and cloud type amounts over the ocean. *NCAR Tech Note* TN-317+STR/DOE Tech. Rep. ER-0406. 42 pp. + 170 maps

Wetzel MA, Vonder Haar TH (1991) Theoretical development and sensitivity tests of a stratus cloud droplet size retrieval method for AVHRR-K/L/M. *Remote Sens Environ* 36:105-119

Wielicki BA, Parker L (1992) On the determination of cloud cover from satellite sensors: The effect of sensor spatial resolution. *J Geophys Res* :(in press)

Wielicki BA, Suttles JT, Heymsfield AJ, Welch RM, Spinhirne JD, Wu M-LC, Starr DO, Parker L, Arduini RF (1990) The 27-28 October 1986 FIRE IFO cirrus case study: Comparison of radiative transfer theory with observations by satellite and aircraft. *Mon Wea Rev* 118:2356-2376

Wylie DP, Menzel WP (1989) Two years of cloud cover statistics using VAS. *J Climate* 2:380-392

5.6 Additional Reading

— Radiation

Davies R (1978) The effect of finite geometry on the three-dimensional transfer of solar irradiance in clouds. *J Atmos Sci* 35:1712-1725

Goody RM, Yung YL (1991) Atmospheric Radiation, Theoretical Basis. Oxford, Oxford Univ.

Hansen JE, Travis LD (1974) Light scattering in planetary atmospheres. *Space Sci Rev* 16:527-610

- Clouds and Climate

Hartmann DL, Short DA (1980) On the use of earth radiation budget statistics for studies of clouds and climate. *J Atmos Sci* 39:431-439

Hartmann DL, Ramanathan V, Berroir A, Hunt GE (1986) Earth radiation budget data and climate research. *Rev Geophys* 24:439-468

Pruppacher HR, Klett JD (1978) Microphysics of Clouds and Precipitations. *D. Reidel*, Dorderecht

Rogers RR (1976) A Short Course in Cloud Physics. *Pergamon Press*, Oxford

Rossow WB, Lacis AA (1990c) Global, seasonal cloud variations from satellite radiance measurements. Part II: Cloud properties and radiative effects. *J Climate* 3:1204-1253

Stephens GL, Webster PJ (1981) Clouds and climate: Sensitivity of simple systems. *J Atmos Sci* 38:235-247

Wang W-C, Rossow WB, Yao M-S, Wolfson M (1981) Climate sensitivity of a one-dimensional radiative-convective model with cloud feedback. *J Atmos Sci* 38:1167-1178

– **Cloud Climatologies**

Hughes (1984) Global cloud climatologies: A historical review. *J Climate Appl Meteor* 23:724-751

Rossow WB, Schiffer RA (1991) ISCCP cloud data products. *Bull Amer Meteor Soc* 72:2-20

Stowe LL, Yeh HTM, Eck TF, Wellemeyer CG, Kyle HL, the NIMBUS-7 Cloud Data Processing Team (1989) NIMBUS-7 global cloud climatology. Part II: First year results. *J Climate* 2:671-709

Warren SG, Hahn CJ, London J, Chervin RM, Jenne RL (1986) Global distribution of total cloud and cloud type amounts over land. *NCAR Tech Note* TN-273+STR/DOE Tech. Rep. ER/60085-H1. 29 pp. + 200 maps

Warren SG, Hahn CJ, London J, Chervin RM, Jenne RL (1988) Global distribution of total cloud and cloud type amounts over the ocean. *NCAR Tech Note* TN-317+STR/DOE Tech. Rep. ER-0406. 42 pp. + 170 maps

– **Cloud Physics**

Hobbs PV (1974) Ice Physics. *Oxford*, Clarendon

Mason BD (1987) Introduction to the METEOSAT operational system. ESA BR-32. 42 pp.

Rossow WB (1978b) Cloud microphysics: Analysis of the clouds of Earth, Venus, Mars, and Jupiter. *Icarus* 36:1-50

Twomey S (1977) Atmospheric Aerosols. Amsterdam, *Elsevier Scientific*

Chapter 6

SATELLITE OBSERVATIONS OF RADIATION AND CLOUDS TO DIAGNOSE ENERGY EXCHANGES IN THE CLIMATE: PART II

William B. Rossow
NASA
Goddard Institute for Space Studies
New York, NY 10025
USA

6.1 Cloud properties

Since the early part of this century, we have had a fairly accurate idea of the variations of cloud cover among different climate zones, except for the more remote portions of the tropical and southern oceans and polar regions (Hughes 1984). Standardization of cloud classification for surface weather observers after the Second World War has also provided a long record of changing cloud morphological types (Warren et al. 1986, 1988). However, quantitative information on the physical properties of clouds has been lacking, particularly for those properties that determine their influence on radiative energy exchanges. Recent analyses of satellite datasets , particularly by the International Satellite Cloud Climatology Project (ISCCP), are providing the first global surveys of some of these cloud properties (Rossow and Schiffer 1991). The results presented below come from ISCCP.

The average cloudiness of Earth is over 60%. Although earlier estimates gave a value slightly larger than 50%; the higher value is obtained when a proper sample of the southern oceans is included and when thinner cirrus clouds are included. (The precise value of "cloud" cover depends directly on how small the optical thickness cut-off value is: a value of 60 – 65% is associated with a cut-off of about

NATO ASI Series, Vol. I 5
Energy and Water Cycles in the Climate System
Edited by E. Raschke and D. Jacob

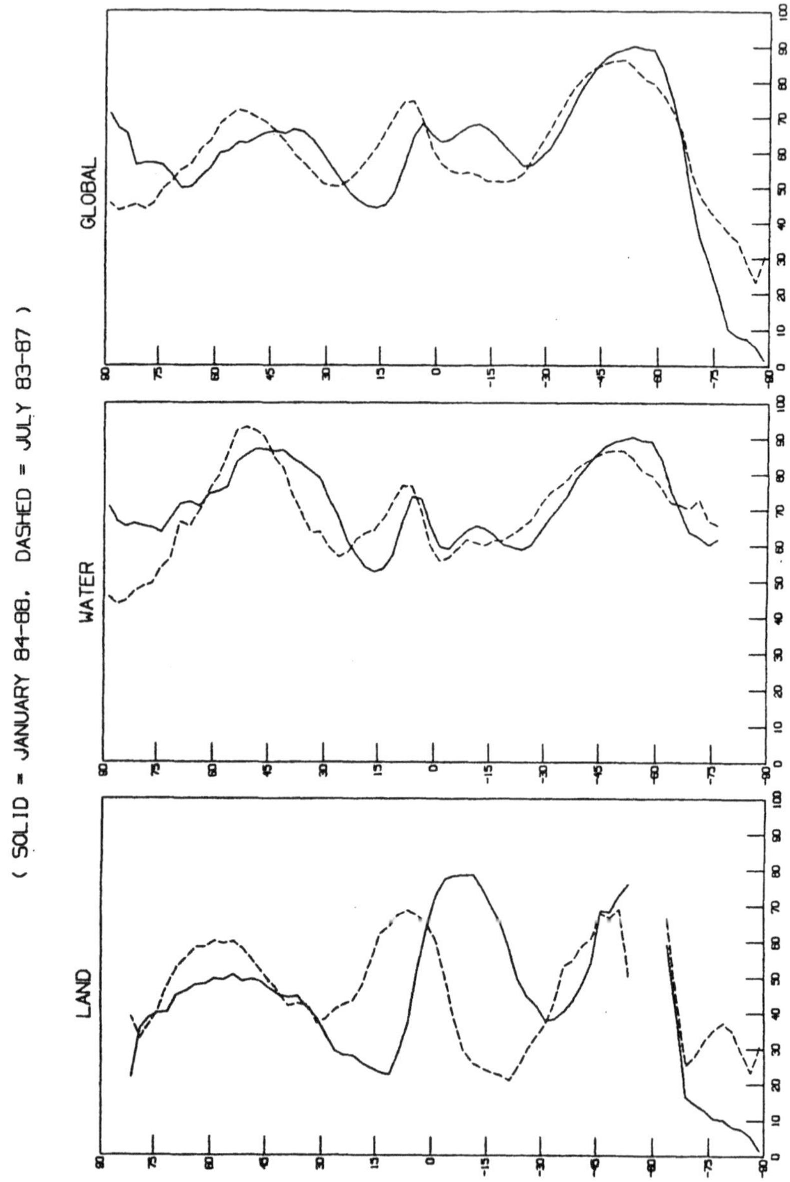

Figure 6.1: *Zonal, annual mean cloud amounts (%) from five years of ISCCP data for January and July.*

0.2. Clouds with smaller optical thicknesses would only alter radiative fluxes by a few percent. If the cut-off were reduced to about 0.1 or slightly less, global cloud cover would be nearly 100% at all times.) Figure 6.1 shows that the latitude zones containing stronger atmospheric motions ("storms") are generally cloudier (60 – 80%) than the more quiescent zones (40 – 50%). The larger scale storm systems in middle latitudes produce somewhat more extensive cloud cover than the smaller scale tropical systems. Drier land areas are generally less cloudy than the oceans at all latitudes, except in the Intertropical Convergence Zone (ITCZ). There are also large ocean "deserts" with little cloud cover. The sea-ice covered polar oceans are very cloudy, but the high altitude ice sheets are more like deserts. Variations in cloud cover with longitude are just as large as they are with latitude (Fig. 6.2).

The distributions of cloud optical thicknesses and cloud top pressures exhibit large variations that also coincide with the classical climate regimes (Fig. 5.1 and 5.2 in Part I). The "storm" zones have optically thicker clouds and more cloud tops at lower pressures (higher altitudes), while the quiescent zones have thinner clouds and fewer clouds in the upper troposphere. Clouds over oceans tend to be optically thinner and to occur more frequently at lower levels than over land (the latter effect is partly due to the higher topography of land areas and to a thicker and drier planetary boundary layer). This difference is also associated with a difference in the average cloud particle sizes for low-level clouds: those over oceans tend to have larger particles than over land (Wallace and Hobbs 1977) and to form precipitation more easily, which may explain their lower optical thicknesses (Tselioudis et al. 1992).

We can consider combinations of cloud radiative properties to represent different cloud types and examine how the relative frequency of occurrence of these cloud types varies with climate regime. One classification can be defined by cloud top pressure and optical thickness (Fig. 6.3(a)); these two properties, together with the frequency of occurrence, determine the primary effect of clouds on the solar and terrestrial radiative fluxes. The quiescent climate regimes over oceans are dominated by a single cloud type that is low-level and optically thin (Fig. 6.3(b)); over land, some thin, low-level clouds occur, but most clouds are thin, high-level cirrus. Midlatitudes have two common cloud types (Fig. 6.3(c)): one with middle to upper level cloud tops and large optical thicknesses, corresponding to storms, and one low-level, optically thin type corresponding to fair weather. Tropical latitudes have three cloud types (Fig. 6.3(d)): low-level, optically thin clouds associated with fair weather and two types associated with convective complexes. The convective "towers" have very high optical thicknesses and very high tops; the mesoscale anvils have moderate optical thicknesses and somewhat lower tops. Cirrus clouds are at similar altitudes to the convective clouds in the tropics but are optically thin. The latter cloud type can also occur alone in both the tropics and midlatitudes.

6.2 Seasonal variations

Low and high latitudes exhibit completely different seasonal variations in cloud properties. At low latitudes seasonal temperature variations are small and the seasons are marked by large changes in precipitation. The "summer" season is the wet season with large storms; the cloud cover becomes much more extensive and the cloud types associated with convective complexes occur much more

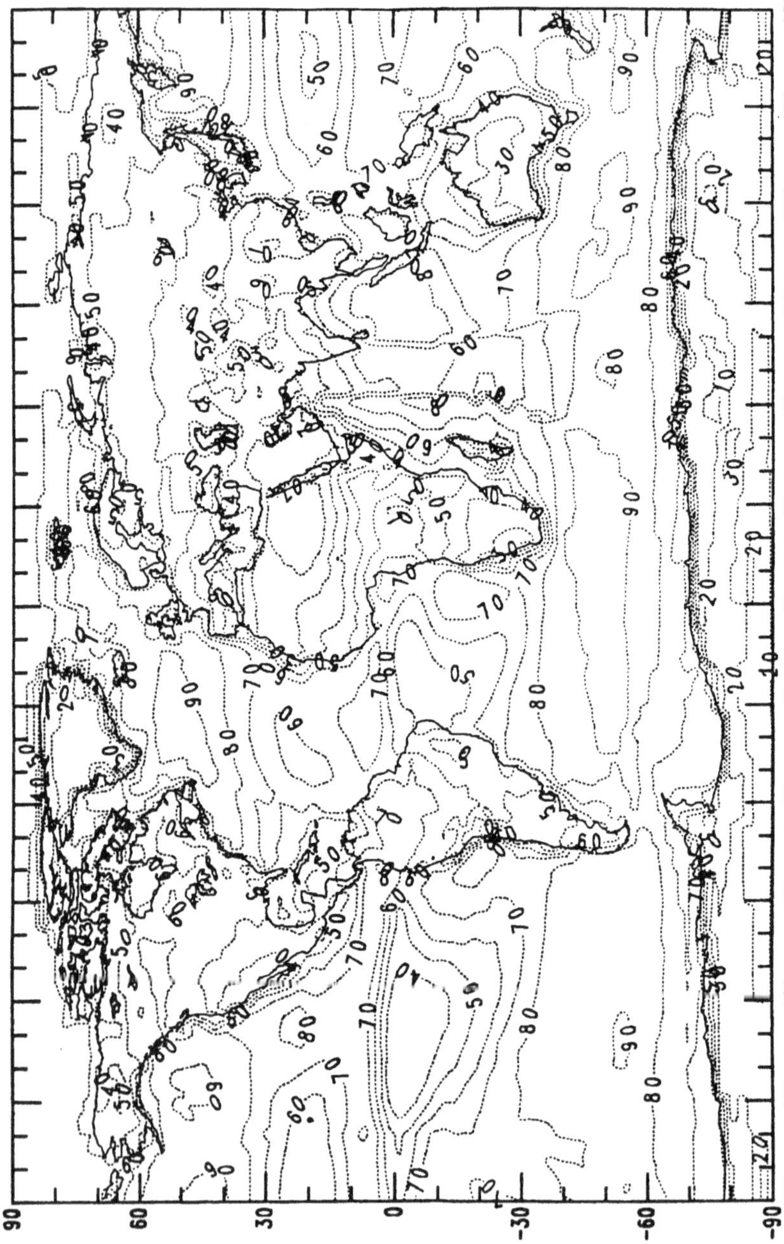

Figure 6.2: *Annual mean cloud amount (%) from five years of ISCCP results.*

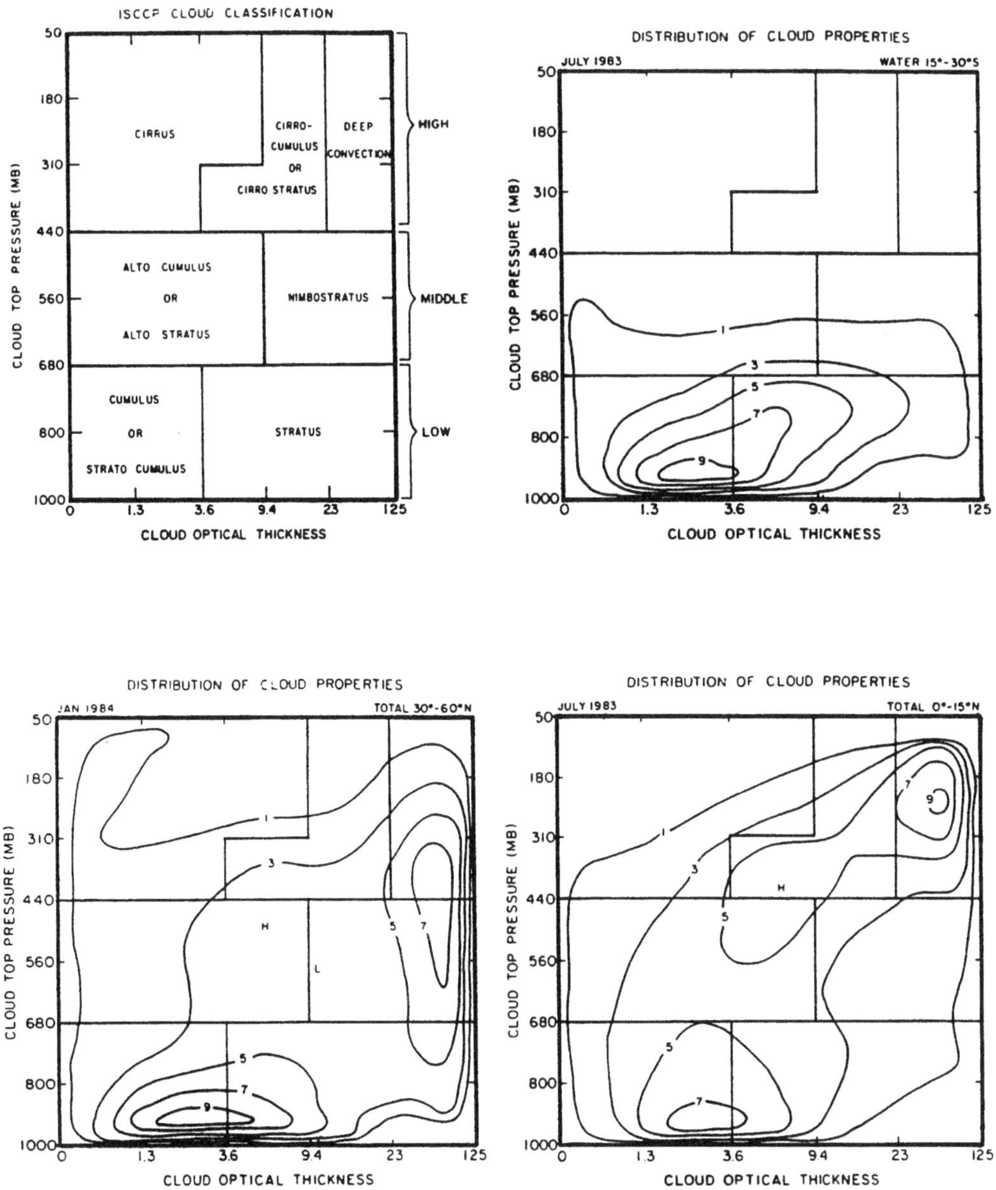

Figure 6.3: *Classification and frequencies of occurrence of combinations of cloud optical thickness and top pressure for different latitude zones from ISCCP.*

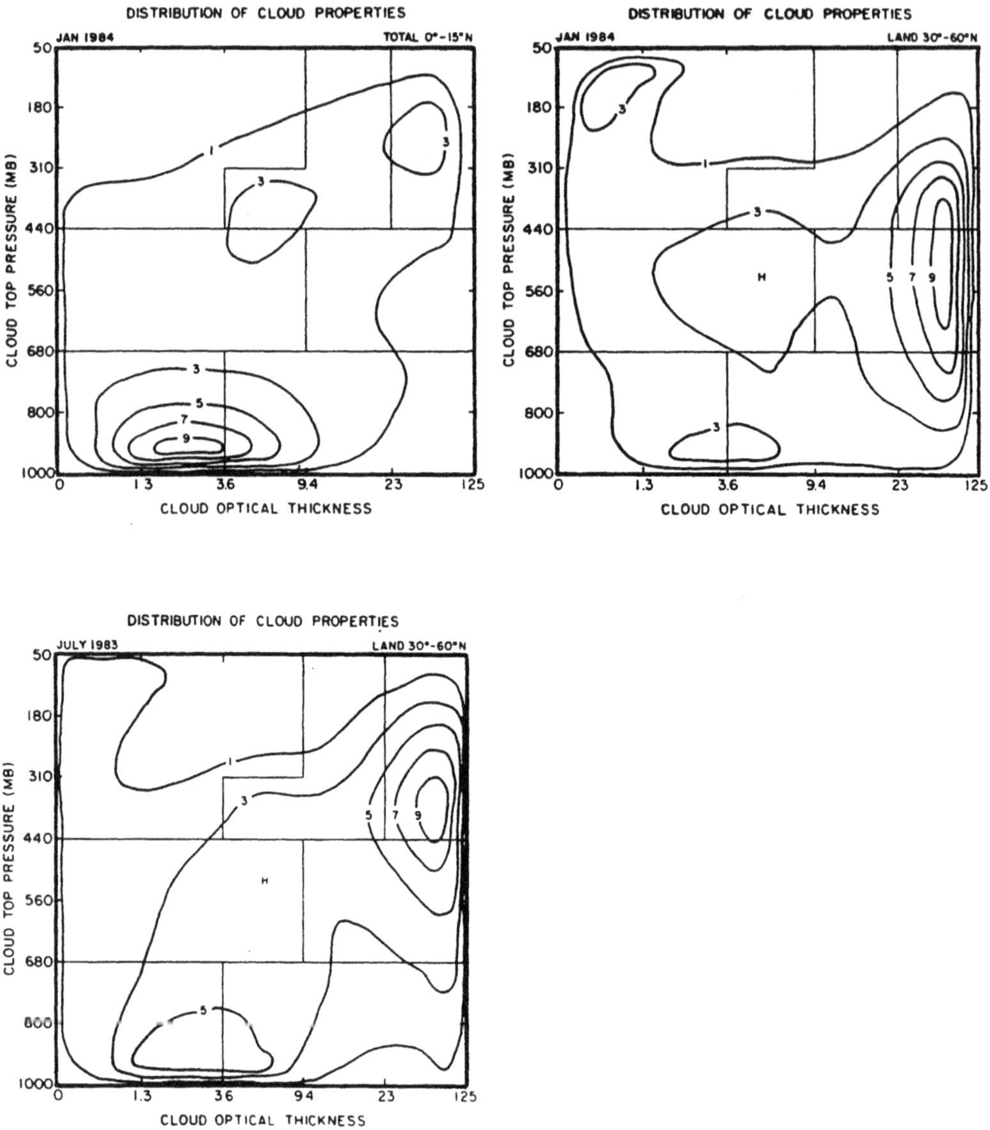

Figure 6.4: *Frequency of occurrence of different combinations of cloud optical thickness and top pressure from ISCCP.*

frequently (Fig. 6.3(d)). ITCZ cloudiness over land shifts latitude dramatically with season, generally occurring near the sub-solar latitude. Over ocean, on the other hand, there is usually evidence of a double structure (Fig. 6.1) that does not move, one part in each hemisphere; the summer part is much more active as evidenced by more frequent convective complexes and generally larger cloud cover. The dry (winter) season is characterized by more low-level, optically thin (fair weather) cloudiness (Fig. 6.4(a)).

At higher latitudes, the winter (colder) and summer (hotter) seasons differ in the relative frequency of occurrence of fair-weather and storm cloud types and in the average properties of the storm clouds. Winter brings more storm clouds (Fig. 6.4(b)). The winter storm clouds over oceans are generally optically thicker with higher cloud tops, (contrast Fig. 6.3(c) and 6.4(b)); over land cloud tops are higher in summer (Fig. 6.4(c)), but average optical thicknesses are larger in winter. Summer also brings somewhat more frequent fair-weather conditions, particularly over land where more cirrus clouds occur (Fig. 6.4(c)).

Although the seasonal temperature variations at low latitudes are quite small, the cloudiness changes are dramatic; whereas, the seasonal temperature changes at middle latitudes are very large, but the changes in cloud properties are more modest. The seasonal variations of cloud properties in these latitude zones are also opposite in sign: summer cloud amounts, optical thicknesses and top heights are larger in the tropics, while they are smaller in middle latitudes, as compared with winter values. In all cases, seasonal variations over land tend to be larger than over ocean. These facts all show that there are not simple relationships between cloud properties and surface climate, usually represented by temperature.

6.3 Diurnal variations

A similar variety of behavior is apparent in the style of daily cloud variations over land and ocean and at lower and higher latitudes. Higher latitude ocean clouds exhibit no significant diurnal variations. Winter clouds over land also show little variation; however, summer cloudiness has larger cloud cover and top heights, but lower optical thicknesses, in the afternoon (Fig. 6.5(a)). Subtropical latitudes show a weak tendency for increased cloudiness in the afternoon over land, but over oceans the maximum cloudiness occurs near dawn (Fig. 6.5(b)). The amplitude of this oceanic cycle is significant despite the very small change in surface temperature it is associated with. In the tropics, a variety of diurnal cycles over land and ocean, depending on the predominant cloud type and local meteorological conditions, combine to produce a more complicated cycle (Fig. 6.5(c)).

6.4 Why do clouds vary?

The names given to the radiative cloud types in Fig. 6.3(a) were selected to correspond, approximately, to the classical morphological cloud types identified by weather observers. A comparison of the frequencies of occurrence of these types shows that this correspondence, while not exact, is very good (Fig. 6.6). A relation between "dynamical" and "radiometric" cloud types is not unexpected, however, since it is the atmospheric motions that cause clouds: the more vigorous these motions are, the more cloud mass is expected to be produced. Going from the "fair-weather" cloud types (Fig. 6.6(b) and 6.6(c)) to the "stormy-weather" cloud types (Fig. 6.6(a) and 6.6(d)) corresponds to an increase in cloud top height and optical thickness (or water content). Even the cirrus amounts are similar (Fig. 6.6(e)).

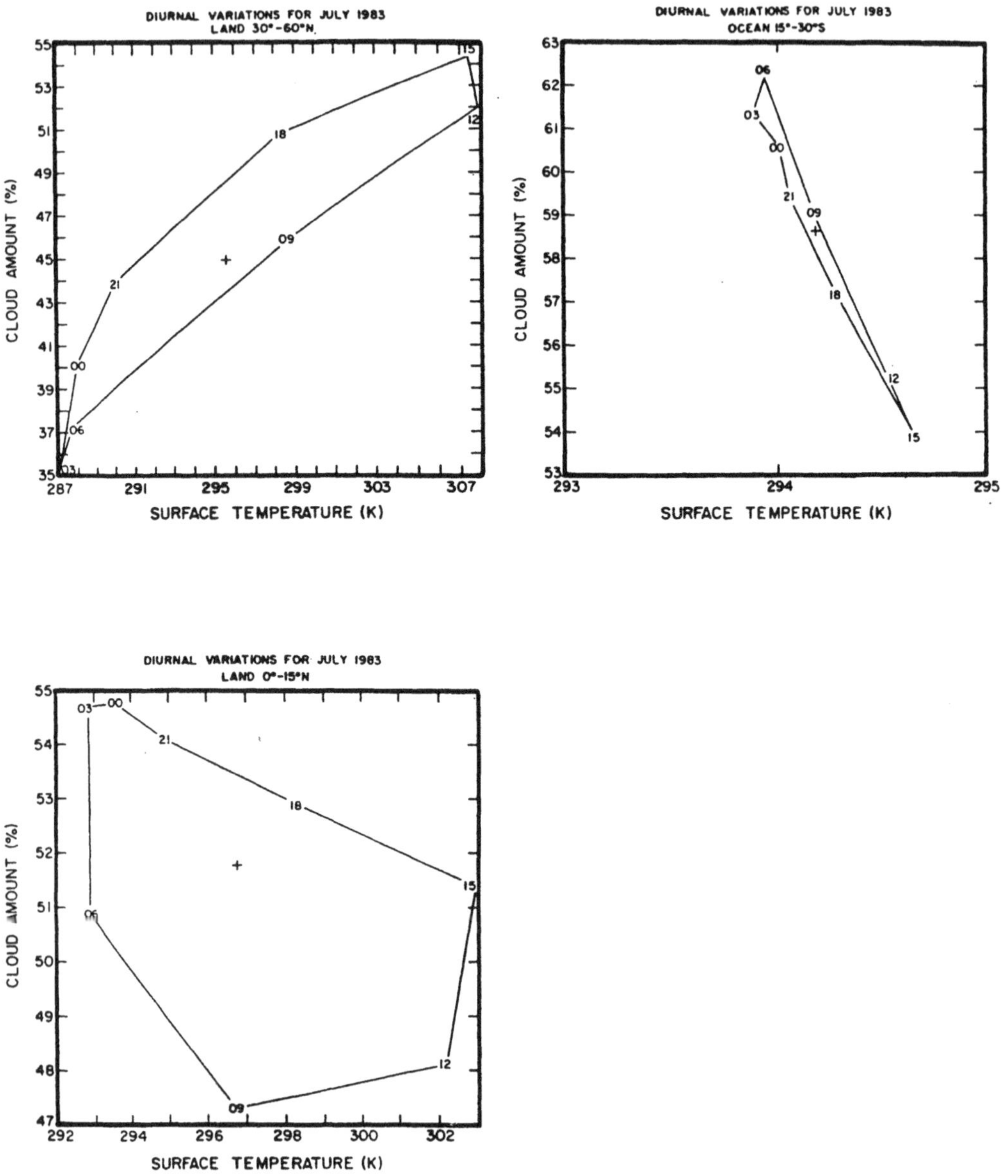

Figure 6.5: *Zonal mean cloud amounts (%) versus mean surface temperatures as a function of local time of day (0-24 hr) from ISCCP.*

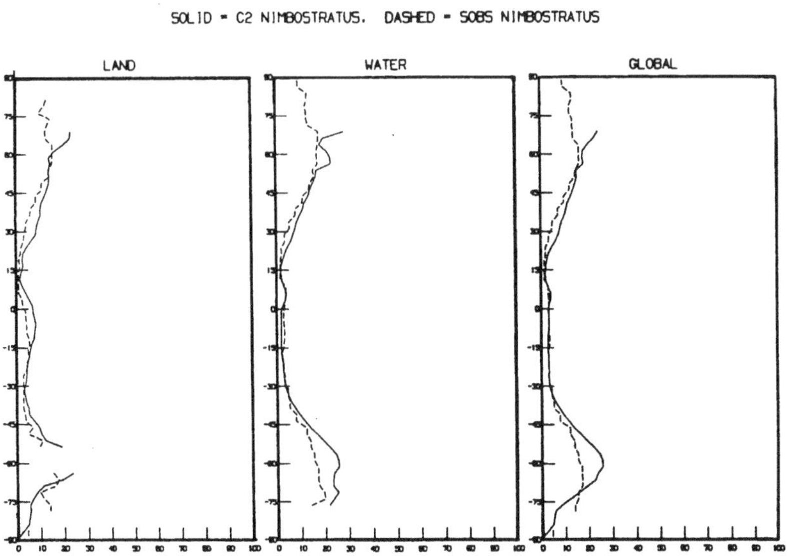

Figure 6.6: *(a) Zonal mean amount (%) of nimbostratus cloud in NH winter from surface observations and ISCCP.*

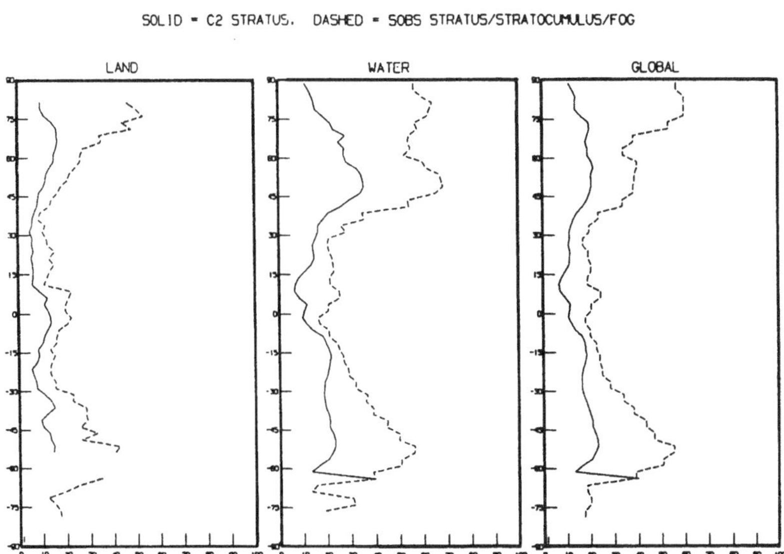

Figure 6.6: *(b) Zonal mean amount (%) of stratus cloud in NH summer from surface observations and ISCCP.*

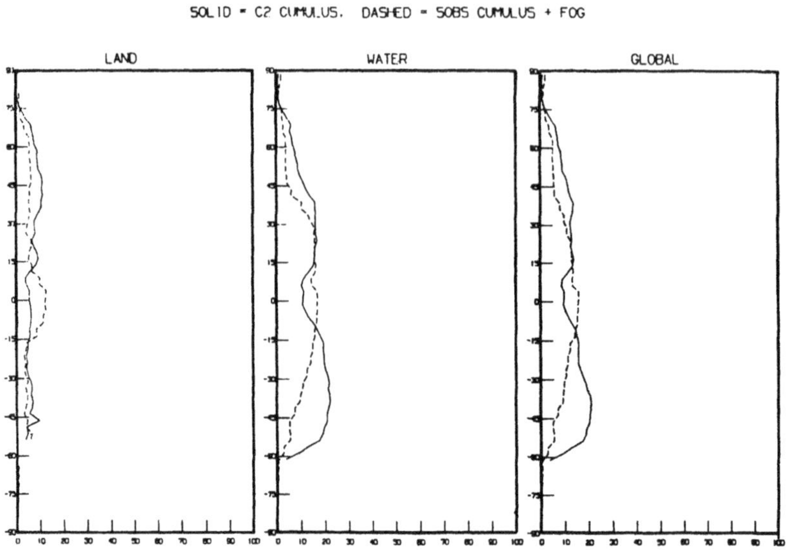

Figure 6.6: *(c) Zonal mean amount (%) of cumulus cloud in NH summer from surface observations and ISCCP.*

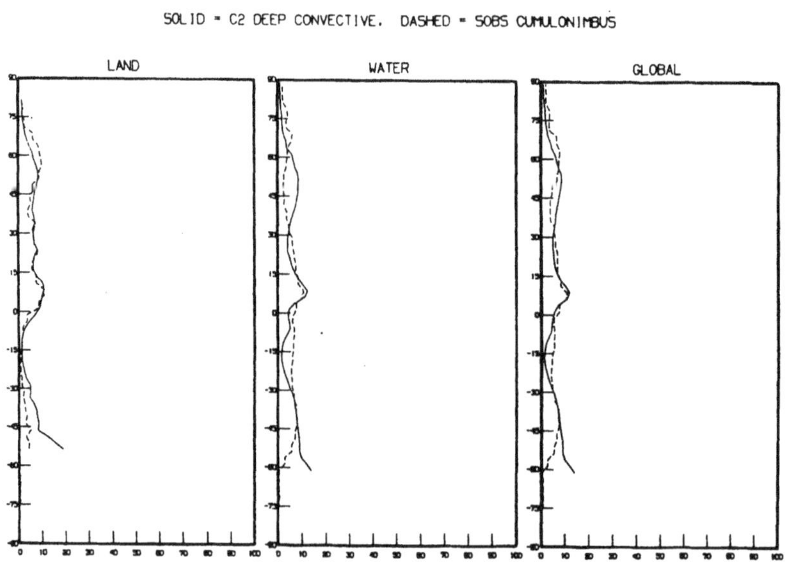

Figure 6.6: *(d) Zonal mean amount (%) of deep convective cloud in NH summer from surface observations and ISCCP.*

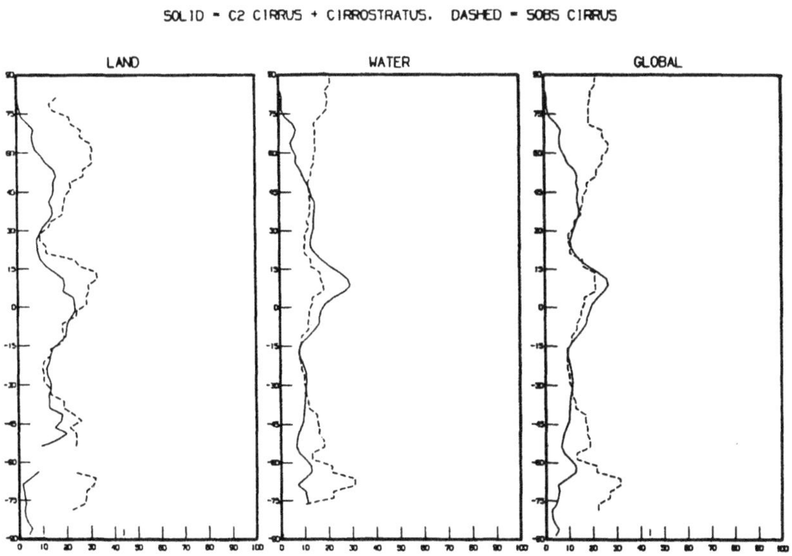

Figure 6.6: *(e) Zonal mean amount (%) of Cirrus cloud in NH autumn from surface observations and ISCCP.*

The results of several preliminary studies help elucidate this connection between cloud properties and atmospheric motions, but much more work is needed to quantify this physical relationship so that it can be used as the basis of improved cloud parameterizations in climate models. One such relationship is that between the amount of water formed in the cloud and the temperature at which the cloud forms. Since the condensation rate increases with temperature, it has been proposed that cloud water amount also increases (Somerville and Remer 1984; Betts and Harshvardhan 1984). However, this behavior also depends on how the precipitation rate changes with temperature, since it is the balance between condensation and precipitation that determines the cloud water amount. The observed behavior of the cloud optical thickness of low-level clouds (Fig. 6.7), which is proportional to water content when the vertical extent cannot change too much, changes sign between temperatures colder than about $-10\,^\circ C$ and warmer temperatures. At the colder temperatures, which occur mostly over land, (and where most of the available aircraft observations were collected), the cloud water amount tends to increase with increasing temperature; but at warmer temperatures, which compromise a majority of the Earth, cloud water amount decreases with increasing temperature. A possible explanation is that precipitation efficiency is higher at warmer temperatures (Tselioudis et al. 1992). This different perspective obtained from a global dataset illustrates one major advantage of satellite observations.

This explanation for the behavior in Fig. 6.7 is also consistent with the facts that the average cloud particle size and the temperature at which the sign of the cloud water - temperature relationship changes differ between land and ocean: the generally larger cloud particle sizes over ocean (Fig. 5.3 in Part I) are associated with a colder transition temperature as if precipitation is more efficient

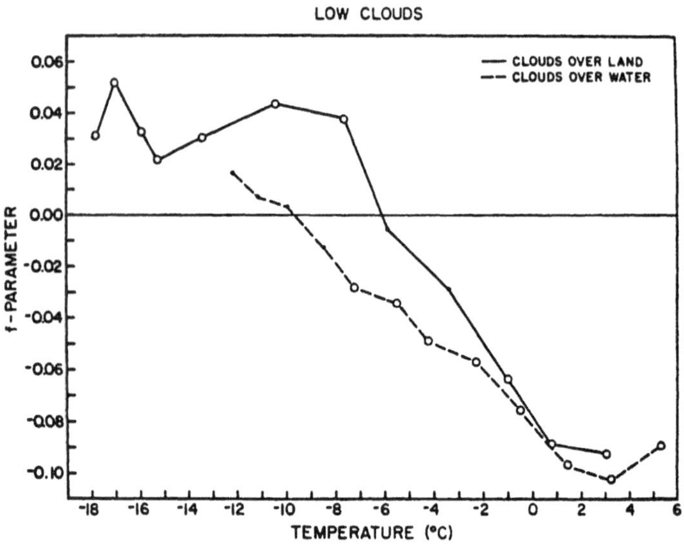

Figure 6.7: *Zonal mean values of $f = (1/\tau)(d\tau/dT)$ for low clouds over land and ocean from 1984 ISCCP results (after Tselioudis et al. 1991).*

in ocean clouds. Further study of the details of this relationship may help identify the key factors governing the formation of clouds and precipitation.

The location and structure of tropical convective complexes are the most highly variable cloud features on Earth. It has been proposed that sea surface temperature (SST) controls the spatial distribution of tropical convection based on observed variations with season and during El Niño events (e.g., Graham and Barnett 1987). Thus, a tight feedback between tropical SST and convection has been postulated, which also implies that the energy and water transports by convection, as well as the radiative flux changes associated with the convective cloud formations, are closely related to SST. A more detailed examination of the data (e.g., Fu et al. 1990) shows that the correlation between changes in SST and convective cloudiness is not actually very simple: Fig. 6.8 shows that the seasonal variations do not occur at the same time, that some areas with very warm surface waters can be free of convection and that some areas with cool surface waters are covered by convection. These results (Fu et al. 1990) show that this linkage cannot be considered as a simple one-dimensional relationship, but rather must be considered as three-dimensional (ie, dynamical), since the vertical structure of the atmosphere above the boundary layer as well as the character of the atmosphere in the boundary layer play equally important roles in determining the location of convective systems.

There is also some evidence that the sizes of convective complexes are related to the average properties of the clouds (Machado et al. 1992). As the systems grow larger, the cloud top heights increase and the average optical thicknesses increase (Fig. 6.9). This might be explained by all of these properties being determined by the magnitude of the updraft velocities in the convective towers. This supposition may also be supported by differences in lightning activity between ocean and land systems (Price and Rind 1991).

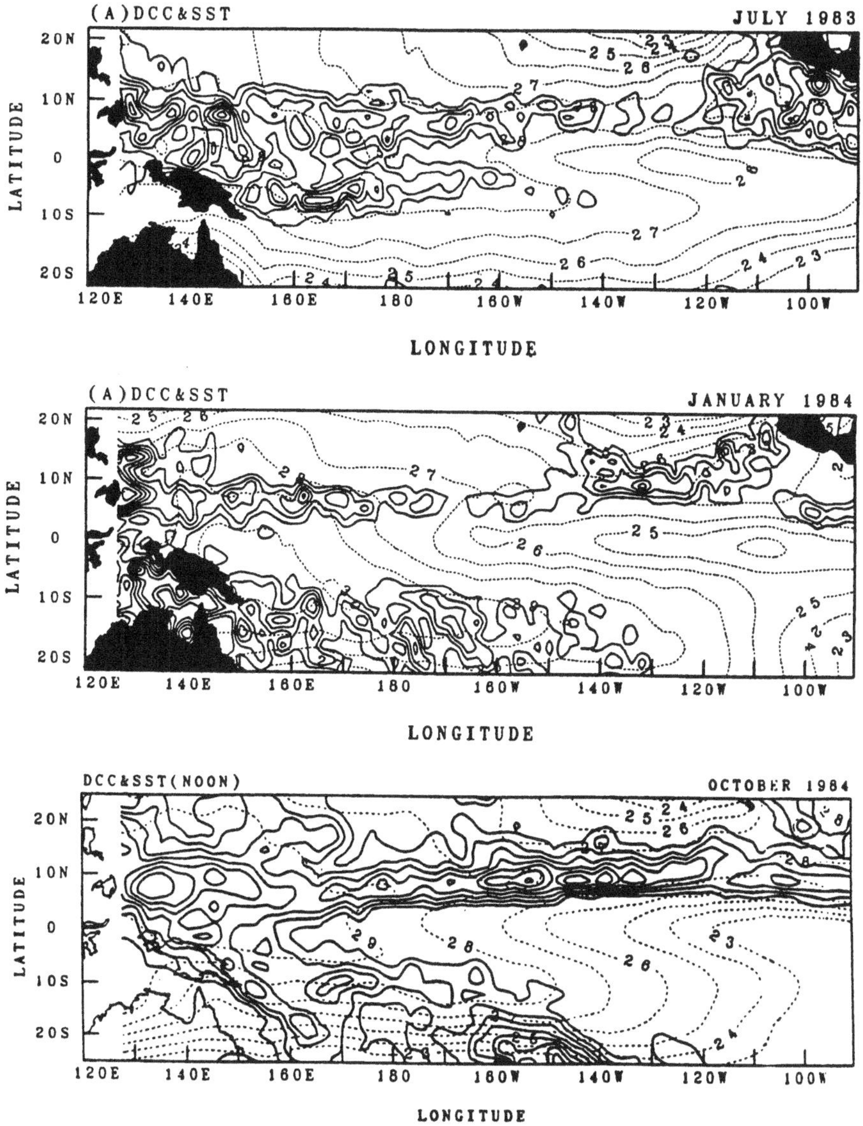

Figure 6.8: *Distribution of intense convection and sea surface temperatures in the tropical Pacific for three months (after Fu et al. 1990).*

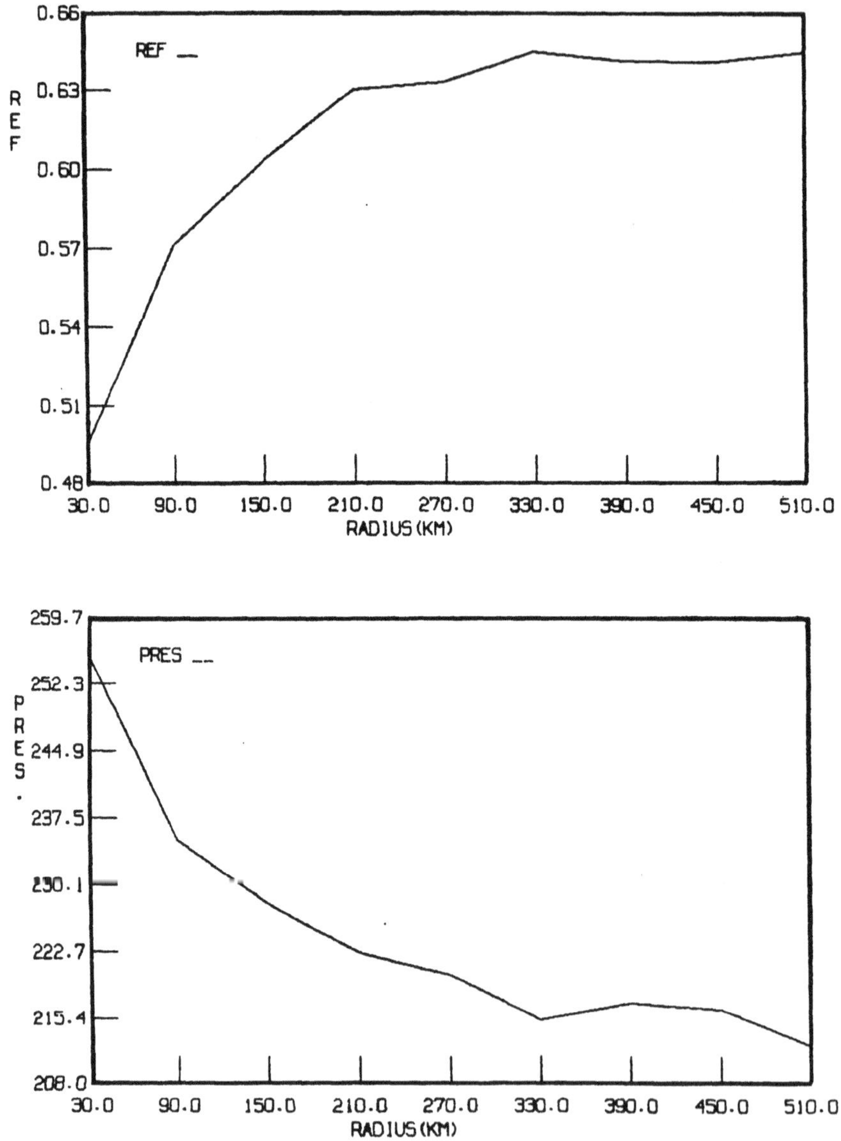

Figure 6.9: *Variation of average cloud reflectivity and top pressure with radius of the high (Temperature < 245 K) cloud clusters (based on Machado and Rossow 1992).*

6.5 Cloud effects on radiative flux variations

The basic radiation balance (Tab. 6.1) shows that over 70% of the solar flux absorbed by Earth is absorbed at the surface; however, the opacity of the atmosphere to terrestrial radiation allows little direct cooling of the surface by emitted radiation. This is the "greenhouse effect" which is due primarily to water vapor in the atmosphere. The net radiative heating of the surface is balanced primarily by the latent cooling associated with evaporation of water from the oceans (there is also some direct transfer of heat from the surface by turbulent atmospheric motions, particularly over land). The atmosphere is heated by the formation of precipitation, closing the water cycle, and emits this extra heat as terrestrial radiation. Thus, the net radiative balance of the atmosphere is a cooling.

The addition of clouds to a planet with the same surface and atmospheric properties as Earth (with clouds) causes a decrease in the amount of solar radiation absorbed by the planet and a decrease in the amount of terrestrial radiation emitted by the planet (Tab. 6.1). Since the former change is twice a large as the latter, there is a net cooling effect (Harrison et al. 1990). If the imbalance caused by sudden removal of all clouds were allowed to continue, the surface temperature would rise. The increasing upward flux of terrestrial radiation from the surface would be absorbed by the atmosphere causing its temperature to rise. This purely radiative effect would probably be aided by increasing latent heat fluxes at higher temperatures. Eventually, as the surface and atmosphere approach a new, higher temperature equilibrium, the downward terrestrial flux from the atmosphere would nearly cancel the upward flux so that most of the surface cooling would be by latent (and sensible) heat transfer. Thus, the nature of the changes of energy exchanges in response to a perturbation away from equilibrium depends on the time scale considered: the most rapid response to changing surface temperature is caused by changes of the terrestrial radiation, but the final (new) balance would be determined by changes in latent and sensible cooling. In this sense, a rapid change in cloudiness results in a change of solar heating at the surface and a change of terrestrial heating of the atmosphere (cf., Stephens and Webster 1981).

Atmospheric and oceanic motions are driven by horizontal and vertical gradients in the heating. The Earth's spherical shape produces the primary contrast of solar heating between the equator and poles, but atmospheric motions transport some heat so that the net flux at the top of the atmosphere has the pattern shown in Fig. 6.10(a). Clouds alter the average latitudinal gradients of the radiative fluxes (Fig. 6.10(b)), serving to decrease the forcing for the general circulation in the tropics and near the poles and to increase it in midlatitudes.

Clouds also alter the vertical distribution of heating/cooling, because of their radiative effects and because of the latent heat exchanges associated with their formation (e.g., Stephens and Webster 1981). The nature of this effect depends on the amount water vapor in the atmosphere. However, we lack detailed enough information about cloud vertical structures to assess this effect quantitatively.

Small systematic variations of clouds over a diurnal cycle alter the radiative forcing that would obtain. The atmospheric response is more complicated than a simple local change in atmospheric turbulence. Over lower latitude oceans, since the surface temperature does not respond much to increased solar heating, it is the small change in atmospheric heating caused by the presence of clouds which controls

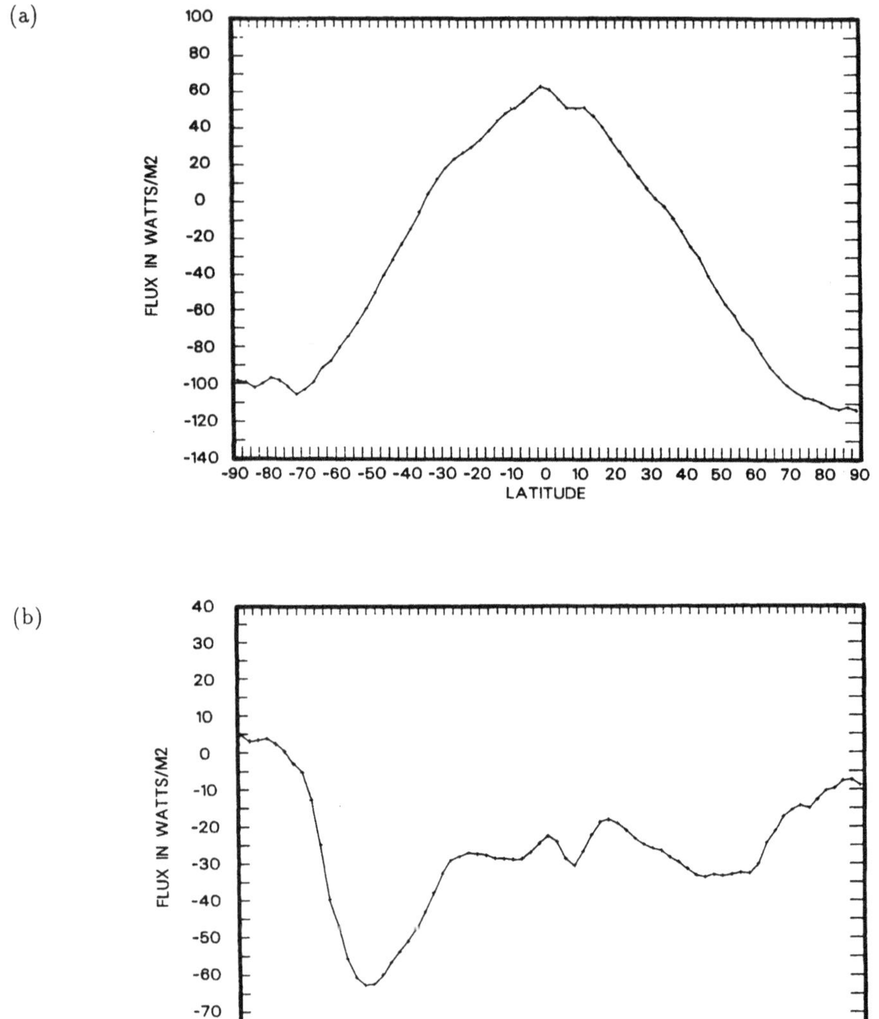

Figure 6.10: *(a) Zonal, annual mean net radiative flux at the top of the atmosphere, calculated from ISCCP datasets, (b) Net changes in the zonal, annual net radiative flux, shown in (a), caused by adding partial cloud cover.*

Quantity	ISCCP Values	
WITH CLOUDS		
Absorbed Solar at TOA	228.0	(232.4)
Emitted Terrestrial at TOA	232.4	
Net Radiation at TOA	-4.4	(0.0)
Absorbed Solar at SRF	162.2	(165.3)
Emitted Terrestrial at SRF	51.4	
Net Radiation at SRF	110.8	(113.9)
Absorbed Solar by ATM	65.8	(67.1)
Emitted Terrestrial by ATM	181.0	
Net Radiation by ATM	-115.2	(-113.9)
WITHOUT CLOUDS		
Absorbed Solar at TOA	280.2	(285.6)
Emitted Terrestrial at TOA	254.6	
Net Radiation at TOA	25.6	(31.0)
Absorbed Solar at SRF	212.7	(216.8)
Emitted Terrestrial at SRF	75.0	
Net Radiation at SRF	137.7	(141.8)
Absorbed Solar by ATM	67.5	(68.8)
Emitted Terrestrial by ATM	179.6	
Net Radiation by ATM	-112.1	(-110.8)
DIFFERENCE PRODUCED BY PARTIAL CLOUDINESS		
Absorbed Solar at TOA	-52.2	(-53.2)
Emitted Terrestrial at TOA	-22.2	
Net Radiation at TOA	-30.0	(-31.0)
Absorbed Solar at SRF	-50.5	(-51.5)
Emitted Terrestrial at SRF	-23.6	
Net Radiation at SRF	-26.9	(-27.9)
Absorbed Solar by ATM	-1.7	(-1.7)
Emitted Terrestrial by ATM	1.4	
Net Radiation by ATM	-3.1	(-3.1)
CONTRAST BETWEEN OVERCAST AND CLEAR		
Absorbed Solar at TOA	-68.6	(-69.9)
Emitted Terrestrial at TOA	-29.3	
Net Radiation at TOA	-39.3	(40.6)
Absorbed Solar at SRF	-67.0	(-68.3)
Emitted Terrestrial at SRF	-38.1	
Net Radiation at SRF	-28.9	(-30.2)
Absorbed Solar by ATM	-1.6	(-1.6)
Emitted Terrestrial by ATM	8.8	
Net Radiation by ATM	-10.4	(-10.4)

Table 6.1: *Global summary of annual average fluxes at the top of the atmosphere (TOA), at the surface (SRF) and in the atmosphere (ATM), with and without clouds, calculated from ISCCP datasets. All values are in Wm^{-2}. Values in parentheses are adjusted by requiring a flux balance at the top of the atmosphere in the annual mean. The third part of the table indicates the changes in the fluxes caused by removing (partial) cloud cover, while the fourth part indicates the differences between completely cloudy and completely clear conditions.*

the strength of boundary layer turbulence. Apparently, the daytime boundary layer is stabilized by heating in the clouds near the top of the layer; consequently, the cloud amount, optical thickness and height decrease throughout the day from a peak near dawn (e.g., Nicholls 1984). These changes shift the peak solar heating towards afternoon, but also shift the minimum terrestrial cooling towards morning. Since water vapor amounts are relatively high, the former effect is more important.

Over midlatitude land, the surface temperature responds quickly to the changing heating, so that it controls the boundary layer dynamics. In this case the larger temperatures near noon appear to enhance boundary layer turbulence and cloud amount, optical thickness and height all increase towards afternoon. Although these cloud changes serve to decrease the solar heating of the surface more rapidly than for clear conditions as sunset approaches, they also decrease the surface cooling, which may be a more important effect for the relatively dry boundary layer air over land.

The radiative effects of seasonal cloud variations are even more subtle. Figure 6.11(a) shows the seasonal extremes of the net radiation at the top of the atmosphere (cf., Fig. 6.10(a) for the annual means). At midlatitudes, cloud amount and optical thickness increase in winter, while cloud top temperature decreases primarily because the atmosphere is colder but also because the cloud top heights increase slightly. Increasing optical thickness and cloud amount enhance the wintertime decrease of solar heating caused by the changing inclination of Earth's rotation axis; but the colder cloud temperatures imply an offsetting decrease in cooling by terrestrial radiation. The clouds switch from a net cooling for summer to almost no net effect in winter (Fig. 6.11(b)). In the tropics, the cloud changes act in the opposite way: cloud amount, optical thickness and cloud height increase in summer. Increasing optical thickness and cloud amount decreases the summertime increase of solar heating; but the colder cloud temperatures imply an offsetting decrease in cooling by terrestrial radiation. These two effects appear to cancel, (Fig. 6.11(b)), leaving the net effect of the clouds about the same - a weak cooling. These changes appear to strengthen the mean atmospheric circulation by increasing the latitudinal contrast of radiative heating.

6.6 Summary - Possible cloud-climate feedbacks

A closer look at the results, assuming they are accurate enough, may suggest much more subtle cloud feedbacks (Rossow and Lacis 1990); however, the "observed" cloud radiative feedbacks on diurnal, seasonal and El Niño changes cannot be used directly to represent the feedbacks on climate. In all these cases that we can observe, we are monitoring the changes in cloud properties and radiative fluxes that occur during "rapid" deviations from equilibrium. They are rapid in the sense that the temperatures do not change much; hence the changes in fluxes that occur will depend on whether the temperatures have had time to change or not. As an example, the cloud effects shown in Table 6.1 are valid only for the case where the temperatures do not change and imply that both the solar and terrestrial net fluxes change most at the surface; but as the surface temperature changes, the largest terrestrial flux change caused by removing the clouds shifts from the surface to the atmosphere. A strong hint that the results may differ when the temperature changes is given by the completely different behavior exhibited by ocean and land areas: over oceans, perturbations which change the clouds produce changes in the radiation balance without significant changes in temperature, whereas over land, surface temperature changes occur almost immediately, producing a change in the terrestrial fluxes in addition to those produced by changing clouds. Thus, the actual radiative effect of a cloud change depends on the time scale on which it occurs. All of the observational

(a)

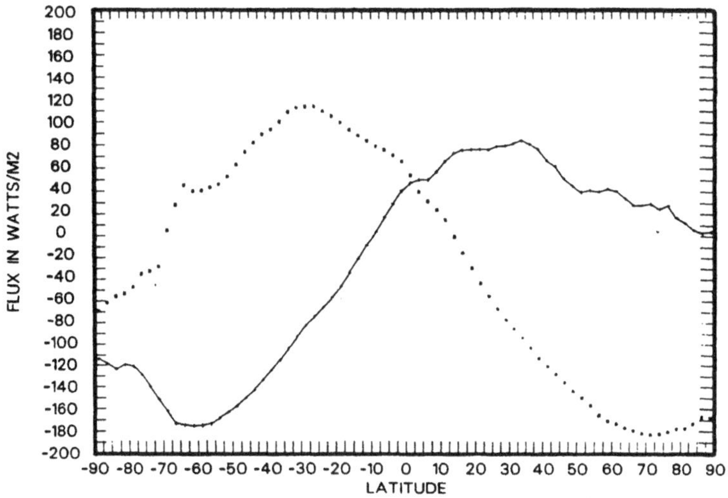

(b)

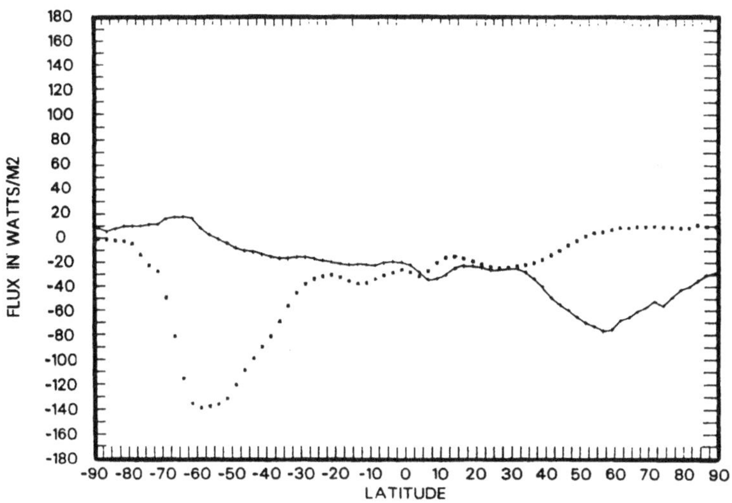

Figure 6.11: *(a) Zonal mean net radiative flux at the top of the atmosphere for January and July, calculated from ISCCP datasets, (b) Net changes in the zonal net radiative flux for January and July, shown in the upper figure, caused by adding partial cloud cover.*

evidence we have seems to suggest that cloud radiative feedbacks are negative for short-time scales (e.g., Hartmann and Short 1980; Rossow and Lacis 1990; Ardanuy et al. 1991), implying that clouds tend to suppress variations. What happens at longer time scales, we do not know as yet.

Despite the large amount of work that has been completed to produce the ISCCP, ERBE and NIMBUS-7 datasets, they are just the beginning of the more difficult studies needed to establish the relationships between atmospheric motions, cloud properties and their effects on radiation. By combining these datasets with other meteorological and surface datasets, some of these relationships can be found. Moreover, these data can be exploited to examine the role of clouds in the water cycle, too. This work is only just beginning.

There are also plenty of satellite (and other) datasets that have not been fully analyzed. In some cases the techniques exist but no one has undertaken the large analysis task. Examples of this kind are systematic analysis of satellite measurements of surface temperature and albedo and use of infrared sounder measurements to infer cloud properties. In other cases the analysis techniques still need to be developed. Examples of this are improvements in the infrared sounder methods and further development of microwave analysis methods to base them more on physical radiation models.

In the next 5-10 years, new kinds of satellite measurements are planned that also require development of new types of analysis methods. The most notable cases are:

1. methods for a simultaneous analysis of information from multiple instruments, each making measurements at many wavelengths, covering the electromagnetic spectrum from UV to microwave,
2. methods for extracting the full information from real spectrometers, and
3. analysis methods that can exploit the information contained in the polarization of reflected sunlight and scattered microwaves.

6.7 References

Ardanuy PE, Stowe LL, Gruber A, Weiss M (1991) Shortwave, longwave and net cloud-radiative forcing as determined from NIMBUS-7 observations. *J Geophys Res* 96: 18537-18549

Betts AK, Harshvardhan (1984) Thermodynamic constraint on the cloud liquid water feedback in climate models. *J Geophys Res* 92:8483-8485

Fu R, DelGenio AD, Rossow WB (1990) Behaviour of deep convective clouds in the tropical Pacific deduced from ISCCP radiances. *J Atmos Sci* 3:1129-1152

Graham N, Barnett TP (1987) Observations of sea surface temperature and convection over tropical oceans. *Science* 238:657-659

Harrison EF, Minnis P, Barkstrom BR, Ramanathan V, Cess RD, Gibson GG (1990) Seasonal variation of cloud radiative forcing derived from the Earth Radiation Budget Experiment. *J Geophys Res* 95:18687-18703

Hughes (1984) Global cloud climatologies: A historical review. *J Climate Appl Meteor* 23:724-751

Machado LAT, Rossow WB (1992) Structural characteristics and radiative properties of tropical cloud clusters. *Mon Wea Rev* (submitted)

Machado LAT, Desbois M, Duvel J-P (1991) Structural characteristics of deep convective systems over tropical Africa and Atlantic Ocean. *Mon Wea Rev* 120: 392-406

Nicholls S (1984) The dynamics of stratocumulus: Aircraft observations and comparisons with a mixed layer and model. *Quart J Roy Meteor Soc* 110:783-870

Price C, Rind D (1992) A simple lightning parameterization for calculating global lightning distributions. *J Geophys Res* 97: 9919-9933

Rossow WB, Lacis AA (1990) Global, seasonal cloud variations from satellite radiance measurements. Part II: Cloud properties and radiative effects. *J Climate* 3:1204-1253

Rossow WB, Schiffer RA (1991) ISCCP cloud data products. *Bull Amer Meteor Soc* 72:2-20

Somerville RCJ, Remer LA (1984) Cloud optical thickness feedbacks in the CO_2 climate problem. *J Geophys Res* 89:9668-9672

Stephens GL, Webster PJ (1981) Clouds and climate: Sensitivity of simple systems. *J Atmos Sci* 38:235-247

Tselioudis G, Rossow WB, Rind D (1992) Global patterns of cloud optical thickness variation with temperature. *J Climate* : in press

Wallace JM, Hobbs PV (1977) Atmospheric Science, An Introductory Survey. New York, *Academic Press*

Warren SG, Hahn CJ, London J, Chervin RM, Jenne RL (1986) Global distribution of total cloud and cloud type amounts over land. *NCAR Tech Note* TN-273+STR/DOE Tech. Rep. ER/60085-H1. 29 pp. + 200 maps

Warren SG, Hahn CJ, London J, Chervin RM, Jenne RL (1988) Global distribution of total cloud and cloud type amounts over the ocean. *NCAR Tech Note* TN-317+STR/DOE Tech. Rep. ER-0406. 42 pp. + 170 maps

6.8 Additional reading

– Cloud Climatologies

Hughes (1984) Global cloud climatologies: A historical review. *J Climate Appl Meteor* 23:724-751

Rossow WB, Schiffer RA (1991) ISCCP cloud data products. *Bull Amer Meteor Soc* 72:2-20

Stowe LL, Yeh HTM, Eck TF, Wellemeyer CG, Kyle HL, the NIMBUS-7 Cloud Data Processing Team (1989) NIMBUS-7 global cloud climatology. Part II: First year results. *J Climate* 2:671-709

Warren SG, Hahn CJ, London J, Chervin RM, Jenne RL (1986) Global distribution of total cloud and cloud type amounts over land. *NCAR Tech Note* TN-273+STR/DOE Tech. Rep. ER/60085-H1. 29 pp. + 200 maps

Warren SG, Hahn CJ, London J, Chervin RM, Jenne RL (1988) Global distribution of total cloud and cloud type amounts over the ocean. *NCAR Tech Note* TN-317+STR/DOE Tech. Rep. ER-0406. 42 pp. + 170 maps

– Cloud Radiative Feedbacks

Cess RD (1976) Climate change: An appraisal of atmospheric feedback mechanisms employing zonal climatology. *J Atmos Sci* 33:1831-1843

Hansen J, Lacis A, Rind D, Russell G, Stone P, Fung I, Ruedy R, Lerner J (1984) Climate sensitivity: Analysis of feedback mechanisms. Climate Processes and Climate Sensitivity. *Geophys Monogr Ser*, Vol. 29.

Harrison EF, Minnis P, Barkstrom BR, Ramanathan V, Cess RD, Gibson GG (1990) Seasonal variation of cloud radiative forcing derived from the Earth Radiation Budget Experiment. *J Geophys Res* 95:18687-18703

Hartmann DL, Short DA (1980) On the use of earth radiation budget statistics for studies of clouds and climate. *J Atmos Sci* 39:431-439

Hartmann DL, Doelling D (1991) On the net radiative effectiveness of clouds. *J Geophys Res* 96:1204-1253

Paltridge GW (1974) Global cloud cover and earth surface temperature. *J Atmos Sci* 31:1571-1576

Rossow WB, Lacis AA (1990c) Global, seasonal cloud variations from satellite radiance measurements. Part II: Cloud properties and radiative effects. *J Climate* 3:1204-1253

Schneider SW (1972) Cloudiness as a global climate feedback mechanism: The effects on the radiative balance and surface temperature of variations in cloudiness. *J Atmos Sci* 29:1413-1422

Stephens GL, Webster PJ (1981) Clouds and climate: Sensitivity of simple systems. *J Atmos Sci* 38:235-247

Stephens GL, Greenwald TJ (1991a) The Earth's radiation budget and its relation to atmospheric hydrology, 1, Observations of the clear sky greenhouse effect. *J Geophys Res* 96:15311-15324

Stephens GL, Greenwald TJ (1991b) The Earth's radiation budget and its relation to atmospheric hydrology, 2, Observations of cloud effects. *J Geophys Res* 96:15325-15340

Wang W-C, Rossow WB, Yao M-S, Wolfson M (1981) Climate sensitivity of a one-dimensional radiative-convective model with cloud feedback. *J Atmos Sci* 38:1167-1178

Chapter 7

OBSERVATION AND ANALYSIS OF GLOBAL RAINFALL

E. Ruprecht
Institut für Meereskunde an der Universität Kiel
D-2300 Kiel
Germany

7.1 Introduction

Precipitation is a most important meteorological parameter for the whole biosphere including human life. The consequences of sufficient or inadequate rainfall are obvious comparing tropical rain forest and desert areas. The latter also can change drastically after a short rain event, vegetation grows immediately, that means the potential for a vegetation cover exists, but cannot be used due to lack of water. Over the oceans, precipitation is the source of freshwater, and therefore the main component of the forcing of the thermohaline circulation. For the atmospheric circulation precipitation stands for the amount of net energy (latent heat) which is liberated during condensation and partly consumed by the evaporation of droplets. This latent heat is for many processes in particular in the tropics the main energy source. The global mean rainfall is about 1 m per year, this corresponds to about 80 Wm^{-2} of latent heat. Comparing with the mean incoming solar radiation of about 240 Wm^{-2} $(= \frac{S_o}{4}(1-\alpha_p)$, when α_p is the planetary albedo) its relative importance for the energy cycle is evident. Since precipitation is very much concentrated in time and space, the energy surplus in certain regions and certain time periods can be larger more than two magnitudes of the incoming solar energy. That is the reason why the hydrological and the energy cycle are to be jointly investigated during GEWEX. Precipitation has a high spatial and temporal variability. This is due to its process of formation which depends on the following atmospheric properties:

- sufficient water vapour, which can condensate;
- rising moist air, that cools the water vapour below saturation point;
- condensation nuclei, to initiate droplet formation.

NATO ASI Series, Vol. I 5
Energy and Water Cycles in the Climate System
Edited by E. Raschke and D. Jacob

It is the vertical velocity, which mainly causes the high variability. Vertical velocity particularly depends on the state (stability, divergence) of a weather system, but it can be largely affected by the conditions at the earth surface. The air is dynamically forced to rise by irregularities at the surface - mountains, hills, houses, trees etc. Different thermodynamic response of the earth surface to the incoming solar radiation - albedo, evaporation, heat storage - can lead to a marked pattern of up- and down-draughts and to a large diurnal variation of precipitation. These surface effects are noticeable in nearly each precipitation event, although they may have different consequences for the precipitation amount and distribution at a certain location, depending for example on the direction of the air flow towards a hill or on the a soil moisture and therefore on the precipitation history. In certain regions, where the synoptic situation changes little during the whole year, the orographic effect dominates the rainfall distribution e.g. on the islands in the trade wind region as Hawaii or Canaries. Here the NE part is moist and covered with vegetation, and the SW part is typically desertlike.

Topography can also affect precipitation measurement and analysis. The representativeness of an observation depends on the location of the gauge and its surroundings. A linear interpolation which is mostly used in the analysis of meteorological data is often not applicable. This fact makes also the analysis of precipitation over the oceans very difficult because coastal and island stations are not representative of the rainfall over the free ocean.

Since the synoptic situation with a certain wind and humidity field mainly determines the formation of precipitation, the rainfall variability decreases with increasing space or time domain over which an average is taken. This leads also to the fact that temporal and areal averages are very similar. Hudlow and Patterson (1979) showed for the rainfall during GATE measured by radar that the hourly rainfall averaged over an area of 28 x 28 km^2 has a very similar statistical distribution as the daily rainfall for a 4 x 4 km^2 area. Many investigations have demonstrated a relationship between rainfall depth, duration and areal extent: rainfall depth decreases nearly exponentially with increasing areal extent, and the slope decreases with increasing rainfall duration (see for example Sumner, 1988). Such relations help to analyse precipitation fields or to calculate spatial average by using point observations.

In the following we describe the state-of-the-art of precipitation measurement. Then problems are discussed involved in the analysis of the observations and the calculation of spatial averages which are needed as one component of the hydrological cycle. It follows a short review about the global precipitation distribution. In particular these published distributions are discussed which are most often used to validate results of atmospheric models. Finally the Global Precipitation Climatology Project is described .

7.2 Measurements of precipitation

7.2.1 Gauge measurements

Precipitation is defined as the amount of (liquid and solid) water particles falling from the atmosphere and reaching the ground. Taking this definition into consideration, its measurement seems to be an easy task. It becomes, however, immediately clear that the measurement poses more questions than expected at first glance. In order to determine the correct precipitation, say in global scale, the mentioned method encounters three general problems:

1. applicability of the method only in populated areas,
2. representativeness of the observation site,
3. accuracy of the gauges.

The first problem excludes most areas of the globe from precipitation measurements e.g. oceanic areas. The second affects network density which therefore is strongly related to the topographic structure of the earth surface. The gauge accuracy depends on three processes: wind effect, wetting and evaporation losses (Sevruk, 1982). Gauges can affect the surrounding windfield so that drops are hindered to fall into the orifice. This is the most serious error in high windspeed regions. The error can be 10 - 50% for rainfall and 200 - 500% for snowfall depending on the gauge type and precipitation structure (Sevruk, 1982). Different gauge designs have been developed and correction coefficients are derived to minimize this error, but a sufficient solution for this problem in particular for snowfall has not yet been attained. Wetting (adhesion of water at the innerwalls of the gauge) and evaporation (of water from the container) losses are small compared to the wind effect, mostly smaller than 0.2 *mm*.

The wind effect is also the largest impediment to precipitation measurement over the ocean by shipborne gauges. Here the windfield is not only disturbed by the gauge but even more by the ship. During GATE 1974 we measured rainfall onboard of the German research vessel F.S. Planet at six different locations with Hellmann type gauges. Fig. 7.1 shows the location of the gauges and the percentage precipitation depth relative to the gauge at the stern, which gave 215.8 *mm* for the period of Sept. 1-20, 1974. Wind direction was mostly from starboard, where the smallest amounts were measured, only about three quarters of the depth at the stern. This distribution agrees in nearly all situations during these three weeks, in 73% of all rainfall episodes the maximum precipitation depth was obtained by the gauge at the stern. A solution to the wind effect may be found by the instrument constructed in Kiel (personal communication with L. Hasse). Besides a normal gauge with a horizontal orifice, rainfall is measured by a vertical column which catches the rain drops with a mainly horizontal velocity component. Calibration of this instrument is in progress.

Rain gauges are the mostly used instruments to measure precipitation, there exist, however, numerous other methods which are developed to overcome the problems with the gauges but not without generating new ones. In general one can say, that today a well-designed gauge network is still needed as basis for verification of all other methods.

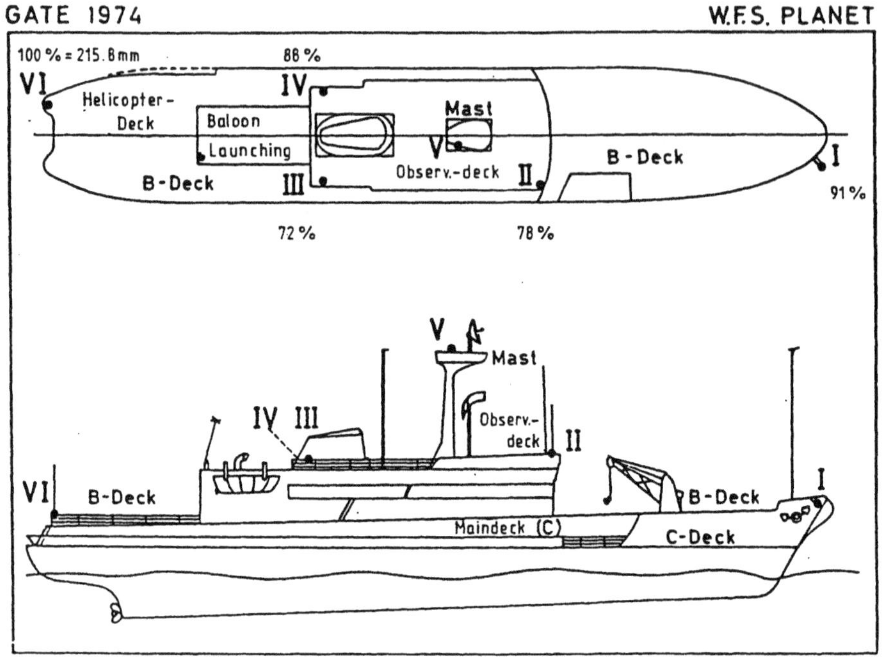

Figure 7.1: *Rainfall measurements at F.S. Planet during GATE (Sep. 1-20, 1974), location of the gauges and its percentage precipitation depth relative to gauge VI at the stern.*

7.2.2 Indirect methods

Observation of precipitation			
in situ - methods		remote sensing methods	
direct	indirect	direct	indirect
gauges	synoptic codes rain frequency salinity	ground based radar spaceborne radar microwave radiometry from satellite	infrared radiometry from satellites visible observations from satellites

Table 7.1: *Methods of observation of global precipitation.*

The methods which are used to derive global distribution are summarized in Tab. 7.1. Since an arrangement of a gauge network is not possible over the ocean, indirect methods have been developed. Tucker (1961) used the synoptic weather reports (ww codes), which describe the present weather. To each of the ww code numbers 50-99, which stand for the different precipitation types (intensities and temporal structures), he assigned a certain rainfall depth and calculated with that monthly mean precipitation. Tucker's method gave less rainfall over the North Atlantic than derived before by using just coastal and island stations. This fact was confirmed for the North Pacific by Reed and Elliott (1973) and by Dorman and Bourke (1978, 1979). The latter derived a correction for

Tucker's method depending on the mean air temperature, since they had found an overestimation of the calculated rainfall in the warm tropics and an underestimation in the cold high latitude regions. Reed (1979) proposed not to use the synoptic weather reports directly but only their frequency distribution, because as he argued in large parts of the oceans the quantity and quality of the ww codes are not sufficient to apply Tucker's method. Already about 80 years ago, Supan (1898) has tried to complete his precipitation distribution over the oceans by using rainfall frequency obtained from ship observations. After the publications of Dorman and Bourke (1978) and Reed (1979) an interesting discussion followed between these two groups about applying the synoptic weather reports or only their frequency to derive the "correct" rainfall over the oceans (Elliott and Reed, 1979; Dorman, 1980; Reed, 1980).

These indirect methods are based on a statistical relationship between qualitative (subjective) observations and a quantitative parameter. Such relationships are only valid if averaging is performed over many situations which actually do not follow a similar relation and may deviate largely, but the deviations should be almost normally distributed. In some regions, a month is sufficient for the averaging period, in others a year is needed. That means only monthly or even annual means can be determined. The situation is very similar for the indirect methods using IR and VIS observations from satellites as we shall see later.

To use the salinity at the ocean surface to determine precipitation was proposed by Wüst (1936). He found a linear relationship between the difference of evaporation minus precipitation and salinity, with the knowledge of evaporation he was able to calculate precipitation. This method, although sometimes discussed among oceanographers, bears too many unknowns to be reasonably applicable. Today such indirect methods are pursued not any further satellite observations are used to derive large-scale rainfall distribution. The methods will be discussed in the next section.

7.2.3 Remote sensing methods

Remote sensing of precipitation has its physical grounds in reflection, scatter or emission of electromagnetic waves by rain drops. The principal methods are listed in Tab. 7.2 together with their physical basis and a few instruments. Only such methods will be included in the discussion, which are designed or were applied for a determination of a large-scale, near-global rainfall distribution.

Methods based on microwave observation

Ground-based radar is after the point measurement by gauges second-most used for rainfall monitoring. Its advantage over gauge measurements is its ability to observe horizontal distribution of precipitation. For the theory of the radar measurement see the text book of Battan (1973). Since the received signal is direct proportional to the reflectivity of the large rain drops, we consider this method as direct. A single-parameter radar measures only the reflectivity:

$$Z = c\sum_i N_i D_i^6 \tag{7.1}$$

Remote Sensing of Precipitation

spectral range	physical basis	instruments
1. active mode		
microwave 3 - 10 GHz	reflection of emitted signal by raindrops	RADAR ground based, spaceborne (planned)
2. passive mode		
microwave 15 - 40 GHz > 80 GHz	emission, scatter of atmosphere and raindrops	spaceborne microwave radiometer (ESMR, SMMR, SSM/I)
infrared 10 - 12 μm	emission of clouds	spaceborne infrared radiometer (e.g. AVHHR and at METEOSAT and GOES)
visible 0.4 - 0.7 μm	reflection of solar radiation by clouds	spaceborne visible radiometer (e.g. AVHHR and at METEOSAT and GOES)

Table 7.2: *Remote sensing methods to observe large scale precipitation.*

N_i number of raindrops with the diameter D_i, c constant coefficient depending on each instrument. Rainfall rate R is estimated by the so-called $Z - R$-relationship:

$$Z = aR^b \tag{7.2}$$

The coefficients a, b have to be empirically determined. This is due to the very variable and mostly unknown drop size distribution in rain clouds. Therefore calibration by an independent method is necessary for different climatic regions and weather situations.

Multi-techniques as dual-polarization and dual-wavelength methods or Doppler-techniques are applied today in research radar and planned for large radar networks e.g. NEXRAD in the USA (Baer, 1991). With such instruments one obtains two measurements which allow to derive two parameters to determine the drop size spectra. This is based on the general accepted assumption that the large raindrops which determine the radar reflectivity and contribute mostly to the rainfall , follow an exponential (or Gamma) distribution. The scatter in the results from the simple Z-R-relationship is certainly reduced, but due to many effects (beam filling, attenuation, variation in dropsize distribution) a "calibration" by other methods (gauges) is also needed for these instruments. Since a ground based radar can monitor rainfall only within areas of about 300 km radius surrounding its site, the vast oceanic regions are excluded from such observations. Therefore spaceborne radars are discussed, for example TRMM (Tropical Rainfall Measuring Mission), but up today no such instrument is in orbit (Simpson, 1988).

The other spaceborne direct method is the microwave radiometry. Here the emission of the atmosphere and surface is measured. Since water has a very low (~ 0.4) microwave emissivity with small variation in contrast to land surfaces ($\sim 0.5 - 0.9$), rain clouds are easily recognized over the ocean as warm islands, but their emission can hardly be distinguished from that of the land surfaces. The measured radiance or brightness temperature, which can be used instead in the microwave spectral range due to the Rayleigh-Jeans-approximation, depends on the drop size distribution. This dependency becomes more dominant for higher frequencies (> $60Hz$) when scattering effects increase. Thus a similar problem is encountered as with the radar methods, empirical coefficients for different climate zones and weather situations have to be determined.

For precipitation estimates up to now, satellite observations are available at frequencies below 40 GHz (ESMR 19.25 GHz; SMMR 18, 21, 37 GHz; SSM/I 19.35, 22, 23.5, 37.0 GHz). An exception is the 85.5 GHz channel of SSM/I, which in fact failed at the first satellite of the DMSP series with SSM/I (F8), but is now available. In the lower frequency range rain clouds are partly transparent. The transparency decreases with increasing rain intensity, so that the radiometer senses more and more the "warm" cloud instead of the "cool" ocean surface and the brightness temperature increases. If the rain intensity increases further the cloud becomes opaque. Then the temperature stratification within the cloud dominates, since then the emission from the upper part of the cloud is measured by the radiometer. That means, brightness temperature is decreasing with increasing rain intensity. The limit of the ambiguity in the rain-rate-brightness-temperature-relationship falls between 20 and 100 $mm\ hr^{-1}$, depending on the cloud thickness, top height or height of freezing level (see Fig. 7.2 and Wilheit et al., 1977).

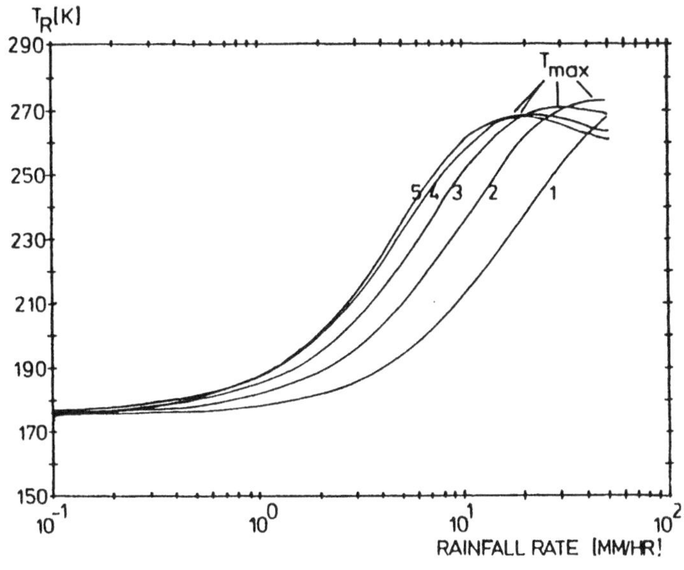

Figure 7.2: *Theoreticcaly derived relationship between microwave brightness temperature (ESMR) T_R and rainfall rate for 5 different rain clouds: rain water (with Marshall-Palmer rain drop distribution) between 0-1 km =1, 0-2 km = 2, 0-3 km = 3, 0-4 km = 4, plus an ice cloud between 4.5-5.5 km = 5 (Medrow at al., 1983)*

For general theory of microwave observations see the 3-volume-textbook of Ulaby et al. (1981, 1982, 1986).

Two other problems occur, the spatial resolution of the microwave observations and their temporal sampling. Due to technical reasons, the spatial resolution of the microwave radiometer is not better than about 15 x 15 km^2, that is in general larger than the extension of a raining cloud. Since the relationship between rain intensity and brightness temperature is not linear, an average over an area larger than the rain cloud in general underestimate the derived rain intensity. The raining part of the microwave pixel has to be determined in advance (Raschke and Ruprecht, 1981). The temporal structure of precipitation cannot be resolved (with one polar orbiting satellite). Such a satellite views the same region depending on its nodal increment and its swath width twice a day or even only twice every ten days. Thus, at least two, better more, satellites are needed to give a sufficient temporal sampling rate. The latter problem does also occur when visible and infrared observations are used, which can be, however, obtained from geostationary satellites. Microwave radiometers for precipitation measurement are not flown on geostationary satellites, the spatial resolution would be too coarse from these far away vehicles.

Methods based on infrared and visible observations

Determination of precipitation by satellite observations in the visible and infrared spectral range must be regarded as indirect methods. It is not possible to derive rainfall or cloud water content directly. One can obtain such cloud properties as optical thickness and top height. Rainfall is then infered from these parameters with the two assumptions:

- high brightness and low brightness temperature imply a large cloud thickness,
- the probability of rain increases with cloud thickness.

Numerous methods have been developed, see the book of Barrett and Martin (1981). One can distinguish between statistical and actual methods: Statistical methods also called cloud indexing methods use a relationship between rainfall amount and cloud cover:

$$R = f(c_i) \tag{7.3}$$

Where c_i is the cloud cover of the i-th cloud type which may be defined by brightness or temperature thresholds. The function $f(c_i)$ is for most methods just a weighting function for the different cloud types for which the weights are predetermined. Arkin (1979) showed that the area covered by cloudy pixels defined by brightness temperature in the infrared spectral range below a certain limit (threshold method) is direct proportional to the rainfall amount of that cloud system. That means, only one cloud type, the thickest one, has to be determined. Similar relationships are widely used today to derive the socalled precipitation index (PI) from the outgoing longwave radiation (OLR) (Arkin and Meisner, 1987; Turpeinen et al. 1987). Kilonsky and Ramage (1976) estimated monthly rainfall with the monthly averaged frequency of highly reflective clouds (large brightness in the visible range). A more detailed cloud analysis is applied by the "Bristol"-method (Barrett, 1970) and the "NESS"-method (Follansbee, 1973). Different cloud types and their coverage are determined. Empirical coefficients (indices), which give the rain intensity of each cloud type, are used as weights.

Actual methods also called cloud history methods need as independent variables not only cloud coverage but also the state of cloud development or the temporal derivative of cloud cover (Stout et al., 1979; Griffith et al., 1978). With such an additional parameter the indirect satellite methods are extended to estimate actual rainfall. All the indirect methods have to be *calibrated* against rainfall amounts given by other methods. Empirical coefficients, mean rain intensity or rain probability of the different cloud types have to be predetermined. Because a statistical analysis is needed to find the correct relationship, the same problem arises as was discussed for the indirect in situ-methods: The rainfall estimation is correct only in a statistical sense. Assuming the calibration was sufficiently carried out, that means the different rainfall situations or systems which are later observed were included in the calibration analysis, then the estimated rainfall agrees the more the larger the rainfall area or the longer the rainfall period, for which the rainfall is to determine. Deviations from the mean relationship can then balance each other.

In summary of the discussion in this section it must be stated that rainfall measurement is a very difficult task. Point measurements by gauges are today the most reliable observations though they are not without problems. How to quantitatively consider the wind effect on the rainfall measurement by gauges is the problem which has to be solved first. Despite of these difficulties the gauges are the standard instruments and all other methods are "calibrated" with the gauge results. Over the oceans, however, this method fails; new instruments for the observations onboard of ships have to be developed. For these vast regions the indirect methods have to be further developed. More microwave observations are needed either in active modes that are spaceborne radars or in passive modes that are microwave radiometers, the observation of which can be combined with observations in the visible and infrared range. Such multispectral methods could yield much better results of the areal precipitation than we have now available at least over the oceans.

7.3 Analysis of Spatial Distribution of Precipitation

The analysis of precipitation encounters a problem which is very similar for other meteorological parameters: observed is a point rainfall (as long as gauge measurements are considered). Required is an areal mean value or the spatial distribution. In the next sections the attempts will be discussed how to solve this problem.

7.3.1 Areal weighted means

The simplest method to derive an areal mean precipitation from point observations is to calculate the arithmetic mean of all available gauge measurements within the considered area. Since the spatial variation of rainfall is very large and is related to the topography, this method works successfully only for a very dense network. Otherwise each gauge measurement is weigthed by its representative area:

$$R = \sum_{i}^{N} \frac{F_i}{F_o} R_i \qquad (7.4)$$

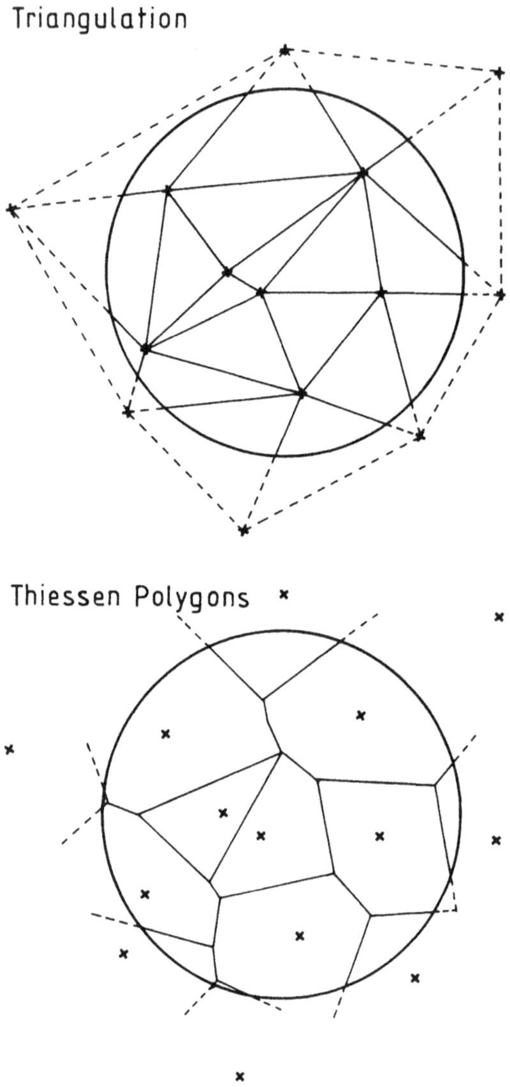

Figure 7.3: *Separation of a catchment area (large circle) by the representative areas of each gauge station (x).*

R_i = rainfall at the ith gauges, F_o = total area, F_i = representative area of the ith gauges, N = total number of gauges within F_o.

The methods which are most often used to determine the areas F_i are: triangulation and Thiessen polygons (Fig. 7.3). For the former method the area is divided into a network of triangles with a gauge site at each apex. The 3 gauge measurements are averaged and then weighted with the area of the triangle. With the Thiessen polygon method a representative area is constructed for each gauge. The boundary of this area is a polygon, each side of which is the perpendicular bisector of the lines joining each pair of adjacent gauges. Stations outside of the region under study (described by the large circle in Fig. 7.3) are included in the analysis. To cover the total region by the representative areas triangles or Thiessen polygons at the boundary have to be construced with outside stations, but only those parts of the triangle or Thiessen polygon areas within the region are used to calculate the means. The results derived by the different methods can largely differ. Sumner (1988) gives an example for these methods. He calculated the areal average of rainfall over an area of 130 km^2 with 10 measurements inside and 7 outside of the region and got the following results:

arithmetic mean	triangulation	Thiessen polygons
15.7 mm	18.2 mm	17.5 mm

Sumner performed also a subjective analysis. He constructed manually the isohyets considering the dependency of rainfall on topography. Integration of the isohyets gave 17.3 mm. The difference of up to 4% between the three latter methods is so small that from this example it cannot be decided which is the more correct one. Also we consider the last as the best estimate. The difference between the simple arithmetic mean and the results of the other methods, however, is large (> 10%). Since the gauge network is dense in that part where little rainfall is observed and less dense in the part of high rainfall the arithmetic mean underestimates the exact areal mean. A sufficient areal weighting can drastically improve the results.

7.3.2 Objective analysis

In order to determine the spatial distribution of precipitation subjective and objective analysis are carried out. Subjective analysis means as mentioned above that isohyets are manually drawn. This is often performed in the precipitation analysis because the experience of the analysist can be thrown in most easily. Even at the Global Precipitation Centre, GPCC (see subsection 7.4.2) a preliminary analysis is carried out in this way (Rudolf et al., 1991).

For many purposes subjective analysis is not sufficient or not possible, so that objective methods have to be applied. For applications in atmospheric sciences a large number of techniques have been developed (see e.g. the book of Thiebaux and Pedder, 1987). In general the methods can be divided into three classes: empirical interpolation, statistical interpolation and function fitting. The latter technique may not be very appropriate for an analysis of precipitation because of the high rainfall variability. A high order surface has to be used. On the other side the topographic influence could be included in such a function which is fitted to the rainfall. Investigations in this direction are not known by the author.

Empirical and statistical interpolations are often applied for precipitation analysis and their advantage and disadvantage are studied (e.g. Bussieres and Hogg, 1989). At the GPCC these techniques are also used.

What is common in both techniques is that a value at a certain grid-point is computed from a distance - weighted sum of the surrounding data. Differences occur, how the weights are determined. In the empirical interpolation technique the weights are predetermined, e.g. exponentially decreasing with distance. In the statistical interpolation technique the ensemble mean of the squared differences between the analysed and the observed field should be a minimum. Then the weights are derived from a covariance relationship (Gandin-technique, see Thiebaux and Pedder, 1987). In order to apply this method the statistical properties of the observed precipitation field has to be known.

Another method, which is often used in hydrology to calculate the weights is the kriging technique (Olea, 1974). It assumes that the field of observation variables can be separated into a drift (spatial trend) and the residuals which are random variates. To apply kriging the variograms of the residuals has to be known together with a constant variance subject to observation errors. Statistical interpolation is a very sufficient technique because topographic effects can easily be included for example by direction-dependent distance weighting.

The high variability of rainfall even for daily averages is reflected by a parameter of the weighting function, the radius of influence. It describes the distance where the weights approach zero or are so small that they can be neglected. Bussieres and Hogg (1989) use values between 100 and 110 km in their comparative study. One can simply interpret this value in such a way that the rainfall at two locations about 100 km apart is uncorrelated. Because the mean density of the gauges is smaller in many parts of the continents, this fact demonstrates very clearly the difficulties which arise for the construction of rainfall maps on the basis of a gauge network. On the other side radar and satellite observations can help to close the gap even if the quantitative results may contain large errors, they can supplement gauge data to construct horizontal distributions.

A new approach, which can help to understand and describe correctly the temporal and spatial variability of rainfall, is the application of the theory of fractals (e.g. Lovejoy and Mandelbrot, 1985). Lovejoy (1982) has shown that rain events have a fractal dimension (about 1.35). If that is in general true, such events can be described by scale-independent fractal models, which comprises the behaviour of actual rain events and of daily or annual averages. In particular hydrological models of catchment areas which need a high spatial and temporal resolution of rainfall, will benefit from this approach.

7.4 Global Distribution of Precipitation

7.4.1 A Review

Precipitation climatology, that is total amount of rainfall on earth and its spatial and temporal distribution, was always of great interest for the meteorologists. Due to the problems of observation and analysis, which have been discussed in the last sections, it is, however, evident that we really do not know the amount of rainfall falling on the earth, and in particular its interannual variability. Today climate modellers are highly interested on global distribution of precipitation to validate their results against. Therefore we shall put most emphasis on those works which are frequently cited by the modellers, to elucidate the data base and analysis methods of these global maps. New global rainfall maps are often not the results of new observations but of a different analysis method of the old data. This is in particularly true for the rainfall over the oceans. Direct observations can be completed by results of other methods or cross-checked by other parameters as evaporation or run-off. Jäger (1976) gives a short overview about the historical development dated back to 1882.

The earlier rainfall maps were based on few observations over the continents, at islands and ships. Thus, a great interpolation effort was necessary, which mostly was carried out by drawing the isohyets by hand and considering for example orographic effects. Wüst (1936) discussed the overestimation of island, coastal and ship stations and proposed a reduction factor of 0.72. He derived this factor by comparing rainfall with evaporation over the oceans. Jäger (1976) has remapped a few older maps to give the same projection and same scale e.g. from 1898 of Supan, 1933 of Schott, 1934 of Meinardus and 1965 of Geiger. In general the rainfall distribution looks very similar, only the amounts change slightly. This reflects the above mentioned facts that often not new observations but new analysis schemes were introduced e.g. using rainfall frequencies as by Supan (1898) or evaporation and salinity as by Wüst (1936).

New and old maps were also indirectly related to each other, since the statistical dependency of rainfall on other meteorological parameters as rainfall frequencies or evaporation was calculated on the basis of the known rainfall. Thus, in some regions the rainfall distribution does change by applying these different parameters, but the overall structure must be maintained. Therefore it is not unexpected that the mean annual global rainfall derived by integrating the different maps published since 1905 varies little (Jäger, 1976): mean value 913 $mm\,year^{-1}\pm$ 10%, lowest 690 $mm\,year^{-1}$, largest 1020 $mm\,year^{-1}$.

A very detailed distribution of annual precipitation was published by Baumgartner and Reichel (1975). They used all global maps of rainfall , evaporation and runoff, available to them cross-checked the three different fields and tried to balance them. If the data did not balance in a certain region, the field which is based on the most sufficient and reliable observations is given first priority and the other fields had to be changed accordingly. Baumgartner and Reichel obtain for the mean annual global rainfall 973 $mm\,year^{-1}$.

The work which is cited most often in recent years is that of Jäger (1976). He published global maps of the monthly mean precipitation. For the continents they are based on rainfall for each month in a 5° x 5° grid. For the ocean he took the annual mean as given by the map of Geiger (1965) and distributed that precipitation over the twelve months proportional to the mean monthly (percentage) frequency of rainy days in a 5° x 5° grid. The so calculated precipitation depths over the ocean were then multiplied by a constant factor to give a mean annual global rainfall of 1000 $mm\,year^{-1}$, which he cited as best value given in the literature for the mean annual evaporation (Fig. 7.4).

Algorithms to determine precipitation amount from satellite observations are not yet sufficient to be applied globally, though a few attempts were carried out to produce global maps of precipitation. Rao et al. (1976) published a "Satellite-derived Global Oceanic Rainfall Atlas". They used the microwave data of the years 1973 and 1974 measured by ESMR flown at NIMBUS 5. Discussing the results in a different paper (Rao and Theon, 1977) they stated: "Several interesting aspects of global climatoloy are disclosed ...". They describe for example an area of extensive rainfall ($>$ 1000 $mm\,year^{-1}$) over the South Atlantic, which "does not appear in any existing map of global rainfall". Their explanation is that only few ships traverse this area. Of course can better observation techniques reveal uncorrect distribution in the earlier global maps in particularly over the southern hemisphere

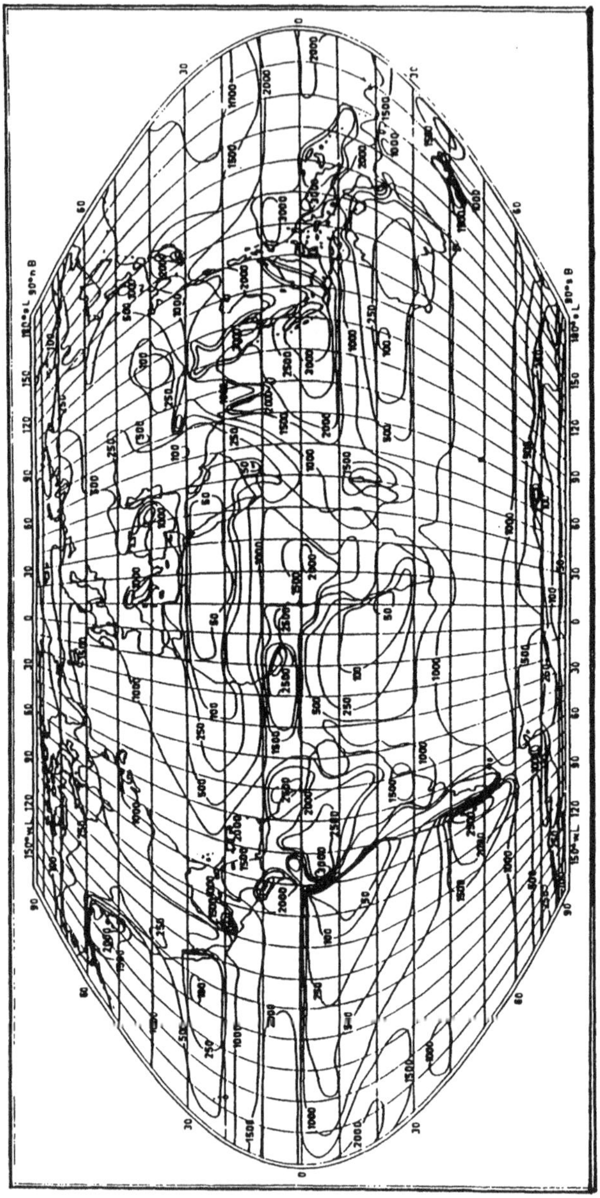

Figure 7.4: *Global distribution of annual mean rainfall (after Jäger, 1976), global mean is 1* $m\,year^{-1}$.

oceans. But satellite-derived rainfall-pattern must be handled with caution when compared with the earlier maps. The latter are normally based on observations of many years (up to 30 years), the former cover only a short period (1-2 years). This fact may explain some of the differences. Second, global applicability of the satellite algorithms is often not valid, and differences can be large in regions where no other data are available to determine empirical coefficents or to verify the results.

In the tropics where convective rain dominates, the outgoing longwave radiation (OLR) is used as precipitation index based on the work of Arkin (1979). WMO publishes each month global maps of OLR (Climate System Monitoring, Monthly Bulletin) with the explanation that in particular in the region between 20°S and 20°N low OLR ($< 200\ Wm^{-2}$) implies intensive rainfall , values $> 260\ Wm^{-2}$ indicate little or no precipitation.

In summary, the gross feature of the global precipitation is known, details may still be obscure. This is in particular true for the interannual variability which is important for climate studies. Based on satellite observations one can certainly disclose qualitative interannual variability, that means a shift of the large raining areas, but quantitative values cannot be derived. That hampers also an answer to the question of the impact of the increasing greenhouse effect on precipitation.

7.4.2 Global Precipitation Climatology Project (GPCP)

Present knowledge of the global distribution of precipitation is not sufficient to satisfy the needs and the requirements of the World Climate Research Programme (WCRP). Therefore the WMO/ICSU Joint Scientific Committee of the WCRP approved 1986 the establishment of the GPCP. Its objectives include

- the production of estimates of tropical precipitation for areas of 2.5° latitude x 2.5° longitude for periods of 5 days using geostationary satellite data ;
- the derivation of rainfall estimates for similar spatial scales and for months from passive microwave radiometric data ;
- the use of these data together with conventional rainfall data to produce analyses of monthly rainfall on a near-global scale;
- the development of a capability to validate and calibrate the various satellite estimates.

In order to carry out these objectives a world-wide organization had to be established (WMO, 1990). Fig. 7.6 shows the organization structure with different data centres and the data flow. The tasks of the centres are

- collect and process conventional rainfall and satellite data;
- develop new global algorithms to derive rainfall from the satellite data;
- develop new methods to measure rainfall especially over the oceans;
- develop analysis schemes for rainfall data;
- merge conventional and satellite-derived rainfall data and produce global maps of the monthly rainfall.

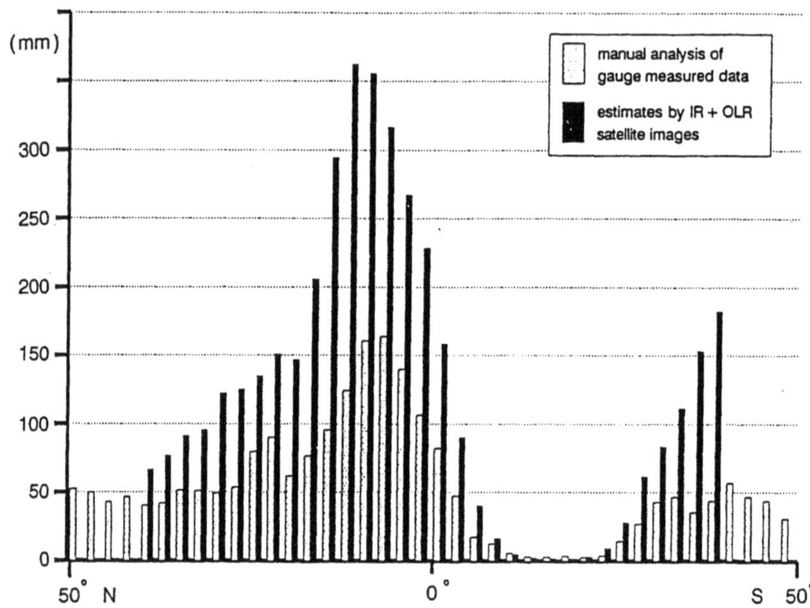

Figure 7.5: *Zonal mean monthly precipitation depth over land for August 1987, comparison between satellite derived (OLR) and station based analysis (Rudolf et al., 1991).*

In order to show the problems which arise during the construction of the global maps, a comparison of station-based and satellite-derived rainfall amounts are given in Fig. 7.5 produced by the Global Precipitation Climatology Centre (GPCC) in Offenbach (Rudolf et al., 1991). In this example the satellite-derived rainfall exceeds in each latitude belt the gauge measurements, in some latitude belts even more than by a factor of 2. Which data set is nearer to the truth could not yet be decided by the GPCC, arguments for both data sets are given. Before global maps are constructed, the problem how to match these data has to be solved. A detailed description about the GPCC, its tasks, routine procedures and first results are published in the new Meteorologische Zeitschrift (1992), its first volume is totally devoted to GPCC.

The GPCC has undertaken a very difficult task, which can successfully be accomplished only with a great afford and by an international cooperation. In a few years we hopefully know much more about the global distribution of precipitation and its variability.

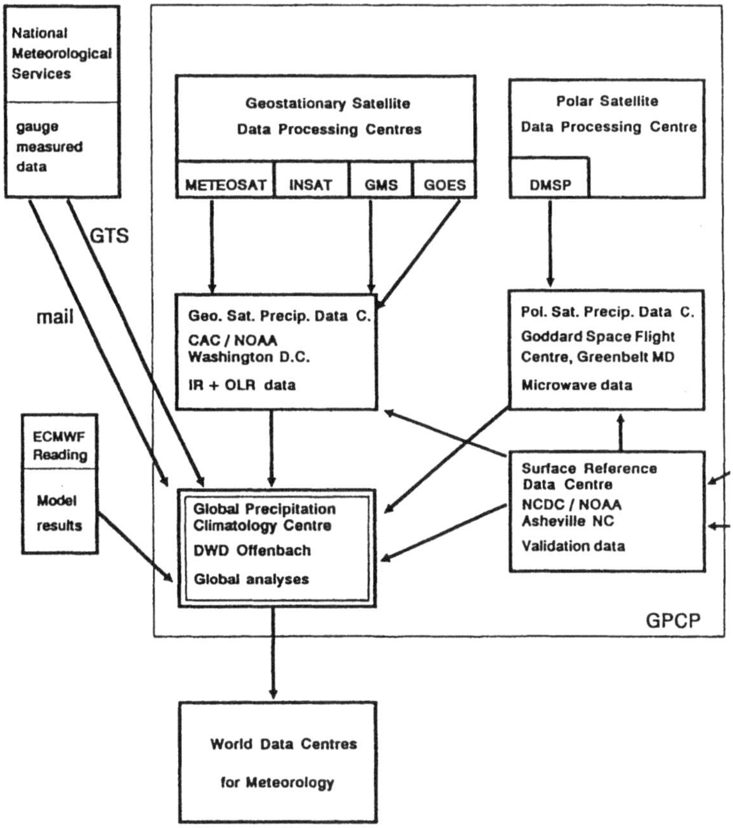

Figure 7.6: *Organization scheme and data flow of GPCP.*

7.5 Outlook

The total amount of precipitation falling at the earth surface and its spatial and temporal distribution is not exactly known today. A value of about 1 $m\,year^{-1}$ for the annual global mean is in general agreed on, the latest estimates do not differ more than 5% from each other, which is reasonable for a longterm (10-30 years) mean. For shorter periods e.g. 1 year this accuracy is not yet attained. Thus, interannual variability cannot be determined yet. This is even more true for the horizontal distribution, in large areas of the globe the rainfall distribution is poorly known.

The reason for this ignorance lies in the observation and analysis methods. A few requirements are listed below, which are needed to improve the situation:

- improve the existing gauge network (representative sites for the gauges, reliability of the observer, correct data transfer,)
- derive correction factors of the wind effect on the different gauge types,

- develop new methods to measure rainfall over the ocean e.g. onboard of ships,
- increase the radar network (installation of radar at coasts and islands to cover part of the ocean),
- launch satellite with active and passive microwave instruments on board,
- improve the algorithms to refer rainfall from satellite observations,
- investigate rainfall statistics with gauge, radar and satellite measurements to support analysis techniques but also to understand differences of the three observation methods,
- improve the parametrization schemes for clouds and precipitation processes in the numerical models,
- develop data assimilation schemes for rainfall data and very important to include satellite observations directly.

Many of the listed requirements are included in the goals of the GPCP. Its results will be of great importance for GEWEX. Therefore a great international support should be given to this work. Other requirements as expansion of gauge and radar network or launch of additional satellites will directly affect the national budgets. If it is agreed on, that water is a valuable but changing natural resource which has to be observed carefully, priorities should be clear. GEWEX will hopefully be so successful that we increase our knowledge about the rainfall on earth, its variability and its possible change due to a changing environment.

7.6 References

Arkin PA (1979) The relationship between fractional coverage of high cloud and rainfall accumulation during GATE over the B-scale array. *Mon Wea Rev* 107: 1382-1387.

Arkin PA, Meisner BN (1987) The relationship between large-scale convective rainfall and cold cloud over the Western Hemisphere during 1982-84. *Mon Wea Rev* 115: 51-74.

Baer VE (1991) The transition from the present radar dissemination system to the NEXRAD Information Dissemination Service (NIDS). *Bull Am Meteor Soc* 72, 29 - 33.

Barrett EC (1970) The estimation of monthly rainfall from satellite data. *Mon Wea Rev* 98: 322-327.

Barrett EC, Martin DW (1981) The Use of Satellite Data in Rainfall Monitoring. *Academic Press* London, New York, 340 pp.

Battan LJ (1973) Radar Observation of the Atmosphere. *The Univ. Chicago Press* Chicago and London, 324 pp.

Baumgartner A, Reichel E (1975) The World Water Balance. *R. Oldenbourg Verlag* München, Wien, 179 pp.

Bussieres N, Hogg W (1989) The objective analysis of daily rainfall by distance weighting schemes on a mesoscale grid. *Atm Ocean* 27: 521-541.

Dorman CE (1980) Comments on "The relationship between the amount and frequency of precipitation over the ocean". *J Appl Meteor* 19: 1131-1133.

Dorman CE, Bourke RH (1978) A temperature correction of Tucker's ocean rainfall estimation. *Quart J Roy Meteor Soc* 104: 765-773.

Dorman CE, Bourke RH (1979) Precipitation over the Pacific Ocean 30° S to 60° N. *Mon Wea Rev* 107: 896-910.

Elliott WP, Reed RK (1979) Comments on the paper "A temperature correction for Tucker's ocean rainfall estimates". *Quart J Roy Meteor Soc* 105: 1082-1083.

Follansbee WA (1973) Estimation of average daily rainfall from satellite cloud photographs. *NOM Techn Memo* NESS 44, Washington DC, 39 pp.

Geiger R (1965) Jährlicher Niederschlag. Wandkarte. *Perthes Verlag* Darmstadt.

Griffith CG, Woodley WL, Grube PG, Martin DW, Stout JE, Sikdar DN (1978) Rain estimation from geosynchronous satellite imagery - visible and infrared studies. *Mon Wea Rev* 106: 1153-1171.

Hudlow MD, Patterson VL (1979) GATE Radar Rainfall Atlas. *NOAA Special Report* U.S. Department of Commerce, 155 p.

Jäger L (1976) Monatskarten des Niederschlags für die ganze Erde. *Berichte des Deutschen Wetterdienstes* Offenbach, 139: 38 pp.

Kilonsky BJ, Ramage CS (1976) A technique for estimating tropical open-ocean rainfall from satellite observations. *J Appl Meteor* 15: 972-975.

Lovejoy S (1982) Area-perimeter relation for rain and cloud areas. *Sci* 216: 185 - 187.

Lovejoy S, Mandelbrot BB (1985) Fractal properties of rain, and a fractal model. *Tellus* 37 A: 209 - 232.

Medrow W, Raschke E, Ruprecht E (1983) Rainfall rates derived from NIMBUS 5 observations analysed against GATE radar rainfall. *Meteor Rundschau* 36: 13-20.

Olea RA (1974) Optimal contour mapping using universal kriging. *J Geophys Res* 79: 695 - 702.

Rao MSV, Abbott III WV, Theon JS (1976) Satellite-Derived Global Oceanic Rainfall Atlas (1973 and 1974). *NASA SP-410* Washington DC, 31 pp.

Rao MSV, Theon JS (1977) New features of global climatology revealed by satellite-derived oceanic rainfall maps. *Bull Americ Meteor Soc* 58: 1285-1288.

Raschke E, Ruprecht E (1981) Microwave radiometry sampling problems demonstrated with Nimbus 5 rain rates versus GATE data precipitation measurements from space. *Workshop Report*

NASA Goddard Space Flight Center, April 28-May 1, 1981. Ed.: D. Atlas and O.W. Thiele, D 84-D 93.

Reed RK (1979) On the Relationship between the amount and frequency of precipitation over the ocean. *J Appl Meteor* 18: 692-696.

Reed RK (1980) Reply to C.E. Dorman (1980). *J Appl Meteor* 19: 1133-1134.

Reed RK, Elliott WP (1973) Precipitation at ocean weather stations in the North Pacific. *J Geophys Res* 78: 7087-7091.

Rudolf B, Hausschild H, Reiß M, Schneider U (1991) Operational global analysis of monthly precipitation totals planned by the GPCC. *Dynam of Atmospheres and Oceans* 16: 17 - 32.

Schott G (1933) Die jährlichen Niederschlagsmengen auf dem Indischen und Stillen Ozean. *Ann Hydrogr Marit Meteor* 61: 1-12.

Sevruk B (1982) Methods of correction for systematic error in point precipitation measurement for operational use. *WMO Operational Hydrology Rep* No. 21 (WMO-No.589), 91 pp.

Simpson J (ed.) 1988: TRMM, a satellite mission to measure tropical rainfall. *Report of the Science Steering Group* NASA, 94 p.

Stout JE, Martin DW, Sikdar DN (1979) Estimating GATE rainfall with geosynchronous satellite images. *Mon Wea Rev* 107: 585-598 .

Sumner G (1988) Precipitation, Process and Analysis. *J. Wiley and Sons* Chichester, 455 pp.

Supan A (1898) Die jährliche Niederschlagsmenge auf den Meeren. *Petermanns geographische Mitteilungen* 44: 179-182.

Thiebaux HJ, Pedder MA (1987) Spatial Objective Analysis. *Acad Press* London, 299 pp.

Tucker GB (1961) Precipitation over the North Atlantic Ocean. *Quart J Roy Meteor Soc* 87: 147-158.

Turpeinen OM, Abidi A, Belhouane W (1987) Determination of rainfall with ESOC precipitation index. *Mon Wea Rev* 115: 2699-2706.

Ulaby FT, Moore RK, Fung AK (1981) (1982) (1986) Microwave Remote Sensing. Vol. I: Microwave Remote Sensing. Fundamentals and Radiometry. Vol. II: Radar Remote Sensing and Surface Scattering and Emission. Vol. III: From Theory to Application. *Addison Wesley Publ Comp* Reading, Massachusetts, 2162 pp.

Wilheit TT, Chang ATC, Rao MSV, Rodgers EB, Theon JS (1977) A satellite technique for quantitatively mapping rainfall rates over the oceans. *J Appl Meteor* 16: 551-560.

WMO (1990) Global Precipitation Climatology Project. Implementation and data management plan. *WMO/TD* 367: Genf, Switzerland.

Wüst G (1936) Oberflächensalzgehalt, Verdunstung und Niederschlag auf dem Weltmeere. Landeskundliche Forschung, *Festschrift N. Krebs* Eds.: H. Louis and W. Panzer, Stuttgart, 347-359.

Chapter 8

DATA ASSIMILATION PROBLEMS

Olivier Talagrand
Laboratoire de Météorologie Dynamique du CNRS
F-75231 Paris Cedex 05
France

8.1 Introduction

The accurate determination of the energy and water cycles of the atmosphere is an ambitious undertaking which will require that the best possible use be made of all the relevant available information, irrespective of its origin, nature and accuracy. The available information will consist first of the *observations*, in the strict sense of the word: in addition to the already existing system of ground-based and satellite observations, a number of new observing instruments, many of which satellite-borne, will be developed in the coming decade, in the context of the World Climate Research Programme and especially of the Global Energy and Water Cycles Experiment. These new instruments will measure many otherwise inaccessible quantities, such as for instance the atmospheric content in precipitable water.

One could argue that such observations, provided they are performed with sufficient accuracy and spatial and temporal resolution, over the whole atmosphere and over sufficiently long periods of time, should in principle suffice for determining the atmospheric energy and water cycles. After all, the determination of these cycles basically only requires the knowledge of the energy and water contents over appropriately defined subvolumes of the atmosphere, together with the associated fluxes through the boundaries of these subvolumes. One could therefore conclude that measurements of the atmospheric content of energy and water under their various forms, together with measurements of wind velocity and of a few other quantities such as precipitation and surface evaporation, should be sufficient for determining the energy and water cycles of the atmosphere. This of course would be a fairly idealised and unrealistic view.

First of all, one would be at odds at precisely specifying what a "sufficient" accuracy and a "sufficient" spatial and temporal resolution could be. More importantly, much information is available which, without consisting of numerical estimates of energy and water contents or fluxes, can be used

NATO ASI Series, Vol. I 5
Energy and Water Cycles in the Climate System
Edited by E. Raschke and D. Jacob

for inferring, through physical or statistical relationships, such numerical estimates. For instance, measurements of the outgoing energy flux emitted by the Earth under the form of electromagnetic radiation are routinely performed by satellite-borne radiometers. These measurements are used, through "inversion" of the radiative transfer equation, for inferring estimates of the atmospheric temperature and of the atmospheric water vapour content. A second source of information, in addition to the observations themselves, therefore consists of the *physical and statistical laws* which link together the various atmospheric fields.

The use of such laws for complementing observations can in some cases be obvious: it is easy to imagine for instance how the principle of conservation of mass can be used for balancing the budget of water mass within a given subvolume of the atmosphere, thereby allowing the determination of quantities, such as water fluxes, which have not been measured. Another, conceptually simple, example is the use of the observed temporal variations of the surface pressure in order to infer values of the vertically integrated divergence of the velocity field. These values can in turn be combined with other sources of information on the velocity field in order to obtain values of the velocity or, at least, to impose bounds on the values of the velocity.

Assimilation of observations is the general procedure by which all the available information (i.e. observations and physical and statistical laws) is processed in order to produce as *accurate* and as *consistent* as possible a *description of the state* of the atmospheric flow. Depending on the particular problem under consideration, the final product of the assimilation may be a description of the state of the flow at a given time, or of the evolution of the flow over a given period of time, or estimates of averaged "climatic" quantities, such as the quantities defining the energy and water cycles of the atmosphere over different spatial and temporal scales.

Because of the extreme diversity of the sources of information to be used in assimilation, which include observations of many different kinds, together with equations (often partial differential equations) expressing the links between meteorological fields, it seems preferable to define a systematic and general approach, capable of treating any source of information, as long as the information is available in quantitative form. In addition, no observation, or no physical or statistical relationship, will ever be absolutely exact, so that there always will be some uncertainty on the result of the assimilation process. An ideal assimilation algorithm should therefore produce, not only the most accurate possible estimate of the state of the flow, but also *an estimate of the associated uncertainty.* As ambitious as the programme of developing such an algorithm may appear, it will be shown in these notes that it can reasonably be based on an appropriately *generalised least-square* approach.

Assimilation of observations is not only the abstract and general problem which has just been stated. It is a reality, implemented in a large number of meteorological centres for the needs of numerical weather prediction (NWP). Indeed, assimilation has been used daily in some meteorological centres for now three decades. An obvious prerequisite, before one can perform a numerical weather forecast, is the definition of appropriate initial conditions. For a model built on the so-called "primitive equations" of meteorology, as most operational NWP models are, the state vector of the model, which must be specified at the initial time of the forecast, essentially consists of the values of temperature, humidity and horizontal wind components at all the grid points of the model.

With the most powerful present global models, which have a horizontal resolution of about 100 *km*, and a vertical resolution of about 1 *km*, the dimension of the state vector is typically $10^6 - 10^7$. The operational meteorological observing system does not produce the synchronous, regularly spaced measurements of all the meteorological fields that the specification of the initial conditions of a forecast would *a priori* seem to require. It is therefore necessary, in order to define initial conditions appropriate for the model which is used, to estimate the values of the various meteorological fields at times and locations at which appropriate observations are not available. To that end, all the relevant information must be used.

The relevant information consists of the observations performed at the initial time of the forecast, but also of observations performed before that time, and of the physical laws which govern the temporal evolution of the flow. In the context of numerical weather prediction, these laws are available in practice under the form of the discretized numerical model used for the forecast, and assimilation of observations, as performed at present, consists in combining in an appropriate way observations distributed in both space and time on the one hand, and a numerical forecasting model on the other. Because of its origins in numerical weather prediction, assimilation of observations possesses a very strongly marked "initial-value-problem", or "dynamical system", aspect, which has dominated its development so far.

A detailed description of the development and of the present state of assimilation methods has been given in a recent book by Daley (1991). Present operational methods for assimilation regularly and dependably produce daily *analyses*, or descriptions of the state of the atmospheric flow, which appear, by all available standards, to be fairly accurate. A number of problems remain, some of which will be discussed in some detail below. In particular, if present operational assimilation methods produce accurate results, these results are not optimal, in the sense that they certainly do not produce the most accurate analyses which could be obtained from the available data and models. Indeed, the numerical cost of producing really optimal analyses would go largely beyond the capabilities of operational numerical weather prediction. But, on the whole, operational assimilation methods work satisfactorily, and improvements in the theory and practical implementation of these methods have largely contributed, in parallel with increase in computing power and improvements in modelling of the atmospheric flow (and, to a lesser extent, improvements in the observing system), to the steady increase observed in the last decades in the quality of operational weather forecasts.

In the last few years, assimilation has progressively extended from its original field of NWP to a broader domain of application. First, the sequences of daily analyses which have now been accumulated, for more than thirty years in some centres, constitute a data base of the greatest climatological interest. For instance, sequences of operational analyses have recently been used for identification of weather regimes and of preferential transitions between these regimes (see, e.g., Vautard, 1990, and references therein). There is no doubt that existing sequences of analyses will be more and more used in the coming years to climatological studies of many kinds, and projects are being developed for "reassimilating", with the present most powerful models and assimilation algorithms, years or even decades of past observations (see, e.g., Bengtsson and Shukla, 1988, or Kinter and Shukla, 1989). It is in this perspective of climatological application of assimilation that the determination of the atmospheric energy and water cycles must be considered.

Secondly, the field of dynamical oceanography, which is closely related to dynamical meteorology, is in rapid expansion because of the development of numerical models of the oceanic circulation and of the implementation of large scale observational experiments, such as the World Ocean Circulation Experiment (WOCE). These experiments will produce an unprecedented amount of data. Appropriate assimilation algorithms will figure among the tools necessary in order to make the best possible use of these new opportunities. A description of the present state of development of assimilation in the context of oceanography can be found in Ghil and Malanotte-Rizzoli (1991). One can reasonably expect that the parallel development of assimilation in the two fields of meteorology and oceanography, which are closely related but also distinctly different in some respects, will prove mutually fruitful.

8.2 Sequential and variational assimilation

Assimilation exists at present under two forms, *sequential assimilation* and *variational assimilation.* In sequential assimilation, which is the only form which has been used so far in operational NWP, the assimilating model is integrated over the time interval over which the observations to be used are distributed. Whenever the model time reaches an instant at which observations are available, the state predicted by the model (that state is often called a *first-guess* in this context) is "updated", or "corrected", with the new observations. The integration of the model is then restarted from the updated state, and the process is repeated until all the available observations have been used. In operational NWP, the state obtained at the end of the assimilation period is taken as the initial state for the ensuing forecast.

The operation which consists in correcting a first-guess at a given time with new observations is called an *analysis* (as already mentioned, the same word, which comes from the pre-computer era, is also used for denoting the output of the correction process and, more generally, for denoting an estimate of the state of the atmospheric flow at a given time). Sequential assimilation is therefore an alternative sequence of *analyses performed at observation times*, and of *integrations of the model* between successive analyses. In all algorithms of sequential assimilation which have been developed so far, there is only one sweep of the model over the assimilation period, so that each individual observation is used once and only once.

One appealing feature of sequential assimilation is the constant updating it performs on the state predicted by the model: each new piece of observation is used for correcting the latest estimate of the state of the atmospheric flow. This feature makes sequential assimilation well adapted to NWP. Numerical forecasts are normally produced at the rate of one forecast every day. In order to define the initial conditions of a new forecast at time t_0, it is very natural, starting from the initial conditions of the previous forecast at time t_0 - 24 hr, to perform a 24-hr sequential assimilation. This approach is implemented operationally in numerous NWP centres, and has now been running continuously in some of them for more than ten years without interruption (but with constant improvement of both the model and the assimilation algorithm itself). However, sequential assimilation also possesses a serious drawback: because precisely of the sequential character of the assimilation, each individual piece of observation influences the estimated state of the flow only at later times, and not at previous times.

There is propagation of the information contained in the observations only from the past into the future, and not from the future into the past. This is of no importance in the case of weather prediction, where one necessarily must at one stage run the model into the future. But it certainly is a disadvantage in the case of assimilation performed for *a posteriori* uses, such as for instance climatological studies. In such cases, it clearly appears preferable to use analyses which have been produced from observations performed both *before* and *after* the analysis time. *Variational assimilation*, on the other hand, aims at globally adjusting a model solution to all the observations available over the assimilation time interval. The adjustment being simultaneous, the adjusted states at all times are influenced by all the observations over the assimilation interval, thereby avoiding the problem mentioned above. In presently existing algorithms for variational assimilation, one first defines a scalar function that, for any model solution over the assimilation interval, measures the "distance", or "misfit", between that solution and the available observations. The distance function will typically be a sum of squared differences between the observations and the corresponding model values, e.g.

$$\mathcal{J} \equiv \sum_j \alpha_j (y_{j,m} - y_j)^2 \tag{8.1}$$

where the y_j's are the observations, the $y_{j,m}$'s are the corresponding model values, and the α_j's are numerical weights reflecting the accuracy of the various observations. One will then look for the model solution which minimizes the distance function. Since a model solution is uniquely defined by the corresponding initial conditions at the beginning of the assimilation period, these initial conditions are taken as *control variables*, i.e. as the variables with respect to which the minimization is effectively performed. The minimizing initial state is obtained through an iterative procedure, each step of which requires the explicit knowledge of the local values of the set of partial derivatives, or *gradient vector*, of the distance function with respect to the initial state. As will be explained below, this gradient can be determined, at a non-prohibitive numerical cost, through use of the *adjoint equations* of the assimilating model.

8.3 Least-square statistical linear estimation. Generalities

Most, if not all, assimilation algorithms which have been used so far, either for research or for operational purposes, and either of the sequential or of the variational type, can be described as more or less simplified forms of *least-square statistical linear estimation*. Even "nonlinear" applications correspond in fact to cases which are very close, in some sense, to linearity. Least-square statistical linear estimation is a classical tool, whose basic principles are very simple, even though practical implementation on large dimension systems can raise many problems. We will first describe the basic principles of least-square statistical linear estimation on elementary examples, before going to present meteorological applications.

Let us consider the following simple estimation problem. We want to determine some unknown quantity x from two known measurements y_1 and y_2 of the form

$$y_1 = x + \epsilon_1 \tag{8.2}$$

$$y_2 = x + \epsilon_2$$

where ϵ_1 and ϵ_2 are "observational" errors. These errors are of course unknown, but we assume that the statistical performances of the instruments which have produced y_1 and y_2 are known. More precisely, we assume that these instruments are unbiased, i.e.

$$E(\epsilon_1) = E(\epsilon_2) = 0 \tag{8.3}$$

where $E(\epsilon_i)$ denotes the statistical mean, and that the statistical variances of ϵ_1 and ϵ_2 are known

$$E(\epsilon_1^2) = \sigma_1^2 \qquad E(\epsilon_2^2) = \sigma_2^2 \tag{8.4}$$

We assume in addition, for the sake of simplicity, that the observation errors are uncorrelated

$$E(\epsilon_1\epsilon_2) = 0 \tag{8.5}$$

This will be the case if, for instance, the two observations have been obtained with different instruments.

We now want to estimate x' as a linear combination of the two observations y_1 and y_2, viz.

$$x' = a_1 y_1 + a_2 y_2 \tag{8.6}$$

where the weights a_1 and a_2 are to be determined. We first want the estimate x' to be statistically unbiased, i.e. to verify the condition $E(x' - x) = 0$. This will be verified if

$$a_1 + a_2 = 1 \tag{8.7}$$

We also want x', among all unbiased estimates, to minimize the statistical variance of the estimation error, viz.

$$\sigma^2 = E[(x' - x)^2] \tag{8.8}$$

The solution to this simple constrained minimization problem (minimize Eq. 8.8 under constraint Eq. 8.7 is easily found to correspond to weights a_1 and a_2 which are inversely proportional to the variances of the corresponding observation errors, i.e.

$$a_1 = \frac{\sigma_2^2}{(\sigma_1^2 + \sigma_2^2)} \qquad a_2 = \frac{\sigma_1^2}{(\sigma_1^2 + \sigma_2^2)} \tag{8.9}$$

In addition, the corresponding minimum of the estimation error variance σ^2 is given by the relationship

$$\frac{1}{\sigma^2} = \frac{1}{\sigma_1^2} + \frac{1}{\sigma_2^2} \tag{8.10}$$

which has a simple interpretation: if one calls "precision" the inverse of an error variance, then the precision of the estimate x' is the sum of the precisions of the observations.

The same estimate x' can be found through a different approach: an acceptable estimate of the exact value x must be close to the observations, at least within the accuracy of the latter. For any value ξ, the "distance" between ξ and the observations can be measured by the following quadratic quantity

$$\mathcal{J}(\xi) \equiv \frac{(\xi - y_1)^2}{\sigma_1^2} + \frac{(\xi - y_2)^2}{\sigma_2^2} \tag{8.11}$$

where the observational error variances σ_1^2 and σ_2^2 account for the accuracy of the observations. Now, the value of ξ which minimizes $\mathcal{J}(\xi)$ is precisely the estimate x' given by Eqs. 8.6-8.9. Minimization of Eq. 8.11 therefore provides another way of determining the estimate x', based on a (very elementary) variational principle. Eqs. 8.6-8.9 generalize in an obvious way to the case of any number m of observations $y_j = x + \epsilon_j$ $(j = 1, \ldots, m)$. A slightly less obvious generalization is to correlated observation errors. But this remains limited to observations y_j which bear on the quantity x to be estimated. This is usually not the case for meteorological observations. As already mentioned, observations are rarely performed at the times and spatial locations at which estimates are sought. In addition, and independently of the problem of space-time location, many observations do not bear on the physical quantities to be estimated. For instance, satellite-borne radiometers measure the radiative flux emitted by the Earth to outer space at different wavelengths, while what one basically wants are estimates of the atmospheric temperature and humidity fields. The measured fluxes are functions of these fields (and of other quantities, such as for instance cloud amount and top-pressure, and surface emissivity) through the *radiative transfer equation.* Indeed, such "indirect" measurements will become more and more numerous in future observational systems, such as those which are planned for GEWEX, and assimilation methods must allow for measurements which are indirect, "complicated" functions of the physical parameters to be estimated. Statistical linear estimation can accommodate such indirect measurements (of course within the limits of linearity, which will be discussed later) as we will now proceed to show.

We assume that what we now want to estimate is a complete vector x, with dimension n and components x_i $(i = 1, \ldots, n)$. That vector can be thought of as consisting for instance of the values of one or several meteorological fields (temperature, wind components, humidity) at a given instant at the points of a two- or three-dimensional regular array. But the developments which follow are very general, and independent of the physical nature or significance of the parameters to be estimated. The vector x to be estimated will be called the *state vector*, since it will in general describe the state of a physical system, and more precisely of the atmosphere. As for the available observations, they make up a vector y, with dimension m and components y_j $(j = 1, \ldots, m)$. We assume that the vector y can be written under the form

$$y = Hx + \epsilon \tag{8.12}$$

where H is an $m \times n$ matrix which defines the link between the parameters to be estimated and the observed quantities. H will be called the *observation matrix.* If the observations bear on the physical field to be estimated, but are performed at points in space-time different from the points at which estimates are sought, H will represent some appropriate space-time interpolation . If the observations bear on "indirect" functions of the parameters to be estimated, H will represent an appropriate linearization of the physical and/or statistical relationship linking x and y. In the simple example (Eq. 8.3) (where $n = 1$, $m = 2$), H is the matrix $(11)^T$ (where the superscript T denotes transposition). As for the m-vector ϵ in Eq. 8.12, its components are the errors affecting the observations. This error vector is of course unknown, but we will assume that it is statistically unbiased, i.e.

$$E(\epsilon) = 0$$

and that the variances-covariances of its components, making up the matrix $E(\epsilon\epsilon^T)$, are known

$$E(\epsilon\epsilon^T) = \Sigma$$

To sum up, the exactly known quantities in Eq. 8.12 are the observation vector y and the observation matrix H, while the observation error vector is known only through its statistical properties, and the vector x, to be estimated, is totally unknown. In order to estimate x, we now proceed as in the example 8.3 above, and look for an estimate x' which is a linear function of y, i.e. is of the form

$$x' = Ay \tag{8.13}$$

where A is an $n \times m$ matrix to be determined. We want the estimate x' to be unbiased, i.e. to be such that $E(x' - x) = 0$. This condition is verified if

$$AH = I_n \tag{8.14}$$

where I_n is the unit matrix of order n. In addition, among all the matrices verifying Eq. 8.14, we want to choose the one which minimizes the variance of the norm of the estimation error, i.e. the matrix which minimizes the trace (sum of the diagonal terms) of the variance-covariance matrix $P = E[(x' - x)(x' - x)^T]$ of the estimation error. The solution to that problem is

$$A = [H^T\Sigma^{-1}H]^{-1}H^T\Sigma^{-1} \tag{8.15}$$

As for the corresponding matrix P, it is equal to

$$P = [H^T\Sigma^{-1}H]^{-1} \tag{8.16}$$

P contains the variances and covariances of the estimation errors on all the components of x. In particular, its diagonal terms are the variances of these estimation errors.

Eqs. 8.15 and 8.16 generalize Eqs. 8.9 and 8.10. As in that previous example, there is a variational formulation to the estimation of x, which now corresponds to minimizing the distance function

$$\mathcal{J}(\xi) \equiv [H\xi - y]^T\Sigma^{-1}[Hx - y] \tag{8.17}$$

where ξ is an n-vector. The meaning of expression 8.17, which generalizes Eq. 8.11, should be clear: for any ξ, the vector $H\xi - y$ is the difference vector between what would be observed if the vector to be estimated was equal to ξ (and if the observations were perfect) and the actual observations y. $\mathcal{J}(\xi)$ is the squared norm of that difference vector, weighted so as to take into account, through Σ^{-1}, the accuracies of the different observations (and also the possible correlations between the various observation errors). The estimate x' is the value of ξ for which that squared norm is minimum, i.e. the value of ξ that would produce, if exactly observed, the values closest to the actual observations.

It must be stressed that the problem of minimizing the statistical variance of the estimation error on x, and of minimizing the distance function (Eq. 8.17), are *a priori* distinctly different problems, even though the fact that they lead to identical results is algebraically obvious in simple cases. In the first problem, one minimizes a quantity defined on the n-dimensional space of *state vectors*, while in the second problem one minimizes a quantity defined on the m-dimensional space of *observations*. It is certainly not a priori obvious that the solutions to these two problems should always be identical.

The vector x' defined by Eqs. 8.13-8.15 is called the *Best Linear Unbiased Estimate*, or BLUE, of x from y. The theory leading to the BLUE is standard, and has been described here in order to stress its generality, and also as an introduction to the various methods for assimilation which, although they often do not explicitly refer to the theory of statistical linear estimation, are almost always, as already mentioned above, more or less simplified applications of that theory. In addition to the estimate x', statistical linear estimation produces the variance-covariance matrix P of the corresponding estimation error, thus fulfilling the goal assigned in the introduction to an ideal assimilation algorithm. It can be noted that the matrix P does not depend on the observation vector y, but only on the observation operator H and on the matrix Σ, i.e. on the nature and accuracy of the observations. In particular, the theory of statistical linear estimation can be used for evaluating the performance of an hypothetical observing system, defined by what that system would observe and with which accuracy, but independently of any actual observations.

Implementation of statistical linear estimation requires, in addition to the knowledge of the matrices H and Σ, the fact that the observation error ϵ be unbiased. Actually, if that error was biased, but the bias was known, it would be sufficient to first subtract the bias from the observation vector in order to obtain a new, unbiased, observation vector. The requirement that the bias is zero is therefore in effect only a requirement that the statistical mean of the observation error is known.

To sum up the results obtained so far: before one can implement statistical linear estimation, one must know what has been observed, in terms of the parameters to be estimated (this is expressed by the observation matrix), and with which accuracy (this is expressed by the mean and variances-covariances of the observation error vector). These requirements may seem extremely demanding, since there will always be instruments, especially newly developed instruments, for which it will certainly be very difficult to assign reliable values to the observational errors. But at the same time, it is obvious that a prerequisite for a rational use of a set of observations is to know what has been observed, and with which accuracy. It is therefore a good thing that the requirement for knowledge of the nature and accuracy of the observations comes out of the mathematics of estimation theory. If the required information is not available, one will then have to compensate for it by as reasonable as possible hypotheses on the observation matrix H and the corresponding error covariance matrix Σ. One advantage of studying assimilation in the perspective of general estimation theory is that it forces one to explicitly formulate hypotheses which are necessarily made in one way or another.

8.4 The sequential form of statistical linear estimation. Kalman filtering

Eqs. 8.13, 8.15 and 8.16 take a particular form, which is extremely useful in many applications, when the "observation vector" y defined by Eq. 8.12 can be decomposed into two components $y = (x_g^T, z^T)^T$, where x_g is a prior estimate, or *first-guess*, of the vector x, and z is an additional set of observations, with dimension p. The first-guess can be written as

$$x_g = x + \epsilon_g \tag{8.18}$$

where ϵ_g is the corresponding error, while z, assumed to be associated with a $p \times n$ observation matrix K, can be written as

$$z = Kx + \eta \tag{8.19}$$

where η is the associated obervation error. Eqs. 8.18 and 8.19 are the analogue of Eq. 8.12, the dimension of the observation vector being now $m = n + p$, the corresponding observation matrix being $H = (I_n, K^T)^T$, and the corresponding observation error vector being $\epsilon = (\epsilon_g^T, \eta^T)^T$. As for the covariance matrix of the observation error, it is defined by

$$\Sigma = \begin{bmatrix} E(\epsilon_g \epsilon_g^T) & E(\epsilon_g \eta^T) \\ E(\eta \epsilon_g^T) & E(\eta \eta^T) \end{bmatrix} \tag{8.20}$$

It is important to stress that the first-guess x_g need not consist of "observations" in the strict sense of the word. The first-guess can for instance be an already known statistical, or "climatological" average of the vector x, or it can be any estimate of x, obtained through whatever means may have been available: processing of previous observations, theoretical or numerical developments. The only important thing is that x_g be numerically known, together with the corresponding statistical variances-covariances of the error ϵ_g.

Eqs. 8.15-8.16 can be applied on the above quantities in order to obtain the corresponding BLUE x' and the covariance matrix P of the associated estimation error. The results can be put into forms which are extremely useful from both the theoretical and the numerical points of view. For simplicity, we will assume that the errors ϵ_g and η are statistically decorrelated, so that the off-diagonal terms in Eq. 8.20 are zero. The matrix $E(\epsilon_g \epsilon_g^T)$ will be denoted P_g, and the matrix $E(\eta\eta^T)$ will be denoted Π. In addition, the corresponding BLUE which, in the meteorological vocabulary, will be an analysis, will be denoted x_a, the associated covariance matrix of estimation error being denoted P_a. With these notations, Eqs. 8.13 and 8.15 and 8.16 can be put in the respective forms

$$x_a = x_g + P_g K^T [K P_g K^T + \Pi]^{-1} (z - K x_g) \tag{8.21}$$

and

$$P_a = P_g - P_g K^T [K P_g K^T + \Pi]^{-1} K P_g \tag{8.22}$$

Considering first Eq. 8.21, we see that it defines the analysed state x_a as the sum of the first-guess $x - g$ and of a correction term. The latter is proportional to the vector $z - Kx_g$, i.e. to the difference between the additional observation vector z and what the observation operator K would produce if it was applied to the first-guess x_g. That difference is therefore essentially the lack of agreement between the first-guess and the new observations. It is obvious that, if that difference happened to be exactly equal to zero, that is if the first-guess happened to agree perfectly with the new observations, there would be no point in performing any correction on the first-guess. And, in the linear theory presented here, the correction to be applied to the first-guess naturally appears as a linear function of the difference vector $z - Kx_g$. The corresponding matrix $\Gamma = P_g K^T [K P_g K^T + \Pi]^{-1}$, which is called the *gain matrix*, is simply the matrix which, taking into account the respective accuracies of the first-guess and of the observations, as defined by the covariance matrices P_g and Π, produces the best estimate, in the sense of the minimum of variance, of the state vector x.

In meteorology, the vector $z - Kx_g$ is called the vector of *residuals*. In general estimation theory, it is called the *innovation* vector. That second denomination is extremely suggestive, because the vector $z - Kx_g$ effectively describes *all the new information* (in a linear sense, at least) contained in the additional observation vector z.

Another remark can be made about Eq. 8.21. Its numerical implementation requires the inversion of the matrix $KP_gK^T + \Pi$, which is of dimension p. The direct use of Eq. 8.15 requires (at least) the inversion of the matrix $H^T\Sigma^{-1}H$, which is of dimension n. If $p \ll n$ (which is usually the case in meteorological problems, where the number of observations available at a given time is normally much smaller than the dimension of the model state vector), use of Eq. 8.21 is much more economical. Also, Eq. 8.21 does not require the covariance matrix Σ to be invertible, contrary to what Eq. 8.15 does (at least apparently). If for instance, the additional observations z are perfect ($\Pi = 0$), Eq. 8.21 can still be used. It is interesting to mention that, in this case of exact observations, the analysed state x_a will be exactly compatible with the observations, in the sense that it will verify the equality $Kx_a = z$.

As for Eq. 8.22, it too has a clear significance. It defines the analysis error covariance matrix P_a as the first-guess error covariance matrix P_g minus a corrective matrix. The latter is symmetric, with non-negative eigenvalues, which implies that the analysis error variance on any parameter is at most equal to the corresponding first-guess error variance. The second term on the right-hand side of Eq. 8.22 therefore represents the gain brought about by the additional observations z on the accuracy with which the state vector x is known.

We mention another expression for P_a, directly obtainable from Eq. 8.16

$$P_a^{-1} = P_g^{-1} + K^T\Pi^{-1}K$$

This expression is generally not of much use for numerical applications, but its analogy with Eq. 8.10 is obvious : it also expresses that the "precision" of the analysis is the sum of the precisions of the first-guess on the one hand, and of the additional observations on the other.

Finally, the variational problem associated with the estimation of x from the first-guess x_g and the additional observation vector z is easily seen from Eq. 8.17 to correspond to the minimization of the distance function

$$\mathcal{J}(\xi) \equiv [\xi - x_g]^T P_g^{-1}[\xi - x_g] + [K\xi - z]^T\Pi - 1[K\xi - z] \tag{8.23}$$

where ξ is, as before, an n-vector. The distance function is the sum of two terms, one measuring the distance to the first-guess x_g, the other measuring the distance to the additional observation vector z. These two terms are weighted by the inverse covariance matrices of the corresponding errors.

The algorithm defined by the "updating" Eqs. 8.21-8.22 is known in meteorology under the name of *optimal interpolation.* It was introduced in meteorology largely through a book by Gandin (1963) and, under more or less simplified forms, is at the basis of most of the operational analysis techniques. It is most usually implemented in the following way: a first-guess x_g produced by the assimilating model for the analysis time is combined with a vector of observations z at the same time through some approximate form of Eq. 8.21. The integration of the model is then restarted from the analyzed state x_a until the next observation time, at which a new analysis is performed. This forms the basis of sequential assimilation, as it is implemented at present, with many variants, in operational NWP. Implementation of Eq. 8.21 requires the preliminary knowledge of the covariance matrices P_g and Π of the forecast and observation errors respectively. Reliable specification of those matrices, especially of P_g, raises a number of problems which will not be discussed here. Let us only mention that P_g is usually modelled on the basis of a number of simple hypotheses on the shape and spatial extension of the corresponding covariance functions. It is also commonly assumed that the forecast errors on geopotential and wind are geostrophically related in middle and high latitudes, which allows unambiguous determination of all required covariances from the knowledge of the covariance of geopotential forecast error only.

A number of simplifications are also made in order to reduce the computational cost of Eq. 8.21 and especially 8.22. Indeed, exact implementation of those equations would be much too costly for operational NWP. Concerning Eq. 8.21, its implementation requires at least solving one linear system of dimension p. In a typical meteorological situation, p can be of the order of $10^4 - 10^5$. This is much too large for the constraints of operational NWP. In order to reduce the corresponding cost, only observations located in the vicinity of a given model gridpoint are used when performing the analysis at that point. This "selection" of observations is certainly legitimate in the sense that observations performed at a large distance of a given point must have a small influence on the analysed fields at that point. However, experience shows that it nevertheless introduces spatial noise in the analysed fields, which must be then filtered out by *ad hoc* smoothing. As for Eq. 8.22, it is not implemented in its totality, but only the diagonal terms of the matrix P_a, i.e. the variances of the analysis error, are usually computed.

Experience also shows that, in spite of the geostrophic link assumed between the wind and geopotential forecast errors, the fields produced by optimal interpolation are still contamined by unrealistic ageostrophic noise. This noise must be filtered out through additional appropriate procedures, designated under the general name of "initialization" (for more information on this aspect, see Temperton, 1988, and references therein).

Optimal interpolation, as implemented in present operational NWP, produces results which are on the whole quite satisfactory. Description of its performances can be found in a number of articles or reports. An article by Lorenc (1981), although it is not very recent, contains a clear description of the basic principles of operational optimal interpolation and of the main properties of the results it produces. Among the most recent developments, we can mention that the National Meteorological Center (Washington, USA) has recently introduced in operations an implementation of optimal interpolation, based not on Eq. 8.21, but on an iterative minimization of the distance function 8.23 (Derber et al., 1991). The results show improvement of the quality of the analysis for a lower computational cost. This is probably largely due to the suppression of the need for selecting observations in the vicinity of each analysis point. It therefore appears that it is numerically more efficient to perform optimal interpolation, not by direct use of Eq. 8.21 and explicit exact solution of one (or several) linear system of equations, but by iterative (and approximate) minimization of the corresponding distance function 8.23.

But there is much more to Eqs. 8.21-8.22 than analysis at a given time, and those Eq. are at the basis of the technique of *Kalman filtering* which, in the linear context considered here, is the appropriate approach to sequential assimilation. Let us first consider the following situation. A vector y of observations, of the form 8.12, has been processed through appropriate equations to produce the corresponding BLUE x' 8.13-8.15 together with the covariance matrix P 8.16 of the associated estimation error. At that stage, an additional vector of observations z, of the form 8.19, becomes available, and one wants to determine the BLUE x'' of x from the combined observation vector $v = (y^T, z^T)^T$. Is it necessary to restart the computations from the beginning, or is it possible to take advantage of the computations which have already been performed and have led to x' and P? It must be clear from the foregoing developments, and it is easy to verify that, if the observation error vector η associated with z is uncorrelated with the estimation error vector $x' - x$, then the BLUE x'' and the associated covariance matrix of estimation error are given by Eqs. 8.21 and 8.22, x_g and P_g being replaced by x' and P respectively. There is no need therefore for restarting the computations from the original y, and one can take advantage of the already performed computations. In particular, if the dimension p of the additional observation vector z is small in comparison to the dimension n of x, the numerical gain of using Eqs. 8.21-8.22 rather than restarting the entire computations is obvious.

Eqs. 8.21-8.22 therefore provide a way for constantly updating with new observations the latest estimate of the state of the system under observation. If the errors associated with the successive observations are mutually uncorrelated, the estimate obtained at any stage of the process will always be the BLUE of x from the observations already introduced into the process, and there will be no loss in the accuracy of the estimate resulting from the sequential character of the procedure.

In the case of assimilation of meteorological observations, an additional complication comes from the fact that the observations to be assimilated are distributed over a time period over which the state of the system is itself varying. In order to take into account the temporal dimension, and in agreement with the linear approach followed so far, we consider a system whose state varies in time according to the linear equation

$$x_{k+1} = Sx_k + \zeta_k \tag{8.24}$$

where x_k denotes the state of the system at time k, while S is a known *transition matrix* expressing the time variation of the state vector between times k and $k+1$. As for the term ζ_k, it represents contributions to the temporal variations of the state vector which are not represented by the transition matrix. One can consider that the transition matrix is the numerical model available for performing the assimilation, while the term ζ_k represents the accumulated effect, on the variation of the state vector between times k and $k+1$, of the various processes not represented by S. The term ζ_k will accordingly be called the model error. It will be considered as an unbiased random vector, uncorrelated in time, with known covariance matrix Ξ at a given time

$$E(\zeta_k \zeta_l^T) = \Xi \delta_{kl} \tag{8.25}$$

where δ_{kl} is the classical Kronecker symbol.

We assume in addition that observations of the general form (Eq. 8.19) are available at the successive instants $k = 0, 1, \dots, N$:

$$z_k = Kx_k + \eta_k \tag{8.26}$$

The corresponding observation errors η_k are supposed as before to be unbiased, to be uncorrelated in time and have covariance matrix P at a given time. This leads to the expression

$$E(\eta_k \eta_l^T) = \Pi \delta_{kl}$$

In addition, the model and observation errors are supposed to be mutually uncorrelated

$$E(\eta_k \zeta_l^T) = 0$$

We now assume that the BLUE of the state x_k of the system at time k from all observations up to time k has been determined, together with the covariance matrix of the corresponding estimation error. The BLUE will be denoted $x_{k,+}$, and the covariance matrix will be denoted $P_{k,+}$ (the subscript $+$ meaning that the observation at time k has been used). The BLUE of the state of the system at time $k+1$ from all observations up to time k can be shown to be equal to

$$x_{k+1,-} = Sx_{k,+} \tag{8.27}$$

As for the covariance matrix of the corresponding estimation error $x_{k+1,-} - x_{k+1}$, it is equal to

$$\begin{aligned} P_{k+1,-} &\equiv E[(x_{k+1,-} - x_{k+1})(x_{k+1,-} - x_{k+1})^T] \\ &= E[(Sx_{k,+} - Sx_k - \zeta_k)(Sx_{k,+} - Sx_k - \zeta_k)^T] \\ &= E\{[S(x_{k,+} - x_k) - \zeta_k][S(x_{k,+} - x_k) - \zeta_k]^T\} \\ &= SP_{k,+}S^T + \Xi \end{aligned} \tag{8.28}$$

where the various non-correlation hypotheses have been used. The first term on the last line represents the error at time $k+1$ resulting from the error at time k, while the second term is the contribution of the model error ζ_k between times k and $k+1$.

At this stage, it is easy to introduce the observation vector z_{k+1} at time $k+1$: one simply has to use Eqs. 8.21-8.22, x_g and P_g being replaced by $x_{k+1,-}$ and $P_{k+1,-}$ respectively, and z being replaced by z_{k+1}. This leads to the BLUE $x_{k+1,+}$ of the state of the system at time $k+1$, from all the observations up to, and including, time $k+1$

$$x_{k+1,+} = x_{k+1,-} + P_{k+1,-}K^T[KP_{k+1,-}K^T + \Pi]^{-1}(z_{k+1} - Kx_{k+1,-}) \tag{8.29}$$

and to the variance-covariance matrix of the corresponding estimation error, viz.

$$P_{k+1,+} = P_{k+1,-} - P_{k+1,-}K^T[KP_{k+1,-}K^T + \Pi]^{-1}KP_{k+1,-} \tag{8.30}$$

The sequential process defined by Eqs. 8.26 to 8.30 is called *Kalman filtering* (Kalman, 1960). At any stage, Kalman filtering produces the BLUE of the state of the system under observation, using all observations up to estimation time. It also produces the variance-covariance matrix of the corresponding estimation error. Kalman filtering has been applied to many different problems. A general description of the theory of Kalman filtering and of its properties can be found in, e.g., Jazwinski (1970).

In the case of assimilation of meteorological or oceanographical observations, one can see that, if one accepts the linear hypotheses which underlie Eqs. 8.24 and 8.26, Kalman filtering fulfills the goal assigned in the Introdution to an ideal assimilation system: namely, to use all the available information in order to produce the most accurate possible description of the state of the flow, together with the uncertainty resulting from the uncertainties on the various sources of information. In the present case, the available information consists, not only of the observations z_k (Eq. 8.26), but also of the model 8.24 (and of the initial estimate $x_{0,-}$ from which the whole process must be started). As for the associated uncertainties, they are defined by the covariance matrices Π and Ξ (and the initial covariance matrix $P_{0,-}$). Kalman filtering consistently combines all these elements in order to produce the BLUE 8.29 and the associated covariance matrix 8.30.

The application of Kalman filtering to assimilation of meteorological and oceanographical observations has been studied by a number of authors, in particular Ghil and collaborators (see, e.g., Ghil, 1989, or Ghil and Malanotte-Rizzoli, 1991). Experiments performed with various linear systems have produced convincing results as to the capability of the method for effectively extracting the information contained in the observations and the model. One major difficulty with Kalman filtering in the context of assimilation of meteorological and oceanographical observations is its numerical cost. Writing the first term on the last member of Eq. 8.29 under the form $S(SP_{k,+})^T$ shows that the corresponding computations require two successive matrix multiplications by S. Now, the multiplication of one vector by S corresponds to one integration of the model between times k and $k+1$ (Eq. 8.27). Implementation of Eq. 8.29 therefore requires $2n$ integrations of the model, where n is, as above, the dimension of the state vector of the model. With values of n of the order 10^6 - 10^7, this goes largely beyond the possibilities of assimilation for operational NWP, or even of *a posteriori* assimilation. In operational NWP, the computation is replaced by a simple multiplication of the variances of the analysis errors by an *a priori* specified coefficient (typically, 1.5 for a $6hr$-forecast), the associated correlations being modelled independently, as already mentioned above. The corresponding numerical cost is negligible, but that procedure amounts to ignoring the influence of the particular meteorological situation under consideration, and especially of the particular instabilities which may develop, on the evolution of the forecast error. This certainly is one of the major weaknesses of present operational assimilation methods, which, in the perspective taken here, can be described as degraded but economical forms of Kalman filtering. Recent comparisons of the results produced by variational assimilation and by algorithms similar to operational algorithms (Rabier, priv. com.) suggest that a more accurate description of the evolution of forecast error might substantially improve the quality of assimilations.

Now, the correlation between forecast errors at points located a large distance apart must be negligible, and a large proportion of the entries of covariance matrices such as $P_{k,+}$ must have zero or negligible values. This should allow to reduce the cost of computation of Eq. 8.29. In addition, it is known that the most rapidly amplifying modes in the evolution of the forecast error are geostrophic modes (see, e.g., Lacarra and Talagrand, 1988), so that it should be possible to restrict computation of Eq. 8.29 to a subset of all the model modes. These ideas have been exploited by several authors (see, e.g., Cohn and Parrish, 1991, or Dee, 1991, and references therein) in order to reduce the cost of computation the Eq. 8.29. However, no fully satisfactory algorithm has been defined so far, and active research is still being done on the problem of defining an algorithm for describing the temporal evolution of the forecast error which is both economical enough for practical implementation, and accurate enough for improving on present operational methods of sequential assimilation.

8.5 The variational form of statistical linear estimation

We will now restrict ourselves to the case where the model is supposed to be perfect, i.e. $\zeta_k = 0$ in Eq. 8.24, so that the exact evolution of the flow reduces to

$$x_{k+1} = Sx_k \tag{8.31}$$

The variational form 8.17 of the estimation problem defined by Eqs. 8.26 and 8.31 leads to considering the distance function

$$\mathcal{J}(\xi) \equiv \sum_{0\leq k\leq N} [K\xi_k - z_k]^T \Pi^{-1} [K\xi_k - z_k] \tag{8.32}$$

where $\xi = (\xi_k^T)^T$ is a sequence of model states at successive times, linked by the model Eq. 8.31. $\mathcal{J}(\xi)$ is the sum of the model-minus-observations squared differences, weighted by the inverse of the observation error covariance matrices. Minimizing the distance function 8.32 under the constraint 8.31 will produce at any time k the BLUE of the real state x_k of the system at time k, from *all the available observations*, i.e. from observations performed before, at, and after time k. In particular, the state at the end of the assimilation period will be the same as the state produced by Kalman filtering (under the assumption of an exact model, i.e. under the condition that $\Xi = 0$ in Eq. 8.29. The variational form of statistical estimation therefore provides a way to globally adjust a model to observations distributed in time.

But we can also note that the assumption of linearity, necessary to establish the link with the theory of statistical linear estimation and with Kalman filtering, is by no means necessary for a variational problem of type 8.31-8.32. One can very well consider the problem of minimizing a distance function of the form 8.32, under a constraint of the form 8.31, where K and S are not matrices, but nonlinear operators. Indeed, numerical models of the atmospheric or oceanic flows are nonlinear, and there are observations which are nonlinearly related to the atmospheric or oceanic variables one wants to estimate. For an already mentioned example, infrared radiances measured by satellites are related to the temperature and humidity profiles of the emitting atmospheric columns through the radiative transfer equation, which is strongly nonlinear. We will therefore drop for the time being the hypothesis of linearity (only to come back to it to later in order to show that it is justified in some sense) and consider the problem of minimizing a nonlinear distance function 8.32 (i.e. a distance function with nonlinear observation operators) under the constraint of a nonlinear model 8.31

It is well-known that there basically exist two methods for solving a constrained minimization problem. The principle of the first method is obvious, and consists in *reducing the constraint*, i.e. in eliminating some of the constrained variables so as to transform the problem into an unconstrained problem. In the present case, one can note that a model solution 8.31 is uniquely defined by the specification of the corresponding initial condition ξ_0. The distance function $\mathcal{J}$ can therefore be considered as a function of ξ_0 only, upon which no constraint is imposed, so that one is led to a problem of unconstrained minimization with respect to ξ_0. The second method, whose principle is much less obvious, consists in associating unknown coefficients, called *Lagrange multipliers*, to the constraints of the problem, and to form the associated *Lagrangian*. In the present case, there are N constraints 8.31, each of which of dimension n, and the set Λ of Lagrange multipliers consists of N vectors $\Lambda_k (k = 0, 1, \dots, N-1)$, each of dimension n. The associated Lagrangian reads

$$\mathcal{L}(\xi, \Lambda) = \mathcal{J}(\xi) + \sum_{0\leq k<N} \Lambda_k^T [\xi_{k+1} - S\xi_k]$$

A well-known theorem (see, e.g., Bertsekas, 1982) then says the minima of the constrained minimization problem 8.31 and 8.32 correspond to the stationary points of the Lagrangian $\mathcal{L}(\xi, \Lambda)$, considered as a function of the independent variables ξ and Λ.

The method of *adjoint equations*, which is a classical tool of control theory (Lions, 1971), seems to be by far the most efficient way for numerically solving the minimization problem 8.31-8.32. Interestingly enough, the method of adjoint equations can be derived by either reducing the constraint 8.31 so as to use only the initial state ξ_0 as independent variable (see, e.g., Le Dimet and Talagrand, 1986, or Talagrand and Courtier, 1987), or alternatively by using the technique of Lagrange multipliers (see, e.g., Thacker and Long, 1988). Assuming for instance that we want to solve problem 8.31-8.32 as a problem of unconstrained minimization with respect to the initial state ξ_0, it is necessary, in order to even start solving the problem, to be able to relate the variations of the initial state ξ_0 to the corresponding variations of the distance function $\mathcal{J}$. For a given initial state, these variations are related through the local vector of partial derivatives, or *gradient vector*, of the distance function with respect to the components of the initial state. In particular, if one is able to numerically compute the gradient for a given initial state, it will be possible to feed that gradient into a standard minimization algorithm which will determine the minimizing initial state through successive iterations. In most situations, it will of course be impossible to establish explicit analytical expressions for the gradient. It is possible to numerically (and approximately) determine the gradient through explicit finite perturbations of the initial state (see, e. g., Hoffman, 1986), but this would be much too costly for practical implementation: it would require to compute the cost function, i.e. to effectively integrate the model over the assimilation period, as many times as there are independent components in the initial state. The method of adjoint equations provides a way for computing the gradient at a numerical cost which is at most a few times the cost of one computation of the distance function. In principle the method is extremely simple. Let us consider a computer code (or part of a code) that, starting from some input vector u with components u_i $(i = 1, \ldots, q)$, produces an output vector v with components v_j $(j = 1, \ldots, r)$. The process can be described by the equation

$$v = G(u) \tag{8.33}$$

where G stands for all the computations which lead from u to v. For a given perturbation δu on the input, the resulting perturbation δv on the output is equal to first order to

$$\delta v = G'\delta u \tag{8.34}$$

where G' is the matrix of local partial derivatives, or *jacobian matrix*, of the components of v with respect to the components of u. Eq. 8.34 is called the *tangent linear equation* to expression 8.33. Let now $\mathcal{J}(v)$ be a scalar function of the output v. The gradient of $\mathcal{J}$ with respect to u is given by the chain rule

$$\frac{\partial \mathcal{J}}{\partial u_i} = \sum_{j=1}^{r} \frac{\partial v_j}{\partial u_i}\frac{\partial \mathcal{J}}{\partial v_j} \qquad i = 1, \ldots, q$$

or, in matrix notation

$$\nabla_u \mathcal{J} = G'^T \nabla_v \mathcal{J} \tag{8.35}$$

where, as before, the superscript T denotes transposition. The adjoint method is based on a systematic use of Eq. 8.35. More precisely, let us suppose that the process G is the composition of a number of more elementary processes, namely

$$G = G_M \circ \ldots G_2 \circ G_1$$

the jacobian G' will be product of the elementary jacobians

$$G' = G'_M \ldots G'_2 G'_1$$

and the transpose G'^T will be the product of the elementary transposes, taken in reversed order

$$G'^T = G'^T_1 G'^T_2 \dots G'^T_M$$

This shows that, in order to numerically determine the gradient $\nabla_u \mathcal{J}$ with respect to the input u, it is sufficient to proceed backwards through the direct computations and, at every step, to perform the corresponding transpose, or adjoint computations. The total cost of one *adjoint* computation Eq. 8.35 will generally be of the same order of magnitude as the cost of one direct computation Eq. 8.33. It can be shown, see e.g., Morgenstern, 1984, that the total operation count of one adjoint computation can be reduced to at most 4 times the total operation count of the corresponding direct computation, this ratio being reduced to 2 if one considers only multiplications and divisions. This is of course much more economical than direct perturbations of the input vector.

In the case of the determination of the gradient of the distance function 8.32 with respect to the initial state ξ_0 of the assimilating model 8.31, the adjoint computations reduce to integrating the equation (see, e.g., Talagrand and Courtier, 1987)

$$\delta'\xi_k = S'^T \delta'\xi_{k+1} + K'^T \Pi^{-1}[K\xi_k - z_k] \tag{8.36}$$

backwards in time, starting from the "final" state $\delta'\xi_{N+1} = 0$. In this equation, S' and K' are the jacobians of the operators S and K. The gradient of the distance function with respect to the initial state ξ_0 will be equal to $2\delta'\xi_0$.

It is seen that the basic model solution ξ_k under consideration explicitly appears in the adjoint equation 8.36 in the quantity $K\xi_k - z_k$, which, except for its sign, is the innovation vector of Eqs. 8.21 and 8.29. This means that the basic solution will have to be computed, and kept in memory, before the adjoint integration can be performed. In the general case of nonlinear operators S and K, the basic solution will also be necessary for determining the jacobians S' and K'. Saving the basic solution in memory may entail large core requirements, which constitute one important feature of the adjoint method.

The method of adjoint equations for performing variational assimilation of meteorological observations seems to have been first suggested by Penenko and Obraztsov (1976), who applied it to a simple, small-dimensional linear problem. Since then, a large number of experiments have been performed on (usually nonlinear) models of increasing complexity and dimensions, and with various types of observations. Experiments have been performed on both meteorological and oceanographical examples (for meteorological applications see, e. g., Lewis and Derber, 1985, Talagrand and Courtier, 1987, Derber, 1987, Courtier and Talagrand, 1987, 1990, Lorenc, 1988, Thépaut and Courtier, 1991, Navon et al., 1991; for oceanographical applications see, e.g., Thacker and Long, 1988, Sheinbaum and Anderson, 1990a and b). The first general conclusion which can be drawn from these experiments is that variational assimilation works in that it is capable of minimizing the distance function. Also, and contrary to what happens in sequential assimilation, there is propagation of information, as should be, both forward and backward in time. However, when the distance function contains only terms measuring the misfit between individual observations and model values, the minimization solution tends to contain unrealistic "noise", often under the form of small-scale oscillations and/or of ageostrophic gravity waves. The minimizing solution is physically realistic only if appropriate terms, measuring the energy contained in the small scales of the flow, or its ageostrophy, are added to the distance function (see, e.g., Courtier and Talagrand, 1990, Thépaut and Courtier, 1991). Indeed, the need for adding terms intended at avoiding unrealistic oscillations in the estimated fields is by no means restricted to meteorological or oceanographical applications, but is almost universal in problems where fields are estimated through a variational method. Such terms are often called "smoothing", "penalizing" or still "regularizing" terms. But it must be stressed that the need for appropriate smoothing is not restricted to variational methods: it is also present in statistical estimation when implemented through Kalman filtering which, as already said, must lead to the same final result as variational algorithms. In present operational optimal interpolation, the requirement for appropriate smoothing is satisfied on the one hand through the presence of the first-guess x_g, which defines what the analyzed field must be in data-void areas, and on the other hand through the "initialization" process, which filters out unrealistic gravity wave oscillations. The most recent experiments on variational assimilation of meteorological observations have been performed with multilevel primitive equation models similar, but not identical yet, to the models used in NWP (Thépaut, 1992). The remaining differences lie in the resolution, which is still significantly coarser in variational assimilation experiments (typically one order of magnitude less points in the horizontal than in operational models), and in the representation of many "physical" processes, such as convection and water phase changes, which are still absent from variational assimilation. The results confirm the results of previous experiments, and show in particular that variational assimilation, because it explicitly uses the evolution equations of the system, is able to propagate the information contained in the observations much more accurately than operational optimal interpolation. Variational assimilation, like Kalman filtering, is therefore able to assimilate observations in a way which is exactly consistent with the dynamics of the system, as described by the model equations. But, as in the case of Kalman filtering, and in spite of the fact that the adjoint equations are by far the most efficient way for computing the gradient of the distance function, the computational price to be paid is heavy: in addition to the necessity of keeping in memory the model solution produced by the direct integration which must be performed before each adjoint integration, minimization of the distance function typically requires from 10 to 30 iterations of the minimization algorithm. Each iteration itself requires one integration of the model over the assimilation period, followed by one adjoint integration. The cost of one adjoint integration is about twice the cost of one direct integration, so that one minimization typically requires the equivalent of between 50 and 100 integrations of the model over the assimilation period. This, although much less than the requirements of Kalman filtering, goes largely beyond the possibilities of present operational NWP (although it would not necessarily go beyond the capabilities of delayed assimilation or *a posteriori* experimentation).

The present situation as concerns assimilation methods is therefore rather clear. In addition to the relatively simple, rather *ad hoc*, but economical and basically satisfactory operational algorithms, there exist two broad classes of algorithms which are capable, in the general framework of statistical linear estimation, to consistently extract the information contained in the observations on the one hand and in the physical laws expressed by the assimilating model on the other : Kalman filtering and variational assimilation. In order to implement these algorithms, one must express the observations under the form defined by Eq. 8.12, i.e. one must know what has been measured, in terms of the parameters to be estimated, and with which accuracy. Now, exact implementation of either of these two classes of algorithms is numerically costly, and a large part of the research being done at present on assimilation is in effect directed at determining the most cost-efficient simplications which can be made on them. This task may indeed be with us for a long time: no end is foreseen to the increase in the power of computers and to the deep modifications in their structures, nor to the development of new observing systems and of more realistic models. Changes in any of these aspects may radically modify any conclusion one may have reached as to the most efficient way to perform assimilation.

Before coming to a number of problems which are more specific to assimilation as it will have to be performed in the context of GEWEX, it may be useful to briefly comment on the relative advantages and disadvantages of Kalman filtering and variational assimilation. As already mentioned, both algorithms will lead to the same final state at the end of the assimilation period in the case of linear observations (i.e. of a linear observation operator H) and of a perfect (i.e. $\zeta_k = 0$) linear model 8.24. A basic difference between the two algorithms is that Kalman filtering carries information only from the past into the future, while variational assimilation carries information in both time directions. On the other hand, variational assimilation, contrary to Kalman filtering, is unable to take into account the fact that the assimilating model, like the observations, will never be perfect and will always contain errors. But it must also be said that these differences are only true of these algorithms as they have been described here and as they have been mostly implemented so far in meteorological and oceanographical applications. As concerns Kalman filtering, there exists a procedure, called Kalman *smoothing* (see, e.g., Anderson, 1979), which allows, once a first pass has been performed over the assimilation period, to proceed backward in time so as to obtain, at any intermediate time k, the BLUE of the state of the system at time k from all available observations, performed before, at or after time k. We will not describe here the theory of Kalman smoothing, which is related to the theory of adjoint equations, and will only refer to Gaspar and Wunsch (1989) as a simple but instructive example of an application of Kalman smoothing to an oceanographical problem. And, as concerns variational assimilation, it can incorporate model errors: it suffices to impose the model equation 8.31 not as a constraint to be exactly satisfied by the sequence ξ_k of assimilated states, but (to use the vocabulary introduced by Sasaki, 1970) as a "weak constraint" to be satisfied only approximately. This can be done by modifying the distance function 8.32 to

$$\mathcal{J}(\xi) = \sum_{0<k\leq N} [K\xi_k - z_k]^T \Pi^{-1} [K\xi_k - z_k] + \sum_{0\leq k<N} [\xi_{k+1} - S\xi_k]^T \Xi^{-1} [\xi_{k+1} - S\xi_k] \tag{8.37}$$

where Ξ is, as in Eq. 8.25 the variance-covariance matrix of the model error. The meaning of the second sum on the right-hand-side of Eq. 8.37 must be clear: it simply expresses that the difference $\xi_{k+1} - S\xi_k$ must not be considered as exactly zero, as it would be if the model was exact, but equal to zero only within the uncertainty defined by the matrix Ξ. Accordingly, the sequence of states must be considered as unconstrained, and the minimization of $\mathcal{J}$ must be performed with respect to the entire sequence $\xi = (\xi_k^T)^T$. This of course multiplies the dimension of the minimization problem by the number of timesteps over the assimilation period. It is not difficult to see from the variational form 8.17 of statistical linear estimation, that the sequence of states which minimize Eq. 8.37 is made up of the BLUEs, at all times k, of the state of the system, from all observations over the entire assimilation period. It results in particular that minimization of Eq. 8.37 must lead to the same sequence of assimilated states as Kalman filtering, followed by Kalman smoothing. Variational assimilation using the model equations as "weak constraints" has apparently not been implemented so far in meteorological or oceanographical applications, but it is among the possibilities which must be studied in the future. Another difference between Kalman filtering and variational assimilation is that the former produces, in addition to the BLUE of the state of the system, the variance-covariance matrix of the corresponding estimation error, while the latter produces only the BLUE. In this sense, only Kalman filtering fulfills the goal assigned to an ideal assimilation system in the introduction to these notes. This of course is obtained at the already mentioned much higher cost of Kalman filtering. Now, it is easy to verify from Eq. 8.16 and 8.17 that the variance-covariance matrix of the estimation error is the inverse of the hessian (matrix of second derivatives) of the distance function. The question therefore arises whether the inverse hessian can be computed in variational assimilation, at least to a sufficient degree of accuracy, at a lower cost than in Kalman filtering. Indeed, some minimization algorithms, of the so-called quasi-Newton type (see, e. g., Gill et al., 1982), do compute an approximate inverse hessian in the course of the minimization. The problem of the determination of the estimation error in variational assimilation is also one of the problems which will have to be studied in the future.

To conclude with the theoretical and methodological aspects of assimilation, we will discuss a point which we have so far left in some obscurity, namely the validity of the linear hypothesis which underlies the theory Kalman filtering and which, although not necessary for variational assimilation, gives it a special significance and facilitates the understanding and analysis of the results it produces. Not only the equations governing the atmospheric and oceanic flows are nonlinear, but their nonlinearity is at the origin of one of the most important properties of these flows, namely their chaotic character. This character imposes stringent limits on the predictability of these flows, and one can legitimately wonder whether a linear hypothesis is legitimate in the context of assimilation. Considering first the equations for Kalman filtering, let us assume that the model operator S and the observation operator K (Eqs. 8.24 and 8.26 respectively) are nonlinear. If the difference $x_{k,+} - x_k$ is small enough, the quantity $Sx_{k,+} - Sx_k$ (Eq. 8.29 can be approximated by $S'(x_{k,+} - x_k)$, where S' is the jacobian matrix of the operator S, taken at point $x_{k,+}$. Eq. 8.29 accordingly becomes

$$P_{k+1,-} = S'P_{k,+}S'^T + \Xi \tag{8.38}$$

Similarly, if the difference $x_{k+1} - x_{k+1,-}$ is small enough so that the innovation vector $z_{k+1} - Kx_{k+1,-} = Kx_{k+1} - Kx_{k+1,-} + \eta_{k+1}$ can be approximated by $K'(x_{k+1} - x_{k+1,-}) + \eta_{k+1}$, calculations show that Eqs. 8.21 and 8.22 can be respectively replaced by

$$x_{k+1,+} = x_{k+1,-} + P_{k+1,-}K'^T[K'P_{k+1,-}K'^T + \Pi]^{-1}(z_{k+1} - Kx_{k+1,-}) \tag{8.39}$$

and

$$P_{k+1,+} = P_{k+1,-} - P_{k+1,-}K'^T[K'P_{k+1,-}K'^T + \Pi]^{-1}K'P_{k+1,-} \tag{8.40}$$

In these equations, K has been replaced by the jacobian K', except in the expression for the innovation vector. The algorithm defined by Eqs. 8.38-8.40 is called *extended Kalman filtering* (see, e.g., Jazwinski, 1970). It is valid whenever the differences between the real and estimated states of the system are small enough so as to allow local linearizations as just described.

A similar argument holds for variational assimilation. For a linear model and linear observation operators, the distance function 8.32 will be a quadratic function of the initial state ξ_0. For a nonlinear model or nonlinear observation operators, the distance function will not be quadratic, but will remain approximately quadratic in a neighbourhood of its minimum. If the initial uncertainty on the state of the system (defined for example by the point from which the minimization process is initiated) is small enough to insure that, at any stage of the assimilation, the estimated state of the system will always lie within that neighbourhood, the theoretical nonlinearity of the distance function will have no practical effect. In particular, the minimizing solution will be the BLUE of the possess the many properties of the BLUE.

The linear hypothesis made previously, and the associated logic of statistical linear estimation, including in particular the equivalence between sequential and variational assimilation, will therefore be valid if the differences between the real and estimated states of the system are always small enough to allow the local linearizations described above. The validity of this so-called *tangent linear approximation* has been checked systematically in a number of situations. For instance, Lacarra and Talagrand (1988) have shown on a barotropic model that, for realistic amplitudes of the error on the initial state of the flow, a linear approximation for the evolution of the forecast error is valid up to about 48 or 72 hours. Similarly, Thépaut and Moll (1990) have shown that, within the uncertainty existing in practice on the atmospheric profiles of temperature and humidity, the tangent linear hypothesis is valid for the radiance observations performed by the TIROS Operational Vertical Sounder (TOVS) carried by the satellites of the NOAA series.

In addition, many results have indirectly confirmed the validity of the tangent linear hypothesis. This means that sequential or variational assimilation can confidently be expected to produce reasonable estimates of the state of the flow. But it certainly does not mean that there do not remain problems. There must exist limitations to the validity of the tangent linear hypothesis, due for instance to the presence, in the physically most realistic models, of processes capable of inducing sharp variations in the model fields. These limitations have so far not been clearly identified, and it is not clear either what will have to be done in circumstances where the tangent linear hypothesis is not valid. But it is very clear at the same time that much development work remains to be done within the context of statistical linear estimation.

8.6 Assimilation and "physical" processes

We have dealt so far with assimilation considered as a general problem and with the general principles underlying assimilation methods, but the more specific problems which will arise in the context of GEWEX have not been much considered. Two classes of difficulties can be identified which, although not absolutely specific to GEWEX, will have to be considered for an accurate estimation of the water and energy cycles. The first of these difficulties has to do with the increasing complexity of future observing systems, including various types of remote sensing instruments, which will produce data linked to the variables to be estimated through "complicated" relationships. The second difficulty is linked to the "physical" processes, associated in particular with phase changes of water, which can introduce sharp variations in the meteorological fields. Such processes must be explicitly represented in realistic models. There exist in particular processes, such as for instance convection or water condensation, which are initiated when the value of a particular parameter reaches a critical value. Such "threshold" processes can introduce discontinuities in the model fields with respect to initial conditions.

As concerns "indirect" observations, it has been shown above that, in order to implement statistical estimation, it is necessary to know, for each individual observation, what has been measured, and with which accuracy. A rather general practice so far, when dealing with satellite observations, has been to first "invert" them to "geophysical variables", such as for instance temperatures and humidities, and then to use the inverted fields as observations in the assimilation algorithm. For instance, the radiance observations performed by the satellites of the NOAA series are routinely inverted by the American National Environmental Satellite, Data and Information Service (NESDIS) to produce the so-called SATEM temperature and humidity "observations". These are used operationally in a number of assimilation schemes. Now, such a preliminary inversion is by no means necessary. It does not avoid the basic need for the definition of an appropriate observation operator and for the specification of the associated observation error. And it usually requires a first-guess which itself depends on the other available observations. This leads to interdependence of the errors associated with the various "observations" used in the assimilation.

The problems raised by such an interdepedence can be solved in the context of statistical estimation, but it certainly seems preferable to avoid them in the first place. For these reasons, and also in order to define a systematic approach to be followed for any type of observations, the tendency is now to avoid as much as possible preliminary processing of the observations before the assimilation, and to introduce raw observations in the assimilation algorithm with an appropriate observation operator. In the case of the radiance measurements performed by the satellites of the NOAA series, the associated problems have been studied, among others, by Eyre (1989a and b) and Thépaut and Moll (1990). The general conclusions are that direct use of the radiance measurements should produce better analyses than use of SATEMs, and several meteorological services are taking steps to directly incorporate radiance measurements in their operational assimilation algorithms. The same general trend is followed for all types of measurements: for instance, radial winds measured by Doppler effect either from ground-based radars or from satellite-borne lidars can be assimilated through an observation operator which reduces to the computation of the wind component along the appropriate direction (see respectively, e.g., Sun et al., 1991, and Courtier et al., 1992). It is certainly the same approach which will be followed for the assimilation of observations made by future observing systems.

As concerns physical processes, and more particularly threshold processes, their effect is more immediately visible in variational assimilation, where they can be at the origin of the existence of multiple minima in the distance function, as shown for instance by Douady and Talagrand (1990) in a rather idealized case. But this effect, which renders the tangent linear hypothesis invalid, is also present in sequential estimation. It is avoided in the practice of present operational assimilation, performed with models which do contain threshold processes, because the approximate description of the temporal evolution of the forecast error smooths out the corresponding discontinuities. The problem of the threshold processes has not been very much studied so far, but it will require due consideration, especially when the time comes to assimilate data, such as precipitation data, which are directly related to these processes.

8.7 Conclusions

Our primary purpose in these notes was to describe the principles which are at the basis of assimilation of observations. These principles are those of statistical linear estimation, and it has been shown that they lead to a generalised least-square approach, amounting to minimizing a measure of the difference between the available observations and the state to be estimated. The word "observations" must be taken here in a very broad sense, to include all information available in quantitative form, and in particular the equations governing the assimilating model. Two classes of algorithms, sequential and variational assimilation, can be used for actually performing the required computations. Both algorithms are costly, and neither of them can be considered at the present stage as intrinsically superior.

The more specific problems associated with assimilation as it will have to be implemented in the context of GEWEX have been considered more briefly, in part because these problems have not yet been studied in much detail. But it is clear that they will have to be solved within the general principles of statistical estimation. In particular, it will always be necessary to specify, for each individual observation, and as precisely as possible, what has been observed, in terms of the variables to be estimated, and with which accuracy. This amounts to expressing the observations under the form 8.12, where the (possibly nonlinear) observation operator H defines what has been observed, and the first and second order statistical moments of the "observation error" ϵ (mean $E(\epsilon)$ and covariances $\Sigma = E(\epsilon\epsilon^T)$) specify the accuracy. Any assimilation requires hypotheses, either explicit or implicit, on what H and the statistical moments of ϵ are.

This leads us to our final remark. Estimating H and the statistical moments of ϵ is by itself an estimation problem, *a priori* no easier than estimating the state of the atmospheric flow at a given time. One difference is that these quantities can be estimated by appropriate statistical accumulation. And, as concerns H, the physics of the measurement process is of course also fundamental. It has been repeatedly mentioned in these notes that statistical estimation, in addition to the BLUE of the state of the sytem, produces the covariances of the associated estimation errors. These covariances depend on H and Σ (Eq. 8.16). Any disagreement between the predicted anlysis-minus-observations differences and the *a posteriori* effectively observed differences must therefore be due to inaccurate estimation of H and/or Σ, and must be usable for improving the corresponding estimates. We refer to Daley (1992) for a recent work along these lines of *adaptive filtering*, which will presumably be the next major step in the development of assimilation methods.

8.8 References

Anderson BDO (1979) Optimal Filtering. Prentice Hall, *Englewood Cliffs* 357 pp.

Bengtsson L, Shukla J (1988) Integration of space and in situ observations to study global climate change. *Bull Amer Met Soc* 69: 1130-1143.

Bertsekas DP (1982) Constrained optimization and Lagrange multipler methods. *Academic Press* London, 395 pp.

Cohn SE, Parrish DF (1991) The Behavior of Forecast Error Covariances for a Kalman Filter in Two Dimensions. *Mon Wea Rev* 119: 1757-1785.

Courtier P, Gauthier P, Rabier F, Flamant P, Dabas A, Lieutaud F, Renault H (1992) Study of preparation of the use of Doppler wind lidar information in meteorological assimilation systems. Final report, *ESA Contract 8850/90/HGE-I* ESA, Paris.

Courtier P, Talagrand O (1987) Variational assimilation of meteorological observations with the adjoint vorticity equation. I : Numerical results. *Quart J Roy Met Soc* 113: 1329-1347.

Courtier P, Talagrand O (1990) Variational assimilation of meteorological observations with the direct and adjoint shallow-water equations. *Tellus* 42A: 531-549.

Daley R (1991) Atmospheric Data Analysis. Cambridge Atmospheric and Space Science Series, *Cambridge University Press* Cambridge, 457 pp.

Daley R (1992) The lagged innovation covariance: A performance diagnostic for atmospheric data assimilation. *Mon Wea Rev* 120: 178-196.

Dee DP (1991) Simplification of the Kalman filter for meteorological data assimilation. *Quart J Roy Met Soc* 117: 365-384.

Derber JC (1987) Variational four dimensional analysis using quasi-geostrophic constaints. *Mon Wea Rev* 115: 998-1008.

Derber JC, Parrish DF, Lord SJ (1991) The new global operational analysis system at the National Meteorological Center. *Wea Forecasting* 6: 538-547.

Douady D, Talagrand O (1990) The impact of threshold processes on variational assimilation. In: Proceedings of International Symposium on Assimilation of Observations in Meteorology and Oceanography, *World Meteorological Organization* Geneva, Switzerland, 486-487.

Eyre J (1989a) Inversion of cloudy satellite sounding radiances by nonlinear optimal estimation. I: Theory and simulation for TOVS, *Quart J Roy Met Soc* 115: 1001-1026.

Eyre J (1989b) Inversion of cloudy satellite sounding radiances by nonlinear optimal estimation. II: Application to TOVS data, *Quart J R Met Soc* 115: 1027-1037.

Gandin LS (1963) Objective analysis of meteorological fields. *Gidrometeor Izd* Leningrad (in Russian), (English Translation by Israel Program for Scientific Translations, Jerusalem, 1965).

Gaspar P, Wunsch C (1989) Estimates from altimeter data of barotropic Rossby waves in the Northwestern Atlantic Ocean. *J Phys Oceanogr* 19: 1821-1844.

Ghil M (1989) Meteorological data assimilation for oceanographers. Part I. Description and theoretical framework. *Dyn Atmos Oceans* 13: 171-218.

Ghil M, Malanotte-Rizzoli P (1991) Data assimilation in meteorology and oceanography. *Adv in Geophys* 33: 141-266.

Gill PE, Murray W, Wright MH (1982) Practical Optimization. *Academic Press* London.

Hoffman RN (1986) A four-dimensional analysis exactly satisfying equations of motion. *Mon Wea Rev* 114: 388-397.

Jazwinski AH (1970) Stochastic Processes and Filtering Theory.*Academic Press* New-York.

Kalman RE (1960) A new approach to linear filtering and prediction problems. *J Basic Eng* 82D: 35-45.

Kinter III JL, Shukla J (1989) Meeting review : Re-analysis for TOGA (Tropical Oceans Global Atmophere). *Bull Amer Met Soc* 70: 1422-1427.

Lacarra JF, Talagrand O (1988) Short-range evolution of small perturbations in a barotropic model. *Tellus* 40A: 81-95.

Le Dimet FX, Talagrand O (1986) Variational algorithms for analysis and assimilation of meteorological observations: Theoretical aspects. *Tellus* 38A: 97-110.

Lewis JM, Derber JC (1985) The use of adjoint equations to solve a variational adjustment problem with advective constraints. *Tellus* 37A: 309-322.

Lions PL (1971) Optimal control of systems governed by partial differential equations. *Springer-Verlag* Berlin, 396 pp.

Lorenc A (1981) A global three-dimensional multivariate statistical interpolation scheme. *Mon Wea Rev* 109: 701-721.

Lorenc A (1986) Analysis methods for numerical weather prediction. *Quart J Roy Met Soc* 112: 1177-1194.

Lorenc A (1988) Optimal nonlinear objective analysis. *Quart J Roy Met Soc* 114: 205-240.

Morgenstern J (1984) How to compute fast a function and all its derivatives. A variation on the theorem of Baur-Stressen. *Report No 49* Laboratoire CNRS 168, Universit de Nice, Nice, France, 5 pp.

Navon IM, Zou X, Johnson K, Derber J, Sela J (1991) Variational data assimilation with an adiabatic version of the NMC spectral model. *Report FSU-SCRI-91-13* The Florida State University, Tallahassee, Florida, USA, 43pp.

Penenko VV, Obraztsov NN (1976) A variational initialization method for the fields of the meteorological elements (English translation). *Soviet Meteorol Hydrol* 11: 1-11.

Sasaki Y (1970) Some basic formalisms in numerical variational analysis. *Mon Wea Rev* 98: 875-883.

Sheinbaum J, Anderson DLT (1990a) Variational Assimilation of XBT Data. Part I. *J Phys Oceanogr* 20: 672-688.

Sheinbaum J, Anderson DLT (1990b) Variational Assimilation of XBT Data. Part II: Sensitivity Studies and Use of Smoothing Constraints. *J Phys Oceanogr* 20: 689-704.

Sun J, Flicker DW, Lilly DK (1991) Recovery of three-dimensional wind and temperature fields from simulated single-doppler radar data. *J Atmos Sci* 48: 876-890.

Talagrand O, Courtier P (1987) Variational assimilation of meteorological observations with the adjoint vorticity equation. I : Theory. *Quart J Roy Met Soc* 113: 1311-1328.

Temperton C (1988) Implicit normal mode initialization. *Mon Wea Rev* 116: 1013-1031.

Thacker WC, Long RB (1988) Fitting dynamics to data. *J Geophys Res* 93: 1227-1240.

Thépaut JN (1992) Variational assimilation of meteorological observations with a multilevel primitive-equation model. *in preparation*

Thépaut JN, Courtier P (1991) Four-dimensional variational data assimilation using the adjoint of a multilevel primitive-equation model. *Quart J Roy Met Soc* 117: 1225-1254.

Thépaut JN, Moll P (1990) Variational inversion of simulated TOVS radiances using the adjoint technique. *Quart J Roy Met Soc* 116: 1425-1448.

Vautard R (1990) Multiple weather regimes over the North Atlantic: Analysis of precursors and successors. *Mon Wea Rev* 118: 2056-2081.

Chapter 9

FORCING THE OCEAN BY HEAT AND FRESHWATER FLUXES

Jürgen Willebrand
Institut für Meereskunde an der Universität Kiel
D-2300 Kiel 1
Germany

9.1 Introduction

The fluxes of momentum, energy, and water through the sea surface constitute the principal coupling between ocean and atmosphere. The momentum flux (wind stress) is a main driving force for oceanic motions, in particular for currents near the surface, along continental margins and in the main thermocline, whereas the fluxes of heat and freshwater drive the slower thermohaline component of the oceanic circulation.

Energy enters the ocean in different ways, and is stored in the ocean in different forms. The kinetic energy flux through direct work by the wind is of the order $10^{-3} Wm^{-2}$. This is a very small number, compared to e.g. the flux of internal energy which occurs via radiative, sensible and latent heat transfer and typically reaches 100 Wm^{-2} and more. Nevertheless, that kinetic energy flux is very important as the wind-induced currents redistribute temperature and salinity and hence influence ocean-atmosphere exchange.

Oort et al. (1989) have estimated the storage of the different forms of energy in both ocean and atmosphere. Very roughly, one finds for the ocean

$$E_k \;:\; E_p^a \;:\; E_p \;:\; E_i \;=\; 1 \;:\; 10^2 \;:\; 10^7 \;:\; 10^9$$

where E_k, E_p^a, E_p, and E_i denote kinetic, available potential, total potential, and internal energies, repectively. The kinetic energy associated with the mean circulation in the ocean is approximately $E_k = 3 \cdot 10^{18} J$. That value which corresponds to an average current velocity of 5 $cm\, s^{-1}$ is

NATO ASI Series, Vol. I 5
Energy and Water Cycles in the Climate System
Edited by E. Raschke and D. Jacob

only a tiny fraction of the total energy in the ocean. The available potential energy E_p^a, i.e. that fraction of the potential energy which can be converted into kinetic energy by adiabatic processes, is two orders of magnitude higher. This is consistent with ocean circulation theory which requires that $E_p^a/E_k = (L/R)^2$ where L is a characteristic horizontal scale of motion and R is the baroclinic Rossby deformation radius (around 30-50 km in mid-latitudes). Most of the available potential energy is indirectly caused by the wind, through geostrophic adjustment of the density field. A certain fraction can however also result from the buoyancy fluxes associated with heat and freshwater fluxes.

Both kinetic and available potential energy are *very* small compared to the overall potential energy and the internal energy. While the interpretation of E_p is somewhat ambiguos as it depends on the choice of a reference state, any reasonable definition will yield $E_p^a \ll E_p$. Most of the potential energy reservoir is therefore not available via adiabatic pathways, it can however be activated when diabatic (i.e., mixing) processes are involved. It is for this reason that mixing plays such a central role for the ocean circulation, and that the correct parametrization of subgrid-scale mixing in ocean circulation models is so important.

The corresponding ratios for the atmosphere are, very approximately

$$E_k : E_p^a : E_p : E_i = 1 : 4 : 600 : 1500$$

and illustrate that kinetic energy in the atmosphere is relatively much more important than in the ocean. The ratio E_p^a/E_k indicates that the scales of motion in the atmosphere are much closer to the deformation radius (which here is around 1000 km). The total kinetic energy in the atmosphere is approximately $E_k = 6 \cdot 10^{20} J$. Thus, both kinetic and available potential energy reside mainly in the atmosphere, whereas the internal energy in the ocean/atmosphere system is almost completely stored in the ocean.

The evaporation over the ocean is a dominant process in the global water cycle. On average, nearly 1 m per year evaporate, corresponding to a global flux of $10^7 m^3 s^{-1}$, or 10 Sv (1 Sverdrup = $10^6 m^3 s^{-1}$ is a convenient unit to measure ocean currents, e.g. the Gulf Stream near Florida transports approximately 30 Sv). Besides controlling the atmospheric latent heat budget, the freshwater flux influences the oceanic salinity and hence density. That influence is often small compared to the density change caused by heat fluxes, but can be fundamentally important for changes in the thermohaline circulation. The amount of water stored in the atmosphere is very small, it corresponds to a water column of only a few centimeters.

The dynamical reaction of the ocean to forcing by heat and freshwater fluxes occurs at several time scales. On monthly and seasonal time scales, the storage of heat in the surface layers is a most relevant process. On time scales of years to a few decades, oceanic motions are mainly forced by the wind. The wind-driven circulation is however strongly affected by the density stratification which in turn is a consequence of the large-scale water mass distribution. On even longer time scales, the dominant mode of motion is the thermohaline circulation which is directly forced by thermohaline fluxes.

The fluxes between ocean and atmosphere are at the heart of the global energy and water cycles. Some aspects of the transport and redistribution of heat and freshwater by ocean currents are discussed in the following. Furthermore, we will consider the oceanic reaction to these fluxes, with the focus being on decadal and longer time scales, and look in particular at the role of freshwater forcing for long-term variations in the ocean-atmosphere system.

9.2 Surface fluxes and large-scale transports

The internal energy flux Q across the air-sea interface is the sum of various contributions,

$$Q = Q_s - Q_i - Q_h - Q_e \tag{9.1}$$

where Q_s and Q_i denote short-wave and long-wave radiation, respectively, Q_h the sensible and $Q_e = L_e E$ the latent heat flux. For the dynamics of the atmosphere, not only the sum of these fluxes but also their individual contribution is important as the atmosphere reacts quite differently e.g. to heating by long-wave radiation or by latent heat flux. In contrast, for the ocean circulation mainly the total sum of all fluxes is relevant as radiation is absorbed in the upper few meters which are normally well mixed by wind stirring. The vertical distribution of radiation is of a certain importance only in low latitudes where radiation can penetrate below the mixed layer and hence destabilize the stratification. The latent heat flux contribution is only insofar individually important as it is proportional to the evaporation E which contributes to the buoyancy flux by changing oceanic salinity.

The various contributions in Eq. 9.1 can in principle be expressed parametrically through sea surface temperature and various atmospheric variables, mainly wind speed, air temperature, water vapour mixing ratio, stability, and cloudiness. These *bulk formulae* involve empirical transfer coefficients which are derived from local air-sea exchange measurements and are subject to considerable uncertainties. For details the reader is referred e.g. to Isemer et al. (1989). Climatological estimates of the net energy transfer have been computed using different parametrizations, e.g. by Hsiung (1985) for the global ocean (except high latitudes) and by Isemer and Hasse (1987) for the North Atlantic. Overall, their accuracy is not better than 30–40 Wm^{-2}, in regions of strong exchange the various estimates differ by more than 50 Wm^{-2}.

The flux of freshwater between ocean and atmosphere is not very accurately known. The main difficulties arise from an accurate determination of precipitation over sea. The global compilation by Baumgartner and Reichel (1975) has long served as a standard reference. A more recent estimate for the North Atlantic has been given by Schmitt et al. (1989) who used precipitation estimates by Dorman and Bourke (1981). The accuracy of such maps is not easy to quantify, 30-50 $cm/year$ for the uncertainty of the difference $E - P$ may still be on the optimistic side. This means that over considerable regions of the ocean even the sign of the freshwater flux is uncertain.

The net heat flux Q and the net freshwater flux $H = E - P$ contribute to the density flux F_ρ according to

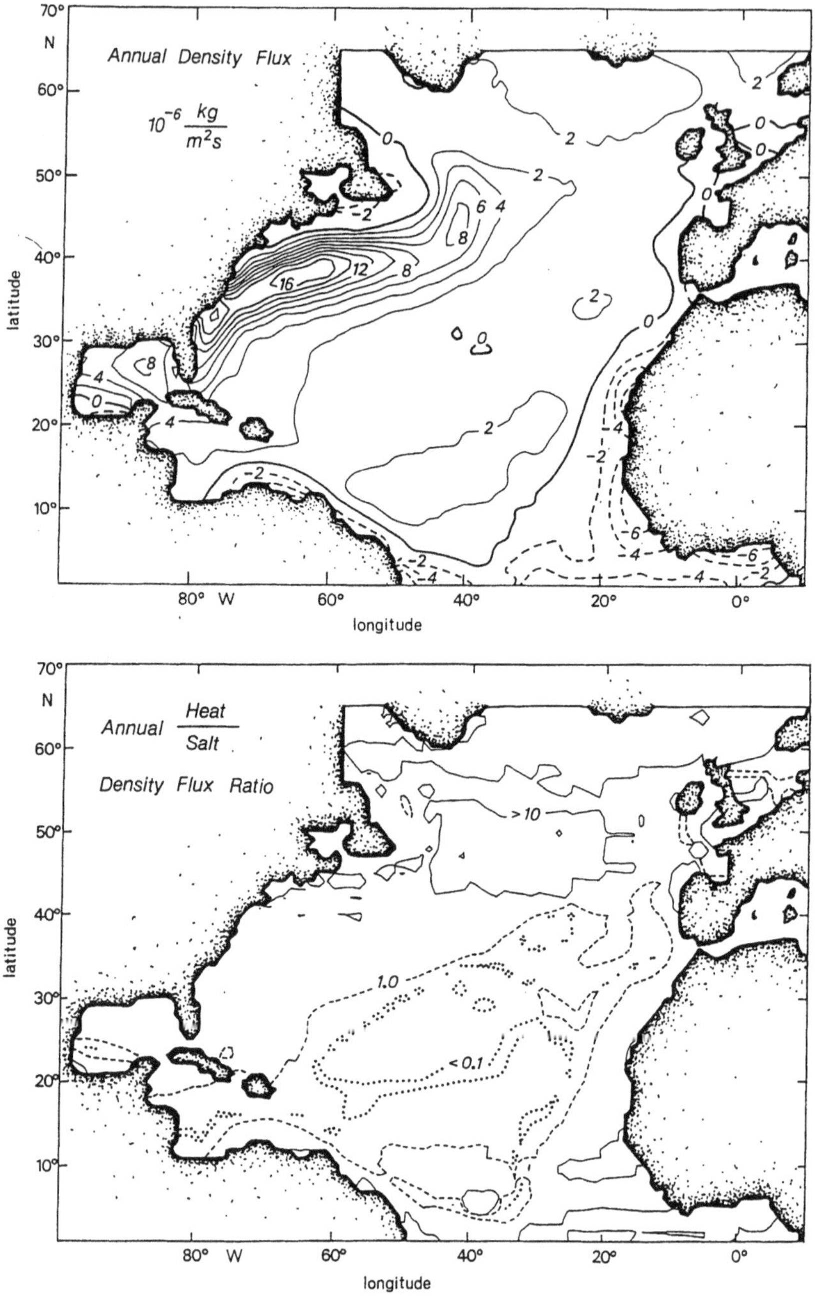

Figure 9.1: *(a) Net density flux based on heat fluxes and evaporation by Isemer and Hasse (1987) and on the precipitation by Dorman and Bourke (1981). Units are* $10^{-6} kg\ m^{-2} s^{-1}$*). (b) Ratio of thermal and haline contributions to the net density flux. Courtesy of K. Speer.*

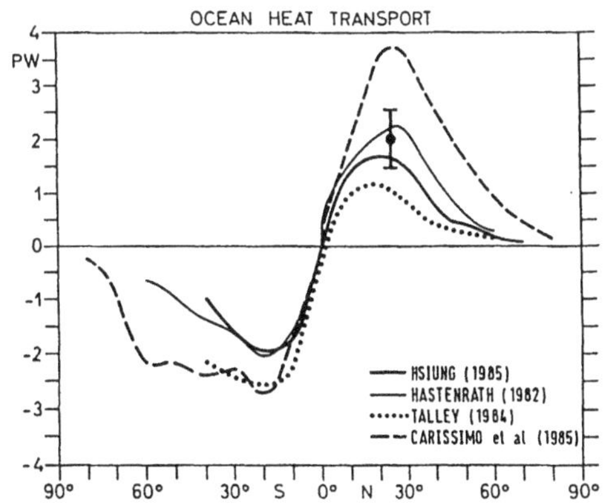

Figure 9.2: *Northward transport of heat by ocean currents, as estimated by different authors using air-sea fluxes (full and dotted lines) resp. as the residual of total and atmospheric transport (dashed). Also indicated is the direct estimate at 24 °N by Bryden et al. (1991).*

$$F_\rho = -\frac{\alpha}{c_p}Q + \frac{\beta\rho S}{(1-S)}H \tag{9.2}$$

where α, β are thermal and haline expansion coefficients, respectively, and ρ and c_p are density and heat capacity of sea water. K. Speer (1991, pers. comm.) has computed the components of Eq. 9.2 for the North Atlantic from climatological data. The density flux is dominated by the heat loss over the Gulf Stream and parts of the North Atlantic Current (Fig. 9.1(a)), regions of net buoyancy gain are found only in the equatorial zone and off the African coast. The ratio of the two terms in Eq. 9.2 is displayed in Fig. 9.1(b) and shows somewhat surprisingly that the haline buoyancy flux exceeds the thermal component over much of the subtropical gyre, a region where the absolute magnitude of the buoyancy flux is however weak. A corresponding analysis which was performed by Schmitt et al. (1989) with different flux parametrizations gave qualitatively similar results, with a relatively stronger thermal contribution. At latitudes north of 45 °, the buoyancy loss due to cooling is an order of magnitude higher than the buoyancy gain due to excess precipitation. The fact that in the climatological average the haline buoyancy flux in high latitudes is small compared to the thermal component, does however not permit the conclusion that salinity has little influence on the circulation. Quite to the contrary, changes in the high-latitude surface salinity can have a profound effect on the circulation, as will be discussed in section 9.4.

The meridional transport of heat by ocean currents has long been recognized as a key parameter of the global climate system, its accurate determination remains a challenging problem in climate research. Various methods have been used so far to estimate that parameter. Assuming steady state, the transport can be found by integrating the net air-sea flux across latitude circles. Three such estimates which are based on the work by Hastenrath (1982), Talley (1984) and Hsiung (1985) are shown in Fig. 9.2. While the resulting global transports show some qualitative agreement, e.g. all exhibit net poleward transport in both hemispheres, the differences are nevertheless quite considerable. Maximum values in the northern hemisphere differ by nearly 100 % (between 1.2 and 2.3 *PW*), and the transports in some regions may even differ in sign.

An alternative determination of the meridional heat transport is based on a direct evaluation of $\int vT dz dx$ along zonal hydrographic sections, with the meridional velocity being estimated geostrophically from the density gradient. This method requires knowledge of the absolute velocity (only the vertical shear can be estimated from the thermal wind relation), and also of the wind stress to determine the Ekman transport in the surface layer. The probably most accurate direct heat transport determination is available at 24 ° N which is near the latitude where the maximum northward transport can be expected. For the Atlantic, the number of 1.2 ± 0.3 *PW* given by Hall and Bryden (1982) has been widely accepted. It has recently been complemented by an estimate for the Pacific, yielding 0.76 *PW* (Bryden et al., 1991). As an interesting aside, it is noteworthy that the transport mechanisms in both oceans are rather different, involving vertical overturning circulation in the North Atlantic and a strong horizontal recirculation in the North Pacific. Combining both numbers results in a global transport of 2 *PW*, with an estimated error of ± 0.5 *PW*. As seen from Fig. 9.2, this value is compatible with two of the three flux-based estimates.

The uncertainty in the parametrizations of the contributions to Eq. 9.1 is the major source of error of flux-based transport estimates. It is in principle possible to combine the two methods, in order to utilize the perceived higher accuracy of direct transport computations to reduce the uncertainty of the parametrizations. This idea was taken up by Isemer et al. (1989) who constructed a method to fine-tune the parameters in Eq. 9.1 so that the resulting transport agrees, in a least-squares sense, with direct estimates. It turned out that it was sufficient to modify several parameters within 50% of their standard deviation in order to reconcile the flux-based transport in the North Atlantic with the value by Hall and Bryden. The single most important parameter was found to be the coefficient for turbulent latent heat exchange resp. evaporation. The direct determination of oceanic heat transport across a number of zonal sections in various ocean basins is one of the main objectives of the World Ocean Circulation Experiment (WOCE). When this data set becomes available, it can be expected that it will significantly contribute to narrowing down the uncertainties of air-sea flux estimates.

An unresolved issue is the apparent discrepancy of all these estimates with indirect evidence from atmospheric and radiation measurements. Carissimo et al. (1985) have estimated the ocean heat transport as the residuum between the total transport as inferred from satellite radiometry, and the atmospheric transport as calculated from the radiosonde network (dashed curve in Fig. 9.2). To close the global heat budget, an oceanic transport reaching nearly 4 *PW* would be required. Such a large value is compatible neither with the estimates from air-sea fluxes nor with direct oceanic determinations.

The transport of freshwater by the oceans has recently been recalculated by Wijffels et al. (1992). Based on the atlas by Baumgartner and Reichel (1975), they found a mean northward transport in the Pacific, and a southward transport throughout the Atlantic. These results seem to be at variance with an earlier calculation by Stommel (1980) who had found equatorward transport of freshwater everywhere except in the South Pacific. The difference results from a different transport through Bering Strait which from direct current measurements is now estimated as $0.8 \pm 0.1 Sv$ (Coachman and Aagaard, 1988). That transport is part of both a freshwater and a salt transport clockwise around the American continent which however does not contribute to the (dynamically relevant) freshwater divergence. The total meridional transport of freshwater (not shown) is equatorward outside an equatorial strip between $\pm 20^\circ$ and reaches approximately 0.8 Sv. It is basically consistent with estimates based on atmospheric water vapour flux divergences.

9.3 The structure of surface boundary conditions

Ocean and atmosphere are coupled by the fluxes of momentum, heat and freshwater through the sea surface. The coupling through the momentum flux is however rather different from that through thermohaline fluxes.

The momentum flux constitutes a main driving force for the oceanic circulation, and is also an important parameter for the atmospheric circulation. Momentum is transferred through wind-generated surface gravity waves, its quantitative description would in principle require a specification of the full wave spectrum. For climatological purposes, a simplified formulation is normally employed whereby the normal component of the momentum flux tensor (wind stress) is expressed in terms of wind velocity and an empirical drag coefficient (see the article by L.Hasse in this book). To a good approximation, the wind stress depends only on atmospheric parameters but not on the oceanic state, and thus the coupling provided by momentum flux is somewhat one-sided. In other words, the atmosphere forces the ocean but not vice versa. A minor exception arises from a weak dependence of the drag coefficient on atmospheric stability and hence sea surface temperature, an effect that can lead to a certain intensification of wind stress over warm ocean currents (Behringer et al., 1979).

With respect to the fluxes of heat and freshwater, the situation is very different. As discussed in the previous section, sensible, latent and radiative heat fluxes and evaporation depend not only on various atmospheric parameters (notably wind, air temperature, moisture, and cloudiness) but also very strongly on ocean surface temperature. Both the net heat flux Q and the net freshwater flux H can hence be expressed as

$$Q = Q(T, \mathbf{a}) \tag{9.3}$$

$$H = H(T, \mathbf{a}) \tag{9.4}$$

where T denotes the sea surface temperature (SST) and $\mathbf{a} = (a_1, a_2, \ldots)$ the set of all relevant atmospheric variables (including radiation parameters and precipitation).

The dependence of Q and H on both oceanic and atmospheric parameters constitutes a true coupling between ocean and atmosphere as the evolution of each system depends on the state of the other. Atmospheric models are frequently run in stand-alone mode using Eq. 9.3 and 9.4, with prescribed ocean surface temperature. This is appropriate for time scales which are much shorter than those relevant for oceanic motions. For ocean models, a corresponding procedure (i.e., specifying the *atmospheric* parameters) is however not adequate as it ignores the rapid atmospheric adjustment to ocean SST. In early ocean models surface temperature and salinity were simply specified and the fluxes necessary to maintain those specified fields diagnosed a posteriori. However, these fluxes were not consistent with Eq. 9.3 and 9.4 implying that the atmosphere would not be in equilibrium with the oceanic state.

A frequently used alternative formulations for the heat flux is

$$Q = \lambda(T^* - T) \tag{9.5}$$

which effectively restores surface temperature to a prescribed value T^*, with a time scale proportional to λ^{-1}. Eq. 9.5 may be considered as a linearization of Eq. 9.3, and both the equivalent temperature T^* and the coefficient λ can in principle be calculated from atmospheric parameters. The equivalent temperature T^* is neither identical with the atmospheric nor with the oceanic temperature, although in practice often the observed SST is used. A linearization of Eq. 9.3 yields typical values of λ around 50 $Wm^{-2}K^{-1}$ which mainly result from the strong temperature dependence of the latent heat flux. That value corresponds to a time scale of 3 months for a surface layer of 100 m thickness.

The main deficiency of the formulation Eq. 9.5 is that it ignores the scale dependency of the ocean-atmosphere exchange. In general λ must decrease for larger scales. On small scales, the atmospheric response to SST anomalies can be removed effectively by the wind, and a large flux can persist. On the global scale, however, the only possible way to remove heat from the atmosphere is long-wave radiation to space, and the value of λ must approach 2 $Wm^{-2}K^{-1}$ as required e.g. by energy-balance climate models. This was already noted by Bretherton (1982) who suggested that Eq. 9.5 should be replaced by a linear but *nonlocal* relation, i.e.

$$Q(\mathbf{x}) = \int \Lambda(\mathbf{x},\mathbf{x}')[T^*(\mathbf{x}') - T(\mathbf{x}')]d\mathbf{x}' \tag{9.6}$$

where the kernel $\Lambda(\mathbf{x},\mathbf{x}')$ could, at least in principle, be determined from a series of sensitivity experiments with atmospheric general circulation models. To this author's knowledge, so far no systematic attempt to determine a relation like Eq. 9.6 has been made.

It is possible to construct local relations between Q and T which reflect the scale dependency of the interaction in a somewhat ad-hoc way. A simple isotropic form is given by

$$Q = \lambda_1(T^* - T) - \lambda_2\nabla^2(T^* - T) \tag{9.7}$$

where $\lambda_1 \approx 2\ Wm^{-2}K^{-1}$ corresponds to the large-scale limit, and λ_2 determines the interaction at the smallest resolved scales. For a climate model with 500 km resolution, an appropriate value would be around $\lambda_2 \approx 10^{13}\ WK^{-1}$. While Eq. 9.7 is not based on any quantitative model for the atmospheric transport, it does capture an essential aspect of the air-sea exchange which is missing in Eq. 9.5, and can be much easier implemented than Eq. 9.6. For problems characterized by one dominant scale L_*, Eq. 9.7 is roughly equivalent to Eq. 9.5 with an effective coefficient $\lambda = \lambda_1 + \lambda_2/L_*^2$.

For the freshwater flux, the situation is different. A formulation analogous to Eq. 9.5, i.e.

$$H = E - P = \lambda_S(S^* - S) \tag{9.8}$$

which very often is used in ocean models, has no physical basis and is inconsistent with Eq. 9.4 as neither evaporation nor precipitation depend on surface salinity. Eq. 9.8 is only insofar defensible as for sufficiently large λ_S it is nearly equivalent to specifying the surface salinity to S^*. The appropriate linear analogy would be to relate the evaporation E to SST, e.g. by

$$E = E^* - \gamma(T^* - T) \tag{9.9}$$

However, another difficulty arises here. Unlike the net heat exchange, the freshwater flux must balance, i.e.

$$\int [E(\mathbf{x}) - P(\mathbf{x})]\, d\mathbf{x} = 0 \tag{9.10}$$

as on long time scales the net atmospheric water content cannot change very much. (If necessary, freezing/melting and river runoff have to be included in the definition of H). The specification of precipitation must hence be such that Eq. 9.10 is satisfied at all times. The most simple solution is to assume that any change in evaporation is compensated *locally* by a corresponding change in precipitation, so that effectively the temperature dependence of evaporation can be ignored and the freshwater flux remains fixed. Other assumptions regarding the distribution of precipitation may however be more plausible, see e.g. Stocker et al. (1991).

From an oceanographer's perspective, the specification of the thermohaline boundary conditions constitutes a model (albeit a very primitive one) of the atmosphere. It should be viewed as equivalent to the simple ocean models (e.g. swamp ocean, or mixed layer ocean) which have been utilized with some success in atmospheric climate models. While obviously inferior to a fully coupled ocean-atmosphere model, a careful choice of flux conditions can help to catch important aspects of ocean-atmosphere interaction. In particular, at the lowest order a combination of a fixed freshwater flux with a heat flux specification according to Eq. 9.5 (often termed as *mixed* boundary conditions) is clearly preferable to the combination of Eq. 9.5 and 9.8 for oceanic climate models.

9.4 Oceanic response to thermohaline forcing

The formation of water masses in the world ocean, their distribution and the associated thermohaline circulation are directly related to the fluxes of heat and freshwater at the sea surface. The thermohaline circulation is a basically nonlinear phenomenon, through the advection of temperature and salinity by a current field that depends geostrophically on the density gradients. For that reason, the thermohaline forcing cannot be considered completely decoupled from the wind forcing as both mechanisms influence the circulation field.

The principal aspects of the dynamics of the thermohaline circulation are reasonably well understood. When forced with observed surface temperatures and salinities, ocean models reproduce at least qualitatively many large-scale aspects of the observed circulation patterns and water mass distribution. Due to their coarse resolution, intense boundary currents and oceanic fronts are however broader and weaker than observed, and only the integral transports can be meaningful. In many simulations the main thermocline has been too deep and too diffuse. In this respect the model by Maier-Reimer et al. (1991) shows a remarkably better performance. Heat transport in models has generally been lower than that inferred from observations, and in most models transports in the northern hemisphere do not exceed 1 PW. Model heat transports and the vertical overturning circulation are very sensitive to several poorly known aspects of circulation models, e.g. vertical diffusion (Bryan, 1987), details of the thermohaline boundary conditions (Maier-Reimer et al., 1991), and even numerical algorithms (Gerdes et al., 1991).

A particularly interesting aspect of the global circulation is the different behaviour of the Atlantic and Pacific Oceans. The northern North Atlantic is warmer (about 3 ° C) and more salty, and is the source region of North Atlantic Deep Water which can be traced through all deep ocean basins. The low surface salinities in the North Pacific prevent sinking and the deep water there is the oldest in the world ocean (C-14 ages which measure the time since a particle was last in contact with the surface layer are 1500-2000 years). Above the main thermocline there is a series of return flows forming what has been called the oceanic *conveyor belt* (Gordon, 1986). In the South Atlantic, this northward flow of warm water causes an equatorward (i.e. up-gradient) heat transport of approximately $0.5\,10^{15} W$ which is transported into the North Atlantic where it is ultimately released into the atmosphere. This remarkable pattern was already known to Wüst (1935) who later attempted to quantify the heat transport in the South Atlantic. Apparently, he was discouraged by the *wrong* (i.e. equatorward) direction of the transport and never published the results.

In recent years, the question of possible long-term changes in the large-scale circulation has become increasingly interesting. The paleoclimatological record indicates that the conveyor-belt pattern has occasionally been disrupted. The best known example is the Younger Dryas Event, approximately 10000 years ago, when a sudden cooling occured on a time scale of 100 years or less. There is evidence that during this period the deep water formation in the North Atlantic, the thermohaline circulation and poleward heat transport were much reduced, causing a significant climate change in Europe and North America. It has been speculated (Broecker et al., 1985) that this event could be interpreted as a transition between different equilibrium states of the ocean circulation.

The clue to understanding the possibility of multiple equilibrium states comes from the way in which heat and freshwater influence the circulation. Temperature and salinity at high latitudes have an opposing effect on density. In the North Atlantic, the buoyancy loss due to cooling clearly exceeds the buoyancy gain due to excess precipitation. The corresponding increase in density causes a sinking to deeper levels and a poleward motion near the surface. The differences in the interaction of heat resp. freshwater with the atmosphere now cause a positive feedback mechanism which tends to amplify the response to haline forcing.

Consider an equilibrium situation where the freshwater flux at high latitudes is initially perturbed. To be definite, we assume that the surface salinity is reduced, say, through excess precipitation or melting. A decrease in salinity corresponds to a decrease in density, this will diminish the sinking motion and ultimately the meridional circulation. Thereby the poleward transport of relatively salty water from lower latitudes is reduced, which in turn further decreases the polar salinity. This positive feedback is visualized as loop 1 in the diagram of Fig. 9.3. The reduction of poleward

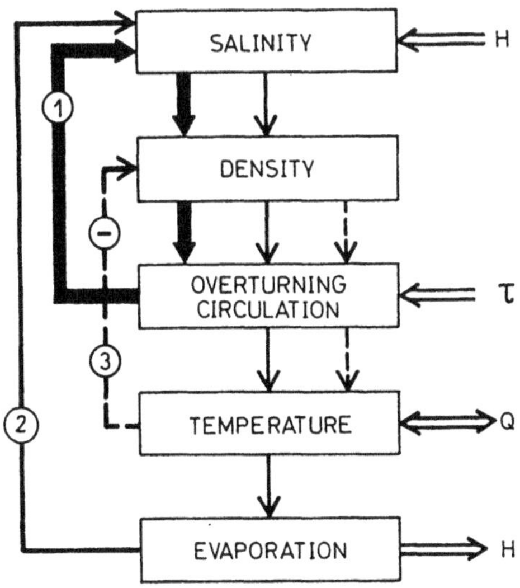

Figure 9.3: *A schematic diagram showing several mechanisms relevant for the thermohaline circulation. The arrows to the right indicate interaction with the atmosphere via momentum flux (τ), heat flux (Q) and freshwater flux (H). The (positive) haline feedback (1) generally dominates the (negative) thermal feedback (3), and is the principal reason for the existence of multiple equilibrium states. The feedback (2) involving the temperature dependence of evaporation is positive, depending on water transport in the atmosphere. It can however be partly or fully neutralized by a corresponding change in precipitation. See text for further explanation.*

advection of warm water will of course also lead to a decrease of temperature, thus providing a negative feedback (loop 3 in Fig. 9.3). However, as discussed in the previous section, the net air-sea heat flux is strongly temperature dependent, it will therefore increase and partly compensate for the temperature reduction. That compensation is not complete, but it is sufficiently strong so that the positive feedback due to salinity altogether dominates.

Another possibly relevant mechanism has been noted by Warren (1983). In the above example, lower surface temperatures in high latitudes will cause less evaporation, and hence lower surface salinity. This can give rise to a positive feedback which is also indicated in Fig. 9.3 (loop 2). The direct influence of temperature on density is weaker than the indirect one through this loop. However, whether or not this loop can become effective depends critically on the atmospheric water budget. A decrease in evaporation must be accompagnied by a corresponding decrease in precipitation *somewhere* as on the time scales considered here no water can be removed from the atmosphere. If that decrease occurs over the same oceanic region, this feedback will be largely neutralized. If, on the other hand, atmospheric water transport between different ocean basins is involved, the feedback can be very important.

It is useful to look at simplified models to illustrate some of these mechanisms. The most simple conceptual model is, in a somewhat modified form, due to Stommel (1961) and consists of two well-mixed reservoirs, one representing the low latitude surface ocean, the other the deep and high latitude ocean in one hemisphere, respectively. The reservoirs are coupled advectively to each other, and by thermohaline fluxes to the atmosphere. Heat and salt conservation can be formulated in terms of the differences in temperature and salinity between equatorial and polar reservoir, ΔT and ΔS, and read

$$V\frac{d}{dt}\Delta T = \Delta Q - \mid m \mid \Delta T \tag{9.11}$$

$$V\frac{d}{dt}\Delta S = \Delta H - \mid m \mid \Delta S \tag{9.12}$$

Here V is an average reservoir volume, and the parameter m is a measure of the advective transport between the reservoirs. It is plausible to assume that the transport depends on the density difference between the reservoirs, a simple linear relation is $m = const(\alpha\Delta T - \beta\Delta S)$.

The character of the system Eq. 9.11 and 9.12 is now determined by the way in which heat and freshwater flux interact with the atmosphere. If both fluxes were specified independent of the oceanic state, Eq. 9.11 and 9.12 could be combined into one equation for the density which then would be uniquely determined by the (fixed) buoyancy flux, a rather uninteresting situation. The discussion in section 9.3 suggests however to use mixed boundary conditions, i.e. to fix only the haline forcing ΔH, and to specify the thermal forcing according to $\Delta Q = \lambda(\Delta T^* - \Delta T)$, with ΔT^* denoting the equator-pole difference of the equivalent forcing temperature. With appropriate scaling (temperature difference by ΔT^*, salinity by its relative density contribution, time scale of advective transport), and omitting all Δ-symbols, the system Eq. 9.11 and 9.12 may then be re-written in dimensionless form as

$$\frac{d}{dt}T = \lambda(1-T) - \mid T-S \mid T \tag{9.13}$$

$$\frac{d}{dt}S = H - \mid T-S \mid S \tag{9.14}$$

and is readily analyzed. The strength of the thermal atmospheric feedback must be stronger than the advection, hence we consider only the case $\lambda \ll 1$. Fig. 9.4 shows the equilibrium solutions for T and S as a function of the strength of the hydrological forcing, H. The most important result is that, in a certain range $0 < H < H_c$, two stable equilibrium solutions exist. One solution is thermally dominated ($T > S$), and has a poleward heat transport and sinking at high latitudes which are relatively warm and saline. The other solution is colder and fresher and without sinking at high latitudes, it has a much weaker circulation and a small equatorward heat transport. In between is a branch with unstable equilibria which can be of some importance during transition states. If the hydrological forcing exceeds the threshold $H_c \approx ((\lambda-1)/2\lambda)^2$, only the haline-dominated solution is possible.

The existence of multiple equilibrium states is thus a direct consequence of the positive feedback described above (loop 1 in Fig. 9.3. Rooth (1982) and Welander (1986) have used similar box models with extended geometrical configuration to include two hemispheres, and found further equilibrium states involving cross-equatorial heat transport. Marotzke (1990) has constructed a model with 7

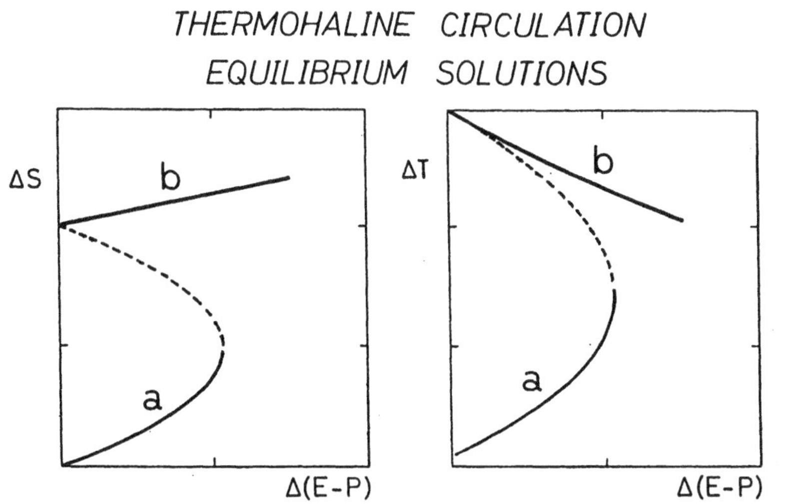

Figure 9.4: *Phase diagram for the equilibrium solutions of the system (3.11,3.12), showing the equator-pole difference of salinity (left) and temperature (right) as a function of the strength of the hydrological cycle. Branch (a) corresponds to a temperature dominated circulation, with a relatively warm and salty polar reservoir, while branch (b) is salinity-dominated, with a colder and fresher polar reservoir. The equilibrium solutions in the dashed region are infinitesimally unstable.*

reservoirs resembling the two major ocean basins, and found a total of 16 stable equilibrium states, essentially consisting of all possible combinations of thermal resp. haline dominated circulation patterns in all 4 hemispheres.

It is not obvious that the results from simple box models should be relevant for the real ocean. So far, a number of models of varying complexity have however confirmed the basic role of the haline feedback mechanism. Bryan (1986) has first demonstrated the existence of multiple equilibrium states in a three-dimensional ocean general circulation model, using an idealized 2-hemisphere configuration. With forcing by specified freshwater fluxes and surface temperatures (mixed boundary conditions), he obtained one solution symmetric to the equator, with poleward heat transport in each hemisphere, and a second equilibrium solution with strong cross-equatorial transport and sinking only in one hemisphere. His experiments have pointed out the important role of convection in high latitudes which greatly increases the sensitivity of the circulation.

From analogy with simple box models, it can be expected that further equilibrium states for the global circulation exist. A series of experiments using a global ocean model with idealized geometry has been carried out by Marotzke and Willebrand (1991), again with mixed boundary conditions. By a systematic variation of initial conditions, they were able to identify at least four stable equilibrium states. (As these equilibrium solutions are constructed by forward integration of time-dependent models, it is possible that not all solutions have been found). The horizontal circulation systems in the various equilibria differ not very much, but the vertical overturning patterns are completely different. One of the solutions represents qualitatively the conveyor-belt circulation characteristic for the present ocean. A second solution corresponds to a conveyor-belt where the respective roles of

Atlantic and Pacific oceans are reversed. Two further states exist where the thermohaline circulation in both major oceans is aligned, i.e. sinking occurs either in both northern or both southern basins. Three of the solutions have a strong poleward global heat transport in the northern hemisphere and weak transport in the southern, whereas one behaves differently with very weak transport in the northern hemisphere. While the surface temperature in these simulations was fixed, it is clear that with more realistic boundary conditions high-latitude sinking will coincide with higher than normal temperatures.

Manabe and Stouffer (1988) found two stable solutions to their global coupled model. Besides the presently observed conveyor belt circulation, a second solution was found which has a much colder (up to 5°C) and fresher (3‰ reduction in salinity) northern North Atlantic and much weaker heat transport, with less significant changes in the Pacific circulation. Corresponding differences occur in the atmospheric climate, in particular over northern Europe. It should be noted that in these experiments, as in many coupled models, a flux-adjustment technique was employed to obtain more realistic precipitation patterns, and the results must therefore be interpreted with some caution. Nevertheless, it appears that the haline feedback also is effective in coupled ocean-atmosphere models with their much higher complexity. Indeed, the first 3-d coupled model by Bryan et al. (1975) showed high salinities in the North Pacific and low salinities in the North Atlantic which in retrospect suggests that an alternate solution resembling a reversed conveyor circulation had been found.

The existence of multiple equilibrium states in the climate system is virtually certain. It is therefore important to consider their stability, i.e. what kind and magnitude of perturbations is necessary to turn the system into a different state. So far we have only some rather preliminary results from uncoupled ocean models. Marotzke and Willebrand (1991) found that changes in the large-scale precipitation patterns of the order of 0.04 Sv were sufficient to cause a transition between equilibrium states. Maier-Reimer and Mikolajewicz (1989) investigated the stability of the North Atlantic circulation to local freshwater input, in an attempt to model the Younger Dryas event as a consequence of increased melt water inflow through rivers. They found a high sensitivity to both the strength and the location of the freshwater input. A freshwater injection of 0.02 Sv through the St Lawrence River was sufficient to trigger an abrupt breakdown of circulation and heat transport within less than 20 years. It corresponds to a transition from a conveyor belt circulation to a state where sinking takes places only in the southern hemisphere.

The short time scale of the response occurs because the fresher surface water suppresses the convection in high latitudes very effectively. The mechanism (termed 'polar halocline catastrophy') was first reported by Bryan (1986). Fig. 9.5 which is based on experiments by Marotzke (1991) shows an example for the rapid breakdown of a circulation state. The fast time scale can be explained neither by advective nor by diffusive processes, and must be associated with wave propagation. It appears that Kelvin waves play an important role in transporting the signal from high to low latitudes along the western boundaries. Two-dimensional models of the meridional circulation have been used with some success to simulate multiple equilibrium states and some aspects of the global water mass distribution (Marotzke et al., 1988, Stocker et al., 1991). They lack however any wave dynamics, and therefore are not able to reproduce the fast changes occuring in 3-d models.

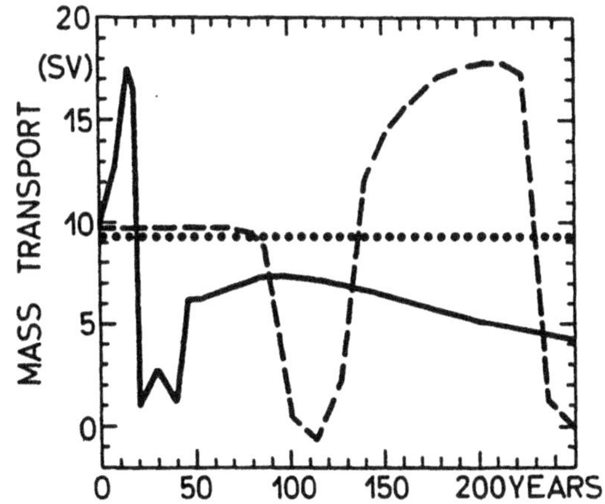

Figure 9.5: *Temporal evolution of the meridional mass transport at 52 °N and 2100 m depth in a 3-dimensional model of the thermohaline circulation, after switching to mixed boundary conditions. The initial (unstable) equilibrium state was obtained with prescribed surface temperature and salinity. The standard case (full line) shows dramatic changes and a complete collapse of vertical overturning within less than two decades. The event is however extremely sensitive to details of the numerical scheme, as demonstrated by the two other curves which were obtained with a different time step (dashed) resp. a different convective mixing algorithm (dotted). After Marotzke (1991).*

Besides by changes in surface salinity, thermohaline circulation patterns can in principle also be perturbed by surface temperature changes. Bryan and Spelman (1985) calculated the response of a coupled ocean-atmosphere model to an instantaneous quadrupling of the atmospheric CO_2-concentration which would lead to a sudden dramatic warming at high latitudes. While this is not a very realistic scenario, the results were nevertheless intriguing. Only 25 years after the increased CO_2 concentration was switched on, the thermohaline cell had almost completely collapsed. A recent study by Cubasch et al. (1992) calculated the climate response to a gradual CO_2 increase in the atmosphere. The warming in high latitudes was strong enough to decrease the transport of the vertical cell, and the reduction of convective mixing actually caused a *cooling* of sea surface temperatures in regions of water mass formation in the North Atlantic.

The apparent possibility that the thermohaline circulation can collapse within a few decades has an obvious relevance for internal variability on decadal and longer time scales. Positive feedback mechanisms such as the one described above can trigger instabilities of the circulation resulting in oscillatory phenomena. Welander (1986) has given an overview on other possible instability mechanisms. Some indications of internal instability mechanisms leading to limit cycle and chaotic behavior have recently been found in experiments with idealized GCM's. Marotzke (1990) found that one of the equilibrium states of a global model exhibited regular self sustaining-oscillations with a period of around 25 years. Both the period and amplitude of the oscillation were highly sensitive to details of the simulation, and some equilibrium states did not show any oscillatory behaviour. Similar results were reported by Weaver and Sarachik (1991), who also found fluctuations involving strongly increased deep ocean ventilation (*flushes*) occurring at scales between 200 and

1,000 years. The mechanisms of these oscillations were investigated by Winton and Sarachik (1992). The fixed freshwater flux condition requires a certain poleward salt transport and hence vertical overturning which can be incompatible with the thermocline structure resulting from that freshwater flux. Furthermore, the poleward salt transport is influenced by the wind, and therefore the existence and form of self-induced oscillations depends, besides on the hydrological forcing, also on the strength of the mean wind forcing.

Another potential source of low-frequency variability is stochastic excitation by high-frequency fluctuations. As shown by Hasselmann (1976), this type of forcing can generate oceanic variability on much longer time scales than that of the forcing, leading to a generally red spectrum of oceanic variables. The forcing may be related to atmospheric weather, or to mesoscale eddy activity in the ocean. Only recently the exploration of long-term variability with ocean circulation models has become feasible. Mikolajewicz and Maier-Reimer (1990) have simulated fluctuations in precipitation and evaporation patterns by adding a random white-noise component to the freshwater fluxes driving their global ocean model. From an integration over more than 3000 model years they found strong variations in the ventilation of the deep Atlantic Ocean on time scales around 300 years, with salinity anomalies propagating southward in the deep ocean and northward in the surface layers. The variations in deep water formation reach nearly 50%, and are associated with heat flux changes of 0.2 PW. Those amplitudes depend on the (somewhat arbitrary) choice of the strength and scales of random forcing which corresponded to a rms value of 16 mm per month precipitation, with lateral scales of a few thousand kilometers. This choice gives a sizable response but is not strong enough to cause a flip to a different steady state. A secondary response was found at time scales of decades, associated with variation in the horizontal gyres. Weaver et al. (1991) arrived at similar conclusions, they found that the structure and amplitudes of the *mean* freshwater fluxes are very critical for the resulting variability.

High resolution models, even when forced with seasonal or annual mean fluxes, exhibit low-frequency variability with time scales of several years. There is some indication that a geostrophic-turbulence cascade is operative producing large scale patterns, modulated by the variation in Coriolis parameter, which give rise to Rossby wave activity, but integration times have so far been too short to clearly identify the cause of this variability. The fluctuations resulting from quasigeostrophic eddies are largely adiabatic, and are therefore less likely to generate variations of the thermohaline circulations on decadal or longer time scales.

9.5 Concluding remarks

The role of the ocean circulation for climate change has long been recognized. Results from a number of quite different model studies are quite robust indicating that the thermohaline circulation may have more than one equilibrium state, a fact which can have serious consequences for long-term climate variability and predictability. The possibility that the ocean may flip from one state to another within a few decades is intriguing, and the issue will become more important as we try to understand climate variability on decadal and longer time scales. It is clear that the interaction of the thermohaline circulation with the hydrological cycle is crucial in determining the stability of the climatic state.

Most of these conclusions have been derived from modeling studies with stand-alone ocean models. While they have illuminated some aspects of the dynamics of the thermohaline circulation on decadal and longer time scales, they have not accurately represented atmospheric feedbacks. With respect to the variability of the ocean-atmosphere system these findings are therefore not conclusive, and systematic variability studies with coupled ocean-atmosphere models are needed to properly model these phenomena. Long-term integrations with coupled models are difficult as the atmospheric component is computationally much more expensive than the oceanic one. Such computations are however beginning to become possible, and we can expect that a close symbiosis between oceanic and atmospheric modelling will be established.

9.6 References

Baumgartner A, Reichel E (1975) The World Water Balance. *Elsevier* 179pp.

Behringer D, Regier L, Stommel H (1979). Thermal feedback on wind stress as a contributing cause of the Gulf Stream. *J Mar Res* 37: 699-709.

Bretherton FP (1982) Ocean climate modeling. *Progr Oceanogr* 11: 93-129.

Broecker WS, Peteet DM, Rind D (1985) Does the ocean-atmosphere system have more than one mode of operation? *Nature* 315: 21-26.

Bryden H, Roemmich DH, Church JA (1991) Ocean heat transport across 24 ° N in the Pacific. *Deep-Sea Res* 38: 297-324.

Bryan K, Manabe S, Pacanowski RC (1975) A global ocean-atmosphere climate model. Part II. The oceanic circulation. *J Phys Oceanogr* 5: 30-46.

Bryan K, Spelman MJ (1985) The ocean's response to a CO_2-induced warming. *J Geophys Res* 90: No. C6 11679-11688.

Bryan F (1986) High-latitude salinity effects and interhemispheric thermohaline circulation. *Nature* 323: 301-304.

Bryan F (1987) Parameter sensitivity of primitive-equation ocean general circulation models. *J Phys Oceanogr* 17: 970-985.

Carissimo BC, Oort AH, Vonder Haar TH (1985) Estimating the meridional energy transports in the atmosphere and ocean. *J Phys Oceanogr* 15: 82-91.

Coachman LK, Aagaard K (1988) Transports through Bering Strait: Annual and interannual variability. *J Geophys Res* 93: 15535-15539.

Cubasch U, Hasselmann K, Höck H, Maier-Reimer E, Mikolajewicz U, Sauter BD, Sausen R (1992) Time-dependent greenhouse warming computations with a coupled ocean-atmosphere model. *Climate Dynamics* (submitted).

Dorman CE, Bourke RH (1981) Precipitation over the Atlantic Ocean, 30 ° S to 70 ° N. *Mon Wea Rev* 109: 554-563.

Gerdes R, Köberle C, Willebrand J (1991) The influence of numerical advection schemes on the results of ocean general circulation models. *Climate Dynamics* 5: 211-226.

Gordon AL (1986) Interocean exchange of thermocline water. *J Geophys Res* 91: 5037-5046.

Hall MM, Bryden HL (1982) Direct estimates and mechanisms of ocean heat transport. *Deep-Sea Res* 29: 339-359.

Hasselmann K (1976) Stochastic climate models, Part I. Theory. *Tellus* 28: 473-485.

Hastenrath S (1982) On the meridional heat transport in the world ocean. *J Phys Oceanogr* 12: 922-927.

Hsiung J (1985) Estimates of global oceanic meridional heat transport. *J Phys Oceanogr* 15: 1405-1413.

Isemer HJ, Hasse L (1987) The Bunker climate atlas of the North Atlantic Ocean. Vol. 2. *Springer Verlag* 252 pp.

Isemer HJ, Willebrand J, Hasse L (1989) Fine adjustment of large scale air-sea energy flux parametrization by direct estimate of ocean heat transport. *J Climate* 2: 1173-1186.

Maier-Reimer E, Mikolajewicz U (1989) Experiments with an OGCM on the cause of the Younger Dryas. In: Oceanography 1988 (Eds.: A Ayala-Castanares, W Wooster and A Yanez-Arancibia). *UNAM Press* Mexico 87-100.

Maier-Reimer E, Mikolajewicy U, Hasselmann K (1991) On the sensitivity of the global ocean circulation to changes in the surface heat flux forcing. *Max-Planck-Inst f Meteorologie* Rep. No. 68 67pp.

Manabe S, Stouffer RJ (1988) Two stable equilibria of a coupled ocean-atmosphere model. *J Climate* 1: 841-866.

Marotzke J (1990) Instabilities and multiple equilibria of the thermohaline circulation. *Ber Inst Meeresk* Kiel 194: 126pp.

Marotzke J (1991) Influence of convective adjustment on the stability of the thermohaline circulation. *J Phys Oceanogr* 21: 903-907.

Marotzke J, Welander P, Willebrand J (1988) Instability and multiple steady states in a meridional-plane model of the thermohaline circulation. *Tellus* 40A: 162-172.

Marotzke J, Willebrand J (1991) Multiple equilibria of the global thermohaline circulation. *J Phys Oceanogr* 21: 1372-1385.

Mikolajewicz U, Maier-Reimer E (1990) Internal secular variability in an ocean general circulation model. *Climate Dynamics* 4: 145-156.

Oort AH, Asher SC, Levitus S, Peixoto JP (1989) New estimates of the available potential energy in the world ocean. *J Geophys Res* 3187-3200.

Rooth CGH (1982) Hydrology and ocean circulation. *Progr Oceanogr* 11: 131-149.

Schmitt RW, Bogden PS, Dorman CE (1989) Evaporation minus precipitation and density fluxes for the North Atlantic. *J Phys Oceanogr* 19: 1208-1221.

Stocker TF, Wright DG, Mysak LA (1991) A zonally averaged, coupled ocean-atmosphere model for paleoclimatic studies. *J Climate* (in press)

Stommel H (1961) Thermohaline convection with two stable regimes of flow. *Tellus* 13: 224-230.

Stommel H (1980) Asymmetry of interoceanic fresh-water and heat fluxes. *Proc Natl Acad Sci USA* 77: 2377-2381.

Warren BA (1983) Why is no deep water formed in the North Pacific? *J Mar Res* 41: 327-347.

Talley LD (1984) Meridional heat transport in the Pacific Ocean. *J Phys Oceanogr* 14: 231-241.

Weaver AJ and E.S. Sarachik (1991) The role of mixed boundary conditions in numerical models of the ocean's climate. *J Phys Oceanogr* 21: 1470-1493.

Welander P (1986) Thermohaline effects in the ocean circulation and realated simple models. In: Large-scale transport processes in oceans and atmosphere (Eds.: J Willebrand and DLT Anderson). *D. Reidel Publ Co* Dordrecht 163-200.

Wijffels SE, Schmitt RW, Bryden HL, Stigebrandt A (1992) Transport of freshwater by the oceans. *J Phys Oceanogr* 22: 155-162.

Winton M, Sarachik SE (1992) Thermohaline oscillations induced by strong steady forcing of ocean general circulation models. *J Phys Oceanogr* (submitted)

Wüst G (1935) Schichtung und Zirkulation des Atlantischen Ozeans. Die Stratosphäre. Wissenschaftliche Ergebnisse der Deutschen Atlantischen Expedition auf dem Forschungs- und Vermessungsschiff 'Meteor' 1925-1927. English translation by WJ Emery, 1978, of Vol. VI, Section 1, The Stratosphere of the Atlantic Ocean. *Amerind Publishing Co* New Delhi 112 pp.

Chapter 10

MODELLING OF OCEANS CIRCULATION

P. Delecluse
LODYC - CNRS
Université Pierre et Marie Curie
75252 Paris - Cedex 05
France

10.1 Physical bases of the ocean circulation

10.1.1 A global hydrographic structure

The ocean covers 70% of the earth surface and it contains 97% of the earth water. It has a considerable "buffer" role in the climate system due to its large heat capacity (2.5 m of ocean water has the same heat capacity than the total atmospheric column above it) and its weight (the density of ocean water is $1.02 \cdot 10^3 kg\, m^{-3}$ as the air density is 1.2 $kg\, m^{-3}$). The sea water has a temperature range within -1.90 to 32 $^\circ C$ for a salinity of open sea varying from 33 to 37 ‰but 75 % of the oceans volume is filled with water in a very narrow range of temperature (between 0 and 4 $^\circ C$) and of salinity (between 34.4 and 34.7 ‰). In enclosed or semi-enclosed seas, much more variability can be found for temperature and salinity values. The vertical profile of temperature is nearly homogeneous in high latitudes and the constrast between surface temperature and deep temperature increases equatorward. In the equatorial band, a strong and sharp thermocline separates the warm reservoir from the cold deep waters. As we move polewards, the thermocline thickens and deepens. These properties are common to the three oceans in winter conditions. The water mass distribution and the ocean circulation are closely linked together and the water mass properties are determined through complex air-sea interactions. It is thus very important to understand and to model the air-sea interactions and how surface properties are transfered into the deep ocean. The ocean communicates with the atmosphere via exchanges in momentum, in water flux and heat flux.

NATO ASI Series, Vol. I 5
Energy and Water Cycles in the Climate System
Edited by E. Raschke and D. Jacob

10.1.2 Interaction with the atmosphere

Seasonal cycle of the mixed layer

The mixed layer is defined as the thin surface layer of the ocean where turbulent mixing is strong enough to ensure a vertical homogeneity of temperature. It is the first receptacle of the atmospheric fluxes. When the ocean is gaining heat from the atmosphere, a thin warm layer forms at the surface. Surface wind stress forcing acts as a source of turbulence and permits the deepening of a homogeneous surface layer. Other sources of intrinsic oceanic turbulence (as the breaking of internal waves) or the shear of ocean currents contribute to maintain it. The gradient of temperature at its bottom is called the seasonal thermocline. As the warm season comes, the seasonal thermocline deepens and gets warmer. In fall, the heat budget reverses and the mixed layer temperature decreases as its depth increases. The maximum depth is usually reached during the winter season and coincides with the position of the main thermocline. The high frequency variability of the ocean-atmosphere system is confined within this depth. The mixed layer depth retreats to the surface level very quickly at the end of the winter season.

Mid-latitude thermocline ventilation

The maximum depth of the seasonal thermocline defines the main thermocline, permanent thermal gradient which separates the warm ocean surface layers from the deep ocean. During the winter season, the main thermocline communicates directly with the atmosphere through a process called the thermocline ventilation. It occurs in areas where the curl of the wind stress is negative. This curl induces important vertical downward motions, allowing the surface information (momentum and density property) to be transmitted to the main thermocline level. This process is located in sub-polar and mid-latitudes. It concerns basin circulation and has a typical time scale of tens of years.

Equatorial thermocline

In the equatorial regions, there is no distinction between the seasonal thermocline and the main thermocline. The thermocline there is shallow and the circulation is vigourous. In all seasons, turbulent mixing is strong enough to homogenize the temperature to the level of the main thermocline which is directly affected by the interactions with the atmosphere. It is an area where ocean and atmosphere interactions rapidly develop a basin wide response (like the ENSO - El Niño Southern Oscillation) within a time scale of several months.

Deep convection and sea ice

In high latitudes, the concept of main thermocline disappears: there is no permanent gradient of temperature between the surface layers and the deep waters. During the winter season, very dense surface waters form at the surface: the temperature gets very cold under strong and rigourous atmospheric conditions and the formation of sea ice significantly increases the surface salinity by rejection of salt during the ice formation. This very dense surface water dives to the deep ocean: it is the process of deep convection. It happens in very limited areas of the world oceans in an intermittent way (a few tens of km^2 during a few weeks). Known areas for formation of the Antarctic bottom waters are the Weddell sea and the Antarctica shelf regions and for the Atlantic deep water the Labrador sea, the Norwegian Sea and the Greenland Sea. Despite the small area of formation, these waters represent the largest proportion of the water masses of the world ocean. They travel very slowly around the world ocean with typical time scales of a few hundreds of years.

10.1.3 Intrinsic turbulence of the ocean

The ocean is far from being a smooth viscous fluid. It is affected by a vigourous turbulent activity which appears clearly on the variability detected by the altimeter (like SEASAT and GEOSAT). This meso-scale turbulence has a typical length scale of 30-50 km for a life time of several weeks. Its contribution to the global circulation is still an open question. In some areas, the amount of heat transported by the eddies is equivalent to the amount of heat transported by the mean circulation. Though the eddies seem to exist everywhere in the oceans, the distribution of the eddy kinetic energy is highly heterogeneous.

10.2 The modelling approach

10.2.1 The modelling dilemma: space or time ?

The first problem faced by the oceanographers is linked to the present limited capacity of the supercomputers. To solve the ocean circulation with the typical length scale of the ocean dynamics requires a grid spacing of at least 10 km. To bring the ocean thermohaline circulation to equilibrium requires time scales of a few hundreds of years: it is impossible to respect both these criteria on present day computer. 100 000 grid-points integrated during 1000 years represent several years of CRAY 2. This is the reason why the regional modelling, favouring process studies with high resolution, has been so popular in oceanography. Nevertheless we will not discussed here the multiple aspects of the dynamical regional modelling which has considerably helped our understanding of the ocean behaviour but we will concentrate our interest on the global ocean circulation. Finally some modelling of the equatorial regions will be discussed in order to highlight the importance of parametrization.

10.2.2 Physical equations

The basic equations used to study the ocean circulation are the primitive equations (Bryan, 1969): Navier Stokes equations plus the assumptions of hydrostatism, Boussinesq and incompressibility. The rigid lid assumption is added in order to filter the external gravity waves which travel at a very high speed compared to the internal waves (200 $m\,s^{-1}$ compared to 1 $m\,s^{-1}$). The momentum equation expressed the conservation of horizontal momentum. It allows the calculation of the two components of the baroclinic speed :

$$\partial_t \mathbf{U}_h = -(\mathrm{curl}_z \mathbf{U}_h + f)\mathbf{k} \times \mathbf{U}_h - w\partial_z \mathbf{U}_h - \frac{1}{\rho_o}\mathrm{grad}(p + \frac{1}{2}\rho_o \mathbf{U}_h^2) + \mathbf{F}(\mathbf{U}) = \mathbf{G} \tag{10.1}$$

where $\mathbf{U}$ is the velocity, $\mathbf{U}_h$ its horizontal projection, p the pressure, $\mathbf{k}$ the local vertical axis, f the Coriolis parameter, ρ_o the reference value for density. The barotropic equation allows to compute the barotropic streamfunction Ψ by solving an elliptic equation (H is the ocean depth) :

$$\mathrm{curl}_z(\frac{1}{H}\mathrm{grad}\partial_t \Psi) = \mathrm{curl}_z \int_{-H}^{0} \mathbf{G}\, dz \tag{10.2}$$

The conservation of heat allows to calculate the evolution of temperature T and the conservation of salt, the salinity S :

$$\partial_t T + \mathrm{div}(T\mathbf{U}) = \mathbf{F}(T) \tag{10.3}$$

$$\partial_t S + \mathrm{div}(S\mathbf{U}) = \mathbf{F}(S) \tag{10.4}$$

Salinity, temperature, barotropic and baroclinic components of velocity are the prognostic variables of the system. Beside them, a few diagnostic variables need to be calculated. Pressure, salinity and temperature are combined together in the equation of state (UNESCO, 1983) in order to compute the density:

$$\rho = \rho(T, S, p) \tag{10.5}$$

The hydrostatic equation permits to compute pressure .

$$\partial_z p = -\rho g \tag{10.6}$$

The assumption of incompressibility allows the computation of the vertical velocity w :

$$\mathrm{div}(\mathbf{U}_h) + \partial_z w = 0 \tag{10.7}$$

In order to express the effect of motion not resolved by the grid size, a parametrization for lateral dissipation and diffusion is added to the equations. The operator is not isotropic in order to take into account the different vertical and horizontal scalings :

$$F(q) = K_H \Delta_H q + \partial_z(K_v \partial_z q) \tag{10.8}$$

Most models are written on a geographic grid on the sphere (in function of latitude and longitude), with the vertical as the third coordinate but it is also possible to work with isopycnal models or σ-models. In isopycnal models, the levels are given in function of density. This is a natural coordinate for oceanic motions which tend to follow isopycnal surfaces with very little diffusion across them. However the treatment of upwelling or disappearing levels is not easy. Another problem comes for the distinct role of salinity and temperature which equally contribute to the density and evolve differently. In σ-models, the vertical coordinate follows the bottom topography. It is stretched above high topography and loose over deep oceans. This is a powerful way to respect the control by mountain ridges but it gives other problems: the pressure gradient is difficult to define in the neighbourhood of slopes as is the turbulent tensor for non homogeneous circulations.

Some models use a distorded grid in order to increase the resolution in regions of interest. The grid is usually defined to respect fluid properties and to limit stability problem. One particular example may be given by the treatment of polar regions in general circulation ocean models. When using geographical coordinates, the zonal grid spacing (Δx) is becoming infinitely small as the pole is approached. This arises numerical stability problem as the first condition for numerical stability is given by the CFL (Courant-Friedrichs-Lewy, 1928) criterion which expresses that: $\Delta x/\Delta t \geq C$ where Δx is the grid spacing in space, Δt is the timestep and C is the maximum phase speed.

The respect of the criterion implies strong restriction on timestep in the vicinity of the pole. In order to avoid it, most models use Fourier filtering or suppress the Arctic ocean (Semtner and Chervin, 1988). This solution is not satisfactory in ocean models because this filter is not acting very well in the presence of coastlines. Another solution is to define a system of coordinates in which the North Pole is no more a singular point. This is possible by removing the singular point into a continental area like Canada or Siberia. This distorded coordinates system is used in the LODYC general circulation model.

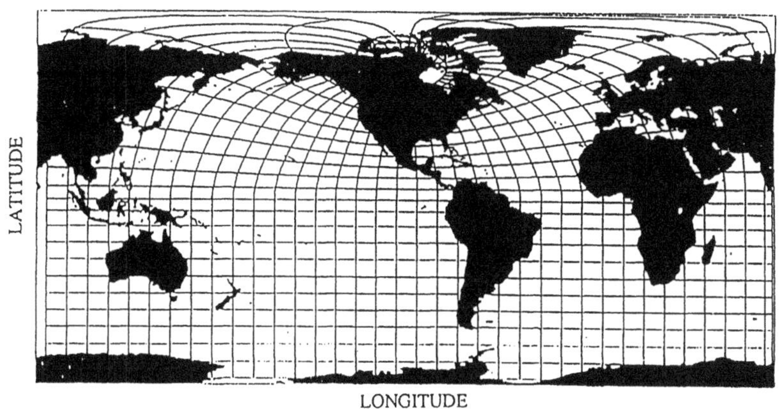

Figure 10.1: *Numerical grid used in the OPA global model (1 line out of 5 is selected)*

Many different coordinate systems permit to treat the general circulation of the oceans. Nevertheless, they all have positive and negative points. The change of coordinates has to respect the fundamental properties of the fluid. When using a variable grid spacing, it is important to respect the conservative properties of the operators from the initial space to the new one. So the correct mathematical way to write the equations is to evidence the vectorial differential operators and to keep their form invariant under coordinates changes.

Let (y^1, y^2, y^3) be a set of orthogonal curvilinear coordinates on the sphere with y^3 parallel to the local vertical and (x^1, x^2, x^3) where $x^i = x^i(y^1, y^2, y^3)$, the cartesian set of coordinates with the origin located at the sphere centre. The local deformation of the curvilinear coordinate system is simply given by the scale factors e_i :

$$e_i^2 = \left(\frac{\partial x^1}{\partial y^i}\right)^2 + \left(\frac{\partial x^2}{\partial y^i}\right)^2 + \left(\frac{\partial x^3}{\partial y^i}\right)^2 \tag{10.9}$$

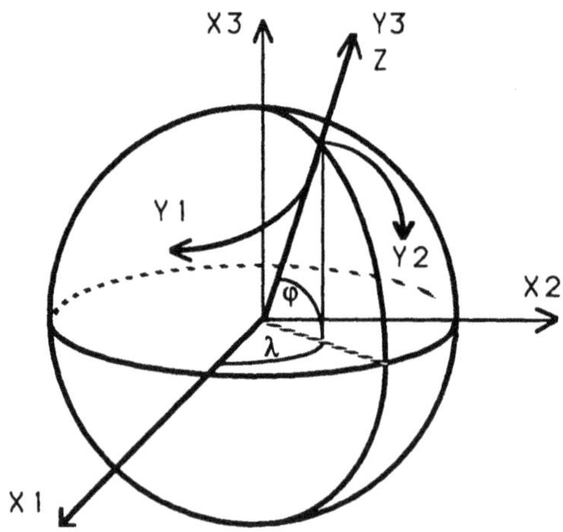

Figure 10.2: *Transformation of orthogonal curvilinear coordinate system on a sphere*

Substituting (y^1, y^2, y^3) by (i, j, k), the scale factors e_i can be expressed as a function of the earth radius a, the latitude $\varphi(i,j)$, longitude $\lambda(i,j)$ and the altitude $z(k)$:

$$\left\{\begin{array}{lcl} e_1 & = & (a+z)\left[\left(\frac{\partial\lambda}{\partial i}\cos\varphi\right)^2 + \left(\frac{\partial\varphi}{\partial i}\right)^2\right]^{\frac{1}{2}} \\ e_2 & = & (a+z)\left[\left(\frac{\partial\lambda}{\partial j}\cos\varphi\right)^2 + \left(\frac{\partial\varphi}{\partial j}\right)^2\right]^{\frac{1}{2}} \\ e_3 & = & \left(\frac{\partial z}{\partial k}\right) \end{array}\right. \tag{10.10}$$

Since the ocean depth is far smaller than the earth radius, $(a+z)$ can be replaced by a in the above sytem. The scalar and vectorial operators can be written in the tensorial form which is invariant in any orthogonal curvilinear coordinate system transformation :

$$\text{grad} f = \begin{pmatrix} \frac{1}{e_1}\frac{\partial f}{\partial i} \\ \frac{1}{e_2}\frac{\partial f}{\partial j} \\ \frac{1}{e_3}\frac{\partial f}{\partial k} \end{pmatrix} \tag{10.11}$$

$$\text{curl}\mathbf{A} = \begin{pmatrix} \frac{e_1}{b}\left[\frac{\partial(e_3A_3)}{\partial j} - \frac{\partial(e_2A_2)}{\partial k}\right] \\ \frac{e_2}{b}\left[\frac{\partial(e_1A_1)}{\partial k} - \frac{\partial(e_3A_3)}{\partial i}\right] \\ \frac{e_3}{b}\left[\frac{\partial(e_2A_2)}{\partial i} - \frac{\partial(e_1A_1)}{\partial j}\right] \end{pmatrix} \tag{10.12}$$

$$\text{div}\mathbf{A} = \frac{1}{b}\left[\frac{\partial}{\partial i}\left(\frac{b}{e_1}A-1\right) + \frac{\partial}{\partial j}\left(\frac{b}{e_2}A_2\right) + \frac{\partial}{\partial k}\left(\frac{b}{e_3}A_3\right)\right] \tag{10.13}$$

where f is a scalar function, $\mathbf{A} = (A_1, A_2, A_3)$ a vector and $\mathbf{b} = e_1e_2e_3$. This formalism is not often used in models and it has been pointed out recently in a paper by Yin and Fung (1991) that variable grid spacings which do not respect this formalism can lose the accuracy of the initial scheme, creating numerical diffusion and artificial upwelling or downwelling motion in case of the vertical advection operator for instance.

10.2.3 Numerical schemes and coding

Discretization in space

The numerical methods to develop the equations are numerous. For atmospheric flows, a widely used method is the spectral technics (the equations are projected on the spherical harmonics). This is not used for ocean general circulation models as the complicated shape of the coastline prevents the use of such a method. In fact a method based on finite elements may be adequate for solving irregular coastlines but most models are based on finite differences and these are the examples we will discuss.

A description of the different properties of numerical schemes can be found in Mesinger and Arakawa (1976). Two different grids have been largely used in ocean modelling. The first one is the B-grid (as in the ocean model developed in GFDL [Geophysical Fluid Dynamics Laboratory], Bryan, 1969) and the second one is the C-grid (as the one used in the model developed in LODYC [Laboratoire d'Oceanographie DYnamique et de Climatologie]- see Chartier, 1985). Multiple studies have been done to exemplify the properties of these two grids. The system of primitive equations is a complex system and it is difficult to find a grid which will keep all the properties of the continuous equations. Certain terms like the energy and the Coriolis term are better written with a B-grid. On the contrary, the C-grid is more natural to calculate the vertical velocity and the curl of the currents.

Extensive studies have been made to describe the properties of these two schemes for the wave propagation in the case of the shallow water approximation (2-D system of equations). Let us show on a simple example how the discretization in different grid-spacing can affect the wave propagation. Considering the one-dimensional approximation of the shallow water equations (only the x-direction is retained), the continuous system of equations is the following :

$$\frac{\partial u}{\partial t} = -g\frac{\partial h}{\partial x} + fv \tag{10.14}$$

$$\frac{\partial v}{\partial t} = -fu \tag{10.15}$$

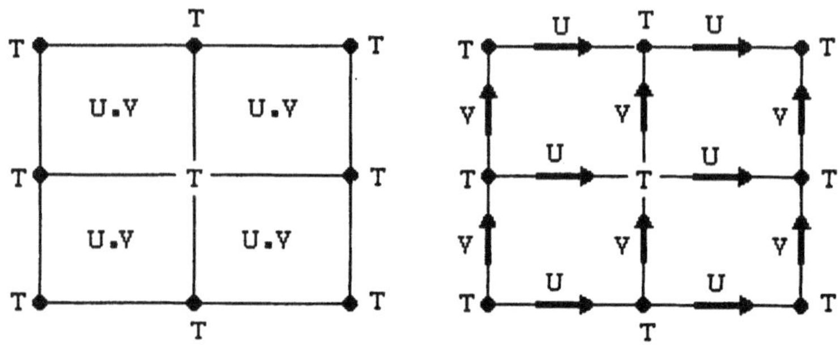

Figure 10.3: *Arrangement of the variables in the B - grid and in the C - grid.*

$$\frac{\partial h}{\partial t} = -H\frac{\partial u}{\partial x} \tag{10.16}$$

where u and v design the zonal and meridional components of velocity, f is the Coriolis parameter, g the acceleration of gravity, H the mean thickness of the layer and h the perturbation of this thickness.

For this system, the relation of dispersion of the inertia gravity waves is the following :

$$\omega^2 = f^2 + gHk^2 \tag{10.17}$$

where ω is the frequency and k the wave number.

When the system is discretized on the B-grid, it becomes :

$$\frac{\partial u}{\partial t} = -g\delta_x h + fv \tag{10.18}$$

$$\frac{\partial v}{\partial t} = -fu \tag{10.19}$$

$$\frac{\partial h}{\partial t} = -H\delta_x u \tag{10.20}$$

and the relation of dispersion for the inertia gravity waves becomes :

$$\omega^2 = f^2 + 4gHk^2\sin^2(k\frac{\Delta x}{2}) \tag{10.21}$$

When the system is discretized on the C-grid and f is defined on the u-point, it becomes :

$$\frac{\partial u}{\partial t} = -g\delta_x h + f\overline{v^x} \tag{10.22}$$

$$\frac{\partial v}{\partial t} = -f\overline{u^x} \tag{10.23}$$

$$\frac{\partial h}{\partial t} = -H\delta_x u \tag{10.24}$$

and the relation of dispersion for inertia-gravity waves becomes :

$$\omega^2 = f^2\cos^2(k\frac{\Delta x}{2}) + 4gHk^2\sin^2(k\frac{\Delta x}{2}) \tag{10.25}$$

For the C-grid as in the B-grid, the discretized relation is not far from the analytical solution when the wave number is close to zero (that is when the long wavelength is well solved by the equations). As the wave number increases, the discretized relation drifts away the analytical one: the numerical waves have a much slower wave speed than the analytical one and for a wavelength equals to the grid spacing, the group velocity is zero : the energy is trapped locally and cannot propagate anymore. This kind of behaviour can easily destabilize a numerical scheme. For this case, C-grid and B-grid are not far from one another but their behaviour is different for other motions like the planetary waves or for 2-dimensional problems.

Discretization in time

Here again it is useful to examine the equations term by term as there is not an only way to differentiate the operators in function of time. The basic scheme for integrating the primitive equations is the leapfrog scheme:

$$u(t+\delta t) = u(t-\delta t) + 2\delta t F(u(t)) \tag{10.26}$$

Three different time steps enter this scheme which is very simple and of second order accuracy. Two following timesteps are necessary to initialize the scheme (or the initial step has to be different). It has one big default: it tends to split the solution between the even and the odd timesteps in a physical solution and a numerical one. So it cannot be used alone. In order to prevent the splitting between two different solutions, it is necessary to use a scheme which will mix the odd and even timesteps. Among the different techniques used, one can cite the Matsuno or Euler-backward scheme which is a two-step damping scheme :

$$u(t+\delta t)^* = u(t) + \delta t F(u(t)) \tag{10.27}$$

$$u(t+\delta t) = u(t) + \delta t F(u(t+\delta t)^*) \tag{10.28}$$

or one can use the Asselin filter (Asselin, 1972) :

$$u_c(t+\delta t) = u(t+\delta t) + \gamma(u_c(t-\delta t) - 2u_c(t) + u(t+\delta t)) \tag{10.29}$$

The leapfrog scheme is used for most terms of the primitive equations except for the dissipative terms which are integrated with different methods. The horizontal dissipation is integrated with a forward timestep which is conditionally stable, implying numerical conditions between the timestep, the grid spacing and the coefficient of horizontal viscous mixing.

$$u(t+\delta t) = u(t-\delta t) + 2\delta t F(u(t-\delta t)) \tag{10.30}$$

Usually the numerical stability conditions implied by the previous scheme present strong constraints for the vertical diffusion term. It may be overcome with a splitting method but very often an implicit scheme is used for its time integration This last scheme is unconditionally stable but diffusive. As it implies a matrix inversion, it is the last term calculated in the equations :

$$u(t+\delta t) = u(t-\delta t) + 2\delta t F(u(t+\delta t)) \tag{10.31}$$

10.2.4 Boundary conditions

Boundary conditions are necessary to close the primitive equations system. Along lateral walls, no flux of heat and salt are permitted (the gradient of these quantities across the boundary is taken to zero). For velocity, different choices are possible. The most commonly used conditions are the no-slip condition ($u = v = 0$ at the boundary) or the slip condition (the velocity normal to the boundary is null but the tangential velocity is conserved). With the first condition, the boundary acts as a source of vorticity. The choice can affect the solution in the vicinity of the boundary where in most cases, the viscous boundary layers are not well resolved by the numerical model.

At the interface with the atmosphere, exchanges are vigourous. Three kinds of energy are exchanged:

- the momentum flux which is expressed as a wind stress imposed on the ocean
- the heat flux which contains the radiative fluxes, the latent heat flux and the sensible heat flux;
- the fresh water flux which corresponds to the budget "evaporation minus precipitation".

The response of the ocean is very sensitive to the specifications of the boundary forcings and it requires much care to define the adequate functions to force the ocean. This point will be developed in Willebrand's lecture.

10.3 Global ocean modelling

In order to highlight different ways to model the global ocean circulation, three different models will be briefly described and compared. They are the following:

TDB the low resolution global ocean model developed in GFDL for large scale climate studies and the references can be found in Toggweiler, Dixon and Bryan (1989 a,b).

SC the high resolution ocean model developed by Semtner and Chervin (1988) for studying the present simulation with a resolution of half a degree. The model is not global but extends to 65°N.

OPA the "medium" resolution global ocean model developed in LODYC to study the large scale ocean circulation.

Neither of these models has a high enough resolution to describe the meso-scale activity within the ocean. SC model solves part of the spectrum but half a degree is too large to explicitly solve the dynamics in mid and high latitudes.

The numerics of these models are different: SC and TDB are written on a B-grid and OPA on a C-grid. This last model uses the tensorial formalism and works on a distorded grid where the coordinate pole is displaced in Canada out of the computational domain (Fig. 10.1). In order to avoid the pole problem, SC stop the model at 65°N and TDB use a Fourier filtering.

For these 3 models, the coastline is defined by an isobath (200 m for OPA) and the land is separated from the ocean by a mask technique. As these models have a very different horizontal resolution, large differences appear in the definition of topography. This is a crucial problem for ocean circulation as basins communicate through very thin passages. The width of the Drake Strait can be doubled for numerical reasons and this has large consequences on the solution: the control of the water volume depends on the bottom specification and water masses properties can be largely affected by the wide extension of a strait. Other difficult areas are the straits in Indonesia between the Pacific ocean and the Indian ocean, or the straits between the Nordics seas and the North Atlantic ocean... The first control exerted over the barotropic flow is given by the contours of f/H : it is important to have a good resolution of ocean ridges and straits. The rigid lid approximation which is used in these three models implies the computation of the barotropic potential on each islands of the domain. This potential may be fixed as a constraint or freely evolving calculated with an elliptic equation (Semtner, 1974). The computation size depends on the number of islands and usually a limited number is selected in order to limit the weight of computation. When an island is linked to coast, the flux between the island and the coast is assumed to be zero. Main connections between the oceans have to evolve freely: Antarctica is a separate land mass as Australia. In OPA, other islands are New-Zealand, Madagascar and the West Indies. In TDB, Madagascar and New-Zealand are not separate islands. TDB consider only two islands and SC three (Antarctica, New zealand, Australia-New guinea). It is also possible to open the Bering strait but this possibility which is not exploited in any of these 3 models requires an accurate topography in high latitudes for this strait is very shallow and narrow.

	TDB	OPA	SC
Δx (in degree)	3.75	1. to 2.	0.5
Δy (in degree)	4.5	1. to 2.	0.5
Levels number	12	20	20
K_{HM} (in m^2s^{-1})	250 000	15 000	1 000
K_{HH} (in m^2s^{-1})	1 000-500	1 000-500	1 000
K_{VM} (in cm^2s^{-1})	20	20	$100/(1+5Ri)^2$
K_{VH} (in cm^2s^{-1})	0.3 to 1.0	0.3 to 1.0	$K_{VM}/(1+5Ri)$
$1/\gamma d$ (in years)	50	1. to 0.15	3 (above 710 m)
Acceleration αg	6 to 180	8. to 11.	7

Table 10.1: *Characteristics and parameters used in three different models of the ocean global circulation.*

A summary of the different characteristics of these models is given in Table 10.1. Although the horizontal resolutions are quite different, the number of vertical levels is similar. These levels are not equidistant but the distance between levels increases with depth. This choice is made to respect the temperature structure in the oceans, with a thin upper layer overlying a stratified region and a thick bottom layer nearly homogeneous.

10.3.1 Horizontal dissipation and diffusion

The value of the horizontal dissipation coefficient of momentum (K_{HM}) is directly linked to the horizontal grid spacing: its value is chosen for numerical stability reasons. All the chosen values are much larger than the molecular viscosity within the ocean and they are supposed to simulate the large eddy activity which is not explicitly solved. A Laplacian operator (or bi-Laplacian) is a poor choice to translate the turbulent activity as a source of dissipation within the ocean. This is certainly an area where major efforts have to be pursued in order to develop parametrizations more adapted to describe the physics. The present value used in TDB is equivalent to the viscosity of tar. On the contrary, the three models use nearly the same value for the coefficient of horizontal diffusion on tracers (K_{HH}). The value is not limited by stability conditions and the order of magnitude is more in accordance with the in-situ large scale observations than with the momentum equation. The choice of a large value for this coefficient (with the same magnitude than in the momentum equation) leads to very irrealistic solutions with a wide boundary layer in the western ocean.

10.3.2 Vertical dissipation and diffusion

The coefficients for vertical diffusion and dissipation (K_{VM} and K_{VH}) are several order of magnitude smaller than they are for the horizontal scale. The choices are similar in TDB and OPA. For the momentum equation, it is constant with depth; for the tracer equation, it is slowly varying with depth, with a larger value in the deep ocean. 0.3 cm^2s^{-1} is a very small value for the surface ocean where large mixing develop in the mixed layer. This choice is made to prevent the slow diffusion of the thermocline. It leads to a very thin surface layer isolated from the deep ocean. As demonstrated in Bryan (1987), the choice of this coefficient is very important for the global balance of the ocean. In fact vertical mixing within the ocean present very different regimes that a linear or constant coefficient cannot simulate. Within the mixed layer, turbulence is very high and, on the contrary, an isopycnal front is very efficient to block the turbulence. The third model (SC) has chosen the parametrization developed by Pacanowski and Philander (1981) which was tested for equatorial simulation. We will come back later to this point. Let us just add that this parametrization allows the vertical diffusion coefficient to be high in regions where the vertical shear of currents is large or the stratification weak, and very weak where the stratification is strong.

10.3.3 Surface boundary conditions

The surface conditions depend on the overlying atmosphere. For the wind forcing, usually a monthly mean climatology is used as, for instance, the data of Hellerman and Rosenstein (1983) which derives from ships data. The three models presented here use this wind stress file. The sampling of ships presents a very irregular coverage in space and time and the bulk formulae which allow to calculate the wind stress from the wind and other parameters of the planetary boundary layer contain large uncertainties. A major source of errors in ocean modelling comes from the uncertainties in the wind stress field. Scatterometer satellites will offer an interesting possibility to increase our knowledge of this quantity.

Heat exchanges between the ocean and the atmosphere are vigourous. The major source of incoming energy in the ocean is the solar flux which is very dependent on the cloud coverage. Its penetration depth depends on the quality of the ocean water (which is linked to the biological activity). In return, the ocean radiates to the atmosphere a long wave heat flux which depends on the sea surface temperature through the Stefan's law. In fact, the long wave exchanges between ocean and atmosphere are complex because of multiple reflections between the ocean surface and the clouds.

The sensible heat flux depends on the temperature difference between the two fluids and on the wind speed. It tends to bring them to the same equilibrium temperature. This flux is usually one order of magnitude smaller than the solar flux and the latent heat flux but it can locally reach very strong value (over the Gulf Stream for instance).

The most important flux involved in the air-sea exchanges is the latent heat flux which depends on the relative humidity, the saturated vapor pressure in air and sea and the wind speed. This term is always a loss for the ocean and it has the same order of magnitude than the incoming solar flux but its distribution is very different. It includes a strong feedback between ocean and atmosphere and is a powerful regulator: it prevents the ocean to get too warm in tropical areas for instance. However this flux is poorly estimated over the ocean and presents a lot of small scale variability which is difficult to parameterize. If the ocean is forced with a prescribed heat flux which is too warm, it tends to increase its stratification in the surface layer and to form a shallow and stable mixed layer which will accumulate the excess of heat. The sea surface temperature can then reach unacceptable values (> 35 C !). There is no possibility for the ocean to get rid of this heat without interactive heat fluxes. This is the reason why in all OGCM's a restoring term is included in the surface heat budget.

Haney (1971) demonstrated that the heat budget at the ocean surface could be written in the form :

$$\frac{Q}{(\rho_o C_p)} = -\alpha(T - T^*) \tag{10.32}$$

where T is the sea surface temperature and T^* a zonal mean equilibrium temperature. ρ_o is the mean ocean density and C_p is the specific heat. α, the restoring coefficient, is the inverse of a time scale (typically 1/30 days). This formulation was recently reviewed by Oberhuber (1988) who proposes to write the heat budget between ocean and atmosphere with the following linearized form :

$$Q = Q_0 + \alpha_0(T - T_0) \tag{10.33}$$

Q_0 is the heat budget determined from observations, α_0 is the derivative $\partial Q_0/\partial T_0$ and T_0 is the observed sea surface temperature. This formulation ensures that, for a simulated sea surface equal to the observed value, the heat exchanged between ocean and atmosphere will be equal to the observed one. It was adopted in OPA.

In SC and TDB, a formulation close to Haney's (1971) was chosen but with some important differences. The heating term was not considered as an external forcing but a restoring forcing. T^* was taken as the observed temperature (from the atlas of Levitus, 1982) and the restoring term was supposed to act over 50 m in TDB and 25 m in SC. With this approach, the heat flux that the ocean wants to simulate the observed sea surface temperature presents large differences with the climatological heat flux from the observations (see the Figure 5 in Sarmiento, 1986) as the computed heat flux should be equal zero if the simulated temperature is equal to the observed one. Such differences are very large in tropical oceans for instance where much too cold water appears at the surface in low resolution models.

The sea surface salinity depends on the evaporation minus precipitation budget (E) at the ocean surface. This term has a very different role than the heat budget as the salt is not visible by the atmosphere i.e. E is independent of the quantity of salt in the ocean ; there is no local feedback between ocean and atmosphere involving salt except in regions where salinity becomes a controlling factor for the mixed layer (like in polar regions during ice formation) but it is important to keep in mind that, on large scale, the salt influences the ocean circulation which in turn modifies the atmospheric circulation. E is the least known quantity exchanged between ocean and atmosphere due to the poor description of the precipitations. It is often specified as a restoring term towards the observed sea surface salinity.

10.3.4 Robust diagnostic modelling

Due to poor parametrization of turbulent mixing and to inadequate grid spacing, global circulation models cannot keep the structure of the thermocline in a long integration. The diffusion of the thermocline is very apparent in intermediate levels in the ocean (between 500 and 1000 m) where there is a spectacular warming (see Fig. 9 in Bryan, Manabe and Pacanowski, 1975 or Figure 19 in TDB, 1989 b). In order to remedy this flaw, a technique was introduced by Bryan and Sarmiento (1982) which was called the robust-diagnostic method. It proposes to take advantage from the knowledge of the temperature field to prevent the ocean drift. A diagnostic term is added to the temperature and to the salinity equation :

$$\frac{\partial T}{\partial t} + \ldots = \gamma_d (T_0 - T) \tag{10.34}$$

γ_d is the inverse of a time scale and it determines the necessary time to reach equilibrium. It is chosen to 50 years in TDB and 3 years in SC to speed up the equilibrium of the deep ocean (below 710 m). In OPA, this parameter is not constant but varies in order to respect some considerations :

- a latitudinal variation is imposed in order to have no restoring force in equatorial regions and to reduce it in polar regions.

  The first point has been discussed in Fujio and Imasato (1990). Considering a geostrophic balance, if there is an error $\Delta\rho$ in the density field, the error in the velocity field will be $(\Delta\rho/f)$. This will give unacceptable large errors close to the equator. It is thus necessary to keep the equatorial regions free from diagnostic forcing.

  In polar regions, the problem is different: the knowledge of in situ temperature is very poor and as most measurements were made when the ocean was ice free, the observed temperature presents a warm bias. It is thus necessary to reduce the impact of the observations in these regions.

- a relaxation to zero was chosen in the vicinity of land point in order to reduce its impact in areas where the dynamics is swift. Moreover the boundary layers are poorly resolved in the Levitus' file where the radius of influence is large.

- a variation with depth is made as in TDB. This variation, proposed by Sarmiento and Bryan (1982), has the purpose to reduce the restoring factor in the deep ocean where the typical time scale is long.

The robust diagnostic technique is very efficient to prevent the temperature drift in intermediate waters but it produces artificial sources of heat within the ocean. In that respect, it cannot be kept to study the climatic evolution of the ocean. It has also undesirable effects on the ocean dynamics: it tends to prevent the deep water formation by convection by stabilizing too much the water columns (see the Figure 4 in TDB-1989a which contrasts the overturning circulation for the global ocean with the prognostic model which has a strong temperature drift but an active convection and the robust diagnostic model which has reduced convective processes). Though it is true that convective processes are strongly limited by the robust diagnostic technique, the quantity of North Atlantic deep water transported to the South can be restored by the diagnostic term. If one compares the overturning circulation in the North Atlantic in the OPA model (Fig. 10.4) and the SC model, the transport of North-Atlantic deep water to the South is very active (more than in TDB) though both

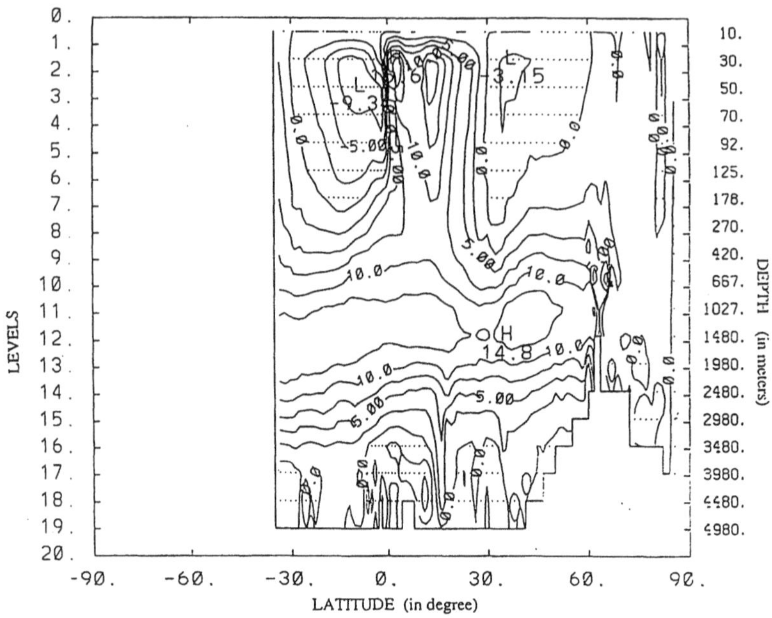

Figure 10.4: *Transport streamfunction of the zonally integrated currents in the North Atlantic in OPA. Dotted areas correspond to negative values. The contour interval is 2.5 Sv (1 Sv =* $10^6\ m^3\ s^{-1}$*)*

these models have a robust diagnostic equation for the temperature and moreover the SC model does not have the Artic ocean. The robust diagnostic technique may be helpful to stabilize the deep ocean to present conditions. It should not be kept in areas strongly interacting with the atmosphere or where drastic changes are suspected in a climate variability study.

10.3.5 Acceleration of convergence

The heat capacity of the ocean is one thousand times larger than the heat capacity of the atmosphere so that the ocean adjusts to a heating anomaly one thousand times more slowly than the atmosphere: one is tempted to speed up the oceanic time integration by a factor of one thousand. However the baroclinic speed which controls the internal adjustment within the ocean is 30 times slower in the ocean than in the atmosphere: the speed up of the ocean may only reach a factor of 30. In order to take advantage of this discrepancy, Bryan (1984) designs a numerical technique which allows to accelerate the spin up to convergence by working with a distorded physics. Two parameters were introduced. The first one α permits to work with a distorded time scale between the momentum equation and the tracer equation. The distorded time t' is equal to: $t' = t/\alpha$. The second parameter γ corresponds to a distorded stratification and it allows the timestep in the tracer equation to vary on a same vertical: $N'^2 = N^2/\gamma$. The consequences of this distorded physics was analyzed for typical ocean motion. They are the followings:

- the free waves have a slower phase speed

- their meridional structure is slightly modified
- the growth rate of baroclinically unstable waves is reduced but there is a wider range of instability.

The technique is efficient for searching an equilibrium state in coarse resolution models. However its application is not suitable for many oceanic problems :

- it cannot be used for transient or time evolving problems (in particular, it is very questionable to keep this technique when studying a mean seasonal cycle)
- it cannot be used in high resolution models where baroclinically unstable processes are important.

The three models use this technique to reach the equilibrium. The different values of the accelerating coefficients are given in Table 10.1.

10.3.6 Results and validation

Analyses and validation of the ocean global circulation model is an important and difficult task. The main difficulty provides from the fact that we are far from being able to describe the global circulation as our knowledge of the oceanic fields contains a lot of blanks. The results can be discussed by comparison with in-situ observations (temperature, currents, salinity) or by confronting simulated fields with theories on ocean circulation. The best observed variable in the ocean is the temperature field. However, as was noticed before, a prognostic integration of the temperature leads to diffuse temperature profiles and sluggish circulation. When using the robust diagnostic technique, temperature is no more available to validate the simulation and the ocean circulation has to be evaluated through the currents distribution. Direct estimations of currents are very scarce and local. Integrated transport (horizontal or vertical) will be calculated and compared to estimations. It is also possible to make an indirect description of the ocean circulation from analyses of passive tracers (like the CFC or the tritium/helium).

Barotropic streamfunction

The balanced barotropic streamfunction (or the streamfunction of the vertically integrated circulation) should verify the model proposed by Sverdrup (1947), in presence of an eastern coastline. The Fig. 10.5 shows the isolines of the barotropic streamfunction simulated from the model currents (a) and calculated (b) from the wind stress curl as proposed by Sverdrup. Over most of the oceans, the agreement is quite good except in the Indian ocean, south of 10°S where the analytical Sverdrup's model cannot take into account the contribution of the throughflow between the Indian and the Pacific oceans. Of course, there is no value from this theory in the Southern Oceans. Close to the western boundaries, the transport values calculated from the Sverdrup's balance are not defined and the values simulated by the model have the same order of magnitude. They correspond to a linear estimation of the circulation along the western boundary and weak values of coastal currents. This is a common drawback of all global ocean models. In the oceanic interior, the flow is basically in geostrophic balance as should be expected.

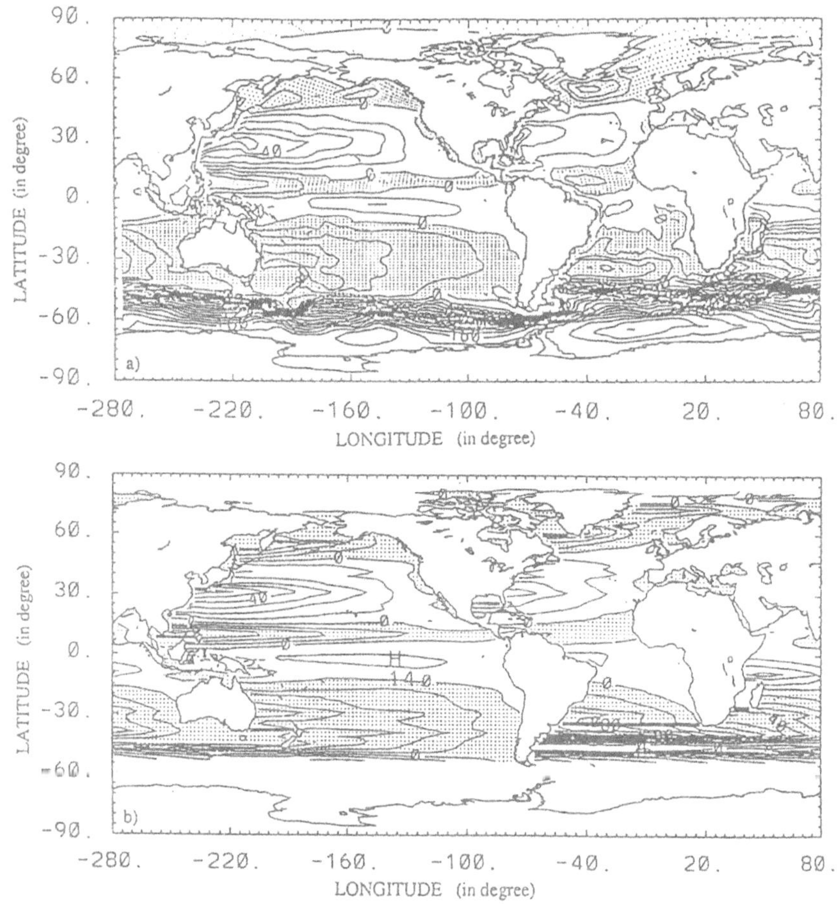

Figure 10.5: *Transport streamfunction of the vertically integrated currents simulated by OPA (a) and from the Sverdrup's balance (b). Dotted areas correspond to negative values. The contour interval is 10.0 Sv (1 Sv =* $10^6\, m^3 s^{-1}$*).*

Vertical overturning

From a global model, one is expected to evaluate the global circulation: the heat, salt and water transport must reach a global balance with the external forcings. A common way to evaluate the models is to look at their streamfunction of the zonally averaged circulation which represents the vertical overturning in the ocean though only a qualitative picture can been drawn from the data. The Fig. 10.6 represents the overturning streamfunction calculated in the OPA model and should be compared to the Figure 4 in TDB-1989a and to Figure 2 in SC. The global surface circulation is marked by an equatorial upwelling which brings up more than 100 Sv ($1\ Sv = 10^6\ m^3\ s^{-1}$) of water to the surface. The main contribution to this upwelled water comes from the Pacific ocean. Nevertheless the magnitude of this upwelling is very large compared to the in-situ estimation (Wyrtki, 1981 - advanced between 37 and 51 Sv for the equatorial Pacific). Upwelled water comes from depths until 600 - 1000 m, especially in the southern ocean. This is very deep and the depth of the equatorial upwelling may be over-evaluated by too much diffusion of the thermocline (at the equator, most models produce too cold waters from a too active upwelling).

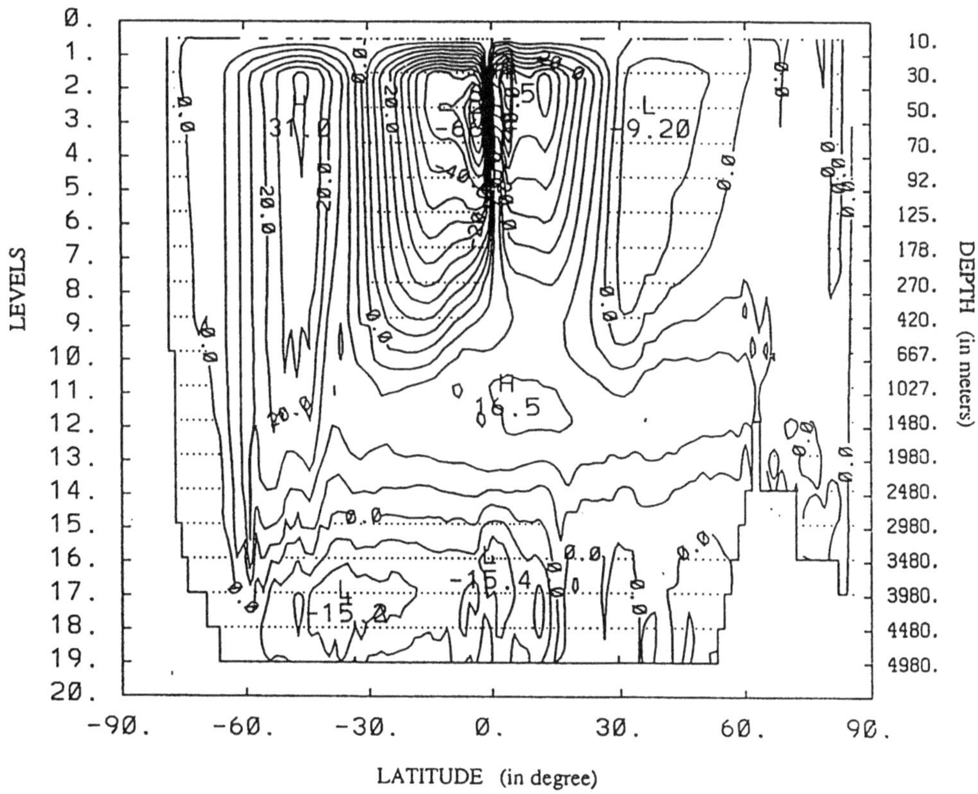

Figure 10.6: *Transport streamfunction of the zonally integrated overturning in the global ocean in OPA. Dotted areas correspond to negative values. The contour interval is 5.0 Sv* ($1\ Sv = 10^6\ m^3 s^{-1}$).

30° N and 30° S mark the position of the convergence where water dives to intermediate levels. Between 1000 and 2000 m the circulation is mainly towards the South and indicates the presence of the North Atlantic deep water. Its contribution is mostly important in the Atlantic ocean where the transport to the south between 1000 and 2000 m can reach 17 Sv. Note that, as discussed before, part of this water is not formed in OPA by convective processes. It is globally form in the northern Atlantic within the water masses in order to adjust the observations. Despite these unphysical mechanism, its path through the world ocean is well marked. Deeper in the ocean, the bottom water created along the Antarctica shelf is slowly moving to the North. It represents nearly 15 Sv of water moving to the North. Once again the source of water formation is not well represented but its effect on the global circulation seems in agreement with estimation. SC model develop a very similar simulation. Its equatorial upwelling may be stronger but the North Atlantic deep water and the bottom water have the same order of magnitude than in OPA. Both of them are stronger than the prognostic model of TDB.

Northward heat transport

The oceans transport as much heat as the atmospheric circulation. It has been evaluated by different methods: as a residual from the heat budget of the overlying atmosphere, from hydrographic sections in the ocean, from ocean models. All these evaluations give a heat transport which is northward in the North Pacific and southward in the South Pacific (between 1 and 3 Petawatts), southward in the Indian Ocean and northward in the whole Atlantic ocean but large discrepancies exist on its magnitude. In the North Atlantic where the in situ data have certainly the densest coverage, the uncertainty on the transport estimation is as large as the transport itself. Nevertheless, its sign is consistent. OPA (Fig. 10.7) and SC have very similar global heat transport. The equator marks the boundary between northward and southward transports. The maximum of transport is reached in subtropical latitudes and attain 1.5 PW (1 Petawatt $= 10^{15}\,W$) for OPA (1.9 PW for SC) in the northern hemisphere and nearly 1.0 PW for OPA (0.6 PW for SC) in the southern hemisphere. There is an increase in the southward transport South of 50 ° S. The major contribution to this transport comes from the overturning term (the vertical circulation). This term is dominant to ensure the equatorial divergence of heat in surface layers and to drop drastically the transport in the convergence areas at 30 ° S and 30 ° N. South of 30 ° S, its contribution is opposite to the mean transport. On the contrary, the gyre circulation (the horizontal circulation) tends to oppose the mean transport in tropical region and to positively contribute to it in southern latitudes. The largest differences between OPA and SC appear in the southern hemisphere where the eddy contribution to the mean transport may be quite differently represented with these models which have different horizontal resolution. The contribution of eddies to the heat transport in the oceans is much debated. In situ measurements (Bryden, 1979 and 1982) indicate that the eddies can locally transport as much heat as the mean circulation. However numerical studies made by Bryan (1986) seem to indicate that the time averaged contribution of the eddies to the heat transport is compensated by modifications of the mean global circulation so that an internal compensation exist in global models : coarse resolution models and high resolution models seem to have nearly the same heat transport (though not for the same physical reasons).

The resolution may not be the most important factor to determine accurately the mean heat transport. In fact, recent studies from Bryan (1987) have shown that the northward heat transport is very sensitive to the parametrization of the vertical mixing. Its value is nearly proportional to the value of the vertical mixing coefficient used in the ocean model.

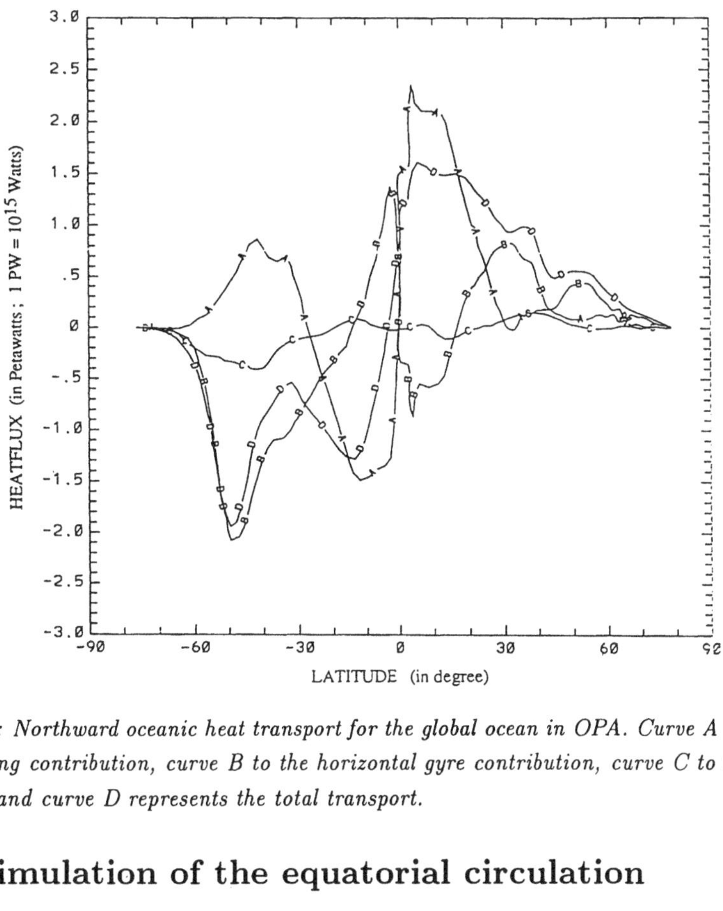

Figure 10.7: *Northward oceanic heat transport for the global ocean in OPA. Curve A corresponds to the overturning contribution, curve B to the horizontal gyre contribution, curve C to the dissipation contribution and curve D represents the total transport.*

10.4 Simulation of the equatorial circulation

As pointed in the first part, regional modelling has been intensely developed in recent years. Here we will give an example of the equatorial modelling and show the progress due to the development of new parameterizations. Equatorial regions present a very specific density profile. The surface layers contain very warm water and are separated from the cold waters below by a very sharp thermocline. The incoming solar radiation is very large in these regions which globally return a large flux in sensible and latent heat fluxes to the atmosphere. These fluxes are important in the trade winds regions and in the regions of gustiness (convective areas). A local heat balance may be reached in warm areas with an equilibrium temperature close to 29 ° C but the equatorial oceans act as a very efficient heat pump in the upwelling areas where the trade winds maintain a divergent ocean circulation. In these regions, cold water goes up to the surface and the latent heat flux comes to zero. These areas gain a tremendous amount of heat which is rapidly transported away from the equator by the divergent circulation. The equatorial oceans can adjust very rapidly to a modification in the atmospheric circulation for the equator is a wave guide where the phase speed is very large: $1\ m\,s^{-1}$ for a Rossby wave, $3\ m\,s^{-1}$ for a Kelvin wave. Both the nullification of the Coriolis force at the equator and the sharp equatorial stratification contribute to define a large scale for the internal radius of deformation

(300 km) and a quick phase speed in these regions. 200 days only are needed to bring the Pacific ocean close to equilibrium. This has to be compared to the adjustment time in mid latitudes basins which is of the order of tens of years! The internal radius of 300 km is nicely solved with a grid spacing of 30 km at the equator and the turbulent activity is explicitly simulated. Moreover it takes a reasonable computing time to bring the ocean to equilibrium. Another favourable factor is that in recent years, there were numerous observational programmes in tropical regions so that the data sampling is more suitable for validation exercises than in mid latitudes. Theoretical advances have provided the needed framework to understand the ocean processes. Moreover the equatorial Pacific undergoes a large scale climatic anomaly each 2 to 10 years which is the most spectacular evidence of the natural and large scale air-sea coupling.

For all these reasons, tropical modelling has experimented a bunch of efforts in recent years. In these regions, progresses have been made in understanding the role of turbulent vertical mixing on the large scale circulation. As pointed above, the use of a constant coefficient for vertical mixing is not adequate to respect the large turbulent mixing in the surface layers and the weak mixing in the thermocline. The spreading of the equatorial thermocline was disastrous in early simulations of equatorial regions. The first effort to remedy this problem was done by Pacanowski and Philander (1981) who introduced a parametrization of the vertical turbulent diffusion which was calculated from the local value of the Richardson number. This formulation which corresponds to a simple diagnostic approach to estimate the local value of the turbulent kinetic energy (TKE) is able to produce large values of K_{VM} (vertical diffusion coefficient of momentum) and K_{VT} (vertical diffusion coefficient of tracer) when the stratification is weak or the vertical shear of currents is large but weak values of K_{VM} and K_{VT} in the thermocline region :

$$K_{VM} = \frac{\nu_0}{(1+\alpha Ri)^n} + \nu_{Mb} \tag{10.35}$$

$$K_{VT} = \frac{K_M}{(1+\alpha Ri)^n} + \nu_{Tb} \tag{10.36}$$

where ν_0 is the maximum value authorized for turbulent diffusion coefficient, α is a parameter chosen to 2 and n another parameter chosen to 5. ν_{Mb} is the background value for the coefficient of vertical turbulence on momentum (equal to its molecular value $10^{-6}\ m^2s^{-1}$) and ν_{Tb} is the background value for the coefficient of vertical turbulence on tracer (equal to its molecular value $10^{-7}\ m^2s^{-1}$). These choices were applied to the tropical version of OPA used to study the tropical Atlantic ocean during the years 1982-84 where measurements were made during the FOCAL (Français Océan Climat dans l'Atlantique équatorial) and SEQUAL (Seasonal Equatorial Atlantic Ocean) experiments (Reverdin et al., 1991). A detailed comparison with observed temperature and currents shows that, though more efficient than a constant value, the Ri parametrization is not able to generate strong enough mixing in the surface layers and the simulated mixed layer is always too shallow. The vertical shear of currents is strong in the surface layers and the impact of the wind stress forcing in the ocean is concentrated in a very thin surface layer: the Ekman divergence is strongly marked in the surface level. In order to improve the simulation, a more sophisticated formulation was introduced in the model. The mixed layer model is given by the equations developed by Gaspar et al. (1990) (hereafter GGL). It is based on a closure assumption for the turbulent length scales which has been tested by Bougeault and Lacarrère (1989) in atmospheric cases. With the following definition for TKE :

$$\overline{e} = 0.5(\overline{u'^2} + \overline{v'^2} + \overline{w'^2}) \tag{10.37}$$

where the various fields are written as a mean value plus a fluctuation part ($x = \overline{X} + x'$), a prognostic equation is developed for TKE

$$\partial_t \overline{e} = -\partial_z(\overline{ew'} + \rho_o^{-1}\overline{p'w'}) - \overline{\mathbf{u'}_h w'}\partial_z \mathbf{U}_h + \overline{b'w'} - \epsilon \tag{10.38}$$

where $B = g\rho_o^{-1}\rho$ is the buoyancy and ϵ is the dissipation. The vertical turbulent fluxes are parameterized with the usual mixing coefficients K_{VM} and K_{VT} according to :

$$-\overline{\mathbf{u}'_h w'} = K_{VM}\partial_z \mathbf{U}_h \tag{10.39}$$

$$-\overline{b'w'} = -g\rho_0^{-1}K_{VT}\partial_z\rho = K_{VT}N^2 \tag{10.40}$$

where N designates the Brunt-Väisälä frequency. The whole vertical flux over TKE, which includes turbulent mixing and pressure effects, requires the introduction of an additional mixing coefficient K_e defined by :

$$-(\overline{ew'} + \rho_0^{-1}\overline{p'w'}) = K_e \partial_z \overline{e} \tag{10.41}$$

Though usually taken equal to K_{VM}, a larger value for K_e has here been chosen ($K_e = 30\ K_{VM}$)) for the tropical model configuration. This choice improves the model stability and allows a convenient TKE homogenization within the surface layers, without affecting the mixed-layer bottom nor the pycnocline. The value of K_e is poorly documented and is likely to depend on the spatial scales resolved by the OGCM. Defining a characteristic dissipation length l_ϵ and a characteristic mixing length l_κ, the dissipation is parameterized following Kolmogorov (1942) and the coefficients K_{VM} and K_{VT} are expressed in function of TKE and l_κ :

$$\epsilon = C_\epsilon \overline{e}^{3/2} l_\epsilon^{-1} \tag{10.42}$$

$$K_{VM} = C_\kappa l_\kappa \overline{e}^{1/2} \tag{10.43}$$

$$K_{VT} = K_{VM}/P_{rt} \tag{10.44}$$

where the constants $C_\kappa = 0.1$ and $C_\epsilon = 0.7$ were designed by GGL to deal with vertical mixing at any depth and where P_{rt} is the Prandtl number. In order to calculate the mixing length, the closure assumption proposed by Bougeault and Lacarrère (1989) for atmospheric models was used: the mixing length is determined by the distance on which a particle will dissipate its TKE in available potential energy. A linearization of their global expression leads to a simple formulation for the mixing length:

$$l_d = l_u = l_\kappa = l_\epsilon = 2^{1/2}\overline{e}^{1/2}N^{-1} \tag{10.45}$$

which is obtained in a stably stratified region with constant values of the Brunt-Väisälä frequency. For tropical simulations, this simplified calculation of the turbulent length scales is a good approximation.

TKE is diagnosed at the surface and the bottom of the ocean. The bottom value for TKE is fixed to a numerical minimal threshold ($e_{\min} = 10^{-11}\ m^2s^{-2}$) while the surface value includes a dependence on the wind stress:

$$\overline{e}_{surface} = 3.75u^{*2} \qquad \text{with} \qquad u^{*2} = \frac{\tau}{\rho_o} \tag{10.46}$$

The time integration of the TKE equation may lead to negative values because the numerical scheme used does not ensure its positivity. A minimum threshold ($e_{\min} = 10^{-11}m^2s^{-2}$) is prescribed to avoid negative values. The computed values of K_{VM} and K_{VT} have to be greater than the molecular viscosity ($10^{-6}m^2s^{-1}$) and diffusivity ($10^{-7}m^2s^{-1}$) respectively.

Let us briefly present the evolution of the results with this new physics. The primary quality of the new physics is to improve the vertical mixing in the surface layers of the ocean. A true mixed layer, with nearly homogeneous temperature profile, exists in the first layers of the simulated ocean. At the thermocline level, the isotherms are pinched and a physical barrier is established across the thermocline. The new distribution of heat and momentum has interesting consequences on the circulation. The Ekman drift is no more concentrated in a thin surface layer but presents a much deeper extension. Consequently, the surface currents are reduced: this is particularly important for the equatorial divergence which is nearly divided by a factor of 2 with this new scheme. Fig. 10.8 contrasts the vertical and zonal streamfunction of the currents integrated in longitude with the physics dependent on the Richardson number and with the new physics. Changes are very important within a few degrees of the equator: the upwelling circulation is drastically reduced and there is a deepening of the overturning cells with TKE. Despite a less intense vertical circulation, upwelled waters are colder and the equatorial upwelling is well marked with an eastward displacement. With the new parametrization, currents are strongly modified. Consequences for the surface divergent circulation have already been mentioned but the most spectacular signature appears on the Equatorial Under Current (EUC). This current is very sensitive to small changes in the equatorial balance. In the western and central parts of the basin, a deeper mixed layer leads to a deepening of the zonal pressure gradient which is a powerful term to accelerate the EUC. The increase of vertical turbulence is more or less compensated by the zonal pressure gradient. As the EUC moves eastward, the thermocline is upwelled to the surface and the vertical mixing is decreasing at the EUC core level: the acceleration of the EUC persists further and maximum speed is reached in the eastern tropical ocean. This result bears a much better agreement with in situ observations than the EUC simulated with the Richardson number dependent physics where the EUC is too sluggish and hardly reaches the eastern part of the basin.

By modifying the distribution of the vertical turbulent fluxes of heat and momentum in the ocean, significant changes are induced in the circulation. Surface currents are slightly reduced with a deeper extension of the Ekman circulation, sub-surface currents are intensified, vertical circulation is reduced but reaches deeper levels. Consequently, a new balance for the surface heat reservoir appears with this new physics and this shows the sensitivity of the ocean response to internal parametrization. The response of a coupled model, where the air-sea feedbacks increase small differences, may evolve very differently with different physical parametrizations.

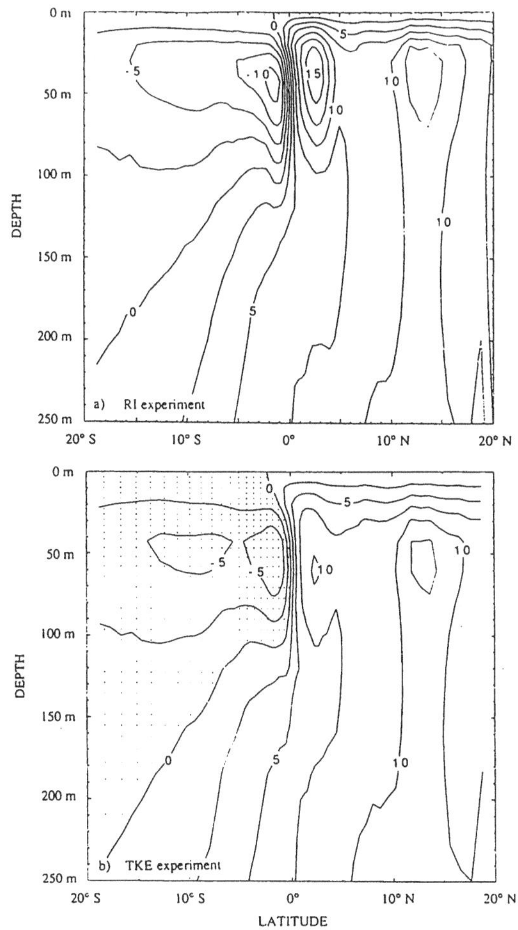

Figure 10.8: *Transport streamfunction of the zonally integrated currents in the tropical Atlantic simulated with the vertical diffusion coefficient depending on the local Richardson number (a) and depending on TKE (b). The contour interval is 2.5 Sv (1 Sv = 10^6 m^3s^{-1}) and dotted areas correspond to negative values.*

10.5 Conclusion

20 years of experiments with numerical models of the ocean circulation based on the primitive equations have considerably helped our understanding of their behaviour and have built increasing confidence in their outputs. Primitive equation models are very successful in regional modelling at high resolution. They are particularly adapted and extensively used to study the equatorial circulation. They are also used in multiple process studies and in theoretical approaches.

Nevertheless, the global approach of the ocean circulation remains a very difficult task. Though the knowledge and the understanding of the simulated ocean circulation is growing, the results are still far from the observed ocean and there is a crucial need to improve many physical processes in the numerical code. Among many problems, one can refer to vertical turbulent diffusion, convection and static stability, topography and continental shelf... and one major problem which is the horizontal turbulent diffusion. Present computers cannot support to bring to equilibrium a global ocean model with the ten degree resolution required to solve the meso-scale activity. It is thus necessary to search for adequate parametrizations but the present understanding of oceanic turbulence makes it difficult. Besides physical improvements, a careful check has to be done on the numerical algorithms used in different codes for the physics depends on the numerical tools used to write it.

Modelling the global ocean still offers a challenge to numerical mathematics and super computers. How can one speed up an ocean code ? These codes require long integration and large memories. Moreover the computations have to be accurate. New architectures like the massively parallel computers may offer an alternative way to accelerate the integrations.

Yet through long term experiments and high resolution modelling, present computers have brought to the oceanographers huge amounts of data which have considerably modified the understanding of the ocean. Regional and scattered observations are no more considered for themselves but also as pieces of a global framework. Global observing and modelling of the ocean is the challenge nowadays.

10.6 References

Asselin R (1972) Frequency filter for time integration. *Mon Wea Rev* 100: No. 6, 487-490.

Bougeault P , Lacarrère T (1989) On the stability of the third-order turbulence closure for the modeling of the stratocumulus-topped boundary layer. *J Atmos Res* 88: 4579-4592.

Bryan F (1987) Parameter sensitivity of primitive equation ocean general circulation models. *J Phys Oceanogr* 17: 970-985.

Bryan K (1969) A numerical method for the study of the circulation of the world ocean. *J Comput Phys* 4(3): 347-376.

Bryan K (1984) Accelerating the convergence to equilibrium of ocean-climate models. *J Phys Oceanogr* 14: 666-673.

Bryan K (1986) Poleward buoyancy transport in the ocean and mesoscale eddies. *J Phys Oceanogr* 16: 927-933.

Bryden HL (1979) Poleward heat flux and conversion of available potential energy in the Drake Passage. *J Mar Res* 37: 1-22.

Bryden HL (1982) Sources of eddy energy in the Gulf Stream recirculation region. *J Mar Res* 40(4): 1047-1068.

Chartier M (1985) Un modèle numérique tridimensionnel aux équations primitives de la circulation générale de l'océan, Thèse de l'université Pierre et Mairie Curie, *CEA Report R-5372* 111 pages.

Courant, Friedrich, Levy (1928) Über die partiellen Differenzengleichungen der mathematischen Physik. *Math. Annalen* 100: 32-74.

Fujio S, Imasato N (1990) Diagnostic calculation for circulation and water mass movement in the deep Pacific. *J Geophys Res* 96: 759-774.

Gaspar P, Gregoris Y, Lefevre JM (1990) A simple eddy-kinetic-energy model for simulations of the ocean vertical mixing: tests at station Papa and Long-Term Upper Ocean Study Site site. *J Geophys Res* 95: 16179-16193.

Haney R L (1971) Surface thermal boundary condition for ocean circulation models. *J Phys Oceanogr* 1: 241-248.

Hellerman S, Rosenstein M (1983) Normal monthly wind stress over the world ocean with error estimates. *J Phys Oceanogr* 13: 1 093-1 104.

Kolmogorov AN (1942) The equation of turbulent motion in an incompressible fluid. Izv. Akad. Nauk. SSSR, *Ser. Fiz.* 6: 56-58.

Levitus S (1982) Climatological atlas of the world ocean, *NOAA Prof paper* 13: Washington D.C.

Mesinger F, Arakawa A (1976) Numerical methods used in atmospheric models. *GARP Publication* No. 17.

Oberhuber J M (1988) An atlas based on the 'COADS' data set: The budget of heat, buoyancy and turbulent kinetic energy at the surface of the global ocean. Max-Planck-Institut für Meteorologie, Hamburg, *Report* No. 15.

Pacanowski R, Philander SGH (1981) Parametrization of vertical mixing in numerical models of tropical ocean. *J. Phys. Oceanogr* 11: 1443-1451.

Reverdin G, Delecluse P, Lévy D, Andrich P, Morliére A, Verstraete JM (1991) The near surface Atlantic in 1982-1984: Results from a numerical simulation and a data analysis. *Prog Oceanogr* 27: 273-340.

Sarmiento JL (1986) On the north and tropical Atlantic heat balance. *J Geophys Res* 91: 11677-11698.

Sarmiento JL, Bryan K (1982) An ocean transport model for the North Atlantic. *J Geophys Res* 87: 394-408.

Semtner AJ Jr (1974) An oceanic general circulation model with bottom topography. *Tech Rep 9* 99 pp., Dep. of Meteorol., Univ. of Calif., Los Angeles.

Semtner AJ Jr, Chervin R (1988) A simulation of the global ocean circulation with resolved eddies. *J Geophys Res* 93: 15, 502-15,552.

Sverdrup HU (1947) Wind-driven currents in a baroclinic ocean: with application to the equatorial currents in the eastern Pacific. *Proceedings of the National Academy of Science* 33: 318-329.

Toggweiler JR, Dixon K, Bryan K (1989) Simulation of radiocarbon in a coarse-resolution world ocean model, 1, Steady state prebomb distributions. *J Geophys Res* 94: 8217-8242.

Toggweiler JR, Dixon K, Bryan K (1989) Simulation of radiocarbon in a coarse-resolution world ocean model, 2, Distributions of bomb-produced carbon 14. *J Geophys Res* 94: 8243-8264.

Unesco (1983) Algorithms for computation of fundamental property of sea water. UNESCO Tech Paper in *Marine Science* 44: 53 pp.

Wyrtki K (1981) An estimate of equatorial upwelling in the Pacific. *J Phys Oceanogr* 11: 1205-1214.

Yin FL, Fung IY (1991) Net diffusivity in General Circulation Models with nonuniform grids. *J Geophys Res* 96: N0 C6, 10773-10776.

Chapter 11

OBSERVATIONS OF AIR SEA FLUXES

Lutz Hasse
Institut für Meereskunde an der Universität Kiel
D-2300 Kiel 1
Germany

11.1 Introduction

Atmosphere and Ocean are parts of a coupled system. However, the relevant scales of the two subsystems are different: The velocity scale of the ocean surface currents is by a factor of roughly 30 smaller, and the time constant of heat storage in the upper mixed layer of the ocean is by the same factor larger than the corresponding scales in the atmosphere. Coupling of ocean and atmosphere is via the fluxes of momentum, heat, and matter (freshwater) through the air water interface. The fluxes of fresh water are not only important for the thermohaline driving of ocean currents but also for driving the atmospheric motions by release of latent heat, and for the moisture balance at the continents. While most water vapour evaporated at the oceans is also precipitated over the sea, the remaining fraction transported over land is of primary importance there. The concern about manmade climate change forces to better determine the fluxes of water vapour from sea to land.

Air sea interaction studies may conveniently be divided into large scale studies like tele-connections and climate investigations, and small scale experimental work. Since all air sea fluxes must go through the interface, and since the surface layer can be seen as the rate determining layer for the turbulent fluxes, larger scale coupling of atmosphere and oceans always implies some small scale interactions. The techniques for experimental measurements of fluxes are expensive in terms of manpower and sophisticated equipment. Thus for larger scale applications some type of parametrization is needed. For a better understanding of problems encountered in air sea interaction it is advantageous to distinguish between measurement, parametrization, and remote sensing. Remote sensing methods will be dealt with in other contributions of this volume, the present paper is concerned with measurements and parametrization at the sea surface.

NATO ASI Series, Vol. I 5
Energy and Water Cycles in the Climate System
Edited by E. Raschke and D. Jacob

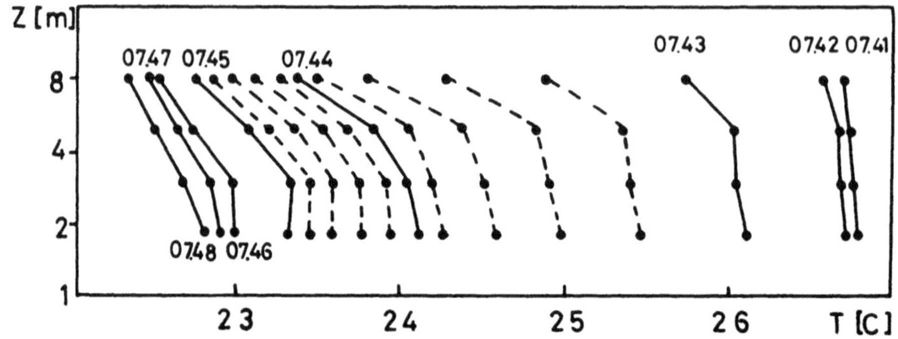

Figure 11.1: *Series of temperature profiles in the surface layer over sea. At each profile the time is indicated by hours and minutes. The series illustrates an event where advection of cold air, supposedly from a cold downdraft, leads to a cooling of the surface layer. The new equilibrium profile is reached after about 6 minutes, in accordance with an estimate of the time constant of the surface layer from energy considerations (Hasse et el. 1978b).*

For the hydrological cycle, evaporation and precipitation are the relevant fluxes. However, for the energy cycle, the other air sea interactions, like sensible heat flux, momentum flux, shortwave and longwave radiation contribute and will be mentioned briefly.

With regard to measurement and parametrization, it is convenient to group the air-sea interactions as follows:

- Turbulent fluxes of latent and sensible heat, and of momentum
- Radiative fluxes
- Precipitation and
- Bubbles and spray.

We may note that in the following we imply the concept of horizontal homogeneity. Considering that internal boundaries, as are induced by a change of roughness or other surface properties, grow roughly with a slope of 1:100 (height to distance), and the height of the surface layer is a few 10 m, this is a good assumption. Also, during GATE, the surface layer has been observed (Fig. 11.1) to establish a new equilibrium profile after a sudden change within about 5 minutes (Hasse et al., 1978b).

11.2 Turbulent fluxes of water vapour, heat and momentum

Methods of measurements are

- eddy correlation or direct method
- profile method
- inertial dissipation or spectral slice method.

This is supplemented by the bulk aerodynamic method that we see as a method of parametrization .

11.2.1 Eddy correlation or direct method

The eddy correlation method gives the vertical fluxes by

$$\begin{aligned} \tau &= -\rho\overline{u'w'} \\ H &= c_p\rho\overline{\theta'w'} \\ E &= \rho\overline{q'w'} \end{aligned} \tag{11.1}$$

from measurements of velocity components u, v, w and of humidity q and potential temperature θ. The dash denotes fluctuations, i.e. deviations from the mean. The necessary frequency resolution is roughly 5 to 10 Hz, and an averaging time of 30 to 60 minutes is advisable in order to obtain stable estimates. The horizontal stress vector τ is parallel to the wind vector. In Eqs. 11.1 τ is the modulus of the stress vector and u designates the downwind component.

A good description of the surface layer is obtained assuming the turbulent fluxes to be constant with z. As characteristic scales of the surface layer one defines u_*, θ_*, q_* by:

$$\begin{aligned} \tau &= \rho u_*^2 \\ H &= -c_p\rho u_*\theta_* \\ E &= -\rho u_* q_*. \end{aligned} \tag{11.2}$$

Typical instruments are: sonic anemometer, hot film anemometer, propellor anemometer, pressure and thrust anemometer for wind components; lyman alpha and infrared hygrometer, microwave refractometer and fast dry and wet bulb psychrometer for humidity; thermocouple or NTC sensors and platinum wire or platinum film for temperatures. For a more extensive listing and detailed discussion of instruments for air sea interaction see Dobson et al. (1980).

Propellor anemometer and psychrometer are in fact somewhat slow for direct measurements and need corrections for frequency response at the high frequency end of the spectra. The difficulty with lyman alpha hygrometer and microwave refractometer is the instability of their calibration. Drifts of calibration of lyman alpha hygrometer appear to happen even within one day. The still best advice is an in situ running calibration, i.e. comparison against a slow response, stable psychrometer (e.g. Lind and Shaw, 1991). Note, that running calibration will work only if there is enough variation of humidity to compare the response of both instruments.

A peculiar problem with the measurement of temperature fluctuations in the marine surface layer is the unavoidable contamination with salt particles at the sensor. This will not only give raise to a change of time constant (see Larsen et al., 1980), but also spurious spikes. These spikes of the temperature trace are due to changes in surrounding humidity. The small salt particles at the sensor adjust to the changes in humidity, and the energy necessary is taken from the salt particles and felt by the sensors. This seriously deteriorates the high frequency part of temperature spectra. This effect is especially felt in the tropics, where the specific humidity is high, and was less pronounced in midlatitudes. Cold/warm spikes would be less important with the direct method, where the cross-correlation matters. However, determination of sensible heat flux from the dissipation method with cold wires or thermocouples is certainly severely biased. The effect of hot/cold spikes is clearly visible in temperature spectra, and is often felt already after 30 minutes of exposure.

With all these instruments the flow distortion of the instrument itself and of its support (ship, platform, buoy) has to be considered (Oost 1991, Wucknitz 1980). Since the fluctuations of the vertical component are so much smaller than the horizontal mean wind, contamination of the vertical components by tilt of the instrument needs to be kept small and to be corrected for. Fixed, bottom - mounted masts and platforms are confined to shallow waters (shallow in the sense that waves would feel the bottom). For open sea, buoys or ships have been used. On ships, of course, one has to deal with considerable difficulties by ship motions and flow distortion. Hence, servo stabilized or stable buoys have been used, where reference is made to some gyro horizon or inertial navigation system.

The influence of waves on the eddy correlation techniques is more subtle. If we think in terms of cross-spectra, then waves, as long as they are linear, have their energy in the quad-spectrum, while the co-spectrum matters for the vertical fluxes. A small phase shift may, however, contaminate the cospectra by contributions of wave-induced quad-spectra.

11.2.2 Profiles

The profile method reflects the experience, that in the presence of gradients of conservative properties, mixing results in transports down the gradient. Hence, the measurement of vertical profiles of atmospheric variables like wind speed, temperature, and humidity will tell something about the turbulent vertical fluxes. The present profile theory developed from the model concept by Prandtl (1932), that the relevant length scale in the surface layer is the distance from the wall. This yields the mixing length $l = kz$, where k is von Karman's constant. Similarity arguments result in:

$$\begin{aligned} \frac{\partial u}{\partial z} &= \frac{u_*}{kz} \\ \frac{\partial \theta}{\partial z} &= \frac{\theta_*}{kz} \\ \frac{\partial q}{\partial z} &= \frac{q_*}{kz} \end{aligned} \tag{11.3}$$

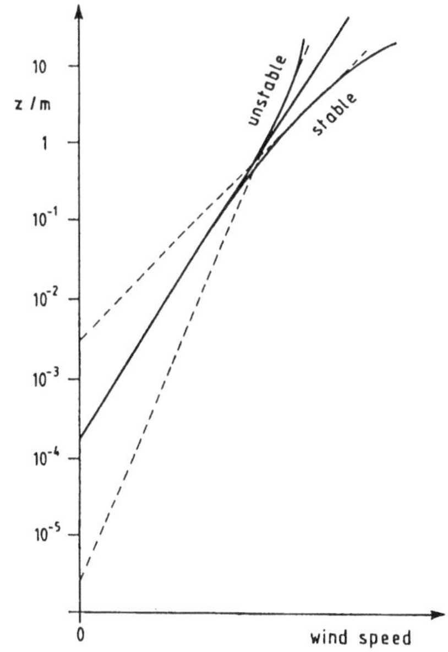

Figure 11.2: *Sketch of influence of stability on profiles in the surface layer. The abscissa is wind speed in arbitary units. Profiles are sketched to have the same friction velocity and roughness length. The curvature of the profiles under diabatic condition is not felt within the experimental scatter when profiles are measured over a limited height range, say 1 m through 10 m. However, the different slope, on linear extrapolation, results in a severely biased apparent z_o.*

for neutral conditions.

Together with the assumption of constancy of vertical fluxes, the logarithmic wind profile is obtained as well as logarithmic profiles of other conservative properties under neutral conditions.

$$u(z) = \frac{u_*}{k} ln\left(\frac{z}{z_0}\right) \tag{11.4}$$

Prandtl (1932) introduced the trick to extrapolate the neutral logarithmic profile to a height z_0, where $u = 0$, called the *roughness length* z_0. Prandtl called this a trick, since very near to the surface turbulence is damped by constraints of continuity, and molecular viscosity takes over (e.g. Hasse and Dobson, 1986).

The original Prandtl theory is applicable only for conditions of neutral stability. It is possible to use the logarithmic profiles to derive fluxes in case of neutral conditions. Unfortunately, within the inevitable experimental scatter, profiles look logarithmic even under non-neutral conditions (Fig. 11.2). There has been repeatedly the misunderstanding that logarithmic appearance has taken as indication of neutral conditions. The resulting estimates of the turbulent fluxes (e.g. expressed as bulk aerodynamic transfer coefficients or roughness length) are severely biased, as impressively shown by Brocks and Krügermeyer (1972). It is advisable to always consider stability for evaluation of profile measurements.

The influence of buoyancy forces on the profiles is expressed by stability dependent flux-gradient relationships ϕ_m, ϕ_H, ϕ_E. For the surface layer, an approach based on dimensional analysis called

Monin-Obukhov theory is widely accepted. The Monin-Obukhov similarity starts from the following set of definitions:

$$\begin{aligned} \frac{kz}{u_*}\frac{\partial u}{\partial z} &= \phi_m \\ \frac{kz}{\theta_*}\frac{\partial \theta}{\partial z} &= \phi_H \\ \frac{kz}{q_*}\frac{\partial q}{\partial z} &= \phi_E \end{aligned} \tag{11.5}$$

The lefthand-side, under neutral conditions, is unity. Under non-neutral conditions these normalized gradients as dimensionless products should depend only on a dimensionless measure of stability. Commonly,

$$\frac{z}{L} = -\frac{g}{T}\frac{H_v/c_p\rho}{u_*^3}kz \tag{11.6}$$

is used where L is the Monin-Obukhov length and the subscript v (virtual) indicates correction for influence of water vapour on buoyancy. Other measures of stability are possible, too. The fluxes are obtained by

$$\begin{aligned} \tau &= \rho k^2 z^2 \left(\frac{\partial u}{\partial z}\right)^2 \frac{1}{\phi_m^2} \\ H &= -\rho k^2 z^2 \frac{\partial u}{\partial z}\frac{1}{\phi_m}\frac{\partial \theta}{\partial z}\frac{1}{\phi_H} \\ E &= -\rho k^2 z^2 \frac{\partial u}{\partial z}\frac{1}{\phi_m}\frac{\partial q}{\partial z}\frac{1}{\phi_E} \end{aligned} \tag{11.7}$$

The stability functions or flux-gradient-relationships are determined from observations and thus are liable to experimental errors. Hence, different formulations are available, that typically deviate from each other by 10%. Despite of this, the following formulations have found a more widespread use (see e.g. Panofsky and Dutton 1984; Dyer, 1974; Webb, 1970):

For unstable conditions

$$\begin{aligned} \phi_m^4 - 16\phi_m^3 \tfrac{z}{L} &= 1 && \text{KEYPS} \\ \phi_m &= \left(1 - 16\tfrac{z}{L}\right)^{-1/4} && \text{Dyer -} \\ \phi_H = \phi_E &= \left(1 - 16\tfrac{z}{L}\right)^{-1/2} && \text{Businger} \end{aligned} \tag{11.8}$$

For stable conditions

$$\phi_m = \phi_H = \phi_E = 1 + 5\tfrac{z}{L} \qquad \text{Webb} \tag{11.9}$$

where the index m stands for momentum, H for heat and E for water vapour. Usually, the stability function for water vapour (and perhaps other conservative admixtures) under unstable conditions is taken to be equal to the one for heat. The KEYPS formulation is preferable since it provides the limiting form for free convection. The Dyer-Businger formulation for momentum may be understood as a rather good approximation for the KEYPS formula. The different exponents in the Dyer-Businger formulations for momentum and heat imply that the ratio of the eddy diffusivities for heat and momentum varies as

$$\frac{K_H}{K_m} = \frac{\phi_m}{\phi_H} = \left(1 - 16\frac{z}{L}\right)^{-1/4} \tag{11.10}$$

for unstable situations. This implies that with increasing instability, the ratio becomes increasingly larger than 1: When z/L is about -1, the eddy diffusivity for heat is about twice as large as the one for momentum. Some difference in the eddy diffusivities is expected since the transfer processes from the eddies to the surrounding air are different for heat and momentum. Pressure forces change the momentum of a particle moving through the surrounding air without material mixing. On the other side, the vertical heat flux is correlated with buoyancy forces that influence vertical motions. It may be noted in passing that profile theory also implies a knowledge of the von Karman constant k. We found good agreement between profile and direct measurements over sea when using $k = 0.4$.

Measurement of profiles of wind speed, humidity, and temperature appears to be simple. However, it is the vertical variation that matters so that small errors become more meaningful. Typical errors are radiation errors of temperature and humidity sensors, and flow distortion errors of wind speed measurements. The influence of masts and spars is felt in the flowfield even at relative distances (distance to radius) larger than 10. It is sometimes assumed that flow distortion does not matter if instruments are mounted on spars on one side of a mast only. This is not true: what you don't see, may still be there (Wucknitz 1977). Personally, I believe that profile measurements at ships are not only useless but rather detrimental: Severely biased measurements are worse than no measurements.

It may be noted that profile measurements of momentum and water vapour at sea appear quite reliable: the unexplained variance of friction velocity squared versus wind speed squared was found to be of order 11% from buoy measurements at the open Atlantic (Dunckel et al., 1974). Water vapour gradients usually are quite large and hence better measured. On the other hand since the air sea temperature difference usually is small, temperature gradients are difficult to measure and C_H as the ratio of two small entities is subject to rather large errors. Substituting C_E for C_H appears to be a reasonable approximation.

For measurements at sea, especially at ships and buoys, one has to consider the waves, both in terms of modulation of air flow above waves and of wave induced motions of instrumentation. For theory of air flow above wave see e.g. Phillips (1977). As a practical rule: the anemometer should be clearly above the crests of the waves when the surface following instrument support is in the wave trough (Hasse et al., 1978a). This is true also for buoys that measure wind speed at one height only.

For application within the profile theory the wind velocity should be determined relative to the sea surface: At a fixed position, the surface drift velocity needs to be considered. At a drifting platform corrections for the drift velocity of the platform are required. Since surface drift velocities are typically of order 2% or 3% of wind speed, these corrections often are neglected.

11.2.3 Inertial dissipation method

The scientific background for this method is the famous Kolmogoroff theory of locally isotropic turbulence:

> *At sufficiently high Reynolds numbers there is a range of high wavenumbers where the turbulence is statistically in equilibrium and uniquely determined by the parameters ϵ and ν. This state of equilibrium is universal.*

and

> *If the Reynolds number is infinitely large, the energy spectrum in the subrange satisfying the condition $k_e \ll k \ll k_d$ is independent of ν, and is solely determined by one parameter ϵ.*
>
> *Since in this subrange the inertial transfer of energy is the dominating factor, this will be called the inertial subrange* (cited from Hinze, 1959).

Here, ϵ is the rate of dissipation, and ν, kinematic viscosity. The condition $k_e \ll k \ll k_d$ means that the wavenumber k of this subrange is large compared to the wavenumber of energy containing eddies k_e and small against the wavenumber of the region of dissipation k_d. According to the second hypothesis, the spectral form is solely determined by ϵ. The theory results in the spectral energy ϕ proportional to minus 5/3 power of frequency or wavenumber. From this, one may work back to the generation of turbulent energy and to the shear stress.

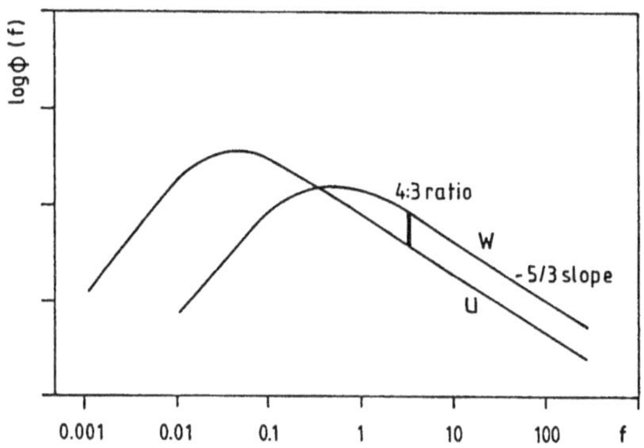

Figure 11.3: *Sketch of spectral variation with frequency in the production range and inertial subrange. In the production range, the longitudinal (or downwind) spectra (marked u) have higher energy than the transversal/vertical spectra (marked w). In the locally isotropic inertial subrange, the spectral energy of transversal and longitudinal components has a ratio of 4:3 and an exponent of - 5/3. The implication is that downwind spectra exhibit a - 5/3 behavior outside the inertial subrange. This shows that one should not estimate dissipation from measurements outside the inertial subrange: A small deviation from the -5/3 law may not be visible within the experimental scatter, but may severely bias the results.*

The reason why the so-called dissipation method is so appealing for air sea interaction experimental work is seen in the spectra of turbulence. Surface waves (and hence any wave induced disturbances) have their peak in the typical production range of turbulent energy. The dissipation range, on the other hand, is at higher frequencies out of the range of the energy containing surface waves and wave induced ship motion (Deacon, 1959; Skupniewicz, 1991). The dissipation method is apparently insensitive to ship or buoy motions and flow distortion errors. However, there are difficulties that are due to the fact that we not really measure dissipation, but rather spectral energy that has to be extrapolated to obtain dissipation (Fig. 11.3). A typical source of bias is the use of too low a frequency range, that is not fully in the inertial subrange (Wucknitz, 1979).

The spectral slice method can be used also for determination of dissipation of water vapour and temperature fluctuations. Therefrom through the budget equation for water vapour and temperature variance the water vapour or sensible heat flux is obtained (Hicks and Dyer, 1972; Fairall and Larsen, 1986). The empirical coefficients that relate spectral density to "dissipation" of water vapour and temperature variances are less well established. From the experimental point of view, this method is not recommended for measuring the sensible heat flux at sea because of the likely contamination with cold/hot spikes.

11.2.4 Parametrization, the bulk aerodynamic method

Often, sophisticated measurements are not possible and recourse has to be taken to more robust parameters as measured with routine weather observations at sea. The so called bulk aerodynamic method, that really is not a method of measurement but rather of parametrization, yields estimates of fluxes by the following equations

$$\begin{aligned} \tau &= C_D \rho \overline{u}^2 \\ H &= -C_H c_p \rho \overline{u} \Delta \overline{\theta} \\ E &= -C_E \rho \overline{u} \Delta \overline{q} \end{aligned} \tag{11.11}$$

The bulk transfer coefficients C_D, C_H, C_E are usually referred to 10 m height and 10 min averaging time. These parametrization can be seen as a result of our experience, that turbulent mixing results in an average transport down the gradient, when a gradient of conservative properties of a fluid exists. The bulk transfer coefficients are thus the result of all the complex processes in the near surface layer above the sea and depend on turbulence intensity as result of wind speed, stability and seastate. Since wind speed and seastate are correlated, the bulk transfer coefficients are often assumed to vary with wind speed and with stability. Their stability dependence may conveniently be obtained by integration of the flux-gradient-relationship from the surface to reference height (Panofsky, 1963). The wind speed and seastate dependence results from experimental determinations. Note, experimental determinations of bulk transfer coefficients are commonly given either as mean over the natural mix of stabilities or reduced to neutral stability. The values of the natural mix are typically of order 10% or 20% higher than neutral coefficients, since unstable conditions prevail at sea.

Similar as with the profile method, the wind speeds should be taken relative to the surface. However, since surface drift velocities are usually small, corrections for drift velocities are generally neglected.

There is a considerable scatter of empirical bulk transfer coefficients. It is important to understand that this is not only due to instrumental and methodical imperfections but also to the fact that sea state depends on fetch and duration and, in shallow waters, on water depth (e.g. Donelan, 1990). Young seas, as are found in conditions were generating waves are not yet in equilibrium with the wind, are usually more steep and hence rougher than fully developed seas.

Despite of the experimental scatter, the bulk transfer coefficients have found to be an indispensible tool in most air sea interaction work, both in data analysis and in modelling. Hence, some choice

U_{10} (m s^{-1})	T_A-T_S						
	>5.0	4.9 – 1.0	0.9 – 0.2	0.1 – −0.2	−0.3 – −1.0	−1.1 – −4.9	<−5.0
< 5	0.05	0.47	0.77	0.95	1.04	1.23	1.42
5–10	0.61	1.03	1.13	1.22	1.26	1.41	1.47
10–15	1.16	1.36	1.42	1.48	1.50	1.58	1.66
15–20	1.54	1.61	1.66	1.71	1.75	1.78	1.83
20–25	1.79	1.82	1.86	1.90	1.91	1.93	1.96
25–30	1.99	2.01	2.03	2.05	2.07	2.08	2.09
30–35	2.20	2.20	2.21	2.21	2.21	2.21	2.21
35–40	2.37	2.37	2.37	2.37	2.37	2.37	2.37
45–50	2.53	2.53	2.53	2.53	2.53	2.53	2.53
>50	2.69	2.69	2.69	2.69	2.69	2.69	2.69

Table 11.1: *Drag coefficients C_D (in units of 10^{-3}) as function of wind speed at 10 m height and air-sea temperature difference (K) applicable for open sea (Isemer and Hasse, 1987).*

U_{10} (m s^{-1})	T_A-T_S						
	>5.0	4.9 – 1.0	0.9 – 0.2	0.1 – −0.2	−0.3 – −1.0	−1.1 – −4.9	<−5.0
< 3	0.06	0.26	0.63	1.15	1.44	1.78	2.19
3– 6	0.19	0.58	0.97	1.17	1.26	1.46	1.75
6– 9	0.60	1.02	1.18	1.25	1.27	1.37	1.56
9–12	0.92	1.18	1.29	1.33	1.37	1.44	1.56
12–15	1.21	1.37	1.40	1.43	1.46	1.51	1.60
15–20	1.38	1.46	1.52	1.57	1.58	1.62	1.69
20–25	1.51	1.56	1.59	1.62	1.62	1.62	1.68
25–30	1.57	1.60	1.61	1.62	1.63	1.64	1.68
>30	1.62	1.62	1.62	1.62	1.62	1.62	1.62

Table 11.2: *Bulk transfer coefficients C_E (in units of 10^{-3}) as a function of wind speed at 10 m height and air sea temperature difference (K). These are optimized by use of an oceanographic constraint (Isemer and Hasse, 1987).*

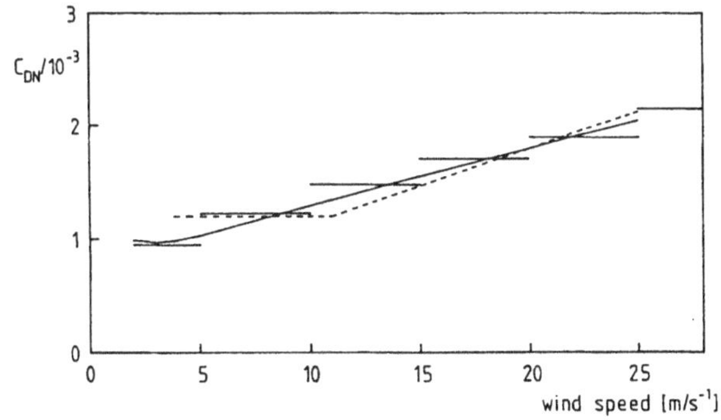

Figure 11.4: *Variation of neutral drag coefficient C_{DN} with wind speed. Full line is from S.D. Smith (1988), dashed line from Large and Pond (1982), step function is from the Bunker atlas (Isemer and Hasse, 1987). These drag coefficients are based on data from the open sea and should be applicable with due correction for non neutral stability, for the natural mix of sea state conditions at open sea. Reference height for wind speeds is 10 m.*

needs to be made. In our evaluation (Isemer and Hasse, 1987) we mainly relyed on direct and profile measurements obtained at open sea. For practical reasons we followed an approach of Bunker (1976), who approximated the transfer coefficients as step functions. The coefficients for a range of stabilities are given as Tables 11.1 and 11.2. The stability dependence used by Bunker, in fact, is based on the Monin-Obukhov similarity following Deardorff (1968).

The drag coefficients for neutral conditions are depicted in Figure 11.4. The bulk transfer coefficient for water vapour or latent heat flux is found near $C_{EN} = 1.2 \cdot 10^{-3}$ from direct measurements at open sea in the wind speed range 4 to 12 $m\,s^{-1}$ under neutral conditions (e.g. Kruspe, 1977; Smith, 1989). Barring the discussion of a possible influence of spray generation on the evaporation for a moment (see below), from the experimental evidence C_{EN} could be taken independent of wind speed. However, since C_E is a parametrization of a turbulent transfer, one would expect a variation with the energy of turbulence. Hence, an increase of C_D with mean wind speed should be reflected in a wind speed dependence of C_E, i.e. C_{EN} should vary as the square root of drag coefficient with wind speed. Since $C_{DN}/10^{-3}$ increases from 1.2 at 8 $m\,s^{-1}$ to roughly 2.0 at 25 $m\,s^{-1}$, one would expect C_{EN} to increase to about $1.6 \cdot 10^{-3}$ at 25 $m\,s^{-1}$ wind speed. This is in good agreement with the coefficients of table 11.2.

11.3 Radiative fluxes

Atmospheric radiation is usually not measured at sea. Some of the ocean weather ships carried solar radiation instruments (Dobson and Smith, 1988). However, it is difficult even for shortwave radiation to maintain an absolute calibration that is better than 5%. Long wave radiation usually can not be measured during daytime and even at night large errors may arise (see Hinzpeter, 1980). A 5% error in radiation appears to be quite acceptable, but is rather large if the energy input into the upper ocean is considered. In principle, the physics of radiative transfer through the atmosphere is well understood and the radiation balance for given atmospheric conditions could be calculated by solving the equation for radiative transfer by existing methods (e.g. Paltridge and Platt, 1976). However, the necessary informations on the vertical structure of the atmosphere are usually not available (Frourin et al., 1988). Hence, parametrizations from ship synop observations are in wide spread use. Both, shortwave and longwave radiation are considerably affected by clouds. The parametrizations of radiation are commonly based on clear sky radiation and a modification by clouds .

11.3.1 Shortwave radiation

Clear sky shortwave radiation is readily calculated as a function of solar height and the atmospheric transmission (for a suitable value see below). The albedo of the sea surface is for practical purposes near 0.06. Higher values are found with clear skies and smooth seas and low solar altitude, a condition not important for air sea interactions. Alternatively, albedo tables by Payne (1972) can be used.

Clouds of different types absorb solar radiation to different degrees. However, it has been found impractical (see Dobson and Smith, 1988) to base a parametrization on type of cloud observation as given in ship synop data. The same accuracy of estimate of insolation in the presence of cloud is obtained through parametrization with aid on mean cloud cover.

For use with climatological data the Budyko-Berliand parametrization , that is non-linear in cloud cover, and has been derived from observations over land, proved to be less suitable for use over sea. At present, a parametrization derived by Reed (1977) that is linear in total cloud cover N, appears to be best suited for use over sea (Dobson and Smith, 1988):

$$S = (1-\alpha)(1 - c_n N + 0.0019h)Q_0 \tag{11.12}$$

Here, S is solar radiation absorbed at the sea surface, Q_0 clear sky solar radiation , α albedo, N total cloud cover (as fraction of 1), c_n empirical cloud cover coefficient, and h solar height at noon (in degrees). For cloud cover 1/8 and 2/8, Reed recommends a constant reduction factor of 0.95 instead of the bracket. In principle, one would expect that with increasing cloud amount the probability of low, dense clouds increases. Consequently a nonlinear relation between extinction by clouds and cloud amount would be expected. As an explanation why a linear relation performs better I may offer that these parametrization are devised to be used with monthly mean cloud covers. The typical frequency distribution of cloud cover is U-shaped. Hence, the climatological mean cloud cover perceived as an index of frequency of sunny or cloudy days and that may well be linearly related to insolation. Eq. 11.12, being linear in cloud cover, can be used with daily averages too.

11.3.2 Longwave or Infrared Radiation (IR)

Usually, upwelling and downwelling IR radiation at sea are not measured, due to contamination by solar radiation during daytime and to the difficulty to obtain an undisturbed exposure. Outgoing IR radiation is calculated according to the surface temperature and Stefan - Boltzmann's law. The counter radiation (downwelling component) with clear skies depends strongly on the vertical distribution of water vapour and its temperature (or rather on the temperature distribution as function of the water vapour coordinate). When the vertical profiles are known the counter radiation can be calculated. However for the majority of data from sea, only ship synop observations are available. Hence, some Ångström or Brunt type parametrization formulæ (like Eq. 11.15 below) are used that relate the counter radiation to σT^4 and a measure of water vapour (note, in many of the old formulæ water vapour is in mm Hg).

Swinbank (1963) has argued that the correlation of clear sky back radiation with the sixth power of temperature is better than with water vapour. The explanation lies in the variation of water vapour with temperature that is imposed through the dependence of saturation water vapour on temperature. With high temperature, there is high water vapour content and hence the counter radiation originates near the surface. With lower temperatures and less water vapour, higher layers are included that have even lower temperatures. The clear sky back radiation I_0 according to Swinbank can well described by

$$I_0 = 0.531 \cdot 10^{-12} (T/K)^6 \; Wm^{-2} \tag{11.13}$$

With view to the fact that

1. there are more ship reports with temperature than with humidity measurements, and
2. it is easier to remotely measure the temperature at the sea surface than water vapour,

it may be worthwhile to test Swinbanks idea again.

The counter radiation, however, is strongly modulated by the presence of clouds. This is accounted for by an equation of the form

$$I = I_0(1 - kN) \tag{11.14}$$

where I is the effective outgoing radiation with clouds , and I_0 with clear skies. N is cloud amount in fractions of 1, and k is an empirical coefficient. For the dependence on cloud amount, a power greater 1 is sometimes recommended, while power 1 is commonly used. The values of k vary with cloud type (see e.g. Sellers, 1965). This variation can be understood as result of a correlation of optical thickness and cloud base temperature with cloud type. However, because there are less informations available of cloud type than of total cloud cover, usually cloud cover only is considered. Isemer and Hasse (1987) used the Efimova formulation that has been found to give good results in an intercomparison with model calculations by Fung et al. (1984):

$$I = \epsilon \sigma T_A^4 (0.254 - 0.00495 e_A)(1 - cN^d) + 4\epsilon\sigma T_A^3 (T_s - T_A) \tag{11.15}$$

Here T_A and T_s are temperature of air and sea, respectively, σ Stefan-Boltzmann's constant, e_A water vapour pressure of the air (in hPa), ϵ emissivity of the sea surface. $c = 0.654$ and $d = 1.1$ are empirical coefficients. The last term, called the jump correction, was added by Bunker to take care of the fact that the outgoing radiation stems essentially from the sea surface (with T_s) while the counter radiation stems from material with a different temperature T_A.

Eq. 11.15 is somewhat unphysical, since upwelling and downwelling radiation are not separated. The value $\epsilon = 0.96$ is part of a parametrization to be used with climatological data. The emissivity of water varies with frequency, it reaches 0.993 in the infrared window, and drops off at longer wave length's (Mikhaylov and Zolofarev, 1970). The emissivity is reduced by surface films. Also, the "flux" emissivity into the hemisphere averaged over all zenit distances is less than the "vertical" emissivity (Paltridge and Platt, 1976, p. 135).

11.4 Precipitation

Inspection of world maps of precipitation discloses that the drawn isolines are educated guesses rather than measurements. It is well known that coastal convergence and orography influences the probability of precipitation at coastal and island stations so that measurements from these can not be extrapolated to sea. Also, the difficulties to measure rain at ships are notorious.

During GATE, most information on precipitation was taken from ship-borne radar measurements. However, radar reflectivity is strongly dependent on drop diameter and consequently the rain to reflectivity relationship is dependent on the type of precipitation physics. It appears to be good practice over land to calibrate actual radar rain estimates by some ground truth. Unfortunately, such is at present not available at sea. The planned rain measuring satellites, that carry a radar type instrument, will need some ground truth, too. The international plans call for seaward looking radars from island stations that in turn are to be calibrated by ship or buoy borne rain gauges. An appealing possibility to measure rain at sea from ships is a vertically looking doppler radar. This could measure the velocity spectra and, interferring the rain drop diameter from the terminal velocity, the rain intensity could be calculated.

The difficulty with rain gauges at ships is seen in a twofold flow distortion: There is the flow distortion around the ship and around the rain gauge itself. These problems are more severe at sea than over land since not only are wind speeds higher at sea, but also wind and ship speed may add to a rather high relative wind. It is thus for good reasons that rain gauges are not routineously used at ships. The flow distortion error by ship superstructure has been investigated during GATE: Austin and Geotis (1980) found the systematic differences from the average of 6 sites at a ship were within 15%, surprisingly low. The result may be somewhat fortuitous since the ship was nearly stationary. The results of their experiment are in agreement with the perception, that high up in the mast, the mean flow distortion will not result in much updraft or downdraft, but the horizontal velocity may be enhanced and a conventional rain gauge would undercatch.

It is evident that ground truth for satellites has to consider the rather patchy structure of rain. However, worldwide there are more than 7000 voluntary observing ships (VOS), of which typically 2000 - 3000 collect synop data four or eight times per day. If the VOS can be outfitted with ship rain gauges, a running calibration of satellite rain estimates could be obtained. Also observed rain amounts can be used to calibrate rain rate predictions from numerical weather-forecast-models.

At the marine meteorology division of IfM, Kiel, we are presently constructing new types of ship borne rain gauges (Hasse et al., 1992). Two types are tested, a mechanical one and an optical. The mechanical has a conventional horizontal and additionally a vertical collecting surface (upper Fig. 11.5). Total rain amount is obtained by combining the two collected amounts with aid of the measured local flow velocity. This ship rain gauge has a smaller cross section than conventional

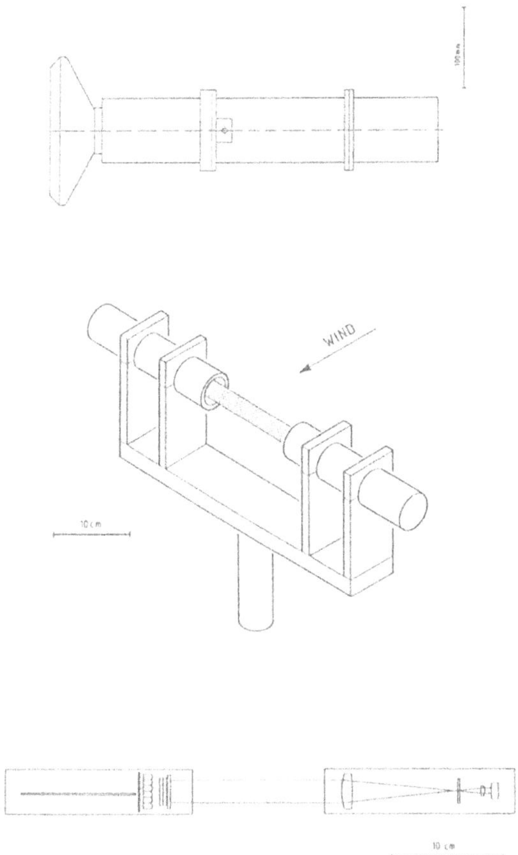

Figure 11.5: *Improved raingauge systems for shipbord use with high relative flow velocities. Mechanical raingauge (top). Rain is collected conventionally at the horizontal surface and at an additional vertical cylindrical surface (125 mm height) above the rim. The water amount from the vertical and horizontal collecting surfaces is recorded separately and, together with the recorded local air velocity, allows calculation of the rain rate. Optical disdrometer, slant view (middle) and vertical cross-section (bottom). When a rain drop passes through the illuminated area (shaded), its size is measured. The rain rate is obtained from the drop-spectra and the fall velocity as function of drop size (Hasse et al., 1992).*

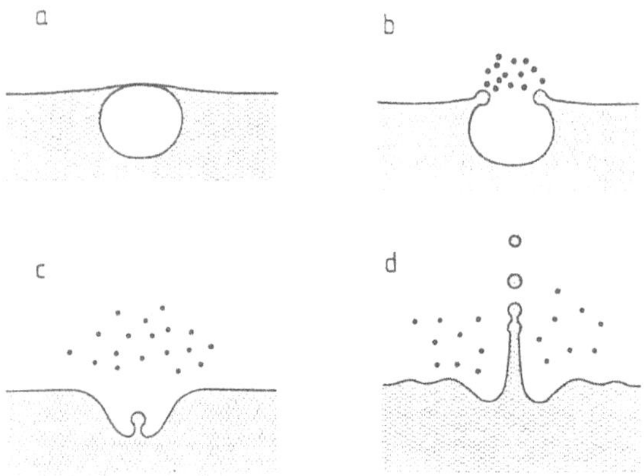

Figure 11.6: *Scheme of bubble bursting. The rupture of the surface produces many small film drops of order 10 μm diameter. The acceleration by surface tension leads to production of a few jet drops of order 100 μm diameter.*

instruments in order to reduce local flow distortion. The optical instrument is of the light extinction type (lower Fig. 11.5). The diameter of each drop is measured by the extinction pulse, when the drop passes through the illuminated volume. Its time in the volume is proportional to the probability to find a drop of the given size in a volume, and the rain rate is obtained assuming terminal fall velocity as function of drop diameter. This type of measurement should be independent of any local up- or downdrafts. Our device is similar to the optical distrometer described by Illingworth and Stevens (1984).

It may be mentioned that there is another method to estimate rain at sea, the so called Tucker method (Tucker, 1961). This is based on the past and present weather coding in synop observations. Rainfall amount is estimated therefrom in analogy to rainfall over land in maritime midlatitude (UK). While this is a reasonable approach, it has not yet been established what the reliability is for other climate regions.

11.5 Bubbles, spray and aerosol generation

Aerosols are of considerable interest in climate studies because of their influence on radiative transfer within the atmosphere. In the context of GEWEX aerosols additionally are interesting since the presence of droplets above the waves may increase the effective water surface where evaporation takes place. We shortly summarize the physical concepts. There are two sources of droplet production at a wind driven sea:

1. Water may be blown by action of the wind from the crests of waves, and
2. air bubbles in the sea rise and burst at the sea surface, ejecting droplets into the air.

The first of these two mechanism is known from the Beaufort scale of wind force: at higher wind speeds (starting with Bft. 8) the wind blows "spin drift" or "spume" from the crests of the waves,

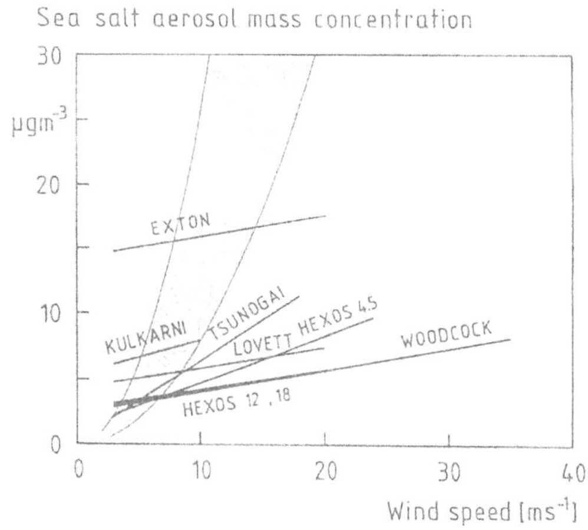

Figure 11.7: *Aerosol content. Measurements by different authors summarized as straight lines by Marks and Monahan (1989). The linear fit through rather scattered data underestimates the increase with wind speed. The shaded area is the compilation of Junge (1965).*

that affect visibility at Bft. 10 and higher. The second mechanism is brought about by the breaking of waves, that entrains air into the water, thus producing white caps, i.e. clouds of air bubbles in the water. When rising to the surface, the bubbles will burst and through the action of surface tension eject droplets into the air (Fig. 11.6). Each bursting bubble produces a few larger and many fine droplets. Bubble generated droplets have a size range of $10 - 100\mu m$ and primarily reach heights of 10 to 15 cm above the water. The bubble bursting mechanism is thought to be the more important one in aerosol generation (Bortkowskii, 1987), at any rate, it operates at much lower wind speeds, and in that sense is more important. There is another factor of spray generation: when deep water waves run into shallower water, the wave form changes, surf developes, and waves finally break. It is likely that the increasing nonlinearity of waves in this process enhances spray and spume generation for a given wind speed, compared to the situation over deep water.

Unfortunately, the aerosol production as a function of wind speed and sea state is not well known. Junge (1965) inferred an increase of sea salt concentration in air approximately with the square of wind speed, with a scatter of about the factor 2 in both directions. Marks and Monahan (1989) fitted a collection of published data of aerosol content over sea by linear increase with wind speed (Fig. 11.7), but found a more rapid than linear increase at higher wind speeds. A possible explanation for the rather wide variation of the experimental data is seen in the fact that near shore measurements will be influenced by surf generated spray.

It is beyond doubt that aerosol will be generated at sea by the action of wind and waves. Its relevance for the enhancement of evaporation is less clear. One reason may be that droplets provide additional surface. Evaporation from their surface would bypass the layer of slow molecular transport near the interface. As becomes evident from Bortkowskii's (1987) work, many details of distributions of bubbles and droplets are not yet known, hence the calculation of the net effect on evaporation is quite uncertain. During the HEXOS - experiment (HEXOS = humidity exchange over sea) evaporation

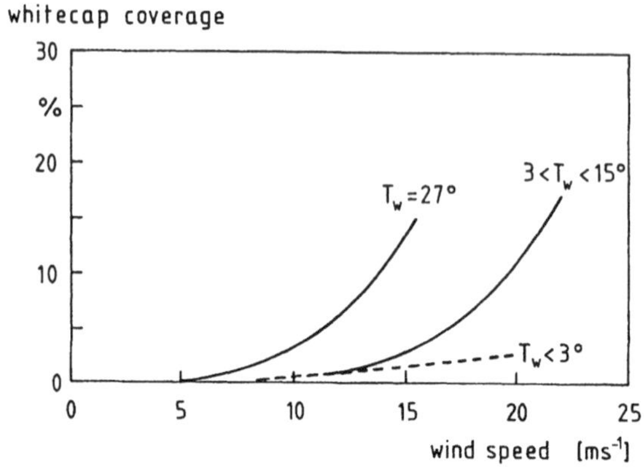

Figure 11.8: *Dependence of white cap coverage on wind speed and water temperature. Ordinate is relative white cap coverage, i.e. percentage of area covered. Water temperature range is indicated at the respective curves. Adapted from the collection by Bortkovskii (1987).*

was measured by an international team at a platform near the dutch coast (Smith et al., 1990) and no strong increase with wind speed was found.

That wave generated aerosol will not enhance evaporation with increasing wind speed, comes as a surprise. It is difficult to find a plausible explanation. Since evaporation at the interface proper will persist, it would need strong evaporation from droplets to provide a noticeable effect. The main argument is given by the energy constraint (Hasse, 1992): evaporation needs the heat of vaporisation. The heat content of a droplet is too small to provide the heat for its own evaporation. In a steady state, this heat has to come from the sensible heat flux. The latter can directly be compared to the latent heat flux. The ratio of sensible and latent heat flux, the Bowen ratio, is typically 0.3 in midlatidudes and 0.1 in the tropics. Since only part of the sensible heat flux is used for evaporation of droplets, the enhancement of evaporation by droplets is limited. At the same time, the sum of latent and sensible heat flux to first order is not influenced by evaporation of droplets in the surface layer.

The white cap coverage has been studied extensively, for a review see e.g. Monahan and O'Muircheartaigh (1986) and Bortkovskii (1987). When bubbles rise to the surface, not all burst immediately. The ensemble of bubbles resulting from a breaking event forms a bubble cloud. These become visible as white caps. White cap coverage increases rather rapidly for wind speeds higher than about 4 $m\,s^{-1}$, roughly with the third power of wind speed. The wind speed where white caps first appear depends on stability, expressed as air sea temperature difference, in the sense that higher wind speeds are needed for stable conditions (Monahan and O'Muircheartaigh, 1986). The white cap coverage (Fig. 11.8) appears to be temperature dependent (Bortkovskii, 1987), as is expected since the relevant molecular coefficients like surface tension are temperature dependent.

Another aspect of white caps in the context of GEWEX is their influence on remote sensing. Whitecaps change the effective emissivity of the sea surface for microwaves. Since whitecap coverage is wind speed and temperature dependent a calibration problem for wind speed or stress algorithms from microwave remote sensing signals is on hand.

11.6 Influence of surface waves

Part of the momentum transferred from the atmosphere to the ocean is used to generate surface waves. In this way, waves are an essential part of air sea interaction. However, in GEWEX we are mainly interested in global aspects. In numerical modelling of ocean circulation, surface waves usually are not considered explicitely: The momentum transferred to the waves is passed on to the currents. Hence, as boundary condition continuity of stress through the air sea interface is assumed

$$\rho_S u_{*S}^2 = \rho_A u_{*A}^2 \tag{11.16}$$

where the subscript A, S stand for air and sea, respectively.

It is not possible in a short contribution, to adequately deal with the physics of waves and their role in air sea interaction. The reader is referred to introductions (e.g. Donelan and Hui, 1990) or to text books like Phillips (1977). A few hints must suffice here. The gravity waves of the sea surface are well described by linear wave theory. The relevant parameter is wave steepness ak, where a stands for wave amplitude, $k = 2\pi/\lambda$ is wave number, λ wave length. Waves can be described formally as small amplitude waves for $ak \leq 1$. The solutions of linear theory are usually applicable to deep water waves. The notion "deep water" requires that the local water depth is deep enough relative to wave length to not influence the orbital motions. Waves are said to be in deep water when $kd \geq \pi/2$ or $d \geq \lambda/2\pi$ where d is water depth, k is wavenumber, λ is wave length. In shallow water, where the relative water depth is much less than the indicated limit, waves become strongly nonlinear. When deep water waves approach a shore and roll in over a self into shallow water, surf developes that finally results in breakers.

Linearity implies that the solutions for monochromatic waves, the Fourier-components, may be superposed. The more interesting processes, however, are with the deviation from linear behavior (see e.g. Phillips, 1977). The equilibrium spectral shape, for example, is determined from weakly nonlinear resonant wave-wave interactions. White caps, as another example, stem from the nonlinear process of wave breaking. GEWEX will rely heavily on satellite remote sensing of sea surface properties. It is necessary to understand the effect of different nonlinearities on the remotely sensed signal (e.g. Brüning et al., 1990). The empirical fact that waves are well described by linear theory does not imply that there are no nonlinearities. Rather, the nonlinear processes like wave breaking work to restore more nearly linear behaviour. Deviations from linear theory grow with wave steepness, wave breaking reduces wave steepness.

In the context of GEWEX we need to deal with the influence of waves in the following areas:

- dependence of drag coefficients on wind speed/sea state
- wave breaking, bubbles, spray and aerosol generation, and their influence on evaporation
- influence of wave induced motions on near surface measurements.

11.6.1 Influence of sea state on stress

The stress at the sea surface is commonly described in terms of a drag coefficient. As mentioned above, this drag coefficient represents all the complex processes that are involved in the turbulent transfer of momentum near the sea surface; e.g. mechanical and buoyant production of turbulent kinetic energy. Hence, drag coefficients are often described in terms of wind speed and stability. Such a description is somewhat strange, since C_D is a dimensionless number and should depend only on other dimensionless products, not on dimensional variables like wind speed. Also, we expect the surface roughness to be somehow related to sea state. Sea state is certainly correlated with wind speed, but it is also true that there is no unique relationship between wind and seas. Waves need fetch and duration in order to obtain equilibrium in a given wave field (e.g. Hasselmann et al., 1973; Phillips, 1977). Hence, some parameter that describe adjustment of waves to the wind field could be relevant. Since the phase velocity of waves is smaller in the beginning and may reach the mean wind speed or even exceed the latter by a few 10%, the ratio phase speed to wind speed c/u is often taken as measure of wave age. However, since in turbulence theory u_* is taken as velocity scale, c/u_* is also used as "wave age". Hence, the drag coefficient is expected to depend on wind speed, stability, and wave age. It is within our practical experience that waves after a sudden increase of wind speed are rougher while waves in equilibrium with the wind appear longer and less rough. The time to reach equilibrium is of order hours or days - longer for higher wind speeds. Thus, with increasing wind speed the frequency of situations where wind and waves are not yet in equilibrium, will increase. That is, in the mean we expect an increase of the effective drag coefficient with wind speed. On the other hand, since wave energy/wave height increases with higher winds, so would the form drag of the wind moving over the waves, again leading to an increase of drag coefficient with wind speed. At present, the experimental evidence is not very convincing (see e.g. Donelan and Hui, 1990).

Part of the remaining uncertainty is due to the difficulty to obtain reliable stress measurements at sea. Caution: repeatedly, some relationships of the type C_D versus wave age c/u_* were presented, that look quite convincing. However, since $C_D = (u_*/u)^2$ and u_* is beset with experimental scatter, there are spurious correlations and such data convey no more information than the common knowledge that wave height and wave speed increase with wind speed.

For practical use, we may argue that the initial adjustment of the wave field to a changing wind is more rapid than the progress to equilibrium sea state. Accordingly, Smith et al., (1992) give C_D as a function of wind speed for "very young", "mature" and "fully developed" seas ($c/u = 0.5, 1.0, 1.2$ respectively), where the mature drag coefficient at 20 $m\,s^{-1}$ is about 8% higher than for fully developed seas. The habit to take the drag coefficient as function of wind speed (and perhaps stability) only is thus seen as a reasonable approximation.

It may be mentioned that instead of the empirical description, that generally is prefered by scientists who work experimentally in air sea interaction, modellers often prefer the so called Charnock relation that gives the roughness length as proportional to u_*^2/g, where the coefficient of proportionality is adjusted to fit the empirical data. Personally, I object to the Charnock relation. The roughness length can not reasonably be related to characteristics of the wave field. Most likely, the conceptional model that underlies the dimensional argument, is not applicable (Charnock, 1981; Hasse, 1986). As an empirical interpolation of data, the Charnock relation is of limited value since it yields unrealistic low values at lower wind speeds (say below 5 $m\,s^{-1}$) and, though formally a relation, is no better than other empirical descriptions.

11.6.2 Wind field above sea waves

In the context of the present chapter the wave field is also of interest because of its influence on measurement techniques of air-sea interactions. Both measurements of wind profiles and of mean wind speed are influenced by the wind field above the waves. Since waves to a good approximation are linear and wave steepness (wave height divided by wave length) is small, typically even less than 0.1, the wind field above the wavy surface can be described, using a wave modulated curvilinear rectangular coordinate system (Brooke Benjamin, 1959). The origin of the system follows the local interface that is assumedly sinusoidal, say, a Fourier component of the wave field. The deformation compared to the Cartesian coordinate system decays with distance from the interface. In this coordinate system, surface layer theory is applied. This is not the full truth, since there are small, but important phase shifts between wave field and pressure field, but sufficient to approximate the wave influence on the wind field. It implies enhanced speeds at the crests of the waves and reduced speeds in the wave troughs (Hasse et al., 1978a). The influence on the wind profile has already been mentioned, but also the influence on measurements of mean wind velocity at low heights from surface following equipment needs to be emphasized. Dobson (1990, personal communication) found that wind speed measurements at about 5 m height on a surface following buoy as used for ground truth of satellite algorithms significantly underestimates the real wind speed in high seas.

11.6.3 Influence of wave induced motions on measurements in the surface layer

The difficulty with direct measurement from ships originates from the influence of flow distortion at the ship and from instrument tilt due to ship motions. The importance of an unbiased exposure of the equipment is seen in the fact that the stress is the crossproduct of fluctuating horizontal and vertical components. Any mean deflection of the wind vector from the horizontal (say, by an angle α) would contaminate the vertical velocity by a contribution $u' \sin\alpha$ and the stress by $u'^2 \sin\alpha$.

Such a mean inclination could be produced by platform induced updraft, by misalignement of the sensor with the vertical, and by asymmetries within the sensor itself. Similarly, fluctuating deviations of the sensor from the vertical (say, by an angle α') would produce spurious vertical velocities $\overline{u} \sin\alpha'$, where $\overline{u}$ is mean wind speed. The appropriate corrections to counteract these influences are rotations of the coordinate system that make the mean vertical velocity component zero (or rather $\overline{\rho w} = 0$) and that removes fluctuating inclinations by reference to a gyro horizon. The mean rotation would also take care of minor misalignements in the sensor geometry.

11.7 Practical considerations

11.7.1 Field experiments

Field experiments at sea are considerably more expensive than measurements in the laboratory, and, even worse, can not be repeated easily. It is advisable to have a high degree of redundancy in field measurements. In research, one commonly has some highly sophisticated instrumentation, that naturally has a higher chance to fail. Such measurements should be backed up by simultaneous records from proven, reliable instrumentation of lesser sophistication, that is suitable to check the more sophisticated instruments and helps to detect failures, biases, or drifts. As third level, standard meteorological observations at hourly intervals are recommended that may lateron help to explain pecularities of a given situation and may provide parameters for estimation of environmental conditions. For example, with our air sea flux measurements we usually paralleled direct measurements with profile measurements. The latter, at the same time, provided mean parameters of wind speed, air and sea temperatures and humidity. Additionally standard observations with hand held instruments were taken and e.g. surface drift velocities determined.

Experience shows that flux measurements are often not reliable enough for determination of Monin-Obukhov z/L. However, it is usually possible to determine mean wind speed and air sea temperature difference that yield a bulk Richardson number. The bulk Richardson number can be used to estimate or check z/L. Also, because of the better sampling statistics of mean parameters compared to the one of covariances, the bulk Richardson number is more stable and less susceptible to experimental errors than z/L from direct measurements. Experimental air sea interaction work is quite demanding. It is not enough to have bright new ideas. It takes experience to bring new ideas to work and to avoid the experimental difficulties and sources of failures, that others have suffered from before. It is necessary to have a group of meteorologists, experimentalists and data reduction experts together for some extended time in order to successfully work in flux measurements at sea. It is also important that our fellow scientists who work in other fields of meteorology and use our products, understand the difficulties with air sea interaction data. There is no sense in averaging unreliable or biased data (like measurement of turbulent fluxes where the wind comes through the legs of the platform rig, or profiles at shore with accelerated flow due to sloping terrain).

The remaining part of this chapter we will devote to practical problems in the use of meteorological information to study air sea fluxes.

11.7.2 Surface winds at sea

There are problems with wind observations at sea that are often overlooked. These problems are interrelated. I may summarize these under the following subheadings: "Wind is not Wind", "Scientific Beaufort Scale", and "Historic Beaufort Equivalents"

Wind is not Wind

When numerical modellers talk about surface wind, this typically is a wind at the lowest computational level, representative of a grid scale area of some 100 km squared and perhaps even geostrophically adjusted. When experimentalists talk about surface wind, they mean 10-minute averages measured at 10 m height above sea level or at least reduced to this height. It is evident that these two winds have different physical properties; however, this is rarely realized by numerical modellers.

Scientific Beaufort scale

Until about 1970 the wind data from sea almost exclusively were estimates according to the Beaufort scale, not measurements. Even today, only up to 40% are measurements, the majority still are Beaufort estimates. Major shipping nations still require Beaufort estimates and for good reasons: Measurements require the error prone vector subtraction of speed and course of ship from relative wind, reduction to reference height, and perhaps corrections for variable flow distortion. Beaufort estimates are based on characteristics of the sea surface and hence are independent of height of observations and of most typical error sources of ship borne measurements. However, with Beaufort estimates, one needs an empirical relationship to go from Beaufort force to wind speed. World Meteorological Organisation rules demand coding of surface wind estimates at sea in terms of knots or $m\,s^{-1}$. Because these data have the dimension of speed, researchers take these as surface wind speeds without even questioning the relevance of the data. It is known that the WMO code 1100 is biased. Hence, the use of the coded data without corrections will give erroneous results. Note that in climate studies often the so-called COADS data are used. Almost all of the older wind data in COADS are Beaufort estimates. Hence, COADS is biased, and air-sea interaction determined from COADS without due correction is biased, too. This again is rarely realized by climate researchers or modellers.

The WMO commission of marine science affairs (WMO, 1970) recommended a scientific scale to be used for scientific purposes, while code 1100 is still in official use for general purposes. We applied a scale derived by Kaufeld (1981) that is similar to the scientific scale. The resulting changes in mean wind speed in certain areas of the world ocean reach 20% compared to use of code 1100. The changes in regional and seasonal variations are even more serious! The relevance for GEWEX is evident by the resulting mean error in latent heat flux or evaporation: For the whole of the North Atlantic Ocean, removal of the bias amounted to an increase of heat flux from the ocean to the atmosphere by 24 Wm^{-2} or a roughly 0.3 m higher annual evaporation (Isemer and Hasse, 1991).

Historic Beaufort equivalents

Originally, the Beaufort scale was defined in terms of sea behavior of a certain type of ship (how much sail it should carry). When steam ships took over, the scale was less well defined. Considerably later, the scale was redefined in terms of visible characteristics of the sea surface. It is obvious that over the time of half a century the scale may have slipped. In fact, in our opinion the Beaufort scale has changed with time, and we can prove this as follows (Lindau et al., 1990): The basic idea is that we can use wind and pressure observations in order to describe wind fields at sea. Pressure observations can be used to derive pressure gradients and hence geostrophic winds. These are frictionless winds and thus differ in their meaning from surface winds. For calibration of the

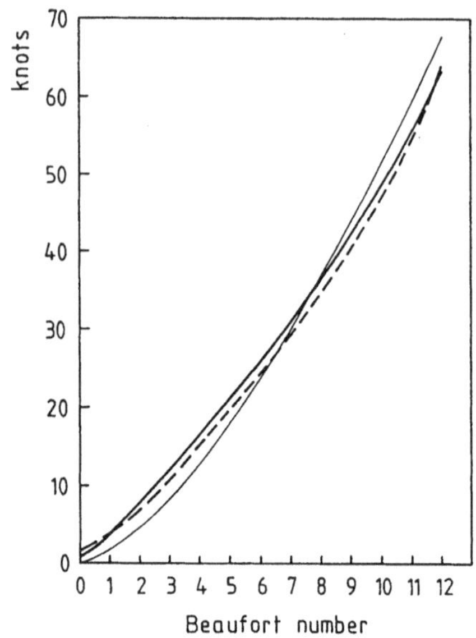

Figure 11.9: *Beaufort force equivalent wind speeds (Kaufeld, 1981). 1) (full line) from comparison of wind speed measurements at ocean weather ships and Beaufort estimates at passing voluntary observing ship (referen height for measurements is roughly 25 m). 2) (thin line) WMO code 1100. 3) (dashed line)* Scientific *Beaufort scale (WMO, 1970).*

historical Beaufort scale we assume that the actual to geostrophic wind relationship (in other words: the actual to geostrophic wind speed ratio and the ageostrophic angle) has not changed over the last century (although we do not consider this relationship in any detail). This allows to compare ship estimates of Beaufort strength against pressure gradients and adjust historical Beaufort estimates to a common basis in terms of pressure gradient. This is done for each Beaufort force separately. The trick in this case is that pressure observations are independent of wind observations and hence can be used for calibration.

The relevance of the correction is demonstrated for a certain $10^\circ \times 10^\circ$ field at the shipping route near the Brazilian coast (Fig. 11.10). The wind speeds from Beaufort estimates, uncorrected, showed a definite and somewhat strange trend. Corrected with aid of pressure gradients, there is nearly no discernible trend of mean wind speed. Most of the apparent trend was spurios!

11.7.3 Consistent parametrization of air-sea interactions from ship data

Large scale fluxes at the ocean surface can be determined from Voluntary Observing Ship data by aid of parametrizations. The selection of best parametrizations is in fact a matter of scientific judgement. Within our studies to determine air sea interaction fluxes from the North Atlantic ocean (Isemer and Hasse, 1987), we used updated parametrizations. In a first step, we obtained "best" parametrizations for short wave insolation, net longwave radiation , latent heat flux, sensible heat flux, and surface

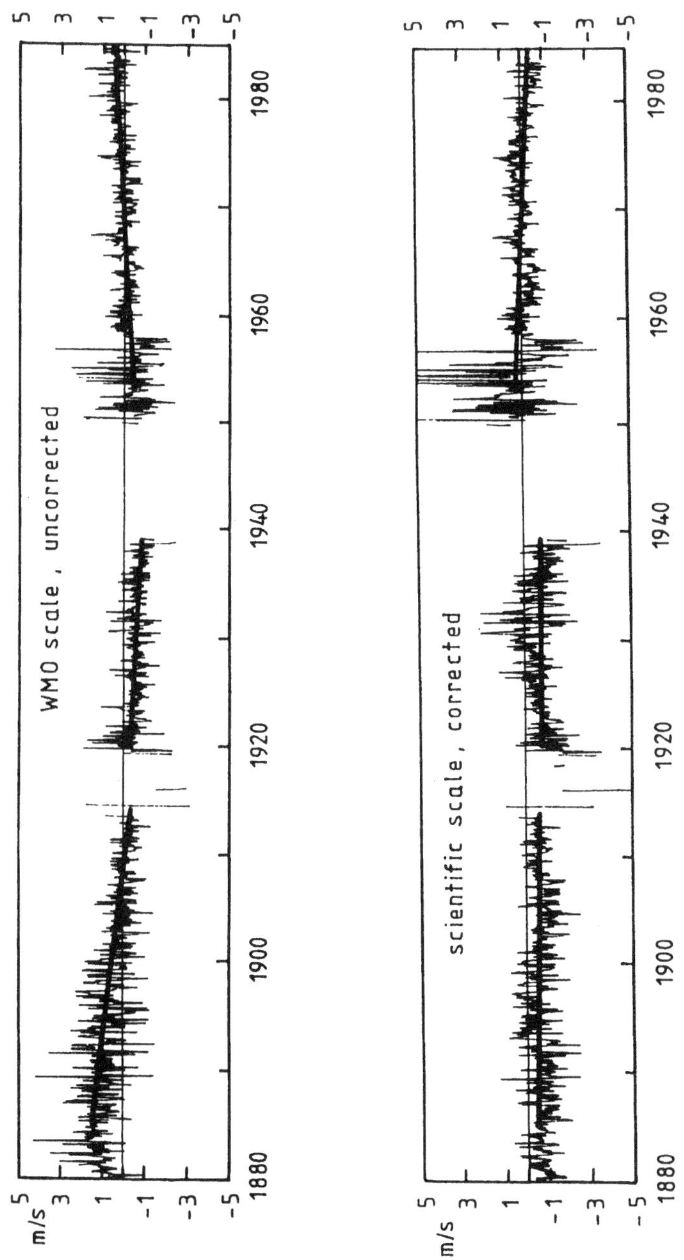

Figure 11.10: *Time variation of mean monthly wind speed at the Brazilian coast between 10° and 20°S. Upper panel, taking code 1100 Beaufort equivalents at face value. Lower panel: Beaufort estimates corrected by aid of pressure gradients for drift of scale. Kaufelds (1981) scale is used to convert from Beaufort force to wind speed (Lindau et al., 1990).*

stress. We tried to use the best available evidence applicable for open ocean conditions. This includes use of the scientific Beaufort equivalent scale. The parametrizations were selected independently; they need not be consistent. Consider the surface energy balance: the major energy supply to the sea surface is by insolation, the major loss by evaporation. A perfectly acceptable accuracy (say 5%) of the two major terms may led to a 50% uncertainty of the net heat balance at the sea surface.

In a second step a fine adjustment was made based on the consideration that an imbalance of heat exchange at the sea surface in the mean must be balanced by heat advection within the ocean. Since some estimates of meridional oceanic heat transport at different latitudes are available, these can be used as constraints for the parametrizations of heat components at the surface. In fact, since the ocean heat transport divergences are small quantities compared to the heat balance components at the surface, a fairly coarse estimate of ocean heat transport provides a valuable constraint (Isemer et al., 1989).

A meridional heat transport of 10^{15} W at 25 ° N and $0.09 \cdot 10^{15}$ W at 65 ° N has been used with an inverse technique. This technique provides an objective method to adjust parametrization coefficients with as small as possible (in r.m.s. sense) change, where the changes are weighed with the respective error bands of each coefficient. It showed that only about 40% of the error bands were used for the adjustments. This points to reasonable description that allows a confident use of these parametrizations. We do not pretend that we achieve perfectly correct parametrizations in this way. Rather, we derived a set of parametrizations where known biases are reduced and that provide consistent estimates of air sea interactions. Since these were derived for the North Atlantic Ocean, that spans a wide range of ocean climates , it is likely that these parametrizations would be valid for other oceans too.

The relevance of these revisions is undervalued, if only mean changes for the North Atlantic Ocean are considered: E.g. the revised parametrizations yield a mean heat loss by evaporation of 139 Wm^{-2} instead of 117 Wm^{-2} according to Bunkers parametrizations. Moreover, the adjustments of regional and saisonal variations of the net energy input are of order 40 Wm^{-2} and present a rather strong change in the driving force of the ocean. It is evident that a more detailed investigation of evaporation at sea is an essential feature of GEWEX.

11.8 References

Austin PM, Geotis SG (1980) Precipitation over the ocean. In: Dobson, Hasse, Davis (eds) Air-sea interaction - instruments and methods. *Plenum Press* New York. pp. 523 - 541.

Bortkowskii RS (1987) Air-sea exchange of heat and moisture during storms (revised engl. ed). *D Reidel Dordrecht* 194 p.

Brocks K, Krügermeyer L (1972) The hydrodynamic roughness of the sea surface. Gordon (ed) Studies in Physical Oceanography 1. *Gordon and Breach* New York. pp. 75-92.

Brooke Benjamin T (1959) Shearing flow over a wavy boundary. *J Fluid Mech* 6: 161-205.

Brüning C, Alpers W, Hasselmann K (1990) Monte-Carlo simulation studies of the nonlinear imaging of a two dimensional surface wave field by a synthetic aparture radar. *Int J Remote Sens* 11: 1695-1727.

Bunker AF (1976) Computations of surface energy flux and annual air-sea interaction cycles of the North Atlantic Ocean. *Mon Wea Rev* 104: 1122-1140.

Charnock H (1981) Air sea interaction. In: Warren BA, Wunsch C (eds) Evolution of Physical Oceanography. *MIT Press* 482-503.

Deacon EL (1959) The measurement of turbulent transfer in the lower atmosphere. *Adv Geophys* 6: 211-228 Academic Press New York.

Deardorff JW (1968) Dependence of air-sea transfer coefficients on bulk stability. *J Geophys Res* 73: 2549-2557.

Dobson F, Hasse L, Davis R (eds) (1980) Air-sea interaction, Instruments and Methods. *Plenum Press* New York 801 pp.

Dobson FW, Smith SD (1988) Bulk models of solar radiation at sea. *Q J R Meteorol Soc* 114: 165-182.

Donelan MA (1990) Air-sea interaction. In: The Sea. Vol 9A: *Wiley* New York 239-292.

Donelan MA, Hui WH (1990) Mechanics of ocean surface waves. In: Geernaert and Plant (eds) Surface waves and fluxes. *Kluwer Dordrecht* 209-246.

Dunckel M, Hasse L, Krügermeyer L, Schriever D, Wucknitz J (1974) Turbulent fluxes of momentum, heat and water vapor in the atmospheric surface layer at sea during ATEX. *Bound Lay Meteorol* 6: 81-106.

Dyer AJ (1974) A review of flux-profile relationships. *Bound Lay Meteorol* 7: 363-372.

Fairall CW, Larsen SE (1986) Inertial-dissipation methods and turbulent fluxes at the air ocean interface. *Bound Lay Meteorol* 34: 287-301.

Frouin R, Gautier C, Katsaros KB, Lind RJ (1988) A comparison of satellite and empirical formula techniques for estimation insolation over the oceans. *J App Met Soc* 27: 1016-1023.

Fung IY, Harrison DE, Lacis AA (1984) On the variability of the net longwave radiation at the ocean surface. *Reviews Geophys and Space Phys* 22: 177-193.

Geernaert GL, Plant WJ (eds) (1990) Surface waves and fluxes, Vol 1: Current theory. *Kluwer Dordrecht* 336 p.

Hasse L (1986) On Charnock's relation for the roughness at sea. Monahan EC, MacNiocaill G (eds) Oceanic Whitecaps 49-56 *D. Reidel Dordrecht* .

Hasse L (1992) On the contribution of spray droplets to evaporation. *Bound Lay Meteorol* 61: 309-313.

Hasse L, Dobson F (1986) Introductory physics of the atmosphere and ocean. *D. Reidel Dordrecht* 126 pp.

Hasse L, Grünewald M, Hasselmann DE (1978a) Field observations of air flow above the waves. In: Hasselmann, Favre (eds) Turbulent fluxes through the sea surface, wave dynamics and prediction. *Plenum Press* New York 483-494.

Hasse L, Grünewald M, Wucknitz J, Dunckel M, Schriever D (1978b) Profile derived turbulent fluxes in the surface layer under disturbed and undisturbed conditions during GATE. *METEOR-Forschungsergebnisse* B 13: 24-40.

Hasse L, Grossklaus M, Isemer HJ, Uhlig K (1992) New instrumentation for measurement of precipitation at sea. *WMO* Geneva In: WMO-TD-No462: 195-198.

Hasselmann K. et al (1973) Measurements of wind-wave growth and swell decay during the Joint North Sea Wave Project (JONSWAP). *Dtsch Hydrogr Zeitschr* Reihe A Ergänzungsheft Nr 12.

Hicks BB, Dyer AJ (1972) The spectral density technique for the determination of eddy fluxes. *Q J Roy Meteorol Soc* 98: 838-844.

Hinze JO (1959) Turbulence. *McGraw-Hill* New York 586 p.

Hinzpeter H (1980) Atmospheric radiation instruments. In: Dobson, Hasse, Davis (eds). Air-sea interaction - instruments and methods. *Plenum Press* New York 491-507.

Illingworth AJ, Stevens CJ (1987) An optical disdrometer for the measurement of raindrop size spectra in windy conditions. *J Atmos Oceanic Tech* 4: 411-421.

Isemer HJ, Hasse L (1987) The Bunker climate atlas of the North Atlantic Ocean Vol 2: Air-sea interactions. *Springer Verlag* Berlin Heidelberg New York 252 p.

Isemer HJ, Hasse L (1988) Consistent parametrisation of air-sea interaction from ship data. In: Scoggins JR (ed) Atmospheric forcing of ocean circulation. *Institute for Naval Oceanography* Stenis 261-272.

Isemer HJ, Willebrand J, Hasse L (1989) Fine adjustment of large scale air-sea energy flux parametrisations by a direct estimate of ocean heat transport. *J Clim* 2: 1173-1184.

Isemer HJ, Hasse L (1991) The scientific Beaufort equivalent scale: Effects on wind statistics and climatological air-sea flux estimates in the North Atlantic Ocean. *J Clim* 4: 819-836.

Junge CE (1963) Air chemistry and radioactivity. *Academic Press* New York 382 p.

Kaufeld L (1981) The development of a new Beaufort equivalent scale. *Meterol Rundsch* 34: 17-23.

Kruspe G (1977) On moisture-flux parametrisation. *Bound Lay Meteorol* 11: 55-63.

Larsen SE, Hojstrup J, Gibson CH (1980) Fast-response temperature sensors. In: Dobson, Hasse, Davis (eds) Air-sea interaction - instruments and methods. *Plenum Press* New York 269-292.

Large WG, Pond S (1982) Sensible and latent heat flux measurements over the ocean. *J Phys Oc* 12: 464-482.

Lind RJ, Shaw WJ (1991) The time-varying calibration of an airborne Lyman-alpha hygrometer. *J Atmos Oceanic Tech* 8: 186-190.

Lindau R, Isemer HJ, Hasse L (1990) Towards time-dependent calibration of historical wind observations at sea. *Trop. Ocean Atmosph Newsletter* No 54: 7-12.

Marks R, Monahan EC (1989) Relationships between marine aerosols, oceanic whitecaps and low-elevation winds observed during the HEXMAX experiment in the North Sea. In: Monahan, van Patten (eds) The climate and health implications of bubble-mediated sea-air exchange. *Connecticut Sea Grant Groton* 83-99.

Mikhaylov BA, Zolofarev VM (1970) Emissivity of liquid water. *Iv Atmosph Ocean Phys* 6: 52 (engl ed).

Moisture (1985) Moisture and humidity, measurement and control in science and industry. *Instrument Society of America Research Triangle Park* 1028 p.

Monahan EC, O'Muircheartaigh IG (1986) Whitecaps and the passive remote sensing of the ocean surface. *Int J Remote Sens* 7: 627-642.

Monahan EC, MacNiocaill G (eds) (1986) Oceanic whitecaps and their role in air-sea exchange processes. *D. Reidel Dordrecht* 294 p.

O'Muircheartaigh IG, Monahan, EC (eds) (1986) Statistical aspects of the relationship between oceanic whitecap coverage, wind speed, and other environmental factors. *D. Reidel Dordrecht* 125-128.

Monahan EC, van Patten MA (1989) Climate and health implications of bubble-mediated sea-air exchange. *Connecticut Sea Grant College Program* Groton CT 170 p.

Oost WA (1991) Flow distortion by an ellipsoid and its application to the analysis of atmospheric measurements. *J Atm Oceanic Tech* 8: 331-340.

Paltridge GW, Platt CMR (1976) Radiative processes in meteorology and climatology. In: Developments in atmospheric science 5. *Elsevier Sci Publ Comp* New York Amsterdam 318 pp.

Panofsky HA (1963) Determination of stress from wind and temperature measurements. *Q J Roy Meteorol Soc* 89: 85-94.

Panofsky HA, Dutton JA (1984) Atmospheric turbulence: Models and methods for engineering applications. *Wiley* New York 397 pp.

Payne RE (1972) Albedo at the sea surface. *J Atmos Sci* 29: 959-970.

Phillips OM (1977) The dynamics of the upper ocean. *Cambridge Univ Press* second edition 336 pp.

Prandtl L (1932) Meteorologische Anwendung der Strömungslehre. *Beitr Phys d freien Atmos* 19: 188-202.

Reed RK (1977) On estimating insolation over the ocean. *J Phys Ocean* 7: 482-485.

Schlesinger ME (ed) (1990) Climate - ocean interaction. *Kluwer Dordrecht* 385 pp.

Sellers WD (1965) Physical climatology. *Univ Chicago Press* Chicago 272 pp.

Skupniewicz CE, Davidson KL (1991) Hotfilm measurements from a small buoy: surface wind-stress estimates using the inertial dissipation method. *J Atm Oceanic Tech* 8: 309-322.

Smith SD (1988) Coefficients for sea surface wind stress. *J Geophys Res* 93: 15467-15472.

Smith SD (1989) Water vapor flux at the sea surface. *Bound Lay Meteorol* 47: 277-293.

Smith SD, Katsaros KB, Oost WA, Mestayer PG (1990) Two major experiments in the humidity exchange over the sea (HEXOS) program. *Bull Am Met Soc* 71: 161-172.

Smith SD, Anderson RJ, Oost WA, Kraan C, Maat N, Delosmo J, Katsaros KB, Davidson KL, Bumke K, Hasse L, Chadwick HM (1992) Sea surface wind stress and drag coefficients: the HEXOS results. *Bound Lay Meteorol* (accepted)

Swinbank WC (1963) Long-wave radiation from clear skies. *Q J Roy Meteorol Soc* 89: 339-348.

Tucker GB (1961) Precipitation over the North Atlantic Ocean. *Q J Roy Meteorol Soc* 87: 147-158.

Webb EK (1970) Profile relationships: the log-linear, and extension to strong stability. *Q J Roy Meteorol Soc* 96: 67-90.

Wexler A (1965) Humidity and moisture; measurement and control in science and industry. *Reinhold* New York, Vol. I-IV.

WMO (1970) The Beaufort scale of wind force. WMO Commssion for Maritime Meteorology. *Marine Science Affairs Report* No 3: 22 pp.

Wucknitz J (1977) Disturbance of windprofile measurements by a slim mast. *Bound Lay Meteorol* 11: 155-169.

Wucknitz J (1979) The influence of anisotropy on stress estimation by the indirect dissipation method. *Bound Lay Meteorol* 17: 119-131.

Wucknitz J (1980) Flow distortion by supporting stuctures. In: Dobson, Hasse, Davis (eds) Air-sea interaction, Instruments and methods. *Plenum Press* New York 605-626 pp.

Chapter 12

SEA–ICE INTERACTIONS IN POLAR REGIONS

J.P. van Ypersele
Institut d'Astronomie et de Géophysique G. Lemaître
Université Catholique de Louvain
B-1348 Louvain-la-Neuve, Belgium

12.1 Introduction

This chapter reviews the state of knowledge about ocean and sea-ice interactions, their role in the climate system and their modelling. It is meant as an introduction to sea-ice geophysics for researchers with an interest in polar phenomena, and not as an exhaustive review for specialists. Therefore, many details will be omitted, and replaced by appropriate references to the sea-ice literature. The reader who wants to start studying sea ice is also referred to the excellent books by Untersteiner (1986a), Washington and Parkinson (1986), and Smith (1990). Additional information on sea-ice modelling may also be found in Hibler (1988) and van Ypersele (1989, 1990). Remote sensing of sea ice is treated, e.g., in the book by Hall and Martinec (1985), and in Shuchman and Onstott (1990). The World Climate Research Programme (WCRP) publishes reports by its Working Group on Sea Ice and Climate (WCRP, 1983, 1984, 1987, 1989, 1990, 1992), where useful information on sea-ice research related to the WCRP objectives can be found.

In the next section, we will describe the basic properties of sea ice at the microscopic scale, starting with the processes governing its formation, then its thermodynamic and mechanical properties. We will also discuss the brine rejection by sea ice, which has such a large influence on the stability of the ocean mixed layer. Sea-ice behaviour at the macroscopic scale will be reviewed, with particular attention given to ice dynamics.

The role of sea ice in the climatic system will be discussed, together with sea-ice relevance to the goals of GEWEX and WCRP. Sea ice represents an obstacle to heat, mass, and momentum exchanges between ocean and atmosphere. Sea ice affects the rate of CO_2 transfer across the air-ocean boundary.

NATO ASI Series, Vol. I 5
Energy and Water Cycles in the Climate System
Edited by E. Raschke and D. Jacob

It also plays a very active climatic role by reflecting solar energy to space and rejecting brine into the ocean. These different effects contribute to a series of positive and negative feedbacks.

The modelling of atmosphere–ocean–sea-ice interactions will receive particular attention. The large variability of sea ice, its effects on atmosphere and ocean dynamics, and its sensitivity to small changes in climate variables makes the inclusion of an interactive sea-ice model necessary in today's climate models. The ultimate sea-ice model now appears to be one in which sea ice is fully coupled, by heat, salt, and momentum fluxes to an ocean and atmosphere model. The elements needed in fully coupled models are discussed, and the principal methods used are described.

The present status and projects of sea-ice monitoring will be briefly reviewed, and the present and future data needs will be discussed.

Perspectives on possible and needed progress in both monitoring and modelling will then be given.

12.2 Sea-ice processes

12.2.1 Sea-ice microphysics

Sea ice is produced by the freezing of a thin layer of polar seawater (of the order of 2 m in the Arctic, 1 m in the Antarctic) and the accumulation of snow on top of it.

Many large-scale characteristics of sea ice originate in its microphysical properties, which are reviewed in Weeks and Ackley (1986). The fact that ice floats on the surface of lakes and seas come from the fact that the density of ice at 0 °C is less than that of water at the same temperature, with the density of water increasing to a maximum at 4 °C. The low density of ice is the result of the tetrahedral coordination of each water molecule by hydrogen bonding: each oxygen atom is located at the centre of a tetrahedron with four other oxygen atoms located at each of the apices (Weeks and Ackley, 1986). This results in a very open crystal structure possessing hexagonal symmetry. The oxygen atoms are concentrated close to a series of parallel planes that are referred to as the basal planes. Fracture along these basal planes is easier than fracture along any plane normal to the basal planes. This helps to explain a number of ice properties: for example, ice cleaves and glides easily on the basal plane, and dendritic sea-ice crystals grow along the directions of the secondary axes in the basal plane (Weeks and Ackley, 1986).

Sea ice is not made out of pure water, and some attention needs to be devoted to the role of salts in seawater. Very few molecules present in seawater have the right size and charge to fit in the ice structure as substitutes for some of the water molecules (Weeks and Ackley, 1986). Therefore, as sea ice forms, essentially all the solute is rejected back into the seawater, which becomes saltier. Young sea ice has a typical (practical) salinity lower than 15, while older sea ice, in which excess salt has had more time to be drained or to migrate downward, has a typical salinity of 5 after one year and 2 after a few years (Weeks and Ackley, 1986) (the practical salinity will be used from now on, as recommended by IAPSO, 1985). This can destabilize the water column, which has profound consequences for the climate system, as we shall see below. The freezing temperature of seawater depends on its (practical) salinity S: it is for example −1.9 °C for $S = 35$ and −1.4 °C for $S = 25$ (Millero, 1978).

The first crystals to form seem to be minute spheres of pure ice (Hobbs, 1974). Growth along the basal plane of the crystal structure rapidly changes these spheres into thin circular discs of about 1 mm in diameter. At some critical radius, the discoidal growth form becomes unstable, breaking up into a hexagonal dendritic star (Weeks and Ackley, 1986). The star-like crystals grow rapidly across the surface of calm seawater until they overlap and freeze together, forming a continuous thin ice skin. Commonly, however, in the open ocean there is some wave-induced turbulence during initial ice formation. This increases abrasive action between crystals and favours discoidal growth. Wind action herds the resultant ice crystals, which are generally termed *frazil*, into agglomerations such as *grease ice*, a soupy layer which exhibits viscous fluid-like properties, and can reach thicknesses of 1 m. *Pancake ice*, roughly circular pieces of new ice with upturned edges, also results from frazil accumulations. Pancakes range from 0.3 to 3 m in diameter. Wave action causes repeated contacts and separations between the floes, pushing the newly formed frazil crystals up to build the characteristic raised pancakes rims. With time, pancakes typically agglomerate with grease ice into loosely joined pancakes and finally to a complete sheet of joined pancakes.

Vertical growth rates in winter are of the order of 10 $cm\,d^{-1}$ for thin ice and less than 1 $cm\,d^{-1}$ for ice thicker than 1 m (Thorndike et al., 1975).

A more detailed and exhaustive review of the growth process in sea ice can be found in Weeks and Ackley (1986).

12.2.2 Sea ice and climate

The climate system is affected by sea ice in several ways. Through its high albedo (0.6 to 0.7 for bare ice, and up to 0.9 for snow-covered ice), sea ice reduces the proportion of solar energy absorbed at the surface; this effect participates in the famous ice–albedo positive feedback (Budyko, 1969; Schlesinger, 1988). It also replaces a relatively warm ocean surface (above −2 °C) by a snow or ice surface well isolated from the ocean heat reservoir, and whose surface temperature can drop to −30 °C or less in winter. This means a reduction of surface infrared radiation to the atmosphere from 300 to 200 Wm^{-2}.

Sea ice represents an obstacle to air–sea exchanges of sensible heat, momentum, water vapour, carbon dioxide, and other gases. For example, Maykut (1978) calculated that the total heat loss through 3 m of arctic ice in winter is only $\approx$ 1.5% of that from the open ocean. The air–ice drag coefficient is similar to the air–water drag over close, smooth sea ice (Banke et al., 1980) but is up to 3 times larger than the latter in the marginal ice zone, where the ice is broken and dispersed by wave activity (Andreas et al., 1984). Sea ice also replaces the wind stirring and wave breaking activity driving the ocean mixed layer by the largely inferior stirring by a slowly moving solid (Semtner, 1984a).

Over the 1973–86 period, the area partially covered by sea ice reached at its maximum (during the austral winter) 28.5 million km^2 of ocean, i.e., 8% of the world ocean surface, as shown by estimates of sea-ice extent based on passive microwave satellite data (Gloersen and Campbell, 1988). Tab.

Southern Ocean		Arctic Ocean	
Minimum	Maximum	Minimum	Maximum
3.6	18.5	8.0	14.6

Table 12.1: *Average sea-ice extent (1978–87)* $10^6\ km^2$. *(Gloersen and Campbell, 1988).*

12.1 shows the seasonal range of sea-ice extent for both hemispheres as obtained from the same data. It can be seen from Tab. 12.1 that arctic sea-ice seasonal variability is largely inferior to the antarctic one. This is mostly due to the fact that a continent is surrounded by an ocean in the Antarctic, while the reverse is true for the Arctic, with ice movement and decay constrained by the continents. This geographical factor also contributes to the different average thicknesses observed in either polar region: being more constrained in its movement, arctic ice has more time to grow before being transported away. Other factors differentiating arctic and antarctic sea-ice processes, in particular the low relative humidity associated with the dry winds from the Antarctic continent and a larger ratio of net radiation to latent heat flux in the Arctic are discussed in Andreas and Ackley (1982).

The various phenomena and effects mentioned above are complicated by the fact that sea ice is not an homogeneous medium: it can be deformed, or fractured in individual floes, which tend to act as rafts under the ocean and wind stresses. These movements can produce areas of ridging, with sea-ice thickness locally larger than 25 m, and can also create open water areas intertwined in the sea-ice cover. If these open water areas are longitudinal they are called *leads*, otherwise they are called *polynyas*. Due to the constant motion of ice in response to varying winds and currents, a certain percentage of leads is present in all but the most compact winter ice cover. Polynyas can be as large as the Weddell Sea Polynya, which had an area of 300 000 km^2 between 1974 and 1976 (Zwally et al., 1983).

Because of the large difference in heat flux from the ocean to the atmosphere over water or over ice of various thicknesses, the ice thickness distribution needs to be taken into account to estimate this heat flux in climate models. Maykut (1982) summed the area weighted contributions to the large-scale heat flux made by 8 thickness categories representing the time-dependent thickness distribution of the Central Arctic. He found that the annually-averaged large-scale turbulent heat loss to the atmosphere is 12 times larger than if a uniform 3 m perennial ice cover is considered. On the other hand, the annual net radiation balance for the region is 2.3 times larger than the perennial ice value, because of the low albedo of leads and thin ice and greater longwave radiation emission by young ice during the winter.

Another category of effects is made by those related to the ocean. As explained in section 12.2.1, when it forms or melts, sea ice affects the upper ocean salt budget, because most of the salt contained in seawater is expelled from the ice by the building of the ice crystals. Sea-ice formation thus increases the density of the under-ice water, which can induce convection and overturning, while sea-ice melting has the effect of stabilizing the water column. Since the entrained underlying water is warmer than the water it replaces below the ice, the ocean–to–ice flux is increased at the time of ice freeze-up. It is interesting to remark that, on the seasonal time scale, the thermodynamic processes of sea ice act to replace the seasonal cycle of sea-surface temperature with one of sea-surface salinity (Semtner, 1984a). Brine rejection is thought to contribute to the maintenance of the cold upper halocline in the Arctic Ocean (Aagaard et al., 1981), and to the formation of dense shelfwater in the Antarctic (Foster and Carmack, 1976), which can influence the deep ocean circulation (Rooth, 1982). Since, at low temperatures, seawater density is controlled by salinity rather than by temperature, the correct representation of salt or freshwater influx is of key importance fo the study of the ocean thermohaline circulation.

These important effects of brine rejection and meltwater flux make the inclusion of the salt conservation equation in sea-ice/ocean coupled models essential. They also justify the need for a much better set of temperature and salinity data to close the energy and freshwater budget ($E-P$) in the ocean, and this has been recognized by GEWEX (WCRP, 1991).

12.3 Modelling

The introduction of sea ice in climate models is progressively taking into account the various processes affecting it. Logically, the first approach is to model the thermodynamic aspects, which govern the creation and melting of sea ice. The main result of these models are the ice thickness and surface heat fluxes. The second step is to incorporate the dynamic processes, which can move and redistribute the sea ice produced by the thermodynamic model. As far as atmospheric climate modelling is concerned, the main variable output by dynamic models is the ice thickness distribution, which is so important to estimate correctly the large-scale heat flux to the atmosphere. The simplest representation of ice thickness distribution is one with only two categories: thin ice (often assimilated to open water) and thick ice. A single additional variable is then introduced: the ice concentration, or percentage of each grid square that is ice-covered. The two steps above are usually done in uncoupled models, where the sea-ice forcing from both atmosphere and ocean is imposed without allowing feedbacks with the sea-ice processes themselves. The ultimate sea-ice model now appears to be one in which sea ice is fully coupled to oceanic and atmospheric models. In the following sections, the basic principles of thermodynamic and dynamic sea-ice modelling are introduced, and the elements needed for coupled modelling are discussed.

12.3.1 Thermodynamic models

Modern sea-ice thermodynamic models are based on the energy conservation law. Their aim is to compute the ice and snow thickness from atmospheric and ocean forcing. When they include a leads parameterization, thermodynamic models also compute the ice concentration, or percentage of ice-covered water. Figure 12.1 shows the various thermodynamic fluxes which affect an idealized slab of sea ice.

Typically, sea ice is assumed to be a horizontally uniform slab of ice, represented by a certain number of layers of equal thickness. A layer of snow can optionally be present. The principle behind any thermodynamic sea-ice computation is to compute at each timestep the heat balance of both upper

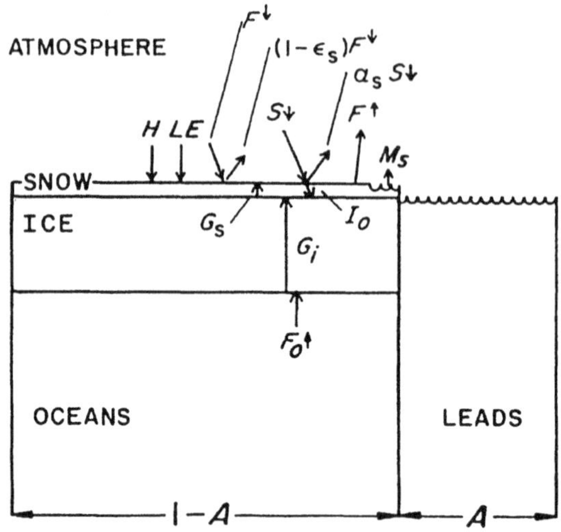

Figure 12.1: *Schematic of a grid square in a thermodynamic sea ice model, with ice and snow layers of uniform thickness, a variable lead fraction, A, and the relevant energy fluxes at the water/ice, ice/snow, and snow/air interfaces. (From Washington and Parkinson, 1986).*

and lower ice surfaces, and compensate any energy deficit or excess by seawater freezing or snow/ice melting, and by changes in upper surface temperature. The initial freezing of ice-free seawater is similarly determined by a heat budget equation applied to the oceanic mixed layer: ice appears as soon as the freezing temperature is reached. By integrating the energy conservation equation in time under a periodic atmospheric forcing, it is possible to obtain seasonal cycles of ice thickness that converge towards an equilibrium cycle after a few years of integration. The fundamental equations of ice thermodynamics are given below, in the form derived by Semtner (1976).

The upper surface flux balance equation is given by

$$Q(1-I_0)(1-\alpha_{surf}) + Q^{\downarrow}_{LW} - Q^{\uparrow}_{LW} + Q_H + Q_{LE} + Q_{cond} = \sum Q$$

and

$$\sum Q = \begin{cases} 0 & , \quad \text{when} T_{surf} < 273\ K \\ -L\left(\rho_i \frac{\partial h_i}{\partial t} + \rho_s \frac{\partial h_s}{\partial t}\right)_{z=0}, & \quad \text{when} T_{surf} = 273\ K \end{cases} \tag{12.1}$$

where Q is the incoming surface solar radiation (Wm^{-2}), I_0 is the fraction of solar radiation that penetrates the slab typically $= 0$ for snow, 0.17 for bare ice, α_{surf} is the surface albedo, $Q^{\downarrow}_{LW}$ and $Q^{\uparrow}_{LW}$ are respectively the incoming and outgoing longwave radiation (Wm^{-2}), Q_H and Q_{LE} are respectively the sensible and latent heat flux (Wm^{-2}), Q_{cond} is the conductive flux through the ice (Wm^{-2}), L is the latent heat of fusion of ice (Jkg^{-1}), ρ is the density 900 and 330 $kg\,m^{-3}$ for ice and snow, respectively, with the s and i subscripts referring from now on to snow and ice, respectively, and h is the thickness.

The temperature within the ice and snow is governed by the one-dimensional heat equation:

$$\rho c \frac{\partial T}{\partial t} = k \frac{\partial^2 T}{\partial z^2} \tag{12.2}$$

where c is the specific heat capacity 2.09 $kJkg^{-1}K^{-1}$ for both ice and snow, T is the thermodynamic temperature (K), and k is the thermal conductivity 2.03 and 0.31 $Wm^{-1}K^{-1}$ for ice and snow, respectively.

At the bottom of the ice, the flux balance equation is:

$$Q_{cond} - Q_{ocean} = L\rho_i \frac{\partial h_i}{\partial t}_{z=h_i} \tag{12.3}$$

where Q_{ocean} is the oceanic heat flux density (Wm^{-2}).

Unless they are given by a coupled model, atmospheric fluxes are typically computed with various parameterizations, including bulk aerodynamic formulae for the latent and sensible heat fluxes and the Stefan-Boltzmann radiation law for the longwave emission. The oceanic heat flux is commonly assumed constant in uncoupled models, and assigned a value of 2 Wm^{-2} in the Arctic and around 20 Wm^{-2} in the Antarctic.

Semtner (1976) produced a model suitable for climate studies by solving Eqs. 12.1 to 12.3 for the surface temperature by forward time-differencing after dividing the ice and snow slab into a certain number of layers, computing the conductive heat flux by assuming linear temperature profiles within each layer, and linearizing the fourth order Stefan-Boltzmann law giving the outgoing infrared radiation.

After this determination of equilibrium surface temperature and conductive heat flux, the surface is allowed to melt if its temperature reaches 273 K, and bottom accretion or ablation is computed. If the model allows, heat storage in brine pockets is predicted (see Semtner (1976) for more details).

A fixed-depth, purely thermodynamic mixed-layer model replaces the ice model when no ice is present, so that the onset of freezing can be computed. Two main versions of the Semtner (1976) model exist, the first one allows for up to 2 layers in the ice and 1 layer for snow, and takes internal heat storage into account by introducing a heat reservoir which represents internal brine pockets. The second version, called the *zero-layer model*, has only one layer for ice and one layer for snow, and parameterizes the effect of heat storage by adjusting the surface albedo and the ice and snow conductivities so that the zero-layer model gives the same annual mean ice thickness than the 3-layer model for standard conditions. Variants of either version are included as the thermodynamic part in most of today's sea-ice models.

It should be noted that, for climate studies, the 3-layer version is recommended by Semtner (1984b) over the zero-layer simplified model. He showed that significant errors in amplitude and phase of the seasonal cycle of ice thickness occurred with the zero-layer model. In particular, the thickness maximum and minimum occur 1 month earlier with the zero-layer model than in the 3-layer Semtner (1976) or the 40-layer Maykut and Untersteiner (1971) model simulations (see Fig. 12.2). This can be attributed to the delaying effect of internal storage of sensible and latent heat in brine pockets. The choice of the three-layer version would be particularly important for correctly assessing the polar regions sensitivity to a CO_2-induced climatic change.

To extend the Semtner model to the horizontal, it is necessary to include a leads parameterization, because sea ice is not a continuum: it breaks with waves and floe-to-floe interactions, forming leads,

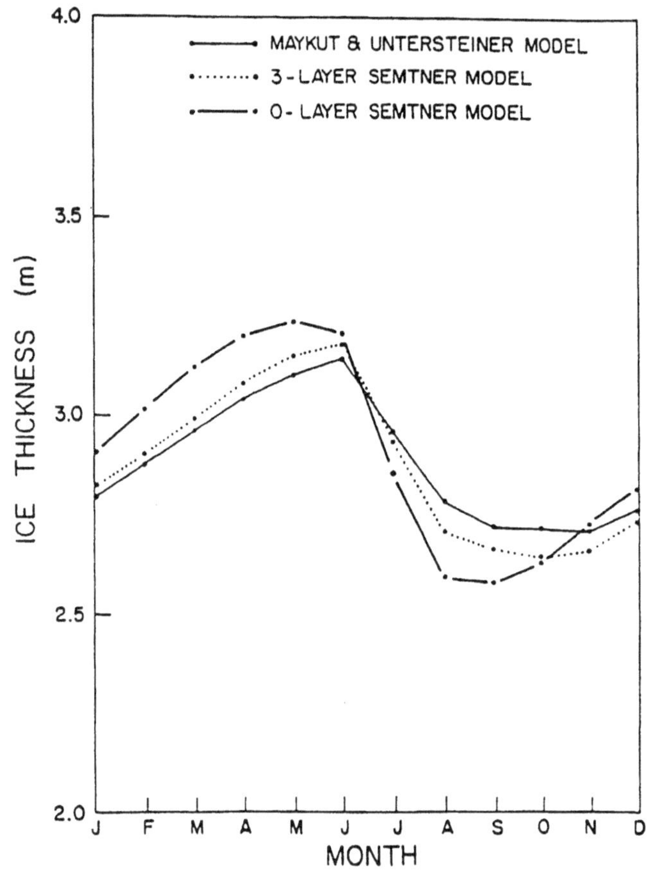

Figure 12.2: *Seasonal cycle of ice thickness simulated for standard arctic conditions with the models of Maykut and Untersteiner (1971) and Semtner (1976). (From Semtner (1976), redrawn by Washington and Parkinson, 1986).*

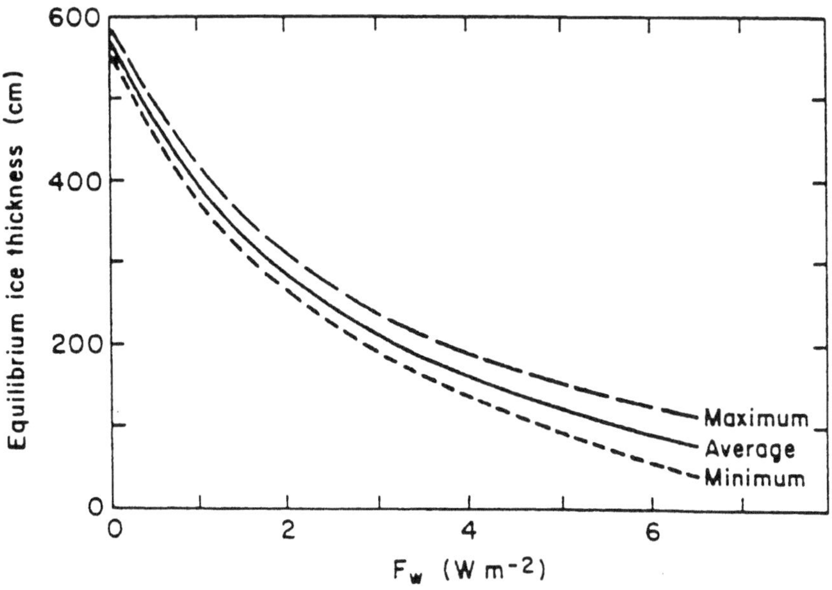

Figure 12.3: *Sea-ice thickness sensitivity to vertical ocean–to–ice heat flux at equilibrium, for arctic conditions. (From Maykut and Untersteiner (1971) redrawn by Maykut, 1986).*

and as soon as its sides are in contact with water, lateral heat exchanges tend to increase or decrease the ice concentration. An important positive feedback is present here in summer: solar energy heats the mixed-layer water in the ice-free area, and this accelerates the sea-ice lateral melting.

A leads parameterization for the zero-layer model was originally suggested by Semtner (1976) and implemented by Washington et al. (1976). Its principle is as follows. An additional variable, representing the amount of open water in leads, is carried at each gridpoint, and the heat balance equation is applied to this fraction of open water. On account of variability in mechanical forcing by the wind, or long-term divergence, a minimum value of 2% is assumed to exist for the fraction of open water. Any surplus of heat absorbed in leads during spring or summer is used for lateral melting of ice at its existing thickness, and tends to increase the fraction of open water. On the other hand, new thin ice is formed in the lead when the heat balance is negative. So as not to require the recording of two species of ice - thin and thick - a new average thickness is computed and the ice concentration is increased to the maximum allowable value. With this parameterization, open areas enlarge rather slowly but freeze over rapidly. The study by Washington et al. showed the improvement in the antarctic seasonal sea-ice cycle produced by the introduction of leads.

Parkinson and Washington (1979) built up on the Washington et al. (1976) model, and used a more sophisticated lead parameterization in which for a positive lead heat balance, the larger the lead area, the larger the energy used directly to warm the lead rather than in melting the ice laterally. For a negative heat balance, the ice fraction is increased, but the ice thickness remains unchanged. Should this result in an ice concentration tendency to exceed an assumed maximum value (99.5% for any grid square in the Arctic and 98% in the Antarctic), the remaining energy deficit is offset by a cooling of the water under the ice, possibly leading to bottom ice accretion. This parameterization might be more realistic than the Semtner scheme, especially for low ice concentrations. More elaborate leads parameterizations are used in dynamic models discussed in the next section. A revised version of Parkinson and Washington (1979) model is described in Crowley and Parkinson (1989).

It should be stressed that those thermodynamic models need appropriate forcing at their upper and lower surfaces. The atmospheric forcing must be derived from observed or modelled variables such as radiation, air temperature, humidity, wind speed, and snowfall. A typical example of parameterizations used to compute the atmospheric forcing is given by the study of Parkinson and Washington (1979). The correct representation of radiation, including the effects of polar clouds, is particularly important for thermodynamic models. Raschke et al. (1991) reviewed the status of polar radiation budget in relation with sea ice, and made appropriate recommendations.

The ocean heat flux into the ice must be specified, as long as no coupling takes place with the ocean. This is very unfortunate because

1. there are virtually no measurements of this flux (see, e.g., Perovich et al., 1989), and

2. ice thickness is very sensitive to it.

Experiments made for arctic conditions by Maykut and Untersteiner (1971) and by Semtner (1976) show that changing the oceanic heat flux from 2 to 0 or 4 Wm^{-2} multiplies the average ice thickness by 2 or 0.5, respectively (see Fig. 12.3). A similar sensitivity exists for the Antarctic (Demuth and van Ypersele, 1987).

This high sensitivity, combined with the fact that this flux is poorly known, gives one of the reasons why coupled sea-ice/ocean models are needed: one of their advantages is that they can compute this heat flux (see section 12.3.3).

12.3.2 Dynamic models

Sea ice is not only subject to heat fluxes: it receives momentum from both the atmosphere and the ocean, and moves in response. To complicate the matter, ice is neither rigid nor a nice compressible material, and ice floes also hit each other, deform, and break. This is the domain of ice dynamics. Ice dynamic processes have an important role in the climate system. They govern the ice motion, which transmits mechanical energy into the ice-covered ocean. Ice motion towards the equator also represents a non-negligible poleward heat transport (Semtner, 1984a). Dynamic processes control the ice concentration, which strongly influences the heat exchange between ocean and atmosphere. They can also export newly-formed ice continuously from a certain area, which causes a net annual average salt influx in that area. This can strongly affect the upper ocean dynamics, and even the deep ocean circulation.

Models that include dynamics seem to be less sensitive to thermal perturbations than models without, and this is important for global warming studies. In regions where the thermodynamics reduces the ice thickness, the ice gets weaker and the dynamics, under favourable conditions, i.e., convergence, can readily increase the ice thickness (by importing ice into the region). In regions where the dynamics reduces the ice thickness (divergence), the thermodynamics, under favourable conditions, i.e., cooling, can easily increase the sea-ice thickness. These interactions produce a negative feedback, which tends to destabilize the solution (Lemke, 1989). On the other hand, ice models that include dynamics are sensitive to the wind forcing (see, e.g., Stössel et al., 1990). Examples of the effect of interaction between ice dynamics, ocean dynamics, and the ice thermodynamics are discussed in Hibler (1984, 1988), Lemke et al. (1990) and van Ypersele (1990).

When Ekman (1905) extracted from Nansen's observations of ice movement in the Arctic a law relating wind and ice velocity, he actually provided the foundation of a first sea-ice dynamic model. Nansen observed that ice drifted at about 2 percent of wind speed, 30 degrees to the right of the surface wind direction in the northern hemisphere. This is also known as the Zubov Law, which can equivalently express the ice velocity as being parallel to the geostrophic wind, with a speed proportional to the pressure gradient (see, e.g., Wadhams, 1986). This simple relation can actually be justified by neglecting the internal ice stress in the full ice dynamics equations (Thorndike and Colony, 1982).

If we consider ice as a continuum, we can write an ice momentum equation, where ice acceleration (measured in a reference frame attached to the Earth) is due to the combined effect of wind stress at the top surface, water stress at the bottom, a force due to the tilt of the water surface, the Coriolis force, and divergence of the internal ice stress.

In the plane of motion of sea ice, the momentum balance is expressed as (e.g., Coon, 1980):

$$m\frac{D\mathbf{V}_{\mathrm{i}}}{Dt} = \boldsymbol{\tau}_a + \boldsymbol{\tau}_w - mf\,\mathbf{k} < \times\, \mathbf{V}_{\mathrm{i}} - mg\nabla H + \operatorname{div}\boldsymbol{\sigma} \tag{12.4}$$

where m is the mass per unit area of ice, D/Dt is the substantial derivative, $\mathbf{V}_{\mathrm{i}}$ is the ice velocity in the horizontal plane, $\boldsymbol{\tau}_a$ and $\boldsymbol{\tau}_w$ are the forces due to air and water stresses, f is the Coriolis parameter, $\mathbf{k}$ is the vertical unit vector, g is the magnitude of the acceleration of gravity, H is the height of the sea surface, $\boldsymbol{\sigma}$ is the Cauchy stress resulting from floe–to–floe stress transmission integrated through the ice thickness.

The complete momentum equation cannot be solved easily, mostly because the relation between internal ice stress and deformation, called the *constitutive law,* is not well known. The constitutive law prevents the ice from excessive accumulation. Typically, the constitutive law has the form

$$\boldsymbol{\sigma} = \sigma_{ij}(\dot{\epsilon}_{ij}, \zeta, \eta, P, \ldots) \tag{12.5}$$

where $\dot{\epsilon}_{ij}$ is the two-dimensional strain-rate tensor, ζ the nonlinear bulk viscosity, η the nonlinear shear viscosity, and P a pressure term dependent on ice thickness and concentration.

Theoretical studies of sea-ice constitutive laws owe much to field studies such as the Arctic Ice Dynamics Joint Experiment (AIDJEX), which took place between 1970 and 1977 (Pritchard, 1980).

AIDJEX provided measurements of the different terms entering the momentum equation, and led to the development of the AIDJEX model. The AIDJEX model (Coon, 1980) consists mainly of a complete momentum balance equation, an *elastic-plastic* constitutive law which relates the internal ice stress to deformation, and an ice thickness distribution law, which relates the dynamic and the thermodynamic terms that change the ice thickness.

Hibler (1979) implemented a model based on the AIDJEX work, in which he proposed a *viscous-plastic* constitutive law, considering the ice as a creeping, linear viscous fluid for small deformation rates and as a rigid plastic for high deformation rates. In general, a *viscous* rheology is defined as one in which stress can only be sustained through non-recoverable dissipation of energy by deformation, while a special case is the *linear viscous* rheology, where the stress depends linearly on the strain rates (Hibler, 1986). A *plastic* material is characterized by a stress independent of deformation rate. Hibler (1986, 1988) documents the need for such a nonlinear rheology, and reviews the studies in which this approach has been used.

Other important aspects of this model are the ice thickness distribution it uses, a simplified version of the Thorndike et al. (1975) formulation, and a relationship between this thickness distribution and ice strength. Hibler (1979) considers only two categories of ice: thin (zero thickness, i.e., open water), and thick ($h > 0.5\,m$), with the fraction of thin ice affected both by deformation and thermodynamics. Freezing conditions decrease the thin ice fraction rapidly, while melting increases it slowly. This approach is different from that of Parkinson and Washington (1979), in which thin ice disappears more slowly than in reality. To allow the ice interaction to become stronger in regions of ice inflow and weaker in regions of outflow, the maximum compressive stress allowable along any principal stress axis, or *plastic strength*, is taken to be proportional to the ice thickness and exponentially related to the ice concentration. Hibler (1980) expanded his 1979 model to consider up to 10 ice thickness categories, but the results showed the more complex ice thickness distribution to give similar results to the two-level distribution.

Hibler's (1979) dynamics was used in a series in simulations to show the importance of including the sea-ice dynamic aspects in climate models. Fig. 12.1 shows for example the results of Hibler and Walsh (1982), which ran an extended version of Hibler's (1979) model subjected to the 1973–1975 atmospheric forcing, with thermodynamics only and thermodynamics + dynamics. It can be observed that the inclusion of dynamics strongly affects the thickness field. Hibler and Walsh found that the interannual fluctuations obtained from the simulation with dynamics correlated better with observations than those obtained with thermodynamics only.

The Hibler (1979) model is too expensive to be used routinely in climate studies, in particular because it requires an iterative solution, and Hibler proposed around 1985 a simplified rheology, called *bulk viscous* below, appropriate for climate time scales, when the forcing varies slowly (Washington and Parkinson, 1986, p. 157–158; Hibler, 1988; Flato and Hibler, 1990). It implies the use of a nonlinear bulk viscosity having a null value when ice diverges, whereby convergence is resisted in a plastic fashion but divergence is allowed to occur without resistance. No shear viscosity exists in this rheology. Hibler (1988) reports, on the basis of a test in the Arctic, that this rheology treats at least the first-order effects of ice dynamics.

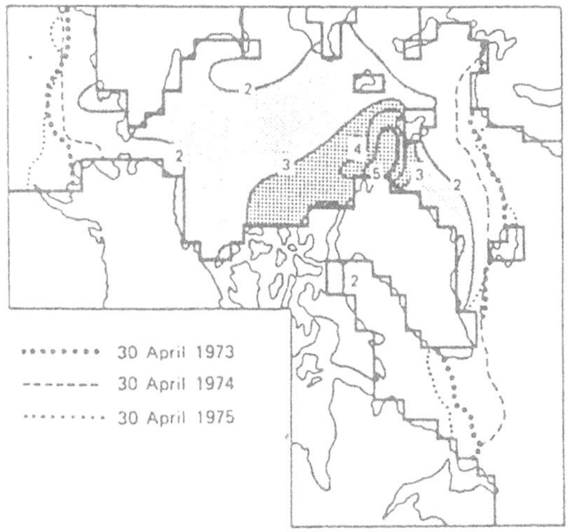

a

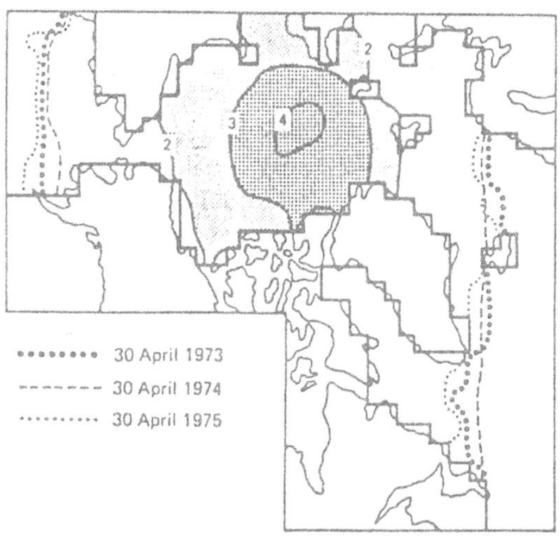

b

Figure 12.4: *Positions of ice edge simulated by (a) a model containing both dynamics and thermodynamics and (b) thermodynamics only. Three-year means of simulated ice thickness are also shown. (From Hibler and Walsh, 1982).*

Semtner (1987) used such formulation in a coupled ocean/sea-ice model of the Arctic, and found results similar to those obtained with the full dynamics of Hibler. Semtner used an explicit time stepping method instead of the relaxation method used by Hibler, and this led to a threefold computer time reduction over the full viscous-plastic approach. The bulk rheology formulation is being progressively implemented in some climate models.

Another, very rough, approach to the modelling of sea-ice dynamics was suggested by Semtner (1984b). It consists in replacing the full momentum equation by a relationship giving the ice velocity as a function of ocean current and geostrophic wind. Thorndike and Colony (1982) analysed data from automatic data buoys deployed in 1979 in the Arctic Ocean, and found that a simple vector relationship could be used to compute the long-term (several months) average ice motion away from the coasts:

$$\mathbf{V_i} \approx C_o \mathbf{V}_{\|\mathbf{g}} e^{-i\alpha} + \mathbf{V_o} \tag{12.6}$$

where C_o is a nondimensional scaling factor, α is a turning angle radians, $\mathbf{V}_{\|\mathbf{g}}$ ms^{-1} is the geostrophic wind, and $\mathbf{V_o}$ is the oceanic current.

The ice velocity derived from Eq. 12.6 enters a snow and ice conservation equation, and a compactness equation used to represent the effect of ice divergence and convergence on the fractional ice cover or ice concentration. This parameterization was used by van Ypersele (1986) in a coupled three-dimensional ocean and sea-ice model of the Weddell Sea area, and by Demuth and van Ypersele (1989) in a coupled mixed-layer and sea-ice model of the same region. Since the internal ice stress neglected in this parameterization is rather low when the ice consists primarily of loose and thin, first-year ice, this approach is probably appropriate for most of antarctic sea ice, but not for the arctic ice, except in marginal ice zones.

In summary, since no measurements of the real internal ice stress in large scale sea ice can be made, and very little data are available to support the choice of a specific form of the ice thickness redistribution function, Untersteiner (1986b) observes that it may prove impossible to find generally applicable equations of state and constitutive laws for sea ice. The ice rheology used in climate models must therefore be adapted to the particular time and space scales and forcing one considers.

12.3.3 Coupled sea-ice, ocean, and atmosphere models

Sea ice affects and is affected by both atmosphere and ocean. Many sensitivity studies have shown that the atmosphere is sensitive to changes in sea ice (e.g., Mitchell and Hills, 1986; Royer et al., 1990). The reverse is also true: sea ice is very sensitive to the net heat flux from the atmosphere. For example, Herman (1986) who discusses the coupling of sea ice to atmospheric general circulation models, showed that a systematic error of 10 Wm^{-2} in this net flux would translate into an error of approximately 9 cm per month in the summer melting ice. On the ocean side, it has been shown that the thermohaline circulation was sensitive to high latitude salinity perturbations, which could be associated with changes in sea ice (e.g., Bryan, 1986; Hoffert, 1990). It has also be mentioned in section 12.3.1 that sea-ice thickness was very much dependent on the oceanic heat flux. The interactive treatment of sea ice is therefore needed in modelling studies that attempt to reproduce the climate behaviour for conditions different from the present ones. Several coupled atmosphere–ocean general circulation models now include a simple thermodynamic representation of sea ice (e.g., Manabe and Stouffer, 1988; Washington and Meehl, 1989). However, they represent only the first attempts to include state-of-the-art ice models in coupled GCMs, as inclusion of dynamics is also necessary.

The following sections describe the main methods used to couple sea ice and other climate components. It concentrates on coupling with the ocean, but most of the basic principles used below can be transposed to the coupling with the atmosphere. The coupling methods described below have been used in several recent studies, including the one-dimensional models by Lemke (1987), Fichefet and Gaspar (1988), and Houssais (1988), and the three-dimensional studies by Hibler and Bryan (1987), Semtner (1987) and van Ypersele (1986) (for a review of ocean/sea-ice coupled models, see van Ypersele, 1989). A common finding of these coupled experiments is that a strong interaction exists between the heat and salt budgets below the ice.

Coupling ocean and sea ice involves ocean model changes that might be categorized in two types:

- type I changes correspond to the *passive* role of ice in the system, seen as an obstacle to the atmospheric forcing of the ocean and affecting the vertical heat, salt, and momentum exchanges;
- type II changes are related to the *active* role of ice; it is limited to the salinity forcing due to freezing and melting.

Three important principles underlying an ideal coupling are that

1. priority should be given to interactive coupling of *sensitive* variables (determined by separate sensitivity analyses of ice and ocean models);
2. physically-based parameterizations of the interactions should be used whenever possible; and
3. conservation laws for mass, energy, and momentum should be respected by the physical and numerical set-up.

Sea ice and ocean must be coupled in three areas: heat, salt, and momentum. Each of the three domains of coupling is discussed below.

Heat

Ocean atmospheric forcing by heat fluxes must be modified when sea ice is present. Below sea ice, the only part remaining from the atmospheric forcing is a small fraction of the solar energy (Grenfell and Maykut, 1977). Therefore, the ice-covered ocean gains or loses energy at its upper surface mostly by latent heat of freezing or melting.

The ocean to ice heat flux can be computed using the hypothesis of thermodynamic equilibrium. If we assume that the ice is in thermodynamic equilibrium with the underlying water, the bottom surface of sea ice must have a temperature close to the freezing temperature of seawater (≈ -1.9 °C). The seawater density dependence on temperature, with the density maximum observed at the freezing temperature (when the salinity > 24.7), is such that the whole mixed-layer temperature must be close to the water temperature below the ice. If advection or convection occurring in the ocean model tends to increase the heat content of this mixed layer, the only way to satisfy the fixed temperature constraint is to pass this heat to the ice, melting some of it. This method allows one to compute the oceanic heat flux that is specified in uncoupled sea-ice studies; it was used by, e.g., Toole (1981), Hibler and Bryan (1987), and Semtner (1987).

An additional problem is posed by the sea-ice movement relative to the ocean. When ice is advected in an area where no ice was present, it must melt proportionally to the difference between the seawater temperature and the freezing temperature. This is also equivalent to an oceanic heat flux.

If the ice cover is not uniform, two additional remarks are in order. First, the above computations must be weighted by the fraction of ice-covered water, keeping energy conservation as a guide. Second, any energy absorbed in areas of open water entrapped in ice (leads, polynyas) may be used to melt ice at the bottom, or for lateral melting. The way this energy must be distributed between bottom and sides is not well known, and different approaches have been used (see the discussion in Harvey, 1990).

Momentum

Coupling sea ice and ocean models means that wind stress over the ocean must be replaced by an ice–water stress when ice is present. The drag of a gas on the wavy ocean surface is replaced by the drag of a solid (the ice floe keel) on the water.

Air–ice drag coefficients are reviewed in Brown (1981), Overland (1985), and McBean (1986), while McPhee (1986) discusses water–ice drag coefficients. Ideally, ice–water drag should depend on the amount of ice ridging, but this kind of refinement is not yet included in coupled ice–ocean models.

As seen above, the ice movement is not simply in the direction of the wind, and the velocity of ice relative to water depends not only on the wind, but also on internal ice factors. When internal ice stress tends to oppose the ice motion, it reduces the turning angle between the ice motion and wind. When a thermodynamic-dynamic model is used, one possible method to compute the surface forcing under the ice cover is to compute explicitly the force due to ice interaction and subtract it from the wind stress. The remainder is used as the stress term transferred into the ocean. The water stress term in the ice momentum equation is a quadratic function of the difference between ice velocity and oceanic surface current. For simplicity, the oceanic surface current can be approximated by the horizontal ocean velocity in the second level of the ocean model. This method is used by Hibler and Bryan (1987). If the ice cover is not continuous, the full wind stress must be allowed in open water areas, and a weighted stress due to air and ice must be used to drive the ocean.

Another method is that used by van Ypersele (1986), who computed the ice velocity directly from the Thorndike and Colony (1982) parameterization (see section 12.3.2), and then used that velocity in an ice–water stress term. This approach gives good results in the Antarctic, where ice resistance is low, but would probably not work well for the Arctic.

For coupling with mixed-layer models, the kinetic energy input into the mixed layer due to wind stress and ice keel stirring is typically proportional to the third power of the ice velocity relative to the ocean below the frictional layer (Lemke and Manley, 1984; Lemke, 1987; Fichefet and Gaspar, 1988).

Salt

As for heat and momentum, ice represents an obstacle to freshwater exchange with the atmosphere, but it is also a source of salt or freshwater in its own right.

Ice as an Obstacle. Evaporation from the ocean to the atmosphere must be set to zero in the ice-covered fraction; this is equivalent to a local increase of freshwater flux into the ocean. On the other hand, snowfall is intercepted by the ice, and the freshwater contained in it is not available to the ocean before it melts, months or years later. This temporary decrease in freshwater flux may introduce a phase shift in freshwater time distribution, or even a net decrease of local freshwater flux if snow-covered ice moves away from the area and is replaced by snow-free ice.

Ice as an Agent. Since most of the salts dissolved in seawater are rejected at the time of sea-ice formation, there is a salt flux proportional to the rate of freezing, which is maximal for thin ice. In addition to that initial process, sea-ice slowly continues to lose salt with time (Untersteiner, 1968; Gow and Tucker, 1990). Sea-ice melting produces the inverse effect, or a freshwater input. Since seawater density is mainly a function of salinity at low temperature, this process tends to destabilize the water column when ice forms, and to stabilize it at the time of melting.

The ice contribution to the surface salt flux can be computed by setting it equal to the growth rate of the ice multiplied by the difference between the water and ice average salinities (weighted by the ratio of ice over seawater density). An ice salinity between 10 and 25 was used by van Ypersele (1986), Lemke (1987), and Fichefet and Gaspar (1988). Hibler and Bryan (1987) assumed that no salt at all is retained when sea ice forms, which has the advantage of being globally conservative for salt.

To illustrate the complexity of ice–ocean interactions, let us examine some consequences of those effects. The salt flux feeds back into the ice thermodynamics, since the convection induced by the brine rejection can upwell warmer water to oppose the freezing process (negative feedback when freezing). On the other hand, the stabilization associated with melting tends to decrease the part of the oceanic heat flux due to deep water convection (negative feedback) but, if leads are present, the stratification helps the water to be heated quickly by the solar energy (positive feedback).

Integrated on an annual cycle, the net salt flux through the upper surface due to the freezing/melting process is equal to zero when ice is stationary. However, the horizontal ice movement can produce positive or negative net salt fluxes. For example, coastal areas subject to off-shore katabatic winds around the Antarctic can be regions of strongly positive net salt influx, since sea ice is continuously formed and evacuated before it can melt.

12.4 Sea ice monitoring and data needs

For a long time, limited accessibility to the polar seas has hampered the systematic monitoring of ice characteristics like ice extent, thickness, and velocity, snow depth, open water fraction, surface albedo or temperature. These parameters were only observed from coasts or during expeditions that were limited in time and geographical coverage. For example, the AIDJEX expedition mentioned in section 12.3.2 provided a tremendous amount of detailed information on sea-ice dynamics, but it only covered a small fraction of the Arctic Ocean for a few years (or even months for the manned observations.) In addition, the lack, until recently, of international standards for sea-ice data archiving helps explain why the state of sea-ice databases remains primitive by comparison with those for meteorological information (for a review, see Barry, 1986). Data on energy fluxes in the polar atmospheric or oceanic boundary layers are also scarce.

The availability of satellites has improved this situation to a certain extent. The first microwave scanning instrument in space was the Electrically Scanning Microwave Radiometer (ESMR) on Nimbus-5 launched in December 1972; it provided the first all-weather (microwaves penetrate clouds), day-or-night, global year-round record of the ice-covered seas (Carsey and Zwally, 1986). Due to the strong contrast in microwave signatures of open ocean and sea ice, its data could be interpreted in terms of sea-ice concentration and ice categories (multiyear/first year). The ESMR operated until March 1983, although with a significant reduction in data quality after October 1976 (Parkinson et al., 1987.) The data collected from this instrument between 1973 and 1976 were analysed and published in the form of atlases of maps of monthly ice concentrations for the Antarctic (Zwally et al., 1983) and the Arctic (Parkinson et al., 1987.) They revealed the formation of a very large polynya in the Weddell Sea in 1974 and several subsequent years. The discovery of this polynya has prompted several modelling studies and cruises, which have increased our understanding of air–sea–ice interactions.

The data from more elaborate sensors have been used since that pioneer period. For example, the Scanning Multifrequency Microwave Radiometer (SMMR), flown on Seasat and Nimbus-7 from 1978, and the Special Sensor Microwave Imager (SSM/I), on board the satellites of the U.S. Defense Meteorological Satellite Program (DMSP), can not only increase the accuracy of ice concentration calculations but also allow determination of such additional variables as ice temperature, snow cover, multiyear ice fraction, and surface melting (NASA, 1984).

In addition, Synthetic Aperture Radar (SAR) data from satellites such as ERS-1 can provide high resolution data on short term variations of sea-ice concentration, floe and lead size distribution, and sea-ice motion over limited areas (WCRP, 1990). Such data can be used for process studies. Comiso (1991) reviews the present and future satellite systems and techniques of interpretation for polar studies.

Even though the satellite data have contributed tremendously to the sea-ice database, several problems remain.

First, changes in satellites, sensors, methods of calibration, of processing or of archiving data reflect the technological progresses, but they also hinder the homogeneity of the long time series that are needed for climate studies. This may also be a result of the fact that more resources are spent for the development of satellite hardware than for the exploitation of the data needed by the climate community. The need for continuity in measurements cannot be overemphasized. In many cases, continuity is more important than the progress that can come from the introduction of a new, different instrument.

Second, (up to now,) satellites cannot routinely provide certain important parameters, such as ice thickness, surface wind over ice, or ocean mixed layer characteristics. For example, ice thickness is the single most important parameter needed to validate the particular rheology used in a sea-ice dynamic model (SINEG, 1990). In the Arctic, it is only available from a few borehole measurements (e.g., Bourke and Paquette, 1989) or from the few upward-looking sonar (ULS) submarine cruises that are unclassified (Keigwin and Johnson, 1992). In the Antarctic, the data are even scarcer, and it is only in December 1990 that 6 ULS were moored 150 *m* below sea level to monitor ice thickness in the Weddell Sea (WCRP, 1992). Some other parameters cannot (yet) be provided by satellites with sufficient accuracy or space and time coverage (e.g., cloud cover over ice, radiation budget, ice motion, ice type discrimination).

Third, insufficient *in situ* measurements by observers near the surface (or at least buoys) are available to validate the algorithms used to extract information from satellite data or to complement them with data they cannot provide. In addition to the experiments recently planned in both the Arctic and the Antarctic (see WCRP (1992) for a recent summary), it would be extremely useful to have access to the data acquired over the years by nuclear submarines from both East and West (Keigwin and Johnson, 1992).

Fourth, the number of observations and studies related to the Antarctic region is scandalously insufficient given its importance in the climate system and compared to the attention given to the Arctic. Since more people live in the vicinity of the Arctic Ocean than around the Antarctic, and since a significant fraction of sea-ice research is funded by military agencies with an interest in the Arctic, this situation can be explained. But in the context of studies of the *global* environmental change experienced by our planet, it is urgent to give a much higher priority to studies related to the Antarctic.

With the above remarks in mind, the reader is invited to refer to the reports of the WCRP Working Group on Sea Ice and Climate (WCRP, 1988, 1989, 1990 and 1992) for additional considerations on the need for sea-ice monitoring and forcing data. Thorndike (1989) details a possible sea-ice monitoring strategy that would assimilate data from various sources in a coherent *data interpretation system*. He pleads eloquently for the need for coordination of the observations and for the development of appropriate analysis procedures. GEWEX should keep these requirements and recommendations in mind in its planning.

Finally, it would also be extremely useful for the understanding of the long term behaviour of sea ice in the climate system to pay more attention to paleo sea-ice data. It is indeed possibe to retrieve information about sea-ice extent from proxy data (see Crowley and Parkinson (1989) for a discussion of the methods used). Reproducing past sea-ice variations with the models used to predict the future climate could help validate these models.

12.5 Conclusions and perspectives

This brief review tried to show the multiple ways by which sea ice affects its surroundings, and how important it is for climate modelling to have sea ice interactively coupled with atmospheric and oceanic models. Progress in sea-ice modelling for climate studies can be made in several directions, some of which are outlined below with references. A condition for this progress is that the sea-ice models need to be validated, and additional measurements of parameters like ice thickness are needed in that respect.

The reader is also referred to WCRP (1983, 1984, 1987, 1989, 1990, 1992) for additional recommendations.

Sea-ice interactions with the climate system cannot be studied without coupling better sea-ice models (including thermodynamics and dynamics) with better ocean models, and forcing them with better air data or coupling them with atmospheric models.

In the area of sea-ice thermodynamics, it is important to represent ice by at least two layers, to allow for internal sensible and latent heat storage, and to take the snow cover into account (use the 3-layer version of Semtner (1976) model). Surface albedo and other optical properties must be parameterized carefully (see Warren (1982) for a review of snow optical properties, Shine (1984) for a study of snow and sea-ice albedo under a cloud cover; for the Arctic, see Grenfell and Perovich (1984), Shine and Henderson-Sellers (1985), Maykut and Perovich (1987); for the Antarctic, see Weller, (1980); Andreas and Ackley, (1982)).

Specification of atmospheric variables such as wind, radiation, cloudiness, and precipitation also needs improvement (see WCRP, (1989) and Raschke et al., (1991)).

In order to compute better energy and mechanical forcing, the representation of both atmospheric and oceanic boundary layers should be improved, over both ice and leads (where strong energy exchanges take place). For the atmospheric boundary layer over ice, see McBean (1986) for a review, and Koch (1988) for a recent example of coupled boundary-layer and sea-ice study. Due to the high sensitivity of sea-ice thickness to the oceanic heat flux, ocean models should be used to compute this flux and allow it to vary in space and time.

It should be noted that the heat and salt budgets interact very strongly, and that climate models which couple ocean and sea ice only by the heat flux, without salt coupling, are missing the very important feedbacks associated with brine rejection or meltwater flux (see section 12.3.3). Mixed-layer models are the cheaper tools one can use to model this process. Detailed mixed-layer studies will also be useful to derive practical heat flux parameterization to be used in other coupled studies. Coupled sea-ice/mixed-layer models could also be used in atmospheric general circulation modelling studies which tend to use oversimplified thermodynamic sea-ice models with a constant oceanic heat flux. Coupled ocean general circulation models and sea-ice models have shown that the large-scale oceanic heat flux was able to determine the position of the ice edge in certain areas. Particular attention should be given in future studies to the treatment of gravitational instabilities, as the convective adjustment scheme might not be appropriate for certain polar areas. For a review of the oceanic mixed layer processes under ice, see McPhee (1986, 1992); for a review of coupled ocean and sea-ice models, see van Ypersele (1989) and SINEG (1990); for coupled model studies with the mixed layer, see Lemke (1987), Fichefet and Gaspar (1988); for coupled model studies with the 3-dimensional ocean, see Hibler and Bryan (1987), Semtner (1987), and van Ypersele (1986).

There is an urgent need to compare the model-generated oceanic heat flux to observations, both in different areas and seasons. The heat flux cannot be measured directly in the ocean, but can be derived in the presence of ice from the difference between heat conduction in the ice and the amount of heat of fusion associated with a change of ice thickness (McPhee and Untersteiner, 1982). This method has not yet been widely applied. Perovich et al. (1989) have tested an alternative method based on thermistor strings attached to a drifting buoy, and found it promising.

To assess the real performance of coupled models, they should be applied to interannual variability studies, which are more stringent tests than equilibrium studies (see Walsh et al., 1985). The simulation of occasional features such as the Weddell Polynya at the proper times also constitutes an interesting test of coupled models. Interannual variability studies imply the need for better forcing data (atmospheric energy fluxes, wind) and accurate validation data in both hemispheres (ice extent, velocity, concentration, thickness distribution, and albedo) for periods of at least 10 years (Lemke et al., 1980). This calls at least for routine, satellite-based, passive microwave measurements, combined with ice buoys observations in *both* hemispheres.

The ice dynamics governs ice movement and deformation, and therefore its thickness distribution. Since the ice thickness distribution affects so much the large-scale heat flux to the atmosphere, and the heat, salt, and momentum budgets of the ocean, ice dynamics should be included in every sea-ice simulation made for climatic purpose. The reduced sensitivity to thermal perturbations of sea-ice models including ice dynamics means that GCMs with thermodynamic sea-ice models might overestimate the climate response to CO_2 increases.

The large-scale dynamic behaviour of sea ice still needs to be better known, especially in the southern hemisphere and in the marginal ice zones. In that respect, drifting ice buoys programs, with measurements of (at least) atmospheric pressure, air temperature, and ice motion, should be expanded (WCRP, 1989, 1992).

It should be stressed that improving the models also means looking for computationally efficient methods, adequate for climate studies. Additional knowledge should also be used, whenever possible, to streamline the description of physical processes involved. An example is given by ice dynamics: the bulk-viscous approach in Semtner (1987) and Flato and Hibler (1990) is more usable in long-term climate integrations with coupled models than the full dynamics approach of Hibler (1979). However, its validity for different circumstances needs further assessment.

A final area in which better coupled sea-ice/ocean/atmosphere models might prove useful is geochemistry. Indeed, for example, an overlooked effect of sea ice is that of a barrier to CO_2 exchanges at the air–sea interface. The inclusion of carbon cycle modelling in coupled ocean and sea-ice models could improve our understanding of global geochemical cycles.

In summary, sea ice is definitely an important element of the climatic system, and GEWEX should pay sufficient attention to it in its planning. Sea-ice modelling requires at least the inclusion of thermodynamic aspects (with internal heat storage taken into account) and some representation of dynamic processes, so that the thickness distribution is also predicted. The numerous feedbacks in which sea ice plays a role, for example the ice–albedo and the salinity–heat flux feedbacks, make the use of coupled ocean/sea-ice models, or atmosphere/ocean/sea-ice models necessary for studies of climatic conditions different from the existing ones. Appropriate forcing and validation data need to be acquired and interpreted in a coherent and continuous way, with coverage of *both* hemispheres.

A final note: at the Glücksburg meeting, a suggestion was made to consider the climatic role played by changes in the abundance of penguins, which modify the sea-ice albedo. A multi-million dollar programme was even advocated by a facetious lecturer. This only illustrates the fact that there exist many more research topics related to sea ice than those reviewed here.

12.6 References

Aagard K, Coachman LH, Carmack EC (1981) On the halocline of the Arctic Ocean. *Deep-Sea Res*,28: 529–545.

Andreas EL, Ackley SF (1982) On the differences in ablation seasons of arctic and antarctic sea ice. *J Atmos Sci*,39: 440–447.

Andreas EL, Tucker III WB, Ackley SF (1984) Atmospheric boundary-layer modification, drag coefficient, and surface heat flux in the antarctic marginal ice zone. *J Geophys Res*,89: 649–661.

Banke EG, Smith SD, Anderson RJ (1980) Drag coefficients at AIDJEX from sonic anemometer measurements. In *Sea Ice Processes and Models,* (ed. R.S. Pritchard) University of Washington Press, Seattle, 430–442.

Barry RG (1986) The sea ice data base. In *The Geophysics of Sea Ice,* Proc. NATO Advanced Study Institute on Air–Sea–Ice Interactions, (ed. N. Untersteiner) Plenum Press, New York, 1000–1131.

Bourke RH, Paquette RG (1989) Estimating the thickness of sea ice. *J Geophys Res*,94: 919–923.

Brown RA (1981) Modeling the geostrophic drag coefficient for AIDJEX. *J Geophys Res*,86: 1989–1994.

Bryan F (1986) High latitude salinity effects and interhemispheric thermohaline circulations. *Nature*,323: 301–304.

Budyko MI (1969) The effect of solar radiation on the climate of the Earth. *Tellus*,21: 611-619.

Carsey FD, Zwally HJ (1986) Remote sensing as a research tool. In *The Geophysics of Sea Ice,* Proc. NATO Advanced Study Institute on Air–Sea–Ice Interactions, (ed. N. Untersteiner) Plenum Press, New York, 1021–1098.

Comiso JC (1991) Satellite remote sensing of the polar oceans. *J Mar Syst*,2: 395–434.

Coon MD (1980) A review of AIDJEX modeling. In *Sea Ice Processes and Models,* (ed. R.S. Pritchard) University of Washington Press, Seattle, 12–27.

Crowley TJ, Parkinson CL (1989) Late Pleistocene variations in antarctic sea ice: Effect of orbital isolation changes. *Clim Dyn*,3: 85–91.

Demuth C, van Ypersele J-P (1987) Sea-ice simulations in the Weddell Sea. *Proceedings of the Belgian National Colloquium on Antarctic Research*, Prime Minister's Office, Science Policy Office, Brussels, 195–212. (Available from the Science Policy Office, 8 Rue de la Science, B–1040 Brussels, Belgium).

Demuth C, van Ypersele J-P (1989) Simulations of the annual sea-ice cover in the Weddell Sea. *Final report, Contract ANTAR/CVT/01, Belgian National Research Programme on the Antarctic*, Prime Minister's Office, Science Policy Office, Brussels, 47 pp.

Ekman VW (1905) On the influence of the earth's rotation on ocean currents. *Ark. Mat. Astron. Fys.* 2: 1–52.

Fichefet T, Gaspar P (1988) A model study of upper ocean–sea ice interaction. *J Phys Oceanogr*,18: 181-195.

Flato GM, Hibler III WD (1990) On a simple sea ice dynamics model for climate studies. *Ann Glaciol*,14: 72–77.

Foster TD, Carmack EC (1976) Frontal zone mixing and Antarctic Bottom Water formation in the southern Weddell Sea. *Deep-Sea Res*,23: 301–317.

Gloersen P, Campbell WJ (1988) Variations in the arctic, antarctic, and global sea ice covers during 1978–1987 as observed with the Nimbus 7 Scanning Multichannel Microwave Radiometer. *J Geophys Res*,93: 10666–10674.

Gow AJ, Tucker III WB (1990) Sea ice in the polar regions. In *Polar oceanography, Part A, Physical science* W.O. Smith (ed.), Academic Press, San Diego, 47–122.

Grenfell TC, Maykut GA (1977) The optical properties of ice and snow in the Arctic Basin. *J Glaciol*,18: 445–463.

Grenfell TC, Perovich DK (1984) Spectral albedos of sea ice and incident solar irradiance in the southern Beaufort Sea. *J Geophys Res*,89: 3573–3580.

Hall DK, Martinec J (1985) *Remote sensing of ice and snow.* Chapman and Hall, London, 189pp.

Harvey LD (1990) Testing alternative parameterizations of lateral melting and upward basal heat flux in a thermodynamic sea ice model. *J Geophys Res*,95: 7359–7365.

Herman GF (1986) Atmospheric modelling and air–sea–ice interaction. In *The Geophysics of Sea Ice,* Proc. NATO Advanced Study Institute on Air–Sea–Ice Interactions, ed. N. Untersteiner, Plenum Press, New York: 713–754.

Hibler III WD (1979) A dynamic thermodynamic sea ice model. *J Phys Oceanogr*,9: 815–846.

Hibler III WD (1980) Modeling a variable thickness sea ice cover. *Mon Wea Rev*,108: 1943–1973.

Hibler III WD (1984) The role of sea ice dynamics in modeling CO_2 increases. In *Climate Processes and Climate Sensitivity*, ed. J.E. Hansen and T. Takahashi, Geophysical Monograph 29, American Geophysical Union, Washington, D.C., 238–253.

Hibler III WD (1986) Ice dynamics. In *The Geophysics of Sea Ice,* Proc. NATO Advanced Study Institute on Air–Sea–Ice Interactions, ed. N. Untersteiner, Plenum Press, New York, 577–640.

Hibler III WD (1988) Modelling sea ice thermodynamics and dynamics in climate studies. In *Physically-Based Modelling and Simulation of Climate and Climatic Change, Part 1,* ed. M.E. Schlesinger, Kluwer Academic Publ., Dordrecht, 509–563.

Hibler III WD, Bryan K (1987) A diagnostic ice-ocean model. *J Phys Oceanogr*,17: 987–1015.

Hibler III WD, Walsh JE (1982) On modeling seasonal and interannual fluctuations of Arctic sea ice. *J Phys Oceanogr*,12: 1514–1523.

Hobbs P (1974) *Ice physics.* Oxford University Press, Oxford, 837pp.

Hoffert MI (1990) Climatic change and ocean bottom water formation: Are we missing something? In *Climate-ocean interaction*, M. Schlesinger (ed.), Kluwer Academic Publ., Dordrecht, 295–317.

Houssais MN (1988) Testing a coupled ice–mixed-layer model under subarctic conditions. *J Phys Oceanogr*,18: 196–210.

IAPSO (1985) *The International System of Units (SI) in Oceanography.* IAPSO Publication Scientifique No. 32, UNESCO technical papers in marine science No. 45, UNESCO, Paris, 124 pp.

Keigwin LD, Johnson GL (1992) Using a nuclear submarine for arctic research. *EOS, Trans Am Geophys Union*,73: 209 and 220–221.

Koch C (1988) A coupled sea ice–atmospheric boundary layer model. Part 1: Description of the model and 1979 standard run. *Beitr Phys Atmosph*,61: 344-354.

Lemke P (1987) A coupled one-dimensional sea ice–ocean model. *J Geophys Res*,92: 13164–13172.

Lemke P (1989) Results from sea ice - ocean modeling experiments in the Weddell Sea. In *Report of the third session of the JSC Working Group on Sea Ice and Climate,* WMO/TD-272, World Climate Research Programme, World Meteorological Organization, Geneva, Annex B.

Lemke P, Manley TO (1984) The seasonal variation of the mixed layer and the pycnocline under polar ice. *J Geophys Res*,89: 6494–6504.

Lemke P, Owens WB, Hibler III WD (1990) A coupled sea ice – mixed layer – pycnocline model for the Weddell Sea. *J Geophys Res*,95: 9513–9525.

Lemke P, Trinkl EW, Hasselmann K (1980) Stochastic dynamic analysis of polar sea ice variability. *J Phys Oceanogr*,10: 2100–2120.

Manabe S, Stouffer RJ (1988) Two stable equilibria of a coupled ocean–atmosphere model. *J Clim*,1: 841–866.

Maykut GA (1978) Energy exchange over young sea ice in the central Arctic. *J Geophys Res*,83: 3646–3658.

Maykut GA (1982) Large-scale heat exchange and ice production in the central Arctic. *J Geophys Res*,87: 7971–7984.

Maykut GA (1986) The surface heat and mass balance. In *The Geophysics of Sea Ice,* Proc. NATO Advanced Study Institute on Air–Sea–Ice Interactions, ed. N. Untersteiner, Plenum Press, New York, 395–463.

Maykut GA, Perovich DK (1987) The role of shortwave radiation in the summer decay of a sea ice cover. *J Geophys Res*,92: 7032–7044.

Maykut GA, Untersteiner N (1971) Some results from a time dependent, thermodynamic model of sea ice. *J Geophys Res*,76: 1550–1575.

McBean G (1986) The atmospheric boundary layer. In *The Geophysics of Sea Ice,* Proc. NATO Advanced Study Institute on Air–Sea–Ice Interactions, ed. N. Untersteiner, Plenum Press, New York, 283–337.

McPhee MG (1986) The upper ocean. In *The Geophysics of Sea Ice,* Proc. NATO Advanced Study Institute on Air–Sea–Ice Interactions, ed. N. Untersteiner, Plenum Press, New York, 339–394.

McPhee MG (1992) Turbulent heat flux in the upper ocean under sea ice. *J Geophys Res*,97: 5365–5379

McPhee MG, Untersteiner N (1982) Using sea ice to measure vertical heat flux in the ocean. *J Geophys Res*,87: 2071–2074.

Millero FJ (1978) Freezing point of seawater. in "Eight Report of the Joint Panel on Oceanographic Tables and Standards," UNESCO Tech. Pap. March. Sci. No. 28, Annex 6. UNESCO, Paris.

Mitchell JFB, Hills TS (1986) Sea-ice and the antarctic winter circulation) a numerical experiment. *Quart J R Met Soc*,112: 953–969.

NASA (1984) Passive microwave remote sensing for sea ice research. Report of the NASA Science Working Group for the Special Sensor Microwave Imager (SSM/I). NASA and Applied Physics Laboratory, University of Washington, Seattle, 55 pp.

Overland JE (1985) Atmospheric boundary layer structure and drag coefficients over sea ice. *J Geophys Res*,90: 9029–9049.

Parkinson, CL, and WM Washington (1979) A large-scale numerical model of sea ice. *J Geophys Res*,84: 311–337.

Parkinson CL, Comiso JC, Zwally HJ, Cavalieri DJ, Gloersen P, Campbell WJ (1987) *Arctic Sea Ice, 1973–1976: Satellite Passive-Microwave Observations.* Special Publication SP-489, NASA, Washington, D.C., 296 pp.

Perovich DK, Tucker III WB, Krishfield RA (1989) Oceanic heat flux in the Fram Strait measured by a drifting buoy. *Geophys Res Let*,16: 995–998.

Pritchard RS (ed) (1980) *Sea Ice Processes and Models,* University of Washington Press, Seattle, 474 pp.

Raschke E, Cattle H, Lemke P, Rossow W (eds) (1991) Sea-ice and climate. Report of a workshop on polar radiation fluxes and sea-ice modelling. WCRP–62, WMO/TD-No 442, World Climate Research Programme, World Meteorological Organization, Geneva, 20 pp + 4 appendices.

Rooth C (1982) Hydrology and ocean circulation. *Prog. Oceanog.,* 11: 131–149.

Royer JF, Planton S, Déqué M (1990) A sensitivity experiment for the removal of Arctic sea ice with the French spectral general circulation model. *Clim Dyn*,5: 1–17.

Schlesinger ME (1988) Quantitative analysis of feedbacks in climate model simulations of CO_2 -induced warming. In *Physically-Based Modelling and Simulation of Climate and Climatic Change, Part 2,* ed. M.E. Schlesinger, Kluwer Academic Publ., Dordrecht, 653–735.

Semtner AJ, Jr. (1976) A model for the thermodynamic growth of sea ice in numerical investigations of climate. *J Phys Oceanogr*,6: 379–389.

Semtner AJ, Jr. (1984a) Sensitivity of ocean circulation to sea-ice processes. In *Report of the meeting of experts on sea ice and climate modelling*, WCP–77, World Climate Research Programme, World Meteorological Organization, Geneva, Appendix D, 1–8.

Semtner AJ, Jr. (1984b) On modelling the seasonal thermodynamic cycle of sea ice in studies of climatic change. *Clim Change*,6: 27–37.

Semtner AJ, Jr. (1987) A numerical study of sea ice and ocean circulation in the Arctic. *J Phys Oceanogr*,17: 1077-1099.

Shine KP (1984) Parameterization of the shortwave flux over high albedo surfaces as a function of cloud thickness and surface albedo. *Quart J R Met Soc*,110: 747–764.

Shine KP, Henderson-Sellers A (1985) The sensitivity of a thermodynamic model to changes in surface albedo parameterization. *J Geophys Res*,90: 2243–2250.

Shuchman RA, Onstott RG (1990) Remote sensing of the polar oceans. In *Polar oceanography, Part A, Physical science.* W.O. Smith (ed.), Academic Press, San Diego, 123–169.

SINEG (1990) *Sea Ice Numerical Experimentation Group. Report of the first session,* WCRP-45, WMO/TD-384, World Climate Research Programme, World Meteorological Organization, Geneva, 20 pp + 4 appendices.

Smith WO (ed.) (1990) *Polar oceanography, Part A, Physical science* Academic Press, San Diego, 406 pp.

Stössel A, Lemke P, Owens WB (1990) Coupled sea ice–mixed layer simulations for the Southern Ocean. *J Geophys Res*,95: 9539–9555.

Thorndike AS (1989) Arctic sea ice: a monitoring strategy. In *Report of the third session of the JSC Working Group on Sea Ice and Climate,* WMO/TD-272, World Climate Research Programme, World Meteorological Organization, Geneva, Annex C, 9pp.

Thorndike AS, Colony R (1982) Sea ice motion in response to geostrophic winds. *J Geophys Res*,87: 5845–5852.

Thorndike AS, Rothrock DA, Maykut GA, Colony R (1975) The thickness distribution of sea ice. *J Geophys Res*,80: 4501–4513.

Toole JM (1981) Sea ice, winter convection and the temperature minimum layer in the Southern Ocean. *J Geophys Res*,86: 8037–8047.

Untersteiner N (1968) Natural desalination and equilibrium salinity profile of perennial sea ice. *J Geophys Res*,73: 1251–1257.

Untersteiner N, (ed.) (1986a) *The Geophysics of Sea Ice,* Proc. NATO Advanced Study Institute on Air–Sea–Ice Interactions, Plenum Press, New York, 1196 pp.

Untersteiner N (1986b) The geophysics of sea ice: overview. In *The Geophysics of Sea Ice,* Proc. NATO Advanced Study Institute on Air–Sea–Ice Interactions, ed. N. Untersteiner, Plenum Press, New York, 1–8.

van Ypersele J-P (1986) *A numerical study of the response of the Southern Ocean and its sea ice to a CO_2-induced atmospheric warming.* Cooperative thesis NCAR CT-99, Université Catholique de Louvain and National Center for Atmospheric Research, Louvain-la-Neuve and Boulder, 135 pp. (NTIS Order No: PB87-163218/GAR).

van Ypersele J-P (1989) Coupled ocean and sea-ice models: review and perspectives. In *Climate and Geosciences*, ed. A. Berger, S.H. Schneider, and J.C. Duplessy, Kluwer Academic Publ., Dordrecht, 253–277.

van Ypersele J-P (1990) Modelling sea ice for climate studies. In *Climate-ocean interaction*, ed. M. Schlesinger, Kluwer Academic Publ., Dordrecht, 97–123.

Wadhams P (1986) The seasonal ice zone. In *The Geophysics of Sea Ice,* Proc. NATO Advanced Study Institute on Air–Sea–Ice Interactions, ed. N. Untersteiner, Plenum Press, New York, 825–991.

Walsh JE, Hibler III WD, Ross B (1985) Numerical simulation of northern hemisphere sea ice variability, 1951–1980. *J Geophys Res*,90: 4847–4865.

Warren SG (1982) Optical properties of snow. *Rev Geophys Space Phys*,20: 67–89.

Washington WM, Meehl GA (1989) Climate sensitivity due to increased CO_2: experiments with a coupled atmosphere and ocean general circulation model. *Clim Dyn*,4: 1–38.

Washington WM, Parkinson CL (1986) *An Introduction to Three-Dimensional Climate Modeling.* University Science Books and Oxford University Press, Mill Valley and Oxford, 422 pp.

Washington WM, Semtner AJ, Parkinson C, Morrison L (1976) On the development of a seasonal change sea-ice model. *J Phys Oceanogr*,6: 679–685.

WCRP (1983) *Report of the WMO/CAS-JSC-CCCO meeting of experts on the role of sea ice in climatic variations,* WCP-26, World Climate Research Programme, World Meteorological Organization, Geneva, 20 pp + 8 appendices.

WCRP (1984) *Report of the meeting of experts on sea ice and climate modelling,* WCP–77, World Climate Research Programme, World Meteorological Organization, Geneva, 34 pp + 10 appendices.

WCRP (1987) *Report of the second session of the WCRP Working Group on Sea Ice and Climate,* WCP–128, World Climate Research Programme, World Meteorological Organization, Geneva.

WCRP (1989) *Report of the third session of the JSC Working Group on Sea Ice and Climate,* WMO/TD-272, World Climate Research Programme, World Meteorological Organization, Geneva, 23 pp + 11 annexes.

WCRP (1990) *Report of the fourth session of the Working Group on Sea-Ice and Climate,* WCRP-41, WMO/TD-377, World Climate Research Programme, World Meteorological Organization, Geneva, 18 pp + 4 annexes.

WCRP (1991) *Global Energy and Water Cycle Experiment (GEWEX); Report of the third session of the JSC Scientific Steering Group for GEWEX.* WCRP-57, WMO/TD–424, World Climate Research Programme, World Meteorological Organization, Geneva, 26 pp + 8 appendices.

WCRP (1992) *Report of the fifth session of the Working Group on Sea-Ice and Climate,* WCRP-65, WMO/TD–459, World Climate Research Programme, World Meteorological Organization, Geneva, 16 pp + 6 annexes.

Weeks WF, Ackley SF (1986) The growth, structure, and properties of sea ice. In *The Geophysics of Sea Ice,* Proc. NATO Advanced Study Institute on Air–Sea–Ice Interactions, ed. N. Untersteiner, Plenum Press, New York, 9–164.

Weller G (1980) Spatial and temporal variations in the south polar surface energy balance. *Mon Wea Rev*,108: 2006–2014.

Zwally HJ, Comiso JC, Parkinson CL, Campbell WJ, Carsey FD, Gloersen P (1983) *Antarctic Sea Ice, 1973–1976: Satellite Passive-Microwave Observations.* Special Publication SP-459, NASA, Washington, D.C., 206 pp.

Chapter 13

THE SOIL-VEGETATION-ATMOSPHERE INTERFACE

W. James Shuttleworth
Institute of Hydrology
Wallingford
Oxfordshire OX10 8BB
UK

13.1 Introduction

The Scientific Plan for GEWEX draws together and enhances a range of climate research activities which have the specific scientific objectives:

- to determine the hydrological cycle and energy fluxes by means of global measurements of observable atmospheric and surface properties;
- to model the global hydrological cycle and its effects on the atmosphere and oceans;

and

- to develop the ability to predict the variations of global and regional hydrological processes and water resources, and their response to environmental change.

In addition to these scientific objectives, there is a technical objective to foster the development of observing techniques and data management and assimilation systems.

Experiments with General Circulation Models (GCMs) have demonstrated that predicted climates are sensitive to surface parameters, in particular to surface albedo (e.g. Charney et al, 1977), moisture availability (Sud et al, 1985), surface roughness (Shukla and Mintz, 1982) and stomatal

NATO ASI Series, Vol. I 5
Energy and Water Cycles in the Climate System
Edited by E. Raschke and D. Jacob

control (Sato et al, 1989). These are in turn determined by surface characteristics, namely vegetation and soil attributes; and by landscape and subsurface attributes, such as topography and underlying geology. The representation of land surface processes is one of the weakest, and most challenging aspects of present global change models, and yet is critical to the accurate prediction of changes in regional climate following *greenhouse warming* and major land use change, such as deforestation and desertification. The impact of climate on hydrological and ecological systems is also of direct human relevance, but here it is the interactive effects of climate-induced feedbacks which require quantification.

Clearly then, the interaction between the atmosphere and the ground is important in determining the hydrology of terrestrial surfaces and the climate to which they are exposed. Describing the processes involved in the soil and the vegetation which make up this interface is vital in modelling the global hydrological cycle, and in accessing the regional consequences of global change. **Observing and modelling the soil-vegetation interface is therefore critical to the fulfilment of all three of the scientific objectives of GEWEX.** But our ability to provide the required understanding is heavily constrained by the grid scale at which global models currently describe continents, and the fact that the vegetation covering the surface, and the topography, is commonly heterogeneous within any single grid square. Seeking to overcome this problem will also foster the technical objective of GEWEX. This paper overviews the basic physical and physiological features involved in the soil-vegetation-atmosphere interface, and the way these have been combined to describe the interaction. The description given here complements the paper by Stricker (1991) in this volume, by focussing on vertical water movement and on above-ground features. *State of the Art* GCM submodels of the soil-vegetation-atmosphere interaction are introduced and described briefly, along with a concise review of observational activity over the last decade under the World Climate Research Programme (WCRP) and the International Satellite Land Surface Climatology Project (ISLSCP), which has sought to provide calibration of such submodels. Finally, the relevant programme of observational and modelling activity proposed under GEWEX is described, some of which will occur in collaboration with the International Geosphere-Biosphere Programme, under the Core Project *Biological Aspects of the Hydrological Cycle* (BAHC).

13.2 Status of the subject

Fundamentally, the purpose of land-atmosphere studies is to provide the understanding necessary to allow simulation of terrestrial global processes by mathematical formulations, and thereby allow their coding as computer algorithms. Nowadays these algorithms are increasingly referred to as Soil-Vegetation-Atmosphere Transfer Schemes (SVATS). SVATS are essentially vertical views of hydrological processes which consider the transports of water and energy just below the surface, across the interface, and within and through the vegetation canopy. They have been recently reviewed in detail (Avissar and Verstraete, 1990), and the most relevant examples are described later in this paper. The demand for such schemes comes on the one hand from atmospheric modelling, to allow a representation of the return of energy and precipitated water to the atmosphere, and on the other hand, from hydrological modelling, to allow calculation of the loss of water from catchments by evaporation.

For many years there have been experimental studies of the soil-vegetation-atmosphere interaction of individual plots of agricultural crop, pasture land and sometimes of forests. The objective has generally been to define their relative water use in response to the weather to which they are exposed. The changing meteorological conditions were regarded as independent variables, to be monitored, and usually considered amenable to statistical description as random or quasi-random variations around an unchanging mean.

Over the past decade the underlying assumptions, the experimental objectives and often the scale, range and complexity of instrumental systems, have all undergone fundamental change. There is now no *a priori* assumption of an unchanging climate; in fact the main experimental objective is usually to improve the models used to predict such change. Moreover, the instrumental and manpower commitment involved in land/atmosphere observational studies these days is often very large indeed, and draws on the resources of many nations (e.g. André et al, 1988; Sellers et al, 1988a).

In the context of GEWEX, with its emphasis on providing global description of the hydrological cycle, it is these larger scale studies which are most relevant. Later in this section we therefore overview past studies of this type, and in a following section, preview such proposed studies in the GEWEX programme. But first we address the SVAT models themselves, and start by considering the basic building blocks from which such models are constructed.

13.2.1 Physical and physiological aspects of the interface

Energy

LATENT HEAT: The molecules in liquid water are held close together by an attractive force. In water vapour they are at least ten times further apart and the intermolecular force is then very much smaller. During evaporation the separation between molecules therefore increases greatly and it is necessary to do work against the attractive force, that is to supply energy. The energy required is called the *latent heat of vaporization of water* λ.

SATURATED VAPOUR CONTENT OF AIR: Natural evaporation occurs by exchange of water molecules between air and a free water surface. This can be inside the vegetation or soil in dry conditions, or on their surface during or just after rain.

According to Shuttleworth (1975), evaporation is the difference between two rates; one is the rate at which molecules leave the surface - determined by the temperature , T; the other is the rate at which they are captured by the surface - related to the vapour pressure, e. If molecules can diffuse away, e can remain low, and the difference between these two rates is finite. If, on the other hand, the air above the water is enclosed, the vapour pressure increases until these rates are equal and there is no more evaporation. The air is then said to be saturated. At a given temperature this equilibrium occurs for a particular vapour pressure e_s. This is called the saturated vapour pressure and is related to temperature; if e_s is in Pascals and T is in degrees Celsius an approximate equation (Randkivi; 1971) is:

$$e_s = 611 \exp\left[\frac{17.27\,T}{237.2 + T}\right] \tag{13.1}$$

It is important in building physically-based models of the soil-vegetation-atmosphere interface that not only is e_s a known function of temperature, but so is Δ the gradient of this function.

SENSIBLE HEAT: A portion of the radiant energy input to the earth's surface is not used for evaporation, rather it warms the atmosphere directly. We speak of the associated flow of energy as *sensible* heat flux because it changes air temperature, a property of air we can sense. The temperature change is proportional to the product of c_p, the specific heat of air at constant pressure, and ρ_a its density.

SURFACE RADIATION BALANCE: The sun is the main source of radiant energy. It is equivalent to a radiator of about 6000 degrees Celsius, but the input is modified by absorption by atmospheric gases (particularly water vapour), through scattering by air molecules in clear sky conditions, and additionally by clouds when these are present.

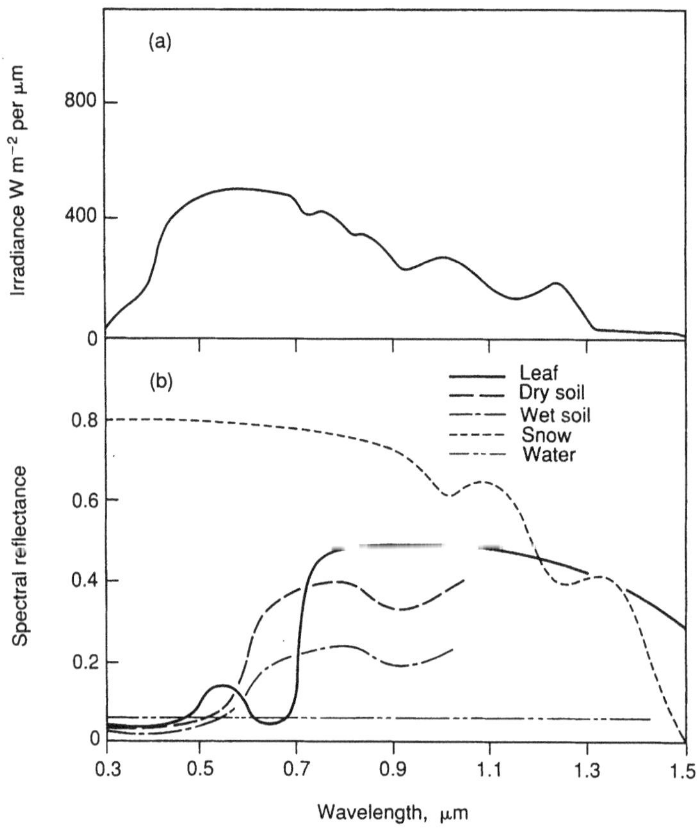

Figure 13.1: *(a) Typical irradiance spectrum for solar radiation for middle latitudes and clear skies. (b) Typical spectral reflectance of different landscape features as a function of wavelength.*

Much of the radiation has short wave lengths, $0.3 - 3.0\,\mu m$, the spectrum varying with the fraction of the total short wave energy input, S_t, reaching the ground in the direct solar beam, see Fig. 13.1(a). Some of the short-wave radiation, S_d, reaches the surface in a diffuse or non-directional form after scattering. This fraction is typically 15 to 25% in clear sky conditions, but approaches 100% in overcast conditions. Part of the short wave radiation is reflected.

The reflection coefficient or *albedo*, α, depends on land cover. Almost half of the solar energy reaches the surface with wavelengths less than $0.72\,\mu m$ (see Fig. 13.1(a)). The spectrum of reflected radiation differs greatly with wavelength, depending on the surface features present. Figure 13.1(b) shows typical curves for snow, water, leaves, and wet and dry soil; there is a very noticeable fall in the reflection coefficient of foliage below $0.7\,\mu m$, which is associated with their use of energy for photosynthesis. The more advanced models of the soil-vegetation system acknowledge this by drawing up radiation balances at the ground for radiation exchange above and below this wavelength, and hence acknowledge the ensuing dependence of area-average albedo on the proportional vegetation cover. Such SVATS also acknowledge, or in some cases model, the influence of the morphology of the vegetation, and in particular the observed decrease with greater vegetation height, due to enhanced trapping of radiation by multiple reflection within the canopy.
There is also a significant exchange of radiant energy between the earth's surface and the atmosphere in the form of radiation at longer wavelengths, i.e. in the range $3-100\,\mu m$. Both the ground and the atmosphere emit *black body* radiation with a spectrum characteristic of their temperature. Since the surface is on average warmer than the atmosphere, there is usually a net loss of energy as thermal radiation. Good models of the interaction at the soil-vegetation-atmosphere interface will seek to represent the exchange of short-wave and long-wave radiation separately, and calculate the outgoing long-wave radiation from the surface temperature of the vegetation. The *net radiation*, R_n is the net input of radiation at the surface; i.e. the difference between the incoming and reflected solar radiation, minus the difference between the outgoing longwave radiation, L_o, and incoming longwave radiation, L_i; thus

$$R_n = S_t(1-\alpha) - (L_o - L_i) \tag{13.2}$$

Nowadays net radiation is comparatively simple to measure, and indirect measurement is increasingly possible using satellite data, (e.g. Choudhury, 1991).

ENERGY STORAGE IN THE SOIL: The heat temporarily stored in the soil can be significant, though it is still typically an order of magnitude less than net radiation. It is considered in detail by Stricker (1991) in this volume.

Conduction is the main mechanism for heat transfer in soils. If the physical properties of the soil constant with depth, the soil temperature obeys an equation of the form

$$\frac{\partial T}{\partial t} = \left(\frac{K_t}{C_s}\right)\frac{\partial^2 T}{\partial z^2} \tag{13.3}$$

where K_t is the thermal conductivity of the soil and C_s is the volumetric heat capacity.

Solving Equation 13.3 in the idealized case of a uniform soil, and assuming the temperature of the soil-air interface oscillates sinusoidally with frequency f (in s^{-1}) and amplitude T (in $^\circ C$), it can be shown (Monteith, 1973) that the heat flux into the surface G is given by

$$G = \sqrt{2}\,T_{ss}^{0}\left[\frac{2\pi\, f\, C_s}{K_t}\right]^{1/2}\sin\left(2\pi\, f\, t + \frac{\pi}{4}\right) \tag{13.4}$$

The flow is therefore at a maximum when the rate of change is greatest, one eighth of a cycle before the temperature, i.e. 3 hours earlier in the case of the diurnal wave and $1\frac{1}{2}$ months in the case of the annual wave. The amplitude of the daily cycle in soil surface temperature can be significantly greater than that in air temperature, with timing mainly linked to radiation input, and G can be large, typically 30% of the net radiation exchange at the soil surface. With dense vegetation little radiation reaches the ground and heat storage in the soil can often be neglected (e.g. Shuttleworth et al, 1984).

Atmospheric diffusion

MOLECULAR DIFFUSION: Individual molecules are in rapid, haphazard motion at normal temperatures. The transfers associated with such motion which are of most relevance to GEWEX are those of water vapour, heat and momentum, and, for convenience, it is common to treat the evaporation rate as the equivalent flow of latent heat λE. To do this, we introduce γ, the so-called *psychrometric constant*, which is defined as the combination $\gamma = (c_p p)/(0.622\lambda)$, where p is atmospheric pressure.

The equations describing the transfer of λE, H and of τ, the momentum transfer per unit area, along an arbitrarily chosen z axis have similar form thus:

$$\lambda E = -\frac{\rho_a c_p}{\gamma} D_V \frac{\partial e}{\partial z} \tag{13.5}$$

$$H = -\rho_a c_p D_H \frac{\partial T}{\partial z} \tag{13.6}$$

$$\tau = \rho_a D_M \frac{\partial u}{\partial z} \tag{13.7}$$

where u is the velocity of the air perpendicular to the z axis. The negative sign in Equations 13.5 and 13.6 is a mathematical convention: when gradients decrease away from the surface the ensuing energy flow is upwards. The molecular diffusion coefficients D_V, D_H and D_M respectively have the values 0.212, 0.181 and 0.133 $(\cdot 10^{-4}\, m^2\, s^{-1})$ at freezing point, and all increase by 0.7% per degrees Celsius.

TURBULENT DIFFUSION: The wind blowing horizontally over the ground is retarded by an interaction with the surface. The interaction gives rise to apparently random and haphazard movement in which portions of air, of varying size, are continuously being created and destroyed, and move in an ill-defined yet coherent way during their transient existence. This phenomenon, the frictional turbulence, is a much more efficient transfer mechanism than molecular diffusion, and is the primary process responsible for exchange between air close to the ground and that at higher levels in the atmosphere. The transfer of latent and sensible heat and momentum by diffusion in turbulent eddies, is conventionally represented in a similar way to molecular diffusion thus:

$$\lambda E = -\frac{\rho c_p}{\gamma} K_V \frac{\partial e}{\partial z} \tag{13.8}$$

$$H = -\rho c_p K_H \frac{\partial T}{\partial z} \tag{13.9}$$

$$\tau = \rho K_M \frac{\partial u}{\partial z} \tag{13.10}$$

in which K_V, K_H and K_M are the relevant eddy diffusion coefficients.

With a so-called *neutral* atmosphere, i.e. one in which there is no atmospheric bouyancy generated by a vertical temperature gradient, the windspeed $u(z)$ at height z above an extensive, horizontal rough surface is described by the equation

$$u(z) = \frac{u_*}{k} \ln \left[\frac{z-d}{z_0} \right] \tag{13.11}$$

In Eq. 13.11, u_* is the *friction velocity*, defined by $\tau = \rho u_*$, and $k = 0.41$ is *Von Karman's constant.* The *zero plane displacement* d and the *roughness length* z are properties of the surface, found by fitting the equation to measured windspeed profiles. If h is vegetation height, typical values are $z \simeq 0.13\,h$ and $d \simeq 0.63\,h$ for *established* agricultural crops (Monteith; 1973), and $z \simeq 0.076\,h$ and $d \simeq 0.78\,h$ for dense forest (Shuttleworth, 1989).

The modification of convection in other than neutral conditions has often been parameterised in terms of a dimensionless parameter R_i, known as *the Richardson number*, which is defined by

$$R_i = \frac{g}{T + 237.2} \frac{\frac{\partial T}{\partial z}}{\left(\frac{\partial u}{\partial z}\right)^2} \tag{13.12}$$

where g is the acceleration due to gravity. In the general situation the (height dependent) eddy diffusion coefficients are defined as

$$K_V = k\,u_*\,(z-d)\,{\phi_V}^{-1} \tag{13.13}$$

$$K_H = k\,u_*\,(z-d)\,{\phi_H}^{-1} \tag{13.14}$$

$$K_M = k\,u_*\,(z-d)\,{\phi_M}^{-1} \tag{13.15}$$

The functions ϕ_V, ϕ_H and ϕ_M are empirical: the best available data (Thom, 1975) suggests

$$\phi_V = \phi_H = \phi_M = (1 - 5R_i)^{-1} \qquad (R_i \text{ positive}) \tag{13.16}$$

$$\phi_V = \phi_H = \phi_M = (1 - 16R_i)^{-1/2} \qquad (R_i \text{ negative}) \tag{13.17}$$

Resistance analogues

It is a useful simplification to apply the diffusion equations (13.5 to 13.17) in integrated form, and this is very straightforward when the flux can be treated as constant. Eq. 13.5 for example, integrated vertically through a layer of constant flux between two points z_1 and z_2, where the vapour pressures are e_1 and e_2, gives:

$$\lambda E = -\frac{\rho\, c_p}{\gamma} \frac{e_2 - e_1}{\int_{z_1}^{z_2} {D_V}^{-1}\,dz} \tag{13.18}$$

This has an obvious similarity to Ohm's Law for electrical current, and it is common to use the name *resistance* for $\left[\int_{z_1}^{z_2} {D_V}^{-1}\,dz\right]$, and for similar entities involving D_H, D_M, K_V, K_H and K_M.

ATMOSPHERIC RESISTANCE: Assuming vertical transfer within and above a vegetation canopy is adequately represented by turbulent diffusion, Eqs. 13.8, 13.9 and 13.10, the equivalent resistance between any two levels for momentum $r_M^{1,2}$, sensible heat $r_H^{1,2}$ and latent heat $r_V^{1,2}$ transfer follows directly, thus

$$r_{M,H,V}^{1,2} = \left[\int_{z_1}^{z_2} K_{M,H,V}(z)\, dz \right]^{-1} \tag{13.19}$$

These are the relevant definitions in a multilayer simulation model of the vegetation-atmosphere interaction. A common requirement is the bulk transfer resistance between an effective source level in the canopy and z_R, the level above at which meteorological variables are measured. The value for momentum transfer in neutral conditions, r_M, is given (Monteith, 1973) by

$$r_{aM} = \left[\ln \frac{z-d}{z_0} \right]^2 / \left[k^2\, u(z_R) \right] \tag{13.20}$$

The bulk transfer resistance for the transfer of sensible heat, r_{aH}, and water vapour, r_{aV}, exceed that for momentum (see below). In neutral conditions, a reasonable estimate is given (Brutsaert, 1982; Thom, 1975) by

$$r_{aH} = r_{aV} = \left[\ln \frac{z-d}{z_0} \right] \left[\ln \frac{5\,(z-d)}{z_0} \right] / \left[k^2\, u(z_R) \right] \tag{13.21}$$

Most present day SVATS use the atmospheric resistance concepts adopted above, which assume *K-theory*, i.e. that Eqs. 13.8, 13.9 and 13.10 provide a valid description of vertical turbulent diffusion. There is now evidence that this is not the case within and just above vegetation canopies (Garratt, 1978; Raupach, 1979; Denmead & Bradley, 1985; Baldocchi and Hutchison, 1987; Raupach, 1989). In practice this seems to have limited consequence on the modelling of whole-canopy exchange. This is partly because other resistances, either at the surface or between the canopy and the atmosphere, tend to dominate; and partly because the consequences of poorly described turbulent transfer on dry canopies is usually in the opposite sense to those on wet canopies (Shuttleworth, 1989).

BOUNDARY LAYER RESISTANCE: Moving air within a vegetation canopy interacts with individual elements, and energy is exchanged through a boundary layer resistance which might be envisaged as the result of molecular diffusion through a layer of laminar flow close to the element. The value, which changes with windspeed and is therefore a function of height in the canopy (Jones, 1984), is approximated by:

$$r_{bH}(z) \simeq r_{bV} \approx 100\ [l_w/u(z)]^{1/2} \tag{13.22}$$

where l_w is a representative leaf width. Integrating this through the height of the canopy (Choudhury and Monteith, 1988) gives the mean boundary layer aerodynamic resistance, r_a^c, often assumed the same for vapour and sensible heat, thus

$$r_a^c \simeq (A/L)\ u_h^{-1/2} \tag{13.23}$$

where u_h is the wind speed at the top of the canopy, L is the total leaf area index. A is a function which is proportional to $w^{1/2}$, but a typical value is $25\, s\, m^{-1}$. Momentum interaction with individual elements is enhanced by the effect of pressure forces, and the boundary layer resistance for momentum, r_{aM}^c, is significantly less than for that for other properties. An approximate relationship (Stewart and Thom, 1973; Thom, 1972) is:

$$r_a^c = r_{aM}^c + 6\, u_*^{-2/3} \tag{13.24}$$

SURFACE RESISTANCE: Sensible heat and momentum can realistically be considered to interact directly with the surface of the individual components in a vegetation canopy, or with the soil surface beneath. However, if there is no surface water on crop or soil, the latent heat flux arises in the first instance as a result of evaporation either from cell walls inside the stomatal cavities within vegetation, or from the surface of soil particles below the soil surface. In each case the water vapour reaches the surface by molecular diffusion through restricted apertures. This additional diffusion can be represented by a resistance, which is not relevant to sensible heat and momentum transfer. Fig. 13.2

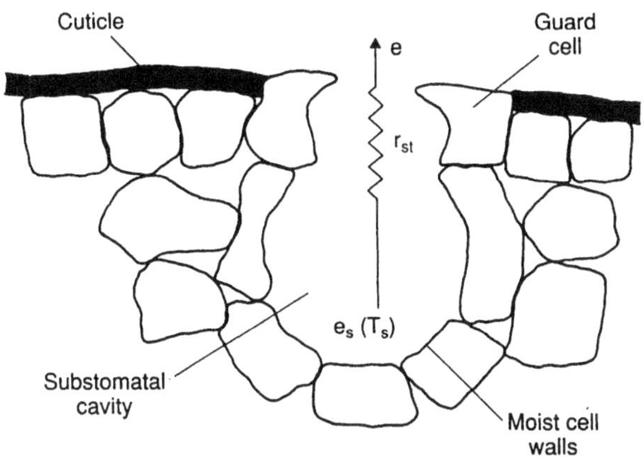

Figure 13.2: *Transpiration by molecular diffusion of water vapour through the stomatal aperture of dry leaves.*

illustrates the stomatal diffusion process. A typical leaf is usually considered small and thin enough for internal temperature gradients to be ignored, and the air inside the stomatal cavity is always close to saturation at leaf temperature. With these simplifications transpiration can be described by the equation

$$\lambda E = \frac{\rho\, c_p}{\gamma}\, \frac{e_s(T) - e}{r_{st}} \tag{13.25}$$

where r_{st} is the stomatal resistance per unit leaf area and T and e are here the temperature and vapour pressure immediately adjacent to the leaf. It is increasingly common to measure r_{st} for different vegetation, usually as r_{ST}, the effective average value for the whole canopy, and to derive empirical functions describing its variation in terms of environmental variables (e.g. Shuttleworth, 1989). In all SVAT models, it is important to include an increase in $r - ST$ as the water in the layer of soil to which the roots have access is reduced, since this is required by basic mass conservation laws. If the substrate beneath the canopy is a free water surface, such as in a paddy field or with wet soil, the substrate surface resistance, r_s^s, is zero. In more normal conditions, the optimum parameterisation of the surface resistance of soil substrate is not yet well defined. Two approaches have been attempted. The first is to define a simple empirical function of near surface soil moisture, an example (Shu Fen Sun, 1982) being

$$r_s^s = 30 + 3.5\, \theta_{w,s}^{-2.3}\, (s\, m^{-1}) \tag{13.26}$$

where $\theta_{w,s}$ is the fraction of water in the top $0.02\, m$ of soil. The alternative approach is to make a simple representation of the physics at the soil surface, e.g. to assume evaporation is occurring from wet soil below a progressively deepening dry layer which is assumed isothermal (Monteith, 1981). This last assumption provides qualitative agreement with measurements from drying soil.

The evaporation process (at plot scale)

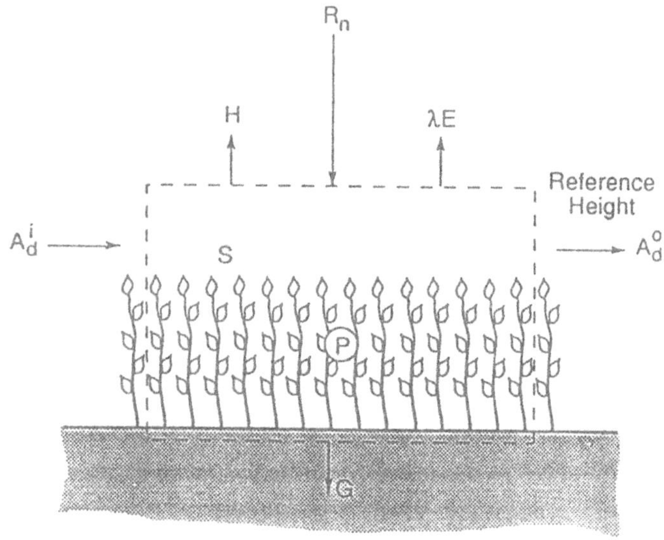

Figure 13.3: *The components of the energy balance for a volume extending from just below the soil surface to the level above the canopy (or bare soil) at which the net radiation balance is determined.*

THE ENERGY BUDGET: Describing the evaporation process for a stand of uniform vegetation, it is usual to draw up an energy budget for volume of defined vertical extent and unit area in the horizontal plane. In the case of the whole crop this volume extends from just below the soil surface, at which level energy lost by heat conduction to the soil is measured or modelled, to that level above the canopy at which the radiation balance between ground and atmosphere is calculated. Figure 13.3 illustrates the components of the whole canopy energy balance, which, in approximate order of their magnitude, comprise:

R_n, net incoming radiant energy;

λE, outgoing energy as evaporation;

H, outgoing sensible heat flux;

G, outgoing heat conduction into the soil;

S, energy temporarily stored within the volume and often neglected except for forests, being proportional to temperature changes in the vegetation, air and shallow soil layer; and to changes in atmospheric humidity;

P, energy absorbed by biochemical processes in the plants, typically just 2% of net radiation;

A_d, the loss of energy associated with near-surface horizontal air movement (Thom, 1975), significant in an *oasis* situation, but generally neglected otherwise ($A_d = A_d^o - A_d^i$ in Figure 13.3).

In the case of more complex, multilayer models of the surface/atmosphere interaction, the canopy is divided into a vertical stack of shallow layers and a separate energy budget drawn up for each. In general there are energy components equivalent to those itemized above, except that layers above the ground have evaporation and sensible heat fluxes entering from below as well as leaving from above. The lowest layer is distinctive since it has heat conduction to the soil.

It is convenient to define an entity A, the available energy, which is the energy available for partition either as latent or sensible heat. For the case of a complete canopy

$$A = \lambda E + H \tag{13.27}$$

and

$$A = R_n - G - S - P - A_d \tag{13.28}$$

Sometimes it is convenient to draw up a separate energy budget at the surface of the substrate (soil) beneath a vegetation canopy. In this case the energy available, A_s, is given in terms of the net radiation reaching the ground R_n^s thus

$$A_s = R_n^s - G \tag{13.29}$$

RESISTANCE NETWORKS: The network of resistances involved in the partitioning of available energy into sensible heat and evaporation is complex for real stands of vegetation, but successful computer models have been written which describe the process in one dimension (Waggoner and Reifsnyder, 1968; Sinclair et al, 1976; Waggoner, 1975). Such models involve the simultaneous, iterative solution of energy balance equations at several heights in the canopy, and require estimates of the resistances operating at each level, i.e. the surface and boundary layer resistances, as well as the effective eddy diffusion resistances operating between levels. Figure 13.4 illustrates the network of resistances commonly used. Typically the model starts by calculating the available energy relevant to each height, and uses the assumed resistance chain to calculate canopy profiles of temperature and humidity (Waggoner and Reifsnyder, 1968). The input data required precludes the predictive use of such models at the global-scale; however, for selected vegetation covers, they may provide a yardstick against which to test the adequacy of simpler models more suitable to predictive application under GEWEX.

In practice it is possible to provide a useful and physically realistic description of the evaporation process without having to resort to the use of a multilayer simulation, though it is still sensible to retain separate description of crop and soil in order to preserve generality and allow description of sparse crops.

Figure 13.5 illustrates a single source model which includes soil evaporation: it has features typical of many SVATS in common use. If D is the vapour pressure deficit above the crop and r_a^a and r_a^s are the aerodynamic transfer resistance shown, the equation (Shuttleworth & Wallace, 1984) describing total evaporation for this network takes the form:

$$\lambda E = C_c PM_c + C_s PM_s \tag{13.30}$$

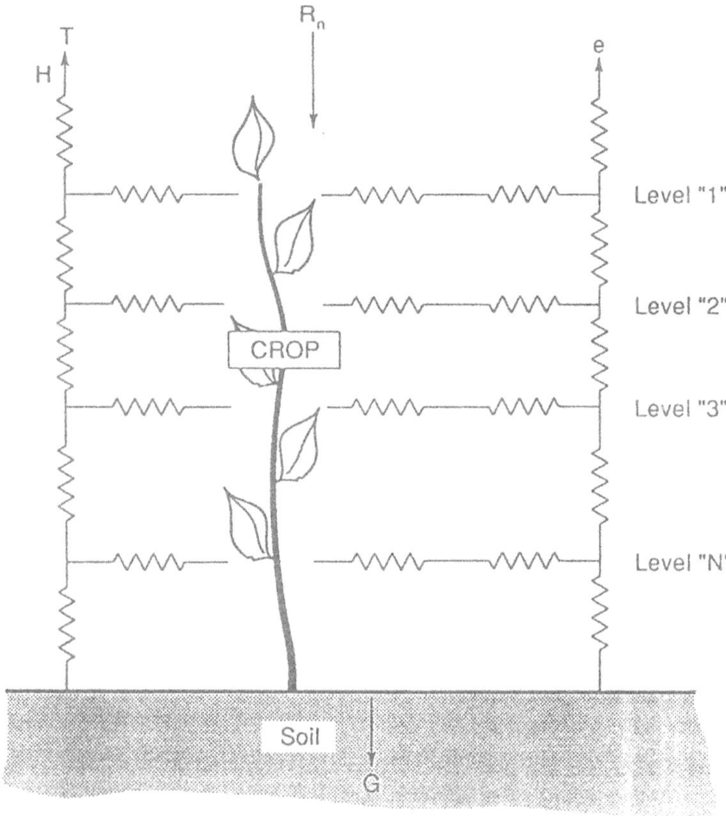

Figure 13.4: *The complex network for resistances used in a typical multilayer, numerical model of a fully developed canopy.*

where PM_c and PM_s have the form

$$PM_c = \frac{\Delta A + \frac{\rho c_p D - \Delta r_a^s A_s}{r_a^a + r_a^c}}{\Delta + \gamma\left[1 + \frac{r_s^c}{r_a^a + r_a^c}\right]} \tag{13.31}$$

$$PM_s = \frac{\Delta A + \frac{\rho c_p D - \Delta r_a^s (A - A_s)}{r_a^a + r_a^s}}{\Delta + \gamma\left[1 + \frac{r_s^s}{r_a^a + r_a^s}\right]} \tag{13.32}$$

and the coefficients C_c and C_s are given by the expressions

$$C_c = \left[1 + \frac{R_c R_a}{R_s (R_c + R_a)}\right]^{-1} \tag{13.33}$$

$$C_s = \left[1 + \frac{R_s R_a}{R_c (R_s + R_a)}\right]^{-1} \tag{13.34}$$

where
$$R_a = (\Delta + \gamma) r_a^a \tag{13.35}$$
$$R_c = (\Delta + \gamma) r_a^c + \gamma r_s^c$$
$$R_s = (\Delta + \gamma) r_a^s + \gamma r_s^s$$

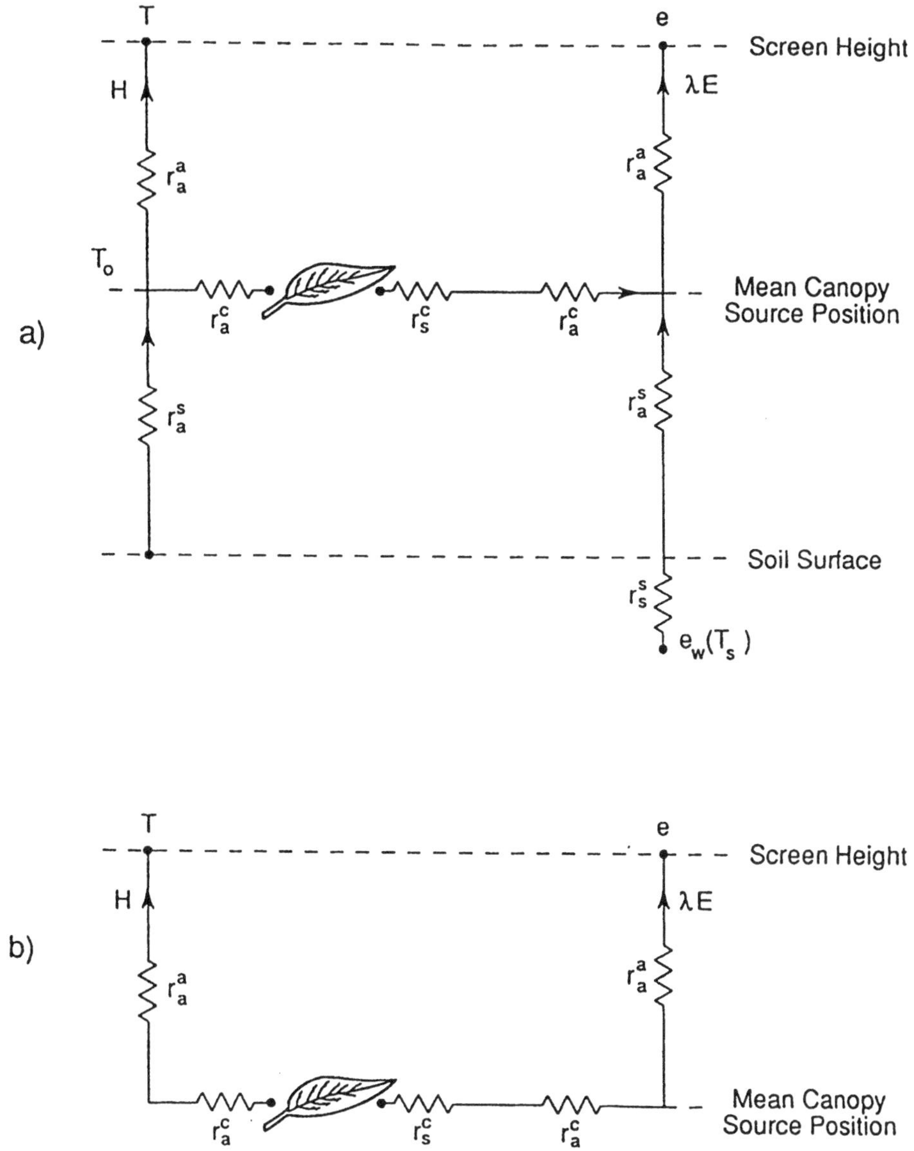

Figure 13.5: *(a) The resistance network which represents the vegetation atmosphere interaction as a single source in the canopy with a substrate interaction below.*
(b) The limit of this model in which soil interaction is neglected, appropriate for well developed, closed canopies.

Formulae for the three aerodynamic resistances, r_a^a, r_a^c and r_a^s, the canopy surface resistance, r_s^c, in Eqs. 13.28 to 13.32 must be made in terms of the extent of canopy cover in the general case see Shuttleworth & Wallace (1984) and Shuttleworth & Gurney (1989).

For the special case of dense, closed canopies, when most of the available energy is partitioned by the vegetation and the substrate interaction is ignored, the resistance network simplifies to Figure 13.5(b), and total evaporation is given by

$$\lambda E = \frac{\Delta A + \frac{\rho c_p D}{r_a^a + r_a^c}}{\Delta + \gamma \left[1 + \frac{r_s^c}{r_a^a + r_a^c}\right]} \tag{13.36}$$

The combined resistance $r_a^a + r_a^c$ is then set equal to the bulk aerodynamic transfer resistance, r_a, given by Eq. 13.19; and r_c^c is set equal to r_s, the bulk surface resistance of the canopy.

Some very simple SVATS ignore the presence of vegetation altogether, and the existence of stomatal control in particular. In this case the evaporation is sometimes referred to as occurring at *potential evaporation* rate, λE, and is estimated by setting $r_s = 0$, to give

$$\lambda E_p = \frac{\Delta A + \frac{\rho c_p D}{r_a}}{\Delta + \gamma} \tag{13.37}$$

This evaporation rate is moderated in conditions of reducing near-surface soil moisture by introducing a moisture-dependent factor β, thus

$$\lambda E = \beta \, \lambda E_p \tag{13.38}$$

The factor β is one when soil water is freshly available, and zero when the soil is completely dry.

Though not universally recognised by meteorologists, the definition of potential evaporation rate given in Eq. 13.37 is *not that used by hydrologists* (e.g. Penman, 1948) and will, in general, provide estimates of evaporation which are too high. In hydrological applications, the unrealistic absence of any allowance for stomatal control in this equation would normally be compensated for by assuming an unrealistically low aerodynamic roughness in r_a (see Thom & Oliver, 1976). A more realistic alternative equation, which yields reasonable estimates of potential evaporation and which is used in some GCMs, is to assume a fixed value of $r_s = 60 sm^{-1}$. Eq. 13.36 then becomes

$$\lambda E_p = \frac{\Delta A + \frac{\rho c_p D}{r_a}}{\Delta + \gamma \left[1 + \frac{60}{r_a}\right]} \tag{13.39}$$

Moisture stores

The size of the several resistances in the above-ground resistance networks overviewed in the previous section *both control, and are controlled by*, the water held in stores in the soil and on the vegetation itself. Both of these stores are filled by precipitation when this occurs, and are depleted by drainage and by evaporation at all times. In SVATS, the amount of water in each store is a so-called *prognostic variable*, its value being updated at each time-step through a continuity equation as the model proceeds.

The vertical movement of liquid water through both the soil and the vegetation canopy can be represented in detail in models by breaking each into many levels, analogous to Figure 13.4, and drawing up a continuity equation for each level. This is particularly relevant if simulation of details of the moisture profiles are of interest. However, in the context of GEWEX, where large-scale applicability is more important than single site detail, representation with a limited number of stores is required, perhaps up to two canopy stores and three or four soil stores.

SOIL MOISTURE STORES: Change in the water content of an individual layer of soil is determined by the balance between drainage into and out of the layer and, where appropriate, the abstraction of water by roots for transpiration. The uppermost layer is unique in that there can be direct infiltration of precipitation, and direct evaporation from the soil surface. Models of the soil-vegetation interface will partition the transpiration loss between layers following an assumed root distribution, but will also include the effect of reducing soil water availability on the subsequent transpiration rate. SVATS which include a vegetation canopy simulate nature by increasing the stomatal resistance of the foliage as soil water reduces, while simpler models merely introduce a soil water related factor, *c.f.* Eq. 13.38.

The soil in the soil-vegetation interface is only occasionally saturated, and vertical movement between soil layers is represented by (the digital equivalent) of the Darcy equation, *viz.*

$$v = -K(\theta)\,\frac{\delta(\psi - z)}{\delta z} \tag{13.40}$$

where v is the macroscopic velocity of water, K is the hydraulic conductivity (which varies with the water content, θ), and ψ is the matrix potential. The minus sign indicates that water movement is in the direction of decreasing potential, while the total potential $(\psi - z)$ includes the effect of gravity. Application of Eq. 13.40 is complicated by the fact that K, ψ and θ are interelated in unsaturated soils, see Fig. 13.6, and that the form of the relationships varies greatly, see Stricker (1991) in this volume for greater detail. In global-scale modelling, it is conventional to assume representative forms for these curves for a restricted number of soil texture classes, typically ten. The classes parameterized by Clapp & Hornberger (1978) are commonly used for this.

CANOPY MOISTURE STORES: The presence of surface water on a leaf effectively *short circuits* the control exerted by the stomata on water vapour flow. When the whole canopy is wet during or immediately after a sufficiently heavy storm, the evaporation is that given by setting the stomatal resistance to zero i.e. setting $r_s^c = 0$ in Eq. 13.31 or 13.36 for example. If the canopy is only partly wet, it is common practice to assume that the rate of evaporation of liquid water from that held on the foliage and stems is reduced in proportion to the fractional fill of a prescribed maximum storage capacity.

The continuity equation for the canopy water storage is given by a balance between the amount of incoming precipitation intercepted (dependent on fractional foliage cover), that evaporated (calculated as above), and drainage from the vegetation to the soil beneath. The rate of such drainage is also related to the fractional fill of the canopy store (e.g., Rutter et al, 1971), though many models simply assume it is zero if the store is incompletely filled, and equal to the intercepted precipitation otherwise, and this is often adequate.

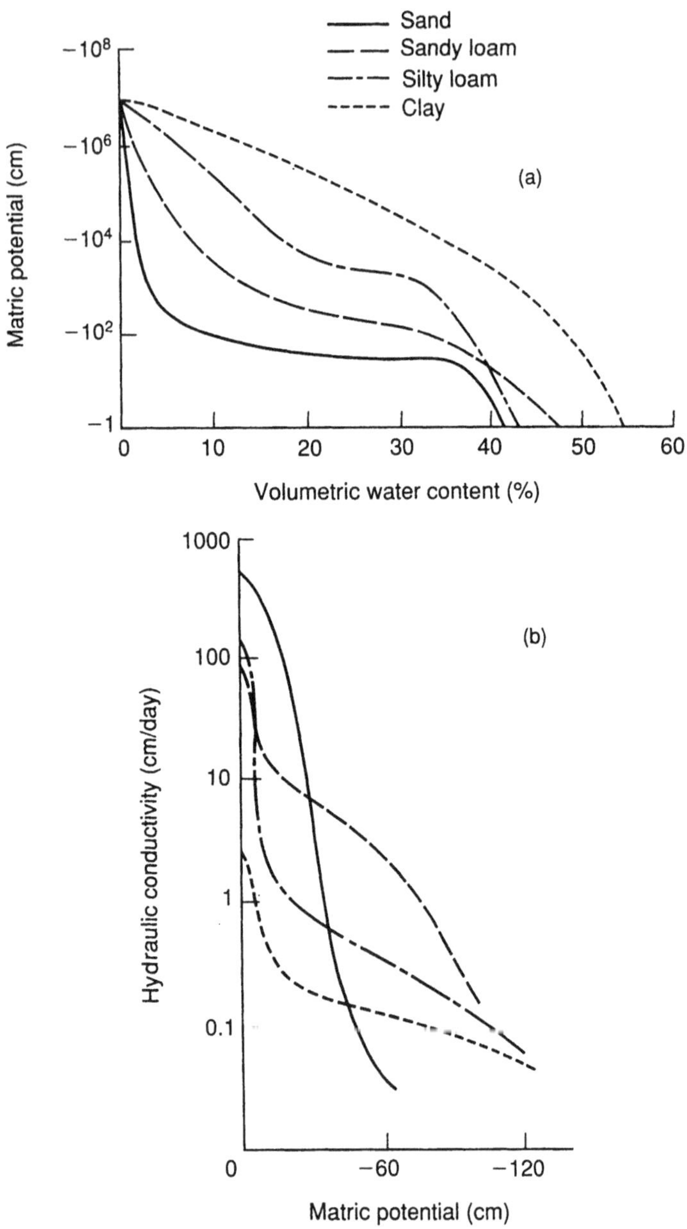

Figure 13.6: *Typical relationships between*
(a) matric potential and volumetric water content, and
(b) hydraulic conductivity and matric potential, for a range of soil types.

13.2.2 Present day Soil-Vegetation-Atmosphere Transfer Schemes

In a recent review of Soil-Vegetation-Atmosphere Transfer Schemes commissioned by the UK's Terrestrial Initiative in Global Environmental Research (TIGER), Geyer & Jarvis (1991) list 57 different examples, with an enormous range of complexity and detail. SVATS have been created for diverse purposes, to seek representation with emphasis on different processes, and at different spatial and temporal scales. Some have up to 20 soil layers described; some have up to 21 canopy layers; while some even include up to 19 atmospheric layers above the canopy. Clearly a great many of these models are not appropriate to the global scale objectives of GEWEX.

There has also been a broad range of complexity in the SVATS actually used in GCMs over the last decade. They extend from simple so-called *bucket models* (e.g. Manabe, 1969), to rather complex though still one-dimensional schemes, in which canopy and soil processes are more realistically represented (e.g. Dickinson et al, 1986; Sellers et al, 1986), along with other schemes of intermediate complexity (e.g. Warrilow, 1985; Noilhan & Planton, 1989). These models seek to provide the global-scale representation of surface processes relevant to GEWEX and therefore merit further discussion here.

Simple Bucket Models

In the absence of vegetation, land surface energy and momentum interactions can be reasonably described by an albedo, an emmissivity, a roughness parameter, and a soil thermal conductivity. These parameters control the absorption of solar radiation, the emission of thermal radiation, the transfer of momentum, and the conduction of heat into the ground. As a first approximation, the emissivity of soil surfaces is high (0.9) and varies little. Soil albedo, however, usually shows a strong dependency on soil moisture.

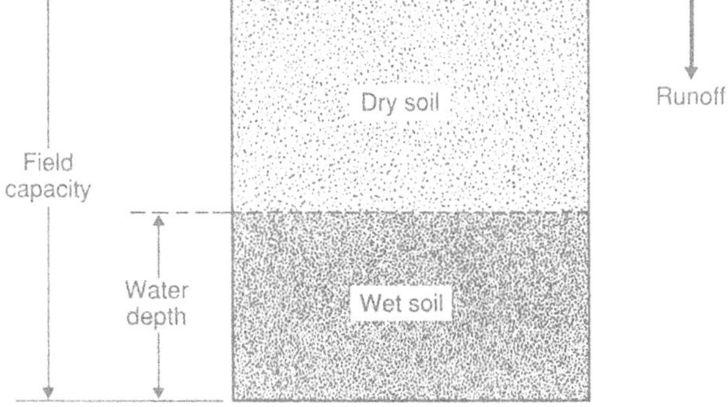

Figure 13.7: *Schematic diagram of a simple* bucket model *of land surface hydrology.*

To partition the energy available at the surface, SVATS must also include at least a rudimentary representation of the surface hydrology. The first numerical scheme to parameterize the hydrological cycle for atmospheric models was introduced by Manabe (1969). His simple concept, known as the *bucket model*, is illustrated in Figure 13.7. The soil had a field capacity of 15 *cm*. The bucket is filled with water when precipitation exceeds evaporation, and after the bucket becomes full, it overflows as runoff. Evaporation occurs at its potential rate as estimated from Eq. 13.37 when the soil is at or close to saturation. When the soil moisture drops below field capacity, the evaporation is proportional to potential evaporation, the proportionality factor, β in Eq. 13.38, being set equal to the ratio of the current soil moisture to field capacity.

Simple models of this type have limited utility in the context of global (and climate) *change prediction* at regional scale, but have continuing relevance in short-term numerical forecasting. Their utility in the context of GEWEX is strongly determined by the availability of relevant calibration data. In a few restricted situations, and in the Mississippi basin in particular (see Section 13.3.2), GEWEX will provide data on surface exchanges (in fulfilment of its first objective) which is extraordinarily rich and extensively available. In this case it will be possible to *fine tune* quite simple SVATS to represent the *status quo* surface interaction very well for weather (as opposed to climate) forecasting purposes. However, it is hard to believe that it will be possible to achieve similar observations worldwide, or to carry out credible long-term *change prediction* without the use of SVATS which provide a more realistic representation. This is because these more difficult tasks will require extrapolation of the calibration of SVATS from place to place, and the inclusion of time dependent change in surface cover.

Complex Soil-Vegetation-Atmosphere Transfer Schemes

Of the SVATS which claim global applicability, and which have been used in GCMs, the two best recognised as providing the most complex, and in this sense realistic description of the land surface interaction are the Biosphere-Atmosphere Transfer Scheme (BATS; Dickinson et al, 1986) and the Simple Biosphere (SiB) Model (Sellers et al, 1986). Fig. 13.8 shows the schematic diagrams normally reproduced in papers on these two models, and overviews their general format and the processes they include.

In many respects the BATS and the SiB model are similar and both require essentially the same surface weather inputs, and provide the same surface flux outputs. SiB is more complex and, in principle, requires more parametric specification. It should be noticed, however, that a great deal of the apparent complexity of the SiB model is only relevant to a pre-processing phase in operation, which is used to derive a reduced number of *whole-canopy* parameters through synthesis from more basic understanding. These less numerous derived parameters are then used in the operational phase.

Both models include a vegetation overstorey (in the case of SiB including both dead and live leaves), which is assumed to be a physiologically uniform, single vegetation type over flat terrain. This canopy is represented by a *big leaf* model in one dimension, with atmospheric turbulence treated through a first order closure (*K-theory*) model. The two models have a similar *two-stream* approximation sub-model for radiation, in which the solar spectrum is partitioned at about 0.7 μm into visible and near-infrared components with diffuse (isotropic) and direct beam contributions, see Section 13.2.1. Radiative transfer calculations are performed for these four components. A similar calculation is

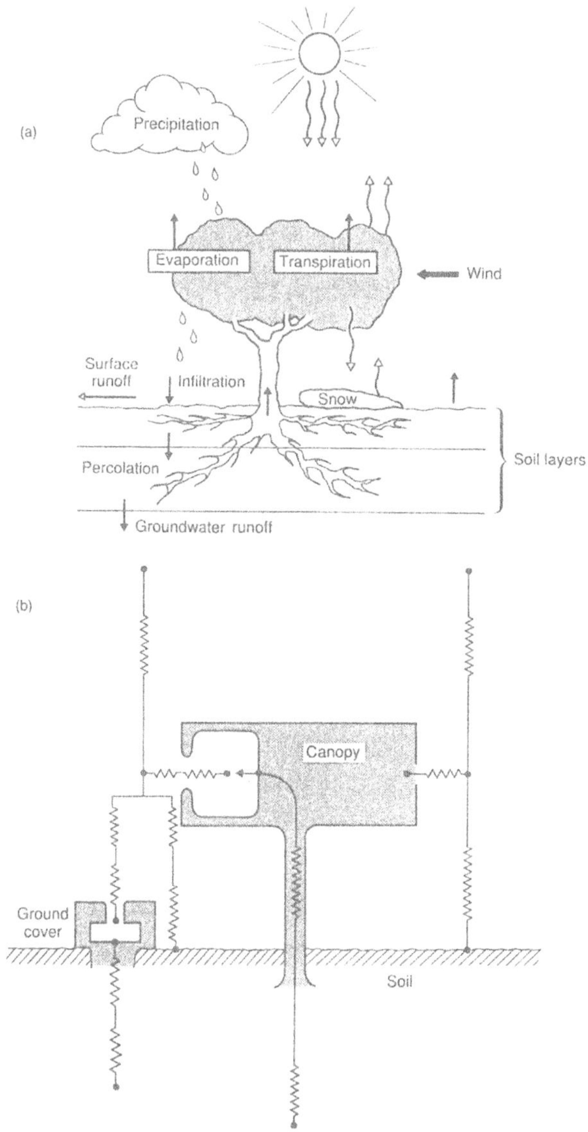

Figure 13.8: *Schematic diagrams of the physical and physiological processes represented in (a) the Biosphere Atmosphere Transfer Scheme [BATS] (Dickinson et al, 1986), and (b) the Simple Biosphere [SiB] Model (Sellers et al, 1986).*

performed to describe the interception of the down-welling longwave radiation in which the radiation flux is assumed to be all diffuse, and emissivities are set to unity.

At first sight, the treatment of momentum exchange is different in the two models, in that SiB synthesizes estimates of the relevant whole-canopy parameters (e.g. roughness length and zero plane displacement) from more basic leaf and canopy parameters during its preprocessing phase. The treatment of the canopy's stomatal response also differs in detail, though both models include allowance for parameterization of a response to light, temperature, vapour pressure deficit, leaf water potential and the moisture content of the soil around the roots. The interception of precipitation by the overstorey is treated in a broadly similar way in both models - with a simple bucket model - but SiB makes some allowance for the effect of sub-grid scale storm size. Both models treat heat and moisture movement into and through the soil with just a few soil layers. In the case of SiB the description of the thin uppermost layer of soil is assumed to include the effect of a short ground cover vegetation.

These two high level models have been adopted and used with considerable success by groups other than those involved in their formulation, and have been successfully incorporated into GCMs, with a significant but acceptable increase in execution time, typically 20% (Sato et al, 1989). Both give improved climate simulation in operation (Dickinson & Henderson-Sellers, 1988; Sato et al, 1989), and both can be provided with (assumed) parameter sets, applicable globally for a restricted range of land cover classes (18 types for the BATS and 12 types for the SiB Model). However, the applicability of these parameters is only established for certain restricted biomes at this time. Neither model currently allows mixtures of the prescribed land covers within individual grid squares.

Intermediate Soil-Vegetation-Atmosphere Transfer Schemes

Not all GCM groups have yet felt the need to incorporate SVATS with the complexity of the BATS or SiB Model, but accept the need for some realism in the description of the land surface they adopt. The SVATS used by the UK Meteorological Office (e.g. Warrilow, 1985), hereafter called UKMO, and the French Centre National de Recherches Meteorologiques (e.g. Noilhan & Planton, 1989), hereafter called CNRM, are useful intermediate examples erring towards simplicity and complexity respectively.

Unlike the BATS and the SiB Model, these intermediate models do not differentiate the solar spectrum above and below $0.72\,\mu m$, and prescribe a fixed albedo to different vegetation types, irrespective of wavelength and time of day. They do, however, have similar (simple) descriptions of heat and water movement in the soil as the BATS and the SiB Model, and also have similar simple descriptions of the canopy interception process. They ascribe a roughness length to different vegetation cover classes, rather than calculating this as in the case of SiB.

One of the main differences between the UKMO SVAT and the CNRM SVAT is the level of complexity employed to describe stomatal control. The UKMO SVAT is the simplest possible in this respect, since it adopts a fixed value of r_s for each vegetation cover to produce a series of equations similar to Eq. 13.39. It (currently) makes no parameterization of the effect of environmental variables on r_s, and no direct allowance for the effect of soil moisture, which it prefers to describe using Eq. 13.38. The CNRM SVAT on the other hand, has a parameterization of r_s which approaches the complexity of that used in the BATS and the SiB Models, and is indeed a hybrid of the two, but with its own representation of the effect of soil water stress.

Despite its plot-scale simplicity, the UKMO SVAT Model has two valuable grid-scale features which distinguish it from many other SVATS, including the complex SVATS described above, and these have considerable merit. This model includes a simple parameterization for the effect of finite convective rainstorm size on the rate of fill of *both* the soil and canopy stores (Warrilow, 1986; Shuttleworth, 1988; Dolman & Gregory, 1992). This can have a significant effect in certain climates. The second outstanding feature in the UKMO SVAT is the provision within the model to include more than one vegetation types at each individual grid square. The value of the (few) vegetation-related parameters at each grid point (e.g. surface resistance) is calculated from very simple aggregation rules, some similar to those tabulated later, see Section 13.2.4. The validity of such simple aggregation procedures is of course not established, and they could well prove over-simple. Nonetheless, in the context of GEWEX, the inclusion of this feature in SVAT schemes is arguably as important as plot-scale detail. It will prove essential as an extrapolation mechanism from those areas of mixed vegetation cover calibrated by GEWEX observations, to other areas of mixed cover where no such calibration is made, see Section 13.5.

13.2.3 Past observational studies

Recently Shuttleworth (1991a, 1991b) has reviewed progress in observational studies carried out during the 1980s. This section provides a summary of these reviews, and the insight they give into the value and reliability of field systems, and into the provision of area-average, or *aggregate*, fluxes and model parameters.

The experiments

Table 13.1 and Figure 13.9 together summarize most of the recent climate-related experiments. Although some are independent, most have been initiatied under international programmes. In the recent past there has been some distinction between the community responsible for the Hydrological Atmospheric Pilot EXperiment (HAPEX), a group largely co-ordinated through the World Climate Research Programme with its priority towards calibrating the hydrology in GCMs, and the International Satellite Land-Surface Climatology Project (ISLSCP), which is more strongly linked to space research agencies (especially NASA) and has priority towards calibrating satellite observation systems. In the course of the next decade this distinction is likely to soften with experiments mainly concerned with the energy and water interaction falling under the umbrella of GEWEX. However, a new distinction may emerge with the development of a more significant experimental component addressing biological interactions and trace gas transfer, fostered under the BAHC. It is important that any rivalry between GEWEX and BAHC does not negatively impact the science, and that experimental and modelling activity is carried out jointly wherever possible.

Experimental systems: Surface interface measurements

SINGLE SITE STUDIES: Table 13.1 includes several patch-scale studies with observations either at a single site, or at several separated sites simultaneously, most of these having been carried out under the ISLSCP programme. Although useful as exploratory exercises for satellite calibration, the direct application of such small-scale activity to GEWEX is problematic. However, patch-scale studies are

LOCATION	EXPERIMENT	AFFILIATION	TIMING OF ACTIVITY	SCALE	DURATION
1	Amazon Region Micrometeorology Experiment (ARME)	Independent	1983-1985	Single Site	2 years, 4 long IFCs
2	Hydrological Atmospheric Pilot Experiment (HAPEX-MOBILHY)	HAPEX	1986	100 km x 100 km	1 year, 1 long IFC
3	First ISLSCP Field Experiment (FIFE)	ISLSCP	1987, 1989	15 km x 15 km	2 seasons, 5 short IFCs
4	'La Crau' Experiment	ISLSCP	1987	Multisite	1 season, 1 short IFC
5	Longitudinal land-surface Transverce Experiment (LOTREX-HIBE88)	ISLSCP	1988	10 km x 10 km	1 season, 1 long IFC
6	Joint Greenland Field Experiment	ISLSCP	1988	Multisite	1 season, 1 long IFC
7	'Niger' Experiment	ISLSCP	1988	Multisite	1 season, 1 short IFC
8	Sahelian Energy Balance Experiment (SEBEX)	Independent	1988-1990	Multisite	3 seasons, 3 long IFCs
9	'Botswana' Experiment	ISLSCP	1989	Multisite	1 season, 1 long IFC
10	Heihe River Experiment	HAPEX	1989	70 km x 90 km	1 year, 4 short IFCs
11	'Sudan' Experiment	ISLSCP	1989	Multisite	1 season, 1 short IFC

Table 13.1: *Summary of past climate prediction-related observational studies*

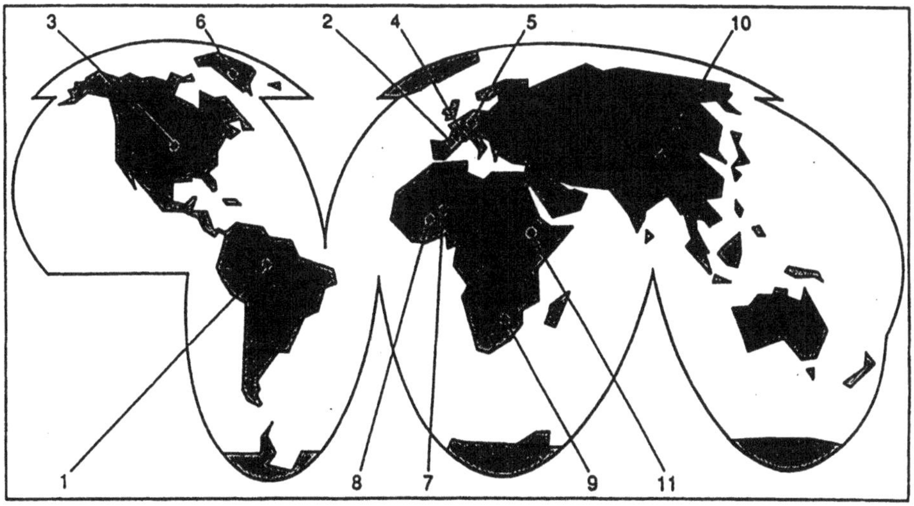

Figure 13.9: *The location of the past climate prediction-related observational studies. The numbers here are those of the experiments listed in the previous table.*

relevant if carried out over land cover which can realistically be considered representative, or at least typical, of one of the extensive, uniform biomes, and particularly so if continuously collected over a long period.

The results from the Amazon Region Micrometeorological Experiment (ARME) are a notable example of effective patch-scale study. Data from this experiment have now successfully calibrated or validated the representation of tropical rainforest in the more realistic SVATS presently used in GCMs (e.g. Sellers et al., 1989; Dickinson & Henderson-Sellers, 1988). Model predictions of the consequences of Amazonian deforestation can now at least have a reasonably realistic description of the atmospheric interaction of the forest prior to clearance.

Of late a subtly different application for these patch-scale data has become apparent, which may well be relevant to GEWEX. It is now clear that in the middle of biomes, single site data can not only be valuable for *calibrations* but also provide a valuable *diagnostic* of GCM performance (Dickinson, 1989a; Shuttleworth & Dickinson, 1989; Dickinson, 1989b). In these locations there is some likelihood that the monthly-average values of weather variables and surface exchange fluxes measured at a point are broadly representative of the monthly-average values for a GCM grid box centred on that point.

LARGE SCALE STUDIES: the data acquisition hardware used in the larger scale experiments which are designed to provide *area-average* calibration of SVATS is extensive and diverse. The fundamental requirement is for simultaneous hourly measurements of near-surface weather variables and the resulting outgoing surface energy fluxes, which are suitable average values over the (say) $10\,km \times 10\,km$ or $100\,km \times 100\,km$ study area, but which are routinely available over long periods, optimally for a season or, better, a year.

No single measurement system can provide data with these characteristics. The hardware systems used in patch-scale studies can provide such data for samples of the vegetation cover in the study area and are therefore an essential component of all area-average studies, but resource limitations necessarily restrict the number of samples which can be taken. Hydrological catchment data (coupled to hydrological models) can, in principle, provide area-average water balance, but the data does not have the necessary time resolution and is limited by the need to estimate the water stored in the soil. Aircraft measurements of atmospheric fluxes should be able to provide occasional *snap shots* of the area-average fluxes, and are frequently used in these studies for this purpose. In addition, speculative techniques to synthesize area-average surface fluxes indirectly from remote sensing measurements or from boundary layer soundings have been investigated. The background philosophy is to provide as wide a range of measurements as possible with at least some of the required characteristics, and to use these to calibrate a mesoscale model of the study area which can then provide the required area-average synthesis of the data actually required.

The network of climatological stations and micrometeorological towers deployed in these large-scale experiments most nearly provides the continuous, hourly-sampled measurements required for the calibration of area-average SVATS. The accuracy and precision of these particular data are therefore important. In the course of past experiments many instrumental systems, some of different manufacture, some based on different measuring princples, have been simultaneously operated in reasonably close proximity, and sometimes at the same site, and this has proved revealing as to the accuracy of the measurements they provide.

Intercomparison and intercalibration studies during FIFE (Field and Fritschen, 1991; Smith et al, 1991) and during HAPEX (Goutorbe, 1991) showed significant differences between net radiometers, typically of the order of 10%, particularly between those of different design. Since surface net radiation is such a critical measurement, the careful selection and intercalibration of the net radiometers is a prerequisite to future observational studies under GEWEX.

The large-scale experiments have so far used three different techniques to make micrometeorological measurements of the energy leaving the surface as latent and sensible heat. The eddy correlation technique worked well in providing the source data for the forested portion of the HAPEX-MOBILHY study area (Gash et al, 1989), and over level sites in the FIFE (Smith et al, 1989). The aerodynamic method, in the form of the so-called SAMER instrument (Itier, 1989; 1981), was used over the other sampled vegetation types in HAPEX-MOBILHY. It provided consistent measurements with minimum supervision, and a long-term comparision with co-located soil water balance measurements gave agreement, better than 10% at the monthly time scale (André et al, 1989). Subsequent station intercomparison at a single site (Goutorbe, 1991) suggests less consistent measurement over shorter time scales. However, Shuttleworth (1990) has cautioned against the use of the aerodynamical method except in situations, like HAPEX-MOBILHY, when evaporation accounts for the major proportion of the available energy, and measurement error associated with technique is least.

Exploratory analysis of the extensive micrometeorological data from FIFE (Shuttleworth et al, 1989, Smith et al, 1991) has revealed much about the experimental systems used, and spatial variability in surface energy exchange across a 15 km square prairie grass site, which is fairly uniform, but which has significant topographic variation. Flux measurements made with the eddy correlation technique are consistent with those made with the Bowen ratio technique within likely experimental error of 15% (Smith et al, 1991). The general result for surface energy partition over this short, prairie grass is that the primary causes of spatial inhomogeneity in surface energy exchange at the $10\,km \times 10\,km$ scale seem either to be of meteorological origin (and transient), caused by differences in cloud cover and occasionally localized rainfall intensity; or they are local, being related to slope and aspect, or to the local albedo.

Experimental systems: Integrating measurements

Large-scale experiments include measuring systems which provide information on the area-average fluxes as a basis for testing aggregation hypotheses. However, no instrumental system exists which can provide the area-average data at the short time resolution required and, in consequence, alternative systems are deployed which provide information which has some of the required characteristics. Such systems have been used in past experiments, and it is timely to review their strengths and weaknesses so revealed.

FIFE data warns against the use under GEWEX of airborne or space- borne measurements of radiometric surface temperature to provide quantitative extrapolation of point measurements of sensible heat flux. The FIFE site is as near an ideal situation in which to apply the technique as could reasonably be expected, having comparatively uniform, short vegetation cover. However, the results show a level of systematic error (a factor two), and spatial and temporal variability in the difference between radiometric and aerodynamic surface temperature (±100% ; Stewart et al, 1989), such that the value of this method for mapping sensible heat flux must be questionable.

Currently one of the most perplexing aspects of those experiments which have deployed flux measuring aircraft (HAPEX, FIFE and LOTREX-HIBE88) is that analysis has revealed a significant, systematic loss in their measurement of energy fluxes, possibly sometimes as big as 40% (Sellers et al, 1988b). This may be of instrumental origin, but early results from LOTREX (Jochum et al, 1990) suggest some flux transfer may be at length scales much longer than those normally associated with the atmospheric boundary layer. At this stage, flux measuring aircraft remain a necessary component of focussed observational studies under GEWEX, though it is probably sensible that their future deployment is orientated more towards supporting mesoscale or boundary layer modelling as the primary source of aggregate flux estimates, rather than attempting to provide these themselves.

Hydrological data from catchments, when synthesized to the scale of the grid mesh used in global models are crucial to validating their long-term average predictions of present rainfall and runoff with existing vegetation cover. The value and role of such data in the so-called GEWEX Mesoscale Field Experiments described later is not as well defined. The primary need in such experiments will be for area-average surface data at the daily, or, better, hourly time scale, very often preferentially during a *drying cycle* when evaporation is mainly from water stored in the soil. Catchment data is of least value in exactly these conditions and calibrating area-average SVATS against the evaporation synthesized from a calibrated catchment-scale hydrological model at such times and time scales is mainly just a model-to-model comparison. Notwithstanding the above, there remains a need to underpin the surface and above-surface measurements carried out in the proposed GEWEX Mesoscale Field Experiments with the best available hydrological model, and to look for sites with a long record of relevant hydrological data. It is essential that the calibrated area-average SVATS ultimately produced to describe the measured short-term response should also be shown capable of describing the long-term response.

The value of boundary layer sounding as an indirect means of estimating surface parameters and surface fluxes has exceeded expectations in past experiments. Satisfactory agreement has, for instance, been reported between the area-average value of the roughness length for the forested portion of the HAPEX-MOBILHY square deduced from balloon-borne and ground-based soundings of the boundary layer, and those deduced by micrometeorological means at the forest tower. There are also encouraging signs that soundings of the daily development of the planetary boundary layer (PBL) between the atmosphere and the ground, or more particularly of the energy stored within it, may provide an indirect measure of the area-average energy input at the ground (Shuttleworth, 1990).

Simple PBL models, such as those of De Bruin (1983) and McNaughton & Spriggs (1986; 1989), have demonstrated firstly, that the development of the boundary is significantly determined by the control on evaporation exerted by vegetation at the surface; and secondly, that the PBL development is fairly robust with respect to initiation. The output of such models can be expressed in terms of the equation

$$\lambda E = \alpha \frac{\Delta}{\Delta + \gamma} \tag{13.41}$$

where A is the energy available for evaporation (approximated by net radiation) and is α a parameter which is a measure of surface energy partition. Shuttleworth (1990) showed that data from FIFE were useful in illustrating the potential of PBLs as an indirect measure of g_s , while work by Munley et al (1991) carried out at the FIFE site tends to confirm the potential of simple energy conservation models of the planetary boundary layer .

Figure 13.10 shows a range of curves for α synthesized using a simple PBL model, taken from McNaughton & Spriggs (1989), with nine different initiation profiles and daily radiant energy inputs from the Cabauw data set (Driedonks, 1981), and for different values of the surface resistance. The mid-day values of a for the FIFE site on 6 June 1987 and 10 October 1987 are also shown, plotted against an estimate of the effective value of r_s at this time. Clearly there is worthwhile agreement, as there is with the equivalent point from the ARME Amazon rainforest data, derived from Shuttleworth et al (1984). However, the data from the forest cover in HAPEX are not consistent. In this case, mesoscale modelling studies (Mascart et al, 1991) show considerable advection of more moist air to the forest from the surrounding agricultural areas and oceans on fine days, thus helping to suppress evaporation.

Estimating area-average surface fluxes

Currently the only means of providing estimates of area-average surface fluxes is by synthesis of aggregate values from multiple samples at representative sites across a study area. Past experiments have provided guidance, though there is as yet no accepted wisdom amongst the scientific community on how best to do this. However, it is the author's belief that the large-scale experiments so far carried out suggest the following:

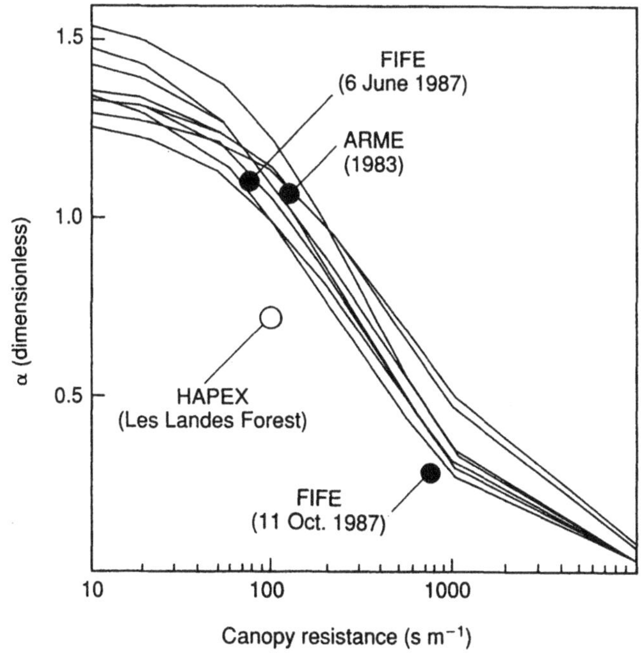

Figure 13.10: *The value of α in Eq. 13.41 as a function of the prescribed value of surface resistance derived from a simple PBL model with 9 different PBL profiles for initiation. Also shown are the equivalent midday values for FIFE on 6 June 1987 and 11 October 1987, a value from ARME for central Amazonia and an average value from HAPEX for Les Landes forest [taken from Shuttleworth (1990)].*

(a) For *uniform study sites* at the 10 *km* × 10 *km* scale, such as in FIFE, it seems that experimental errors probably dominate any persistent spatial variability, and to the accuracy required, simple averaging of the surface fluxes over many sites provides an adequate measure of surface energy exchange.

(b) For *study sites covered in mixed vegetation which occurs in disorganized plots at length scales of less than* 10 *km* [type 'A' in Shuttleworth (1988)], HAPEX and subsequent mesoscale modelling of the data from this experiment (e.g. Pinty et al, 1989), suggest that it might be sufficient to use simple averages of the sampled vegetation covers weighted by their proportional extent. This approach is not yet properly tested, however, and there is a considerable need for detailed tests under GEWEX using coupled surface/atmosphere models, operating with 1 *km* grid mesh or less, to investigate the significance of crop to crop advection, and the differences between the momentum exchange and surface-energy exchange processes.

(c) For land cover which is organized at a bigger scale, with 10 *km* × 10 *km* plots and above [type 'B' in Shuttleworth (1988)], there is evidence that mesoscale advection in the atmosphere may complicate the synthesis of area-average fluxes (Pinty et al, 1989; Mascart et al, 1991). However, mesoscale meteorological models seem able to capture the atmosphere's response and, in principle, coupled hydrological/meteorological models, operating with (say) a 10 km grid mesh, could be used as an integrating mechanism to derive time series of synthetic area-average fluxes and weather variables against which to calibrate area-average SVATS relevant to a larger grid mesh.

It seems that we may be able to synthesize area-average surface (energy) fluxes, either as simple averages, perhaps weighted by the proportion of each vegetation cover, or through the use of mesoscale models operating at various scales.

13.2.4 Area-average model parameters: Strawman Aggregation Rules

Arguably the most difficult challenge in the context of the second objective of GEWEX, *viz.* to model the global hydrological cycle, is to provide values for the parameters in area-average SVATS which are adequately capable of representing the energy and water interactions of land surfaces. This is most obviously the case for those widespread areas of the globe where terrestrial land cover is variable. Most of the present day SVATS used in GCMs, and certainly the most advanced (see Section 13.2.3) ignore this need, and opt instead to select the most typical land cover and to apply its parameterization over the whole grid square. This removes the incentive to address the problem.

Here we meet the aggregation issue head on, and postulate *Strawman* Aggregation Rules. The author's philosophy is that although such simple rules might well not stand the test of time, once in place any critical appraisal of them, based on observation and modelling, can only result in their replacement by superior alternatives, and progress will ensue. Accordingly, we start the process from the simplest assumptions consistent with physical intuition, but first we seek to define the key parameters for which aggregation rules are sought.

It is helpful to seek the essential elements in the description of the energy/water interaction of continental surfaces, the minimum (and it might prove the maximum) features which must be represented. The keyword is *availability*; it is those characteristics which determine the magnitude and accessibility of the energy and water *available* to the atmosphere which are important. To become available, energy and water must be first captured by the surface and held for a while in stores there. They are then released at a rate which is also dependent on certain key surface parameters.

Clearly the area-average albedo is important, since this largely determines the surface-energy captured and available: the surface emissivity must also be averaged but its heterogeneity is less. Equally, the area-average infiltration of water to the soil is important too, since water lost to streams and rivers is generally much more concentrated and less available to the air above. [*Note:* the effect of topography on the availability of water to the atmosphere, through changes in area-average infiltration and soil moisture, is not recognized in present generation SVATS].

Once energy and water have been captured, the *effective size* of the reservoirs in which they are held is certainly important - since they determine the magnitude available for return - but so too are the turbulent exchange characteristics of the surface. These last influence the rate of return of heat and water to the air above, and indeed the capture of momentum from it. In practice and in general, at least two types of stores must be represented. The first is the energy and water stored in comparatively easily accessible form in, or for water on, the vegetation. The second store is the energy and water held in less accessible form in the soil, to the depth where it is still available to the atmosphere. The control exerted by the vegetation in determining the *rate of access* to the available soil water store is also relevant, this commonly being expressed in terms of the surface resistance, or its reciprocal, g_s, the surface conductance.

In summary then, the primary land surface characteristics for which area-average effective values are required are the albedo α; the surface emissivity, E; the maximum surface infiltration rate, I; the aerodynamic roughness, z_0; the surface resistance, or its reciprocal the surface conductance, g_s; and, for both energy and water, the size of two stores, one easily accessed in the vegetation cover, one less accessible in the soil.

SVATS are required to describe area-average surface fluxes. *The required area-average parameters are those which most nearly compute these area-average fluxes*, either their integrated value over a given period in the case of stores, or the rate of flow into and out of the stores. In general, the simplest *aggregation rule* for parameters, and the one which is here advocated pending experimental evidence to the contrary, is that *the effective area-average value of land surface parameters is estimated as a weighted average of the component cover types, through that function involving the parameter which most nearly gives a linear relation with the associated surface flux.*

Table 13.2 interprets this general aggregation rule for a grid square with i component covers of fractional area w_i in terms of the primary land surface characteristics defined above. Although the aggregation rules given in Table 13.2 are only plausible in the case of overhead atmospheric conditions which are constant over the grid square, there is now some preliminary evidence from modelling the HAPEX-MOBILHY data that they may be more generally valid (Noilhan et al, 1992). But systematic correlations with certain sub-grid features such as cloud cover and, particularly, precipitation may well have a significant effect.

In the case of the precipitation captured on the canopy and re-evaporated quickly, Dolman & Gregory (1992) have developed a suggestion of Shuttleworth (1988), to indicate the need to describe enhanced canopy drainage, related to the areal extent of storms, which is different for *convective* and *large-scale* precipitation patterns. Moreover, if the systematic correlation between surface cover and overhead cloud noticed on occasions for the HAPEX-MOBILHY site (e.g. Shuttleworth, 1988) proves to be numerically significant, some adjustment of the effective values of w_i for area-average albedo may be necessary to allow for it.

13.3 Future study of the Soil-Vegetation-Atmosphere Interface

Future land surface observational studies (McBean and McCarthy, 1990) will proceed under the World Climate Research Programme, as part of GEWEX, and under the IGBP, especially through the Core Project *Biological Aspects of the Hydrological Cycle* (BAHC). There are some distinctions between

PARAMETER	'STRAWMAN' AGGREGATION RULE	COMMENTS
Albedo (α)	$\bar{\alpha} = \sum_i w_i \alpha_i$	
Maxmimum Infiltration Rate (I)	$\bar{I} = \sum_i w_i I_i$	Neglects impact of topography on infiltration
Emissivity (E)	$\bar{\epsilon} = \sum_i w_i t_i$	
Aerodynamic Roughness (z_o)	$\ln^{-2}[z_b/\bar{z}_o] = \sum_i w_i \ln^{-2}[z_b/(z_o)_i]$	Averages momentum flux; z_b is a blending height
Surface Conductance ($g_s = r_s^{-1}$)	$\bar{g}_s = \Sigma w_i (g_s)_i$	Assumes aerodynamic transfer is very efficient
or	$\bar{g}_s^{-1} = \Sigma w_i (g_s)_i^{-1}$	Assumes aerodynamic resistance dominates
		(Note the average of these two might be a compromise)
Canopy Stores Heat (C_c)	$\bar{C}_c = \sum_i w_i (C_c)_i$	Neglects subgrid-scale structure in radiation (cloud) distribution
Water (S_c)	$\bar{S}_c = \sum_i w_i (S_c)_i$	Neglects subgrid-scale structure in precipitation distribution
Soil Stores Heat	$\overline{(K_t C_s)^{1/2}} = \sum_i w_i (K_t C_s)^{1/2}$	Assumes cyclic exchange process
Water (S_s)	$\bar{S}_s = \sum_i w_i (S_s)^{1/2}$	Neglects impact of topography on soil water storage

Table 13.2: Strawman *aggregation rules for area-average estimates of the primary parameters in regional-scale soil-vegetation-atmosphere transfer schemes.*

these two international initiatives beyond that associated with the disciplinary mix of the scientists propagating them. GEWEX is more strongly directed towards better description of the present earth system, and toward shorter term prediction. The Core Projects of IGBP are more concerned with global change, and with longer term prediction. Notwithstanding this, the two programmes have jointly planned some of the observational studies in support of the improved formulation and calibration of SVATS. The planned programme (IGBP 1990; WMO/ICSU 1990a) envisages a series of so-called *Mesoscale Field Experiments* selected to take place at locations where global change is most critical; and *Continental-Scale Studies* in which the existing hydrometeorology of major river systems is to be quantified and modelled, the initial study being of the Mississippi basin.

13.3.1 Mesoscale representation of land/atmosphere processes

The *Joint IGBP/WCRP Working Group on Land Surface Experiments* meeting in Wallingford, in January 1990 specified priority topics for the period 1990 to 1997, the experimental locations, and time schedule for a series of Mesoscale Field Experiments (WMO/ICSU, 1990b). These are illustrated in Figure 13.11.

The objectives of this programme, through a judicious mixture of experimental and modelling studies, are to:

- provide quantitative understanding of energy, water and trace gas interactions between land-surfaces and the atmosphere for a range of important biomes;
- create methodologies for the synthesis of area-average descriptions relevant to increasing spatial scales (viz. Soil-Vegetation-Atmosphere Transfer Schemes (SVATS));
- improve the representation of such land-atmosphere interactions within predictive models of global change;
- develop methods to infer the parameters required in such representations from satellite data.

The programme draws on previous experience with experiments in the Hydrologic-Atmospheric Pilot EXperiment (HAPEX) series, the International Satellite Land Surface Climatology Project (ISLSCP), and the International Biosphere Programme (IBP). Future experiments will be essentially interdisciplinary, involving hydrologists, meteorologists and ecologists, along with specialists in trace gas exchange - and will require the co-ordinated resources of several nations.

The proposed experimental format illustrated in Figure 13.12 involves long-term monitoring over several years at each of the study sites including a major experiment lasting one year, preceded by a trial experiment with a lower level of activity. The proposed experimental and associated modelling activities are described in detail by Shuttleworth (1991b).

Studies to investigate desertification through *greenhouse warming* will take place in central Spain (EFEDA) and in Niger (HAPEX-Sahel). The Boreal Forest Study (BOREAS) will be in Canada and the proposed Tundra study will probably be in Alaska or northern Russia. The investigation of

Priority Topic	Location (Experiment)	Provisional Timing						
		1991	1992	1993	1994	1995	1996	1997
Desertification	Spain (EFEDA)	—	····	····	++++	····	····	
	Niger (HAPEX-Sahel)	····	++++	····	····	····	····	····
Boreal Forest Interaction	Canada (BOREAS)			——	++++	····	····	····
Tropical Forest	Brazil (?)	····	····	——	····	····	++++	····
Tundra Interaction	*(to be defined)*	*(to be defined)*						

Key: — Small-scale (single site) experiment
···· Monitoring effort
++++ Large-scale (multiple site) experiment

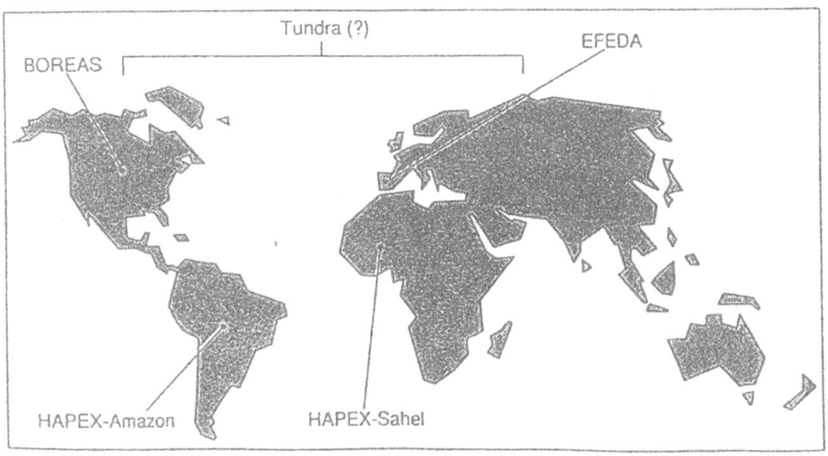

Figure 13.11: *(a) The priority topics, proposed locations and provisional timing of the Mesoscale Field Experiment programme proposed by GEWEX and IGBP-BAHC.*
(b) Proposed locations of the study areas on the globe.

tropical deforestation will occur in Amazonia (HAPEX-Amazonia), and will probably build on the Anglo-Brazilian Study (ABRACOS) already underway there.

In these studies, several typical $10,000\,km^2$ areas will be defined, these being sufficiently large to include mesoscale meteorological activity. The variations in land cover and soil conditions over the large area will require careful definition because of their influence on exchange processes. Since the research issues are related to global change, the sites selected will either include existing ecoclimatological gradients, as in Niger and Canada, or man-made land-use change, as in Amazonia and Spain. Subsidiary river catchments, typically with an area of $100\,km^2$ will be defined within this larger area, and long-term ecological and climate monitoring will take place for at least 5, preferably 10,

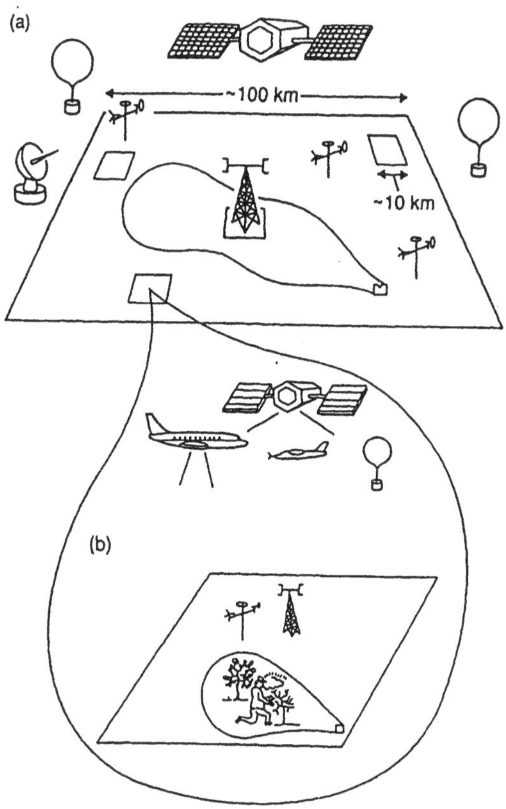

Figure 13.12: *Experimental format for WCRP/IGBP Mesoscale Field Experiments.*
(a) For a site typically $100\,km \times 100\,km$, *long-term (5-10 year) monitoring using a mesoscale meteorological network, rain radar and satellite system; for several subsidiary sites, long-term ecological monitoring; and for at least one of these subsidiary sites, preferably the most representative in terms of surface cover, long-term monitoring of surface fluxes and climate, hydrological catchment data and vegetation characteristics.*
(b) For two or more subsites nested in the larger area, monitoring lasting up to a year, perhaps in short, intensive sessions, of near-surface fluxes and climate, canopy and soil processes and atmospheric profiles, with airborne measurements of boundary layer fluxes and intensive satellite and airborne remote sensing.

years, encompassing the year of major study. These areas also require careful selection to define and replicate variations in surface characteristics. The long-term effects of land-use change can be incorporated into such studies by space-for-time substitution. If the ecoclimatological gradients demand it, as in the case of Boreal Forests, the size of the study area may have to be extended.

Long-term data collection for the larger $10,000\,km^2$ study area will use a network of automatic weather stations, rain radar and satellite systems. At least one, the most representative, subsidiary site will be chosen for long-term measurements of surface flux exchange and for catchment data collection (Figure 13.12).

In each study, during the main experimental year, data collection for the full study site will increase to include boundary layer sounding, with radiosondes and ground-based stations. Surface flux measurements will be made at each of the subsidiary sites for the characteristic classes of surface cover present there, and the hydrological data collection extended to cover all such subsidiary areas. For selected periods of the year these subsidiary sites will be overflown by aircraft carrying remote sensing instruments, and to provide airborne flux measurements. Experimental activity in the precursor experiment, will be broadly of the same nature and intensity but is likely to be restricted to just one subsidiary site.

The experimental methodology is to provide sufficient simultaneous data of surface flux measurements for characteristic vegetation covers, boundary layer soundings and the surface weather measurements to both verify, and provide boundary conditions for, coupled hydrological/atmospheric models. These will operate at a mesoscale with (say) a 10 *km* grid net, and will be used in the development of parameterizations to allow for heterogeneities in the atmospheric variables (e.g. precipitation), and in the surface characteristics. The long time-series of data from the surface meteorological network covering the site, the rain radar system (and associated ground calibration), and meteorological data from the surrounding area will provide the boundary conditions for the mesoscale model syntheses. The surface representations for different land cover types will be calibrated against the micrometeorological flux measurements and associated catchment data, while the coupled hydrological/atmospheric model will itself be validated against the atmospheric sounding and aircraft flux data, and against the catchment data.

Once calibrated and validated in this way, the mesoscale model can then be used to provide synthetic macroscale data as area-average surface fluxes, area-average near-surface weather variables, and surface and subsurface runoff. These can then be used to calibrate grid scale representations. The inclusion of sampled ecoclimatological gradients, or man-made changes within the study area will allow the synthesis of equivalent grid-scale parameterizations for a whole range of hypothetical mixes and spatial distributions of surface vegetation cover. The modelling methodology and the typical calibration for representative cover types can, moreover, be applied to provide simulation of changes elsewhere in the sampled biome, the relative mix of vegetation cover being then determined from remote sensing data.

A second important modelling focus in these studies will be to investigate also the reliability of the surface representations applied at the 10 *km* scale, using finer scale models operating with a grid mesh of 1 *km* or less. Such models will be used to evaluate the significance or otherwise of atmospheric advection as air moves from one plot of vegetation cover to the next, so refining the first stage of the *scaling up* process from the field scale.

13.3.2 Linked hydrological/meteorological models at large basin scales

Over the last few years there has been a major effort to improve the parameterization of land surface processes, but the focus has been on the vertical energy and water fluxes at the land surface/atmosphere interface within each GCM grid (e.g. FIFE and HAPEX experiments). Little attention has been given to the spatial redistribution of water and energy from surface or groundwater runoff within each grid cell and to transfer between cells.

Hydrological models have usually been applied to catchments smaller than a typical GCM grid square. A few modelling studies have examined the relationships between runoff and climate at a large-basin scale or at a continental scale but these have not been linked to atmospheric models. Even where measurements of the distribution of run-off at a local, regional and large basin scale are available, their use in the validation of GCMs on the basis of a water-balance approach has been limited due to inadequate representation of the land surface hydrology at the relevant scale. Improved descriptions of hydrological models are required both as an integral component of the GCM and to predict the response of a catchment to climatic inputs. These responses include the changes in water storage and rates of surface water and groundwater flow within a catchment, including the exchanges of water with the atmosphere or runoff to the sea.

To advance land-atmosphere modelling, a GEWEX Continental Scale International Project (GCIP) has been established. The emerging Science Plan (IGPO, 1990) establishes the need for a large-basin study to provide the framework for the incorporation of the hydrological models into a GCM. The Mississippi basin has been selected from candidates including the Nile, Amazon, Rhine and Danube, since it has long time series of detailed hydrological and meteorological data. These are required so that models can be rigorously validated.

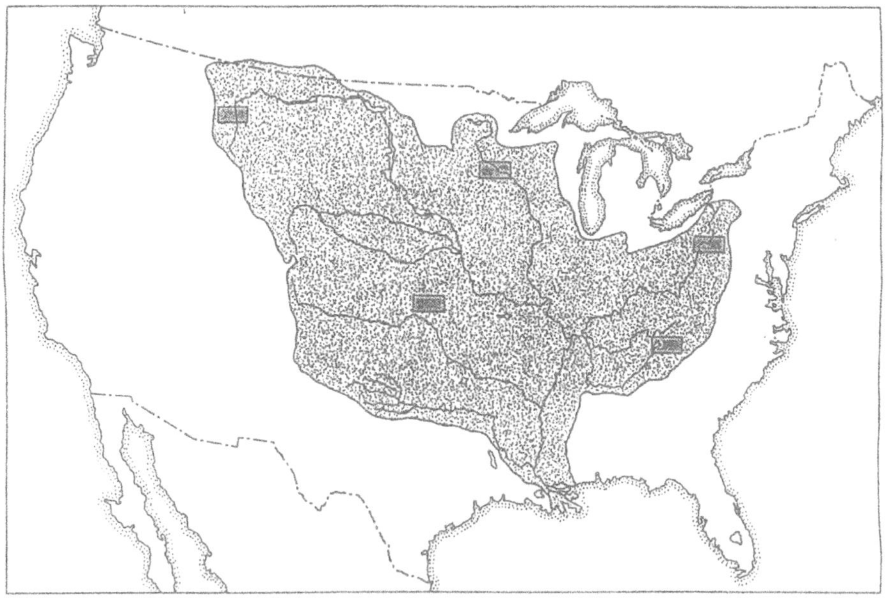

Figure 13.13: *Schematic illustration of the GEWEX Continental-scale International Project (GCIP) to occur in the Mississippi basin, with focussed attention on subsidiary areas (precise location as yet unspecified) to generate the understanding necessary to enhance the global relevance of the study.*

Such data will not exist for many basins in the world and stress will be placed on the development and use of hydrological models that are physically based so that they can be applied to ungauged basins. It is therefore likely that the study will in due course need to include focussed attention on selected areas within the Mississippi, though the location of these is not yet defined, see Figure 13.13.

The principal goal of the study is to produce models which are able to represent both within-grid cell variability and between-cell transfers, and can be applied in a continent without the need for calibration using site-specific historic data (but with, of course, catchment conditions defined by local data). The models can be developed to run both within and independently of a climate model.

Model development will require comprehensive and integrated data management, for example combining remotely sensed data with thematic data bases and digital terrain models. The availability of such data and techniques for their handling will have an important influence on the forms of model to be used and model parameterization. The continental-scale hydrometeorological model will be designed to complement the SVATS currently included within climate models, and will provide periodic updates to the SVATS boundary conditions.

13.4 Concluding summary and comments

As stated in the introduction, it is clear from sensitivity studies with GCMs that the terrestrial hydrological cycle is significantly determined by the interaction at the soil-vegetation-atmosphere interface, and that observing and modelling this is therefore critical to the fulfilment of the scientific objectives of GEWEX. On the basis of the state-of-the-art review given earlier in Section 13.2.1 of this paper, it is also clear that past research over many decades has provided a very substantial understanding of this interaction at the patch-scale.

Present day syntheses of this understanding has resulted in the diverse range of SVATS overviewed in Section 13.2.2, some of which have been specifically created to have the global relevance required by GEWEX. Most of these SVATS, including the most advanced, nonetheless are essentially still patch-scale models applied at the scale of the global grid, with no explicit provision for representing the mixed vegetation which covers much of the Earth, or a defined methodology to allow their calibration to do this.

Past observational studies, reviewed in Section 13.2.3, have begun to generate the attention and commitment of the international scientific community towards addressing the need to provide area-average calibration of SVATS over extended regions of mixed land cover. Section 13.2.4 is a deliberate attempt to force the issue in this regard, and seeks to provide a new generation of young scientists, some participating in this NATO Advanced Study Institute, with a simplistic set of aggregation rules on which they will want to improve. The land surface observational studies to occur under GEWEX, previewed in Section 13.3, will provide significant opportunity to do this. They will generate the data against which SVATS can be calibrated, and improved aggregation rules tested.

Meanwhile it is the author's opinion that GEWEX will force a reappraisal of the modelling priorities used in those SVATS to be applied at global scale. GEWEX is a strongly data-driven programme, which as its first objective focusses on the better quantification and description of the global hydrological cycle in its present form. Such a programme will inevitably tend to encourage the application of simpler SVATS, which are in any case appropriate to numerical weather prediction. Why use a complex scheme when there is enough data to tune the empiricism in simpler schemes?

But GEWEX will exert a subtler pressure on the format and calibration of SVATS. Since so much of the most reliable GEWEX data will be for regions of the developed world where the observational systems are concentrated, it will tend to provide calibration data for regions of mixed land cover. If the experiment is to address its second and third objectives effectively, i.e. to look towards global calibration and prediction, the balance of need within GEWEX-related SVATS will switch away from an emphasis on detailed, single cover representations, towards less detailed, multiple cover representations. The development of aggregation and disaggregation methodologies will gain yet further priority: without them it will be difficult to extrapolate GEWEX's detailed measurements of surface exchanges over areas of mixed land cover, to other unmeasured areas with covers of different mix.

13.5 References

André J-C, Goutorbe J-P, Perrier A, Becker F, Besselmoulin P, Bougeault Ph, Brutsaert Y, Carlson T, Cuenca R, Gash JHC, Gelpe J, Hildebrand P, Lagouard P, Lloyd CR, Mahrt L, Mascart P, Mazaudier C, Noilhan J, Ottle C, Payen M, Phulpin T, Stull R, Shuttleworth WJ, Schmugge T, Taconet O, Tarrieu C, Thepenier RM, Valencongne Ch, Vidal-Madjar D, Weill A (1988) HAPEX-MOBILHY: First results from the special observing period. *Ann Geophys* 6(5): 477-492.

Avissar R, Verstraete MM (1990) The representation of continental surface processes in atmospheric models. *Rev Geophys* 28: 35-52.

Baldocchi DD, Hutchison BA (1987) Turbulence in an almond orchard: vertical variations in turbulent statistics. *Bound Lay Met* 40: 127-146.

Brutsaert W (1982) Evaporation into the atmosphere. *D Reidel Publ. Co* Dordrecht, Holland.

Charney JG, Quirk WJ, Chow SH, Kornfield J (1977) A comparative study of the effects of albedo change on drought in semi-arid regions. *J Atmos Sci* 34: 1366-1385.

Choudhury BJ (1991) Multispectral satellite data in the context of land surface heat balance. *Rev Geophys* 29: 217-236.

Choudhury BJ, Monteith JL (1988) A four-layer model for the heat budget of homogeneous land surfaces. *Quart J Roy Met Soc* 114: 373-398.

Clapp RB, Hornberger GM (1978) Empirical equations for some soil hydraulic properties. *Water Resour Res* 4: 601-604.

De Bruin HAR (1983) A model for the Priestley-Taylor parameter. *J Appl Met* 22: 572-578.

Denmead OT, Bradley EF (1985) Flux gradient relationship in a forest canopy. In: The forest-atmosphere interaction (eds. B A Hutchison, BB Hicks), *D Reidel Publ Co* Dordrecht, Holland: 421-442.

Dickinson RE (1989a) Implications of tropical deforestation for climate: A comparison of model and observational descriptions of surface energy and hydrological balance. *Phil Tran R Soc Lond* B324: 423-431.

Dickinson RE (1989b) Modelling the effects of Amazonian deforestation on regional climate: a review. *Agric For Meteorol* 47: 339-347.

Dickinson RE, Henderson-Sellers A, Kennedy PJ, Wilson MF (1986) Biosphere-Atmosphere Transfer Scheme (BATS) for the NCAR Community Climate Model. *NCAR Technical Note* NCAR/TN-275+STR: 69pp.

Dickinson RE, Henderson-Sellers A (1988) Modelling tropical deforestation: a study of GCM land-surface parameterizations. *Quart J R Met Soc* 114: 439-462.

Dolman AJ, Gregory D (1992) The parameterization of rainfall interception in GCMs. Submitted to *Quart J Roy Met Soc* (In press)

Driedonks AGM (1981) Dynamics of the well-mixed atmospheric boudary layer. *Scientific Report W.R.81-2* KNMI, de Bilt, The Netherlands.

Field RT, Fritschen L (1991) Net radiation observation, calibration and evaluation in FIFE. *J Geophys Res* (Submitted)

Garratt JR (1978) Flux profile relations above tall vegetation. *Quart J Roy Met Soc* 104: 199-211.

Gash JHC, Shuttleworth WJ, Lloyd CR, André J-C, Goutorbe J-P, Gelpe J (1989) Micrometeorological measurements in les Landes Forest during HAPEX-MOBILHY, *Agric For Meteorol* 46: 131-147.

Geyer B, Jarvis PG (1991) A review of models of the Soil-Vegetation-Atmosphere Transfer Schemes (SVATS): A report to the TIGER III Committee, available from *Institute of Hydrology* Wallingford, UK.

Goutorbe J-P (1991) A critical assessment of the SAMER network accuracy. In: Land Surface Evaporation, (eds. T J Schmugge and J-C André), 171-182, *Springer Verlag* New York.

IGBP (1990) The International Geosphere-Biosphere Programme: a study of global change. The initial core projects. *IGBP* Stockholm (IGBP Report No. 12)

IGPO (1991) GEWEX Continental-scale International Project (GCIP). Report of the first GCIP planning workshop, Reston, Virginia, USA, IGPO-No.1, *GEWEX Project Office* Washington.

Jochum AM, Entstrasser N, Fimpel H-P, Mörl P, Rösler F, Willeke H (1990) Evaporation and energy fluxes in the atmospheric boundary layer during LOTREX. *Symposium on FIFE* Feb. 7-9, 1990, Anaheim, Calif. *Pub Amer Met Soc* 177-180.

Jones HG (1984) Plants and Microclimate. *Cambridge University Press* New York.

Penman HL (1948) Natural Evaporation from open water, bare soil and grass. *Proc Roy Soc Lond* A193: 120-145.

Manabe S (1969) Climate of the ocean circulation. Part I. *Mon Wea Rev* 97: 739-774.

Mascart P, Taconet O, Pinty J-P, Mehrez MB (1991) Canopy resistance formulation and its effect in mesoscale models: a HAPEX perspective. *Agric For Met* (in press)

Mason P J (1988) The formation of areally-averaged roughness lengths. *Quart J Roy Met Soc* 114: 281-295.

McBean G, McCarthy (1990) Narrowing the uncertainties: a scientific action plan for improved prediction of Global Change. In: *Climate Change. The IPCC Scientific Assessment* (eds. JT Houghton, G J Jenkins, JJ Ephraums)

McNaughton KG, Spriggs TW (1986) A mixed-layer model for regional evaporation. *Bound Layer Met* 34: 243-263.

McNaughton KG, Spriggs TW (1989) An evaluation of the Priestley and Taylor equation and complementary relationship using results from a mixed-layer model of the convective boundary layer. *IAHS Publ* No. 177: 89-104, Wallingford, UK.

Monteith JL (1973) Principles of Environmental Physics. *Edward Arnold* London, UK.

Monteith JL (1981) Evaporation and surface temperature. *Quart J Roy Met Soc* 107: 1-27.

Munley WG, Hipps LE (1990) Estimation of regional evaporation for a tall grass prairie from measurements of properties of the atmospheric boundary layer. *Water Resources Res* 27: 225-230.

Noilhan J et al. (1992) (Private communication), CNRM, Toulouse, France.

Noilhan J, Planton S (1989) A simple paramaterization of land surface processes for meteorological models. *Mon Wea Rev* 117(3): 536-549.

Pinty J-P, Mascart P, Richard E, Rosset R (1989) An investigation of mesoscale flows induced by vegetation inhomogeneities using an evaporation model calibrated against HAPEX-MOBILHY data. *J Appl Met* 28: 976-992.

Raudkivi AJ (1971) Hydrology. *Pergaman Press* Oxford.

Raupach MR (1979) Anomalies in flux-gradient relationships over forest. *Bound Layer Met* 16: 467-486.

Raupach MR (1989) Stand overstorey processes. *Phil Tran R Soc* London, B324: 175-190.

Rutter AJ, Kershaw KA, Robins PC, Morton AJ (1971) A predictive model of rainfall interception in forests. I: Derivation of the model from observations in a plantation of Corsican Pine. *Agric Met* 9: 367-384.

Sato N, Sellers PJ, Randall DA, Schneider EK, Shukla J, Kinter III L, Hou Y-T, Albertazzi E (1989) Effects of implementing the Simple Biosphere Model in a General Circulation Model. *J Atmos Sci* 46(18): 2757-2782.

Shukla J, Mintz Y (1982) Influence of land-surface evapotranspiration on the earth's climate. *Science* 215: 1498-1501.

Shuttleworth WJ (1975) The concept of intrinsic surface resistance: Energy budgets at a partially wet surface. *Bound Layer Met* 8: 81-99.

Shuttleworth WJ (1988) Macrohydrology - the new challenge for process hydrology. *J Hydr* 100, 31-56.

Shuttleworth WJ (1989) Micrometeorology of temperate and tropical forest. *Phil Tran R Soc Lond* B324: 299-334.

Shuttleworth WJ (1991a) Insight from large-scale observational studies of land/atmosphere interactions. In: Land surface atmosphere interactions: parameterization and analysis for climate modelling (ed. EF Wood) 3-30, *D Reidel* New York.

Shuttleworth WJ (1991b) The Modollion Concept. *Reviews Geophys* 29(4), 585-606.

Shuttleworth WJ, Dickinson RE Comments on "Modelling tropical deforestation: a study of GCM land-surface paramaterizations", RE Dickinson, A Henderson-Sellers. *Quart J Roy Met Soc* 115: 1177-1179.

Shuttleworth WJ, Gash JHC, Lloyd CR, Moore CJ, Roberts JM, Marques A de O, Fisch G, Silva Filho V de P, Ribeiro MNG, Molion LC B, Nobre C, de Sa LDA, Cobral OMR, Patel SR, Mordas JC (1984) Observations of radiation exchange above and below Amazonian forest. *Quart J Roy Met Soc* 110: 1163-1169.

Shuttleworth WJ, Gurney RJ (1990) The theoretical relationship between foliage temperature and canopy resistance in sparse crops. *Quart J Roy Met Soc* 116: 497-519.

Shuttleworth WJ, Gurney RJ, Hsu AY, Ormsby JP (1989) FIFE: the variation in energy partition at surface flux sites. *IAHS Publ* No. 186: 67-74.

Shuttleworth WJ, Wallace JS (1984) Evaporation from sparse crops - an energy combination theory. *Quart J Roy Met Soc* 111: 839-855.

Sinclair TR, Murphy CE, Knoerr KR (1976) Development and evaluation of simplified models simulating canopy photosynthesis and transpiration. *J Appl Ecol* 13: 813-829.

Sellers PJ, Mintz Y, Sud YC, Dalcher A (1986) A Simple Biosphere (SiB) for use within General Circulation Models. *J Atmos Sci* 43: 505-531.

Sellers PJ, Hall FG, Asrar G, Strebel DE, Murphy RE (1988a) The first ISLSCP Field Experiment (FIFE). *Bull Amer Met Soc* 69(1): 22-27.

Sellers PJ, Hall FG, Strebel DE, Kelly RD, Verma SB, Markham BL, Blad BL, Schimmel DS, Wang JR, Kanemasu ET (1988b) First ISLSCP Field Experiment, April 1988, *Workshop Report* Available from Code 623, NASA/GSFC, Greenbelt, Md 20771, USA.

Sellers PJ, Shuttleworth WJ, Dorman JL, Dalcher A, Roberts JM (1989) Calibrating the Simple Biosphere Model for Amazonian tropical forest using field and remote sensing data. Part I: Average calibration with field and remote sensing data. *J Appl Met* 28: 727-759.

Shu Fen Sun (1982) Moisture and heat transport in a soil layer forced by atmospheric conditions. M S Thesis. Dept. of Civil Engineering, *University of Connecticut* 72 pp.

Smith EA, Hsu AY, Crosson WL, Field R, Fritschen LJ, Gurney RJ, Kanemasu ET, Kustas W, Nie D, Shuttleworth WJ, Stewart JB, Verma SB, Weaver H, Wesley M (1991) Area-average surface fluxes and their time-space variability over the FIFE experimental domain. Submitted to *J Geophys Res*

Stewart JB, Shuttleworth WJ, Blyth K, Lloyd CR (1989) FIFE: a comparison between aerodynamic surface temperature and radiometric surface temperature over sparse prairie grass. Proc 19th Agric For Meteorol and 9th Conf Biometeor Aerobiology, March 7-10 1989, Charleston, S. Caroline. *Publ Amer Met Soc* 144-146.

Stewart JB, Thom AS (1973) Energy budgets in pine forest. *Quart J Roy Met Soc* 99: 154-170.

Stricker H (1991) Observing and modelling the terrestrial hydrological cycle. In: Energy and Water Cycles in the Climate System (ed. Ehrhard Raschke), *NATO ASI* (this volume)

Sud YC, Shukla J, Mintz Y (1985) Influence of land-surface roughness on atmospheric circulation and rainfall: A sensitivity experiment with GCM. *NASA Tech Mem* 86219, NASA/GSFC, Greenbelt, Md. 20771, USA.

Thom AS (1972) Momentum, mass and heat exchange of vegetation. *Quart J Roy Met Soc* 103: 345-357.

Thom AS (1975) Momentum, mass and heat exchange of plant communities. Section 3 in Vegetation and the Atmosphere (ed. JL Monteith), *Academic Press* London: 57-109.

Thom AS, Oliver HR (1977) On Penman's equation for estimating regional evaporation. *Quart J Roy Met Soc* 103: 345-357.

Waggoner PE (1975) Vegetation and the Atmosphere, Volume 1. (ed. JL Monteith). *Academic Press* London: 205-228.

Waggoner PE, Reifsnyder RW (1968) Simulation of the temperature humidity and evaporation profiles in a leaf canopy. *J Appl Met* 7: 400-409.

Warrilow DA (1985) Indications of the sensitivity of European climate to land use variations using a one-dimensional model. *Proc. ISLSCP Conference* Rome, Italy, 2-6 December 1985. ESA SP-248 159-166 (May 1986)

WMO/ICSU (1990a) Scientific plan for the Global Energy and Water Cycle Experiment, *WMO/TD* No.376, WMO, Geneva.

WMO/ICSU (1990b) Report on the first meeting of the WCRP-GEWEX/ IGBP-CP3 Joint Working Group on Land-Surface Experiments, *WMO/TD* No.370, WMO, Geneva.

Chapter 14

OBSERVING AND MODELLING THE PLANETARY BOUNDARY LAYER

Søren E. Larsen
Department of Meteorology and Wind Energy
Risø National Laboratory
Denmark

14.1 Introduction

Characteristics of the atmospheric planetary boundary layer (PBL) are important for the atmospheric energy and water cycles because the fluxes of momentum, heat, and water vapour between the atmosphere and the surfaces of the earth all pass through the PBL, being carried and modified by mixing processes here. Since these mixing processes mostly owe their efficiency to the mechanisms of boundary layer turbulence, a proper quantitative description of the turbulence processes becomes essential for a satisfying description of the fluxes between the surface and the atmosphere.

Estimating the water and energy fluxes through the atmospheric boundary layers necessitates that almost all types of the flows, that occur there, must be considered. There are very few combinations of characteristic boundary layer conditions that are not of significant importance for the flux of energy and water between the surface and the atmosphere, at least for some parts of the globe.

14.2 Simple pictures and some basic statistical tools

Fluxes across the planetary boundary layer, PBL, are dominated by turbulent motion. Many pages have been filled in an effort to define turbulence. Here we shall just notice that motions of systems, that can be described by the nonlinear fluid equations, tend to show strongly varying stochastic components, the turbulence, as well as more smooth and predictable characteristics. In the PBL the wind speed as well as temperature and humidity show this stochastic behavior on all spatial and

NATO ASI Series, Vol. I 5
Energy and Water Cycles in the Climate System
Edited by E. Raschke and D. Jacob

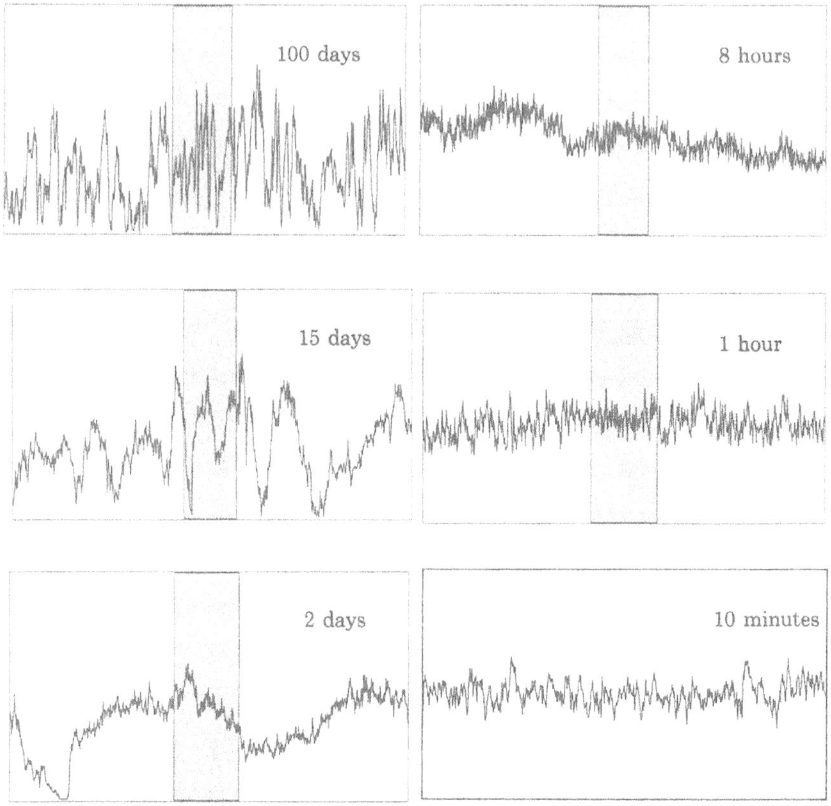

Figure 14.1: *Wind speed measured 30 m above flat homogeneous terrain in Denmark from Troen and Petersen (1989). The data were obtained from a one-year time series recorded with 16-Hz resolution. Each graph shows the measured wind speed over the time period indicated. The number of data points in each graph is 1200, each averaged over 1/1200 of the time period indicated. The vertical axis is wind speed, 0-20 ms^{-1}.*

temporal scales of variation. In Figure 14.1 this is illustrated by a measured time series of the wind speed observed through different time windows.

We shall mostly consider fluctuations on time scales of the order of and less than one hour, because the turbulence processes here carry most of the vertical fluxes, we will try to estimate.

On these time scales the main mechanism for producing turbulence is the vertical gradient of the mean wind. In Figure 14.2 we show typical vertical variations of wind speed, humidity, and temperature between their surface values and values at the top and above the PBL. Temperature and humidity can both increase and decrease with height depending on whether their surface values or values in the free atmosphere are the larger. However, the wind speed will always increase with height from zero at the ground to its value in the free atmosphere just above the PBL. This vertical wind shear gives rise to overturning of the air, producing the turbulence (Tennekes and Lumley,1982). This provides a formidable mechanism for carrying the vertical fluxes compared to the molecular transport mechanism that would have been an alternative. For example, a temperature gradient of

2 K between the surface and 10 meter height with a wind speed at 5 ms^{-1} gives rise to a heat flux of about 0.5 mKs^{-1}. If for the same situation the flux had to be carried by molecular diffusion only, the result would be 4×10^{-6} mKs^{-1} only.

The temperature structure of the PBL strongly influences the turbulence production through its influence on the density of the air. If the air is warmer and thereby lighter close to the ground, it will enhance the production; if it is cooler at the ground the production will be reduced. To a lesser extent the humidity has similar, although smaller effects because also admixture of the water vapour changes the density of the air.

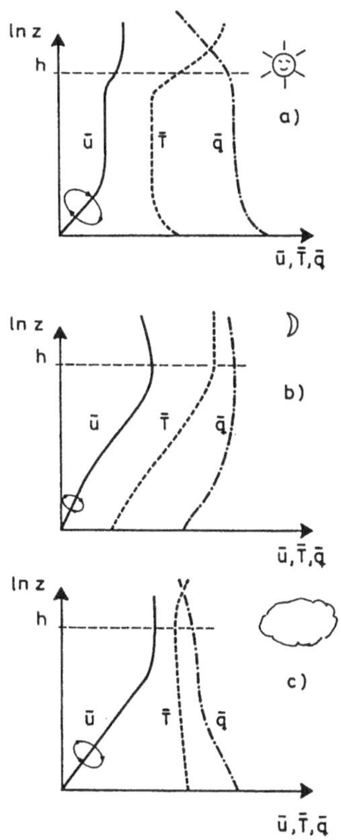

Figure 14.2: *Characteristic height variation (profiles) of the mean values of the wind speed,* $\bar{u}$, *temperature,* $\bar{T}$, *and humidity ,* $\bar{q}$ *from the ground to the top of the PBL, indicated by* h. *Also shown by the arrows on the u-profile is the overturning of the flow induced by the vertical velocity gradient. The profiles are shown for the following characteristic situations: (a) thermally unstable, e.g. a sunny day, (b) thermally stable, e.g. a clear sky night, and, (c) thermally neutral, e.g. a high-wind overcast situation.*

Based on the above discussion we can now specify the planetary boundary layer as being the layer through which the atmospheric variables change between their values in the free atmosphere and their values at the surface, the transition being mostly controlled by turbulent motion and mixing.

In Figure 14.2 the turbulence production is envisioned as a swirling motion induced by the shear. This whorl constitutes a volume of localized vorticity which we shall denote an *eddy*. This picture of turbulence, as a soup of intertwining spaghetti-like eddies, has been very useful in the study of turbulence in spite of its extreme simplicity. Each eddy can be associated with a size, or spatial scale, and a lifetime. When the eddies have been produced they will remain coherent for some time, creating their own smaller shear regions. By the same process as for the mean shear these both erode the larger eddies, due to the associated mixing, and create smaller eddies. This mechanism repeats itself transferring energy to smaller and smaller eddies until they become small enough for the molecular viscosity to dissipate their motion into heat.

In view of the above discussion we shall next consider a few of the statistical tools used to describe the turbulence.

As the simplest we break down the signal into its mean value and its fluctuating part, e.g. for the variable x

$$x = \bar{x} + x' \tag{14.1}$$

where we use an overbar to signify the mean value while the symbol $'$ indicates the fluctuating part. This expansion is called the Reynolds expansion. Note that it implies $\overline{x'} = 0$. The averaging processes used to obtain $\bar{x}$ depend on the type of signal available and the purpose of averaging. In statistics one often talks about ensemble average over an ensemble of realizations of the signal x. This is the type of averaging used for most theoretical manipulations. In practice one will often use time or space averages, either because the signals available are a function of these parameters as e.g for measurements of time series from meteorological towers, or because of the objectives of the study as for example to obtain the area averages often being the goal of hydrological studies. The averaging procedures employed are limited by the signals available but are also a matter of choice. As an example, we can take the signal in Figure 14.1 that would lend itself to time averaging, since it is a time signal, but also to ensemble averaging using ensembles of data from similar days or hours.

Typical height variation of mean quantities as $\bar{u}, \bar{q}$, and $\bar{T}$ is illustrated in Figure 14.2 and the associated discussion. To further study the behavior of x', the simplest measures applicable are:

$$\begin{aligned} \text{the variance} &= \overline{x'^2} \\ \text{the covariance} &= \overline{x'y'} \end{aligned} \tag{14.2}$$

where y is another signal similar to x.

The variance is used to describe the fluctuation intensity of x, while $\overline{x'y'}$ can be used to study the relation between the two signals. Having considered the variances and covariances one can consider higher-order moments, and the distribution functions of the signals to study different aspects of their behavior. However, since it will not be much used here, we shall proceed to the tools used to identify the scales of variation.

Here the Fourier transform is often used to reveal the relevant scales. To analyse a time series like the one used in Figure 14.1, it is convenient to use the following Fourier expansion:

$$x'(t) = \int_{-\infty}^{\infty} dZ_x(\omega) \exp(i\omega t) \tag{14.3}$$

where $x'(t)$ is assumed to be a statistically stationary function of time. The Fourier expansion applied in Eq. 14.3 is the Fourier-Stieltje expansion that can be used even when $x(t)$ cannot be Fourier expanded in the standard sense, since it is not integrable (Lumley and Panofsky, 1964). The Fourier modes are here denoted $dZ(\omega)$ and are stochastic functions as well, being a function of the frequency ω[rad./sec]. Squaring and averaging, one finds from Eq. 14.3 that

$$\overline{x'^2} = \int_{-\infty}^{\infty} S_x(\omega) d\omega \tag{14.4}$$

where $S_x(\omega)$ is denoted the power density spectrum, or shorter simply the power spectrum of the $x(t)$, and it is seen to describe the contribution from the different frequencies to the variance of $x(t)$. This illustrates how the Fourier transform can be used to identify the important scales of variation, in this case the frequency scales. We present in Figure 14.3 the power spectrum of the one-year wind speed record used in Figure 14.1. In Figure 14.3 the spectrum is plotted versus the logarithm of the frequency because of the many decades of frequency scales of interest in geophysical time series. To obtain a proper impression of the contributions of the different frequency decades to the total variance, the spectrum is multiplied by the frequency, $f[Hz]$ because

$$\omega S(\omega) d(\ln(\omega)) = fS(f) d(\ln(f)) = S(\omega) d\omega \tag{14.5}$$

The strong intensity of the spectrum between the annual and the diurnal frequencies comes from the motion of the weather systems across Denmark. Therefore, it is different in other parts of the earth with different climatology as are of course the intensities of the diurnal and annual cycles. The contribution from the boundary layer turbulence described above is represented by the small bump from about 1 hour and out. Around 1 hour is the famous gap between what in relation to the boundary layer turbulence can be considered as the *mean flow* and the turbulence. This is mainly produced by the vertical shear in the mean flow, as discussed above.

There have been some discussion about the existence of this gap. This is because some convection clouds actually create eddies with about the time scale of the gap, see Fig. 14.4, and also since the spectra so far used to illustrate its existence often have been composited from different time series used to compute different decades of the total spectrum. From the point of view of modelling it is advantageous to use grid sizes with frequency and spatial scales within the spectral gap, because the absence of intensity here shows that only few independent processes create variability in this scale region. This in turn means that it becomes simpler to decide if a particular process must be parameterized or resolved explicitely by the model.

As stated above the spatial variability is most important for the structure of the boundary layer turbulence. Therefore, we shall finish this section by considering the wave-number spectrum and the quite general insight that can be drawn from the form of this spectrum.

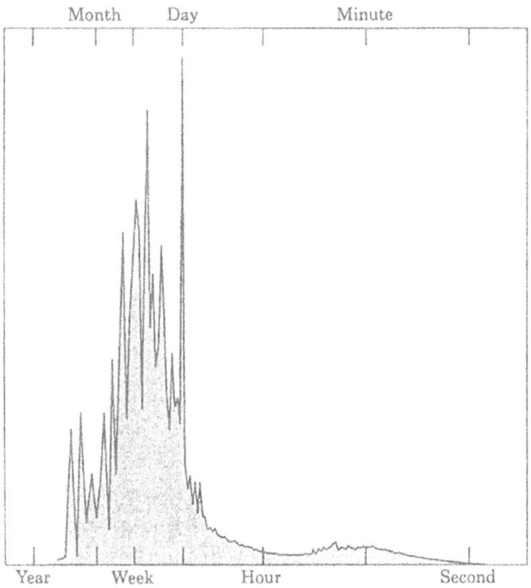

Figure 14.3: *The power spectrum of the one-year time series of wind speed versus the logarithm of the frequency, presented as the corresponding time periods (Courtney and Troen, 1990; Troen and Petersen, 1989).*

For a signal $x(t, \mathbf{r})$, being a stochastic statistically stationary and homogenous function of both time and the three spatial coordinates, $\mathbf{r}$, the spatial Fourier transform corresponding to Eq. 14.3 can be written:

$$x'(t, \mathbf{r}) = \int_{-\infty}^{\infty}\int_{-\infty}^{\infty}\int_{-\infty}^{\infty} dZ_x(t, k)\exp(i\mathbf{k}\cdot\mathbf{r}) \tag{14.6}$$

where $\mathbf{k}$ is now the wave-number vector. As previously we square the equation to obtain

$$\overline{x'^2} = \int_{-\infty}^{\infty}\int_{-\infty}^{\infty}\int_{-\infty}^{\infty} S_x(\mathbf{k})d\mathbf{k} \tag{14.7}$$

which shows how the variance distributes in the three-dimensional wave number space. Note that the time dependency disappears due to the stationarity assumption. For boundary-layer turbulence, strict use of Eq. 14.7 is not possible because even for the most homogeneous of all boundary layers there must be inhomogeneity in the vertical direction by the definition of a boundary layer. However, recall the picture of the eddies being created by mean shear and cascading the energy to smaller and smaller scales. We now assume that the spectral intensity at wave number k is an expression of the contribution to the total variance from eddies with linear size scale, $\ell \sim k^{-1}$. Furthermore, we assume that a range in k (or ℓ) exists where eddies have lost their memory about the orientation of the mean shear, being important only for the largest eddies, but still do not directly feel the effect of dissipation. For these scales the flow can be considered not only homogeneous in all directions, but also isotropic. The latter means that the spectrum is only a function of the scalar wave number,k, and the final dissipation of the variance of x. From this very simple idea one obtain the well known -5/3-law spectra (e.g. Tennekes and Lumley, 1982):

$$S_u(k) = a_u \epsilon^{2/3} k^{-5/3} \tag{14.8}$$

for velocity where ϵ symbolizes the dissipation of velocity fluctuations by molecular viscocity. For any of the scalars, x, like temperature, T, humidity, q and others, the expressions become

$$S_x(k) = a_x \epsilon^{-1/3} N_x k^{-5/3} \tag{14.9}$$

where N_x is the dissipation of scalar x, and the appearance of ϵ reflects the dominating importance of the velocity turbulence. The fluctuations of the scalars are considered to be the result of this carrying along as passive tracers by the air motion.

The spectral formulations in Eqs. 14.8 and 14.9 are called the inertial forms, and the empirical coefficients in front, a_u and a_x, the Kolmogorov constants for velocity and scalars, respectively. There are strong arguments and some empirical evidence that a_x is the same for all passive scalars (Hill, 1989).

When the spectral forms above apply, it is seen that the turbulence intensity for a given scale is characterised by its scale ($l \sim k^{-1}$) and the dissipations only.

The relation between the spatial structure of the turbulence and the time variation of signals measured by stationary sensors are usually handled by appeal to Taylor's hypothesis of frozen turbulence. This hypothesis states that the temporal evolution seen by the sensor in average is due to the spatial variation of the turbulence field which is advected past the sensor by the mean wind, while the spatial field itself shows comparatively little variation. This means that an eddy of size ℓ will give rise to a variation in the time signal of a stationary probe over a time equal to $\ell/\bar{u}$ where $\bar{u}$ is the mean speed. Translated to the language of spectra this transforms to the following:

$$\begin{aligned} \omega = 2\pi f &= \bar{u} k_1 \\ k_1 S_1(k_1) &= \omega S(\omega) = f S(f) \end{aligned} \tag{14.10}$$

where the transformation characteristics of the spectral density function have been utilized and where k_1 is the component of the k-vector along the mean wind direction. The S_1-spectrum is found by integrating $S(k)$ over the other two k-components:

$$S_1(k_1) = \int_{-\infty}^{\infty} \int_{-\infty}^{\infty} S(\mathbf{k}) dk_2 dk_3 \tag{14.11}$$

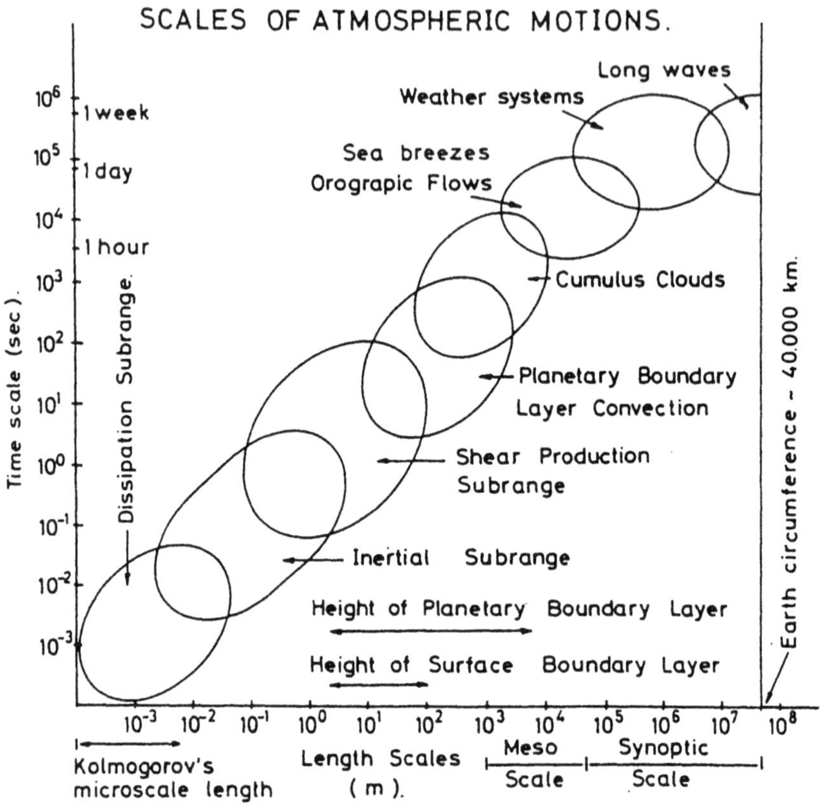

Figure 14.4: *Time and spatial scales for the processes influencing the flow in the atmospheric boundary layer (Busch et al., 1979).*

The spectral formulations and the Taylors hypothesis are based on quite simple ideas mainly of statistical nature and have been found to be so broadly valid that they are extremely useful in both experimental and modelling work. The practical limitations to the use of Taylor's hypothesis show when there is too much variation in the velocity relative to the mean flow, either due to turbulence or due to large vertical wind shear. This can influence the very low frequency, large-scale turbulence (Powell and Elderkin, 1974), and small-scale high frequency measurements (Wyngaard and Clifford, 1977, Mizuno and Panofsky, 1975). The limitation to the validity of the inertial subrange forms of the spectra is found when the assumption behind their validity breaks down, in the high-frequency end by the direct influence of the dissipation and in the low-frequency end through the direct showing of the production scales and the nearness of the surface.

Figure 14.4 presents a more general, but also more qualitative description of the spatial and temporal scales for the different processes influencing the flow in the atmospheric boundary layer. The figure directly shows the scale regions for the inertial subrange, while Taylor's hypothesis usually is used for time scales between 0.01 s and 1 h if the wind speed is not too small.

14.3 Basic equations, problems, and closures

The equation of state is described by the Ideal Gas Law:

$$p = \rho(R_0/M)T \tag{14.12}$$

where T is the temperature, ρ the density, M the molecular weight of the air and R_0 the universal gas constant. The atmosphere is a mixture of many different constituents. We shall concentrate on the most important gases, the *dry air* and the vater vapor. The water vapour and the dry air each exerts their partial pressure according to Eq. 14.12, thus:

$$p = \rho_d(R_0/Md)T + \rho_w(R_0/Mw)T \tag{14.13}$$

where ρ_d and ρ_w are the densities for the dry air and the water vapour, respectively.

Rearranging Eq. 14.13 we obtain

$$p = \rho R_d T_v \tag{14.14}$$

where R_d is the gas constant for dry air, $R_d = 287\ JK^{-1}kg^{-1}$, and Tv is a (the) virtual temperature, i.e. the temperature to be used in connection with the gas law to mimic the influence of water vapour if one wants to use the gas constant for dry air. It is defined by

$$T_v = T(1 + 0.61\,q) \tag{14.15}$$

where $q = \rho_w/\rho$ is the mixing ratio of water vapour in the air. The numerical coefficient derives from the ratio between the mole weights of dry air and water vapour.

Note that we have neglected the influence of the liquid water on the air density. This means that we cannot handle saturated air and thereby the boundary layer clouds that for some conditions can be quite important. Our excuse is that inclusion of the liquid water does complicate the equations quite a bit and that the role of the clouds is treated elsewhere in this book.

Having considered the influence of humidity we now turn to the height variation of the pressure, and introduce the potential temperature, θ, defined as the temperature an air parcel would attain if brought from its height, z, to some reference level usually taken at the sea surface. To a good approximation the vertical pressure distribution is controlled by the hydrostatic equation:

$$\partial p/\partial z = -\rho g \tag{14.16}$$

where g is the acceleration due to gravity. Since the pressure decreases with height, rising air parcels expand and sinking parcels compress. If the resulting vertical mixing takes place without heat exchange, this equation can be combined with the first law of thermodynamics for adiabatic processes, $dp/dT = \rho c_p$ to yield the vertical temperature variation that would be in equilibrium with such a mixing. This gradient is normally called the adiabatic lapse rate for dry air and is denoted Γ and is found to be

$$-\,\partial T/\partial z = \Gamma = g/c_p \sim 0.01\,\mathrm{K/m} \tag{14.17}$$

Combining with the gas law, Equation 14.13, we can now solve the potential temperature at reference pressure, p_0

$$\theta/T = (p_0/p)^{R/c_p} \tag{14.18}$$

and

$$\partial\theta/\partial z = (\partial T/\partial z + \Gamma)\theta/T \tag{14.19}$$

If we use the surface as the reference level, then within the boundary layer $p \simeq p_0$ and $\theta \simeq T$. The vertical variation of θ is therefore given by

$$\theta(z) = T(z) + \Gamma z \tag{14.20}$$

Note that humidity has not been mentioned in this discussion meaning that R and c_p both depend on the water vapour mixing ratio, q and its height variation. Strictly speaking, therefore, the integration for humid air leading to Eq. 14.18 cannot be carried out. However, since R and c_p roughly vary the same way with q, their ratio is not very sensitive and one tends to use Eq. 14.20 for most boundary layer work except of course when clouds become involved. Strictly speaking however, Γ is seen to depend on q, compare Eq. 14.17 also for non-saturated air.

Neglecting the humidity influence on R/c_p in Eq. 14.18, we can define a virtual potential temperature as

$$\theta_v = \theta(1 + 0.61\,q) \tag{14.21}$$

corresponding to equation Eq. 14.15.

Next we proceed to the continuity equation:

$$\frac{\partial\rho}{\partial t} + \frac{\partial}{\partial x_j}\rho u_j = 0 \tag{14.22}$$

where summation over repeated indices is understood here and throughout the text. The coordinate system will be Cartesian with x_1, x_2, x_3 as the coordinates with the corresponding velocities u_1, u_2, u_3.

Within boundary layer meteorology the continuity equation is normally used in the incompressible form:

$$\frac{\partial u_j}{\partial x_j} = 0 \tag{14.23}$$

As seen this form corresponds to the assumption that the relative variations in density are much smaller than the variations in velocity and actually take some profound arguments about the relevant scales of variations for the variables (Businger, 1982). This will mainly be true for flows in the atmospheric boundary layer with scale lengths less than 10 km.

The conservation of momentum is normally used in the following form for the velocity component i in the atmospheric boundary layer:

$$\frac{\partial u_i}{\partial t} + u_j \frac{\partial u_i}{\partial x_j} = -g_i - f_c \epsilon_{ij3} u_j - \frac{1}{\rho}\frac{\partial p}{\partial x_i} + \nu \frac{\partial^2 u_i}{\partial x_j^2} \tag{14.24}$$

where f_c is the Coriolis parameter accounting for the motion to take place on a rotating planet. It is given as $f_c = 2\Omega \sin\phi$ where Ω is the angular frequency of the earth's rotation and ϕ is the latitude. At middle latitude f_c is of the order $10^{-4}s^{-1}$; g_i is zero for $i \neq 3$ and equal to the acceleration of gravity for $i = 3$. ϵ_{ijk} is the socalled alternating unit tensor being +1 if the subscripts are cyclic, -1 if not, and zero if two subscripts are equal. The kinematic viscocity is denoted ν, and as before p is pressure and ρ is the density. The equation shown in 14.24 has already been fairly strongly simplified from the basic Navier-Stokes equation in a rotating coordinate system. Equation 14.23 has been used repeatedly, we have neglected the effect of the rotation in the vertical component as well as other simplifications. It is easily recognized as Newton's second law for an air parcel: the right-hand side is the acceleration, while the left-hand side in the order of appearance is the forcing due to gravity, the effect of the rotating coordinate system, the forcing due to the pressure gradient, and the friction due to molecular forces.

At the same level of approximation as in Eq. 14.24 we can write the conservation of heat as

$$\frac{\partial \theta}{\partial t} + u_j \frac{\partial \theta}{\partial x_j} = k_T \frac{\partial^2 \theta}{\partial x_j^2} - LE/\rho c_p \tag{14.25}$$

where θ is the potential temperature defined in Eq. 14.20, k_T is the thermal diffusivity, E is the production of water vapour from water drops in the air, and L is the heat of evaporation. Relative to Eq. 14.24, Eq. 14.25 is seen to be vastly simpler, missing both the pressure and the Coriolis term. It does, however, couple to the budget for liquid water and humidity through the last term, both through E and through the fact that ρ and c_p pertain to the humid air. To further illustrate this coupling, we present below the budgets for the mixing ratios of liquid water, q_L, and for water vapour, q:

$$\frac{\partial q}{\partial t} + u_j \frac{\partial q}{\partial x_j} = k_q \frac{\partial^2 q}{\partial x_j^2} + E/\rho$$

$$\frac{\partial q_L}{\partial t} + u_j \frac{\partial q_L}{\partial x_j} = S_{q_L}/\rho - E/\rho \tag{14.26}$$

where k_q is the diffusivity of water vapour, E is the vapour production from evaporation, and S_{q_L} is additional local sources of liquid water by convergence of, e.g. spray produced from a water surface or rain drops falling down through the volume.

All of the equations are seen to possess one basic complication: they are nonlinear due to the advection term on the left-hand side. This nonlinearity is a general feature of fluid equation and is the origin of most of the complications one faces when trying to use these equations.

In section 14.2 the meaning of average values were discussed. As pointed out there, the ensemble averages were easiest to use when manipulating the equations. Based on the governing equations established above, we shall now establish the equations for the mean quantities. This is done by expanding all the variables in Eq. 14.23 through Eq. 14.26 into their mean values and fluctuating part, corresponding to Eq. 14.1. Inserting these expansions into the equation, averaging yields the equations for the mean quantities. When averaging the equations, all the fluctuating parameters will disappear except when entering nonlinear. The two important types of nonlinearity are the advection term on the right-hand side of all the equations and those involving the air density, ρ. Before generating the mean equations we shall therefore apply the Boussinesq approximation to remove the last of these nonlinearities. This approximation means letting $\rho\prime/\bar{\rho} \simeq 0$, except for the vertical component of the momentum equation, because here the mean pressure gradient and the mean density times gravity approximately cancel each other through the hydrostatic equation 14.16. From Eq. 14.24 with $i = 3$, we isolate the pressure and the gravity term and multiply $\bar{\rho}$ as:

$$\bar{\rho}(1 + \rho'/\bar{\rho})g - \frac{\partial}{\partial x_3}(p + p') \simeq \rho' g - \frac{\partial p'}{\partial x_3} \tag{14.27}$$

The equation of state (14.14) can be expanded for small variations in the parameters as:

$$\frac{p'}{\bar{p}} = \frac{\rho'}{\bar{\rho}} + \frac{T_v'}{\overline{\overline{T_v}}} + \frac{\rho' T_v'}{\bar{\rho}\overline{T_v}} \tag{14.28}$$

Considerations of the order of magnitude of the different terms, e.g. Businger (1982) and Stull (1988), yield:

$$-\frac{\rho'}{\bar{\rho}} = \frac{T_v'}{\overline{\overline{T_v}}} \simeq \frac{\theta_v'}{\overline{\overline{\theta_v}}} = \frac{\theta'}{\overline{\overline{\theta}}} \tag{14.29}$$

where the last step uses the approximative relations between between θ_v and T_v within the boundary layer, compare Eq. 14.19 through Eq. 14.21.

Based on this discussion, we can now replace ρ by $\bar{\rho}$ everywhere in Eq. 14.24 through Eq. 14.26 and rewrite the momentum equation as

$$\frac{\partial u_i}{\partial t} + u_j \frac{\partial u_i}{\partial x_j} = -g_i \left(1 - \frac{\theta_v'}{\bar{\theta}_v}\right) - \frac{1}{\bar{\rho}} \frac{\partial p}{\partial x_i} - \epsilon_{ij3} f_c u_j + \nu \frac{\partial^2 u_i}{\partial x_j^2} \tag{14.30}$$

Establishing the conditions for the validity of the approximations used above is not straightforward. It takes some pretty profound discussions of the scale of variability for the various terms in the equation. It is usually described as the shallow motion approximation or a somewhat more restrictictive shallow convection approximation (Mahrt,1986). It will *mostly* be fulfilled for scales of variability contained vertically within the boundary layer and horizontally within 15 *km*.

We now revert to the equations of the mean flow to obtain (Busch, 1973):

a) the continuity equation:

$$\frac{\partial \overline{u_i}}{\partial x_i} = 0 \tag{14.31}$$

b) the momentum equation:

$$\begin{aligned} \partial \bar{u}_i / \partial t + \overline{u_j} \frac{\partial \overline{u_i}}{\partial x_j} &= -g_i - f_c \epsilon_{ij_3} - \frac{1}{\bar{\rho}} \frac{\partial \bar{p}}{\partial x_i} \\ &\quad + \nu \frac{\partial^2 \overline{u_j}}{\partial x_j^2} - \frac{\partial}{\partial x_j} \left(\overline{u_i' u_j'}\right) \end{aligned} \tag{14.32}$$

c) the heat or temperature equation:

$$\partial \bar{\theta} / \partial t + \overline{u_j} \frac{\partial \bar{\theta}}{\partial x_j} = k_T \frac{\partial^2 \bar{\theta}}{\partial x_j^2} - \frac{L\bar{E}}{\bar{\rho} c_p} - \frac{\partial}{\partial x_j} \left(\overline{u_j' \theta'}\right) \tag{14.33}$$

d) the equation for the water vapour mixing ratio:

$$\partial q / \partial t + \overline{u_j} \frac{\partial \bar{q}}{\partial x_j} = k_q \frac{\partial^2 \bar{q}}{\partial x_j^2} + \frac{\bar{E}}{\bar{\rho}} - \frac{\partial}{\partial x_j} \left(\overline{u_j' q'}\right) \tag{14.34}$$

Equations 3.21 through 3.23 show the socalled closure problem, i.e. the equations to derive the mean flow contain terms involving the statistics of the fluctuating components as well. Here $\frac{\partial}{\partial x_j}\overline{u_i' u_j'}$, $\frac{\partial}{\partial x_j}\overline{u_j' \theta'}$ and $\frac{\theta}{\partial x_j}\overline{u_i' q'}$. Physically these terms are gradients of turbulent fluxes of the different parameters. The covariance terms involving a velocity component describe fluxes as can be seen by considering the consequences for the flux of the x-parameter in the j-direction, of $\overline{u_j' x'}$ having a positive, negative or zero value. To close the set of equations, one must either prescribe the higher-order terms or relate them to the mean parameters already in the equations. One talks about closures of different order.

In zero-order closure the terms are prescribed by appeal to physical hypotheses. An example is the different similarity formulations to be described later.

In first-order closure the flux terms are characterized through the local gradient and a *turbulent* diffusion coefficient, normally denoted K. For example:

$$\begin{aligned} \overline{u_i'\xi'} &= K_{i\xi}\frac{\partial\overline{\xi}}{\partial x_i} \\ \overline{u_i'u_j'} &= K_{ij}\left(\frac{\partial\overline{u_i}}{\partial x_j}+\frac{\partial\overline{u_i}}{\partial x_i}\right) \end{aligned} \tag{14.35}$$

The K-diffusion concept is borrowed from the molecular gas theory that has given inspiration to how to determine K as well. In the molecular theory the diffusivity, k is found from expressions like (e.g. Chapman and Cowling, 1970):

$$k \sim \langle \text{ mean speed of the molecules } \rangle \cdot \langle \text{ free path between molecular encounters} \rangle.$$

Following this idea we need a length and velocity scale for the turbulence to build a diffusivity. This is normally done by some combination of physical hypotheses and additional equations that can be used to determine these scales. The first efforts concentrated on the physical hyphotheses to prescribe K throughout the boundary layer. As an example of this approach can be mentioned the work of Businger and Araya (1974). When additional equations are carried to determine K, the simplest method used is to determine the equation for the variance of the turbulence velocity. Normally, this is called the energy equation because of the relation between the turbulence velocity and the turbulent kinetic energy. The standard deviation is then used as the velocity scale, and the length scale is derived using this velocity scale and the dissipation of turbulence energy, see Eq. 14.8, derived from a specially derived equation, see e.g. Arpaci and Larsen, 1984. This type of closure is normally called the $K-\epsilon$ closure and has been extensively used in connection with fluid modelling.

Strictly speaking, the $K-\epsilon$ closure is a type of higher-order closure because an equation for a second-order moment, the turbulence energy, is introduced to determine the diffusivity used to relate the covariance terms (second-order moments) in Eq. 14.32 through Eq. 14.34. The $K-\epsilon$ closure is often referred to as one and a half-order closure. A more strict type of higher-order closure is obtained by deriving equations for the covariances $\overline{u_j'u_i'}$, $\overline{u_j'\theta'}$ and $\overline{u_i'q'}$.

The equations for the fluctuating quantities can be generated by subtracting the equations for the mean flow, Eq. 14.32 through Eq. 14.34 from the basic equations, Eq. 14.24 through Eq. 14.26. To generate the equation for for a covariance say, $\overline{u_j'u_i'}$, we produce the equation for each of the fluxtuating quantities and note that

$$\frac{\partial}{\partial t}\overline{u_i'u_j'} = \overline{u_j'\frac{\partial}{\partial t}u_i'} + \overline{u_i'\frac{\partial}{\partial t}u_j'} \tag{14.36}$$

When producing the equations for the second-order moments so as to solve the equations for the mean flow, one will note that the equations generated contain third-order moments meaning that the closure problem will not go away. For these third order moments one can continue and generate equations for the higher order moments as previously, or one can close with a diffusivity hypothesis, e.g.

$$\overline{u_i'u_j'\theta'} = K\left(\frac{\partial}{\partial x_i}\overline{u_j'\theta'} + \frac{\partial}{\partial x_j}\overline{u_i'\theta'}\right) \tag{14.37}$$

Solving the equations for both the mean values and the second-order moments has become quite a well-established technique, while including also the third-order equations is more rarely done. This is because of the difficulty in handling the number of equation, but also because many of the terms that appear in the higher-order equations are very difficult to relate to conceptually and even more uncertain to measure, see e.g. Wyngaard (1973, 1982).

We can summarize the disadvantages and advantages by the trying to close the equations by generating equations for the higher-order moments.

Disadvantages:

1. An ever increasing number of equations with associated boundary and initial conditions.
2. Eventually, one falls back on using the local K-diffusivity closure at the final order of equations. As stated above, the K-diffusivity for turbulence is an useful hyphothesis that is known not always to work, and it does not become more valid by being applied rather to higher-order moments than to second-order moments. Discussions and tests of validity just become more intransparent and difficult.

Advantages:

1. As results, one often wants the higher-order moments, e.g. all of the turbulence fluxes of momentum, heat, and water vapour are second-order moments.
2. By applying the K-diffusivity closure at ever higher-order moments, one might hope for larger generality because the higher-order moments get more and more determined by the structure of the small scale turbulence. Recall the simple picture in section 2 where the eddies gradually lose the memory of how they are produced when moving to smaller scales.
3. In spite of the principal criticisms that can be raised against the higher-order closure, it is a rational approach to solving the equations of motion that in extensive literature on the method and its results has shown its value.

Some of the shortcomings of the local closures have long been well-known: during the afternoon unstable boundary layer, the potential temperature can often be constant with height over the major part of the boundary layer. Under such conditions the vertical heat flux can be very strong, and therefore obviously does not follow a parameterization like Eq. 14.35. Correspondingly, it is well-known within air pollution meteorology that in its inital phase, the dispersion of a plume from a point source is inconsistent with the assumption of a diffusivity being a function of flow conditions only. It would have to be a function of the distance from the source as well (Troen et al, 1980). This knowledge has led to efforts of repairing or replacing the local closure. In one approach, i.e. the transient turbulence theory (Stull,1988), the locality of the closure is given up, and the system simply specifies an exchange between non-neighbour air parcels according to certain rules. Another approach postulates that the diffusivity should be defined in Fourier space so that a diffusivity, $K(k)$, corresponds to each eddy-size or wave number, see Berkowicz and Prahm (1979), Troen et al (1980). Both systems work quite well with numerical codes and can be used when mean profiles of parameters are such that local closure gives wrong results. As the only caution one could say that they do not obviously give correct answers under all conditions. There is less empirical knowledge on how they perform in different situations than for the other closure systems that have been used much more extensively.

Finally, we turn to a modelling principle that is denoted Large Eddy Simulation or LES. Here the ensemble average, being the basic type of average when manipulating the equations above, is dropped for a spatial average more precisely defined. This has some advantages and some disadvantages as will be seen below.

The decomposition of signals is formally defined in the same way as in Eq. 14.1:

$$x = \{x\} + x' \tag{14.38}$$

where, however, the average value $\{x\}$ is now defined through some kind of spatial average over a volume that will typical correspond to a grid size, see e.g. Wyngaard (1982) :

$$\{x\} = \int_{-\infty}^{\infty}\int_{-\infty}^{\infty}\int_{-\infty}^{\infty} G(\mathbf{r}-\mathbf{r}')\,x(\mathbf{r}')\,d\mathbf{r}' \tag{14.39}$$

The equations for the average, $\{\cdot\}$, flow look very much like those of Eq. 14.31 through Eq. 14.34, but the second-order moments are now defined by:

$$\{x'y'\} = \{(x-\{x\})(y-\{y\})\} \tag{14.40}$$

with a closure equation of $\{u_i'y'\} = K_{iy}\frac{\partial\{y\}}{\partial x_i}$.

The main advantage of the LES modelling is that the scales of motion involved in $\{x\}$ and x' are much better defined than the normal Reynolds decomposition, where in principle all scales influence x'. The LES has a number of mathematical complexities, however. Both $\{x\}$ and x' are now stochastic functions, meaning that also K_{iy} is stochastic. A further estimation of $\{x'y'\}$ is more complex because $\{x\}$ and x' are now correlated through Eq. 14.36. In spite of these difficulties the techniques is becoming more and more used. This is because this model type can directly output the time and spatial variation of the larger eddies, those that are resolved by the model, while the models operating with ensemble averages by their nature can produce only ensemble averages of the different moments.

14.4 Simple Boundary Layers

In the former sessions we have generally used the coordinate system x_1, x_2, x_3 with associated wind speeds u_1, u_2, u_3. The only specifications of these coordinate systems have been that the 3-axis has been vertical. As we move down into simple boundary layers we shall more and more switch to a x, y, z coordinate system with associated wind speeds u, v, w. Here, x is along the mean wind close to the surface, meaning that u is the only wind component with a mean value; z is vertical and y is the other horizontal direction perpendicular to x.

To simplify the equations for the mean flow, Eq. 14.31 through Eq. 14.34, we define the geostrophic wind.

$$U_g = -\frac{1}{f_c\bar{\rho}}\frac{\partial\bar{p}}{\partial y} \qquad V_g = +\frac{1}{f_c\bar{\rho}}\frac{\partial\bar{p}}{\partial x} \tag{14.41}$$

Note that U_g is the component of the geostrophic wind along the mean wind direction at the surface. Inserting Eq. 14.5 into Eq. 14.29 through Eq. 14.32 and recalling that in this approximation $\bar{\rho}g \simeq \partial\bar{p}/\partial z$, the equations for the mean flow come out as:

$$\begin{aligned}
\bar{w} &= 0 \\
\frac{\partial\bar{u}}{\partial t} + \overline{u_j}\frac{\partial\bar{u}}{\partial x_j} &= -f_c(V_g - \bar{v}) - \frac{\partial}{\partial x_j}\overline{u_j'u'} \\
\frac{\partial\bar{v}}{\partial t} + \overline{u_j}\frac{\partial\bar{v}}{\partial x_j} &= f_c(U_g - \bar{u}) - \frac{\partial}{\partial x_j}\overline{u_j'v'} \\
\frac{\partial\bar{\theta}}{\partial t} + \overline{u_j}\frac{\partial\bar{\theta}}{\partial x_j} &= -L\bar{E}/\bar{\rho}c_p - \frac{\partial}{\partial x_j}\overline{u_j'\theta} \\
\frac{\partial\bar{q}}{\partial t} + \overline{u_j}\frac{\partial\bar{q}}{\partial x_j} &= \bar{E}/\bar{\rho} - \frac{\partial}{\partial x_j}\overline{u_j'q'}
\end{aligned} \tag{14.42}$$

where we have neglected the molecular terms.

Equation Eq. 14.43 shows that if the left-hand sides are constant, then the two scalars will vary linearly with height, provided of course that we can neglect the E-terms.

The next step in simplifications comes when the left-hand side of Eq. 14.43 is assumed to be zero. In physical terms this corresponds to calling the flow statistically stationary and horizontally homogeneous. Eq. 14.43 now takes the form:

$$\begin{aligned}
\overline{w'\theta'} = \text{const.}\ \overline{w'q'} &= \text{const.} \\
-f_c(V_g - \bar{v}) - \frac{\partial}{\partial z}\overline{u'w'} &= 0 \\
f_c(U_g - \bar{u}) - \frac{\partial}{\partial z}\overline{v'w'} &= 0
\end{aligned} \tag{14.43}$$

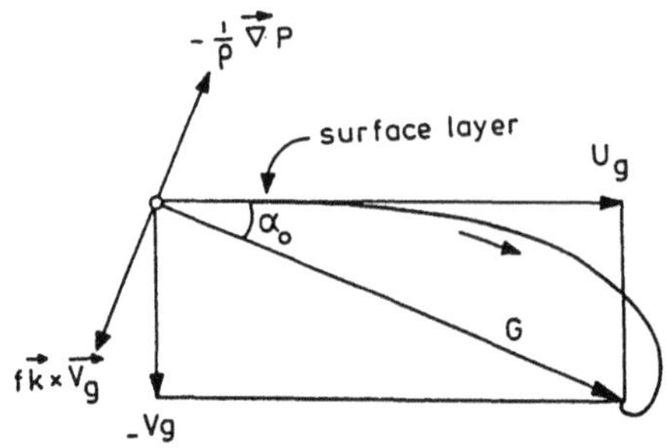

Figure 14.5: *The variation of the wind with height, taken from Tennekes (1982).*

The two equations in Eq. 14.44 can be closed with the type of K-diffusivity already discussed above, and one obtains a spiral-like variation of the wind with height as shown in Figure 14.5. It is often called the Ekman spiral, after the first proposer of this solution (Ekman, 1905).

The exact form of the spiral in Figure 14.5 will depend of course on the details of the closure used. In practice, however, one will often/mostly see a large deviation from the figure because the real world does not satisfy the stationarity and horizontal homogeneity assumption. Often one will find that horizontal temperature gradients induce vertical changes in the geostrophic wind, because under such conditions the hydrostatic equation (14.16) can not in generally be fulfilled with one constant $\bar{\rho}$. Also the curvature of the isobars, instationarity, and inhomogeneity will introduce changes in the equations, see e.g. Hasse (1976).

For the height, z, approaching zero, v goes to zero (because by definition only u can have a mean component close to the ground). From Eq. 14.44 we therefore conclude that:

$$\begin{aligned} \partial\overline{u'w'}/\partial z &= -f_c V_g, \text{ or : } \overline{u'w'} = \overline{u'w'_0} - f_c V_g z \\ \partial\overline{v'w'}/\partial z &= f_c(U_g - u(z)) \end{aligned} \tag{14.44}$$

where subscript 0 indicates surface value.

A closer inspection of the scales involved and the fact that the friction very close to the surface must be in the opposite direction of the mean speed implies that close to the surface we have:

$$\begin{aligned} -\overline{u'w'} &= \overline{u'w'_0} = u_*^2 = \tau_0/\bar{\rho} \\ \overline{v'w'} &= 0 \end{aligned} \tag{14.45}$$

where we have defined the surface stress, τ_0, and the associated scale velocity u_*.

From the above we conclude that for stationary and horizontally homogeneous situations these exist close to the surface a layer, where also the turning of the wind and the change of stress with height can be neglected. The charactestics of this layer can be further illuminated by considering the equations for the variances of the turbulent variations. Specifically the turbulent kinetic energy, $\bar{e} = \frac{1}{2}\overline{u_i'^2}, \overline{\theta'^2}$ and $\overline{q'^2}$. In section 3 we have already discussed how to produce equations for the covariances. The procedure is the same for the variances, just a bit simpler, because more terms can be contracted: For $\bar{e}$ we start with (3.19), and obtain after some manipulation:

$$0 = \frac{\partial \bar{e}}{\partial t} + u_j \frac{\partial \bar{e}}{\partial x_j} = -\overline{u'w'}\,\partial \bar{u}/\partial z + \frac{g}{\theta_v}\overline{w'\theta_v'} - \frac{\partial}{\partial z}\left(\overline{w'\left(e + \frac{1}{\bar{\rho}}p'\right)}\right) - \epsilon \qquad (14.46)$$

The $\overline{\theta'^2}$ and $\overline{q'^2}$ equations derive more easily from Eqs. 14.25 and 14.26, neglecting the E-terms.

$$\begin{aligned} 0 = \frac{\partial \overline{\theta'^2}}{\partial t} + \overline{u_j}\frac{\partial \overline{\theta'^2}}{\partial x_j} &= -2\overline{\theta' w'}\partial\bar{\theta}/\partial z - \frac{\partial}{\partial z}\overline{w'\theta'^2} - 2N_\theta \\ 0 = \frac{\partial \overline{q'^2}}{\partial t} + \overline{u_j}\frac{\partial \overline{q'^2}}{\partial x_j} &= -2\overline{q' w'}\partial\bar{q}/\partial z - \frac{\partial}{\partial z}\overline{w'q'^2} - 2N_q \end{aligned} \qquad (14.47)$$

Here, ϵ, N_θ and N_q are the dissipations we have already met in section 14.2, and which we can now define from the equations as:

$$\epsilon = \nu\overline{\left(\frac{\partial u_i'}{\partial x_j}\right)^2}, \; N_\theta = k_T\overline{\left(\frac{\partial \theta'}{\partial x_j}\right)^2}, \; N_q = k_q\overline{\left(\frac{\partial q'}{\partial x_j}\right)^2} \qquad (14.48)$$

In Eqs. 14.46 and 14.47 the terms on the right-hand side reflect the different sources and sinks for the turbulence. The first terms describe the production, P, of turbulent fluctuations through interaction between the turbulence and the shear of the mean flow. Dependent on its sign, the second term, B, in the equation for the kinetic energy describes the production or destruction of turbulent kinetic energy due to buoyancy. The last term in all of the equations has already been described in Eq. 14.48 as the dissipation of turbulent variations by molecular forces. Likewise, in all of the equations the last but one terms are called the divergence terms, D. They can be shown to redistribute the energy between different components of u_i as well as between different points in space.

Considerations of Eq. 14.45 through Eq. 14.47 have led to the idea that to describe the flow in the near-surface layer, one needs only to know the stress, heat flux, and water vapour flux together with the buoyancy parameter $g/\bar{\theta}_v$ and measuring height, z. Here, the latter is to account for the vertical gradients appearing in the equations. In the Monin-Obukhov scaling system these quantities have been combined to a number of scales that together describe the flow in the surface layer. They are:

$$\begin{aligned} u_* &= \left(-\overline{u'w_0'}\right)^{1/2} & \theta_* &= -\overline{\theta' w_0'}/u_* \\ q_* &= -\overline{q'w_0'}/u_* & g/\bar{\theta}_v & \text{ and } z \end{aligned} \qquad (14.49)$$

Note, that $\overline{w'\theta_v'} = \overline{w'\theta'} + 0.61\bar{\theta}\overline{w'\theta'}$, according to Eq. 14.21, θ_{v*} is defined accordingly. Some of these scales are combined to the stability scale, the Monin-Obuckov stability length, given by:

$$L = +\frac{\bar{\theta_j}}{g\kappa} u_*^2/\theta_{v*} \qquad (14.50)$$

where κ is the socalled von Kármán constant that will later be introduced more rationally. As seen from Eq. 14.46 L can be considered as a ratio between the mechanical turbulence production and the buoyancy force. If the heat flux is positive, as on a sunny day with a warm suface, L is negative, and the situation is called unstable. On the other hand, if the heat flux is downward, as on a clear sky night with radiational cooling of the ground, L is positive. For both situations L increases with wind speed. L becomes largest, when the heat flux is very small, i.e. when the air and the ground are of roughly the same temperature. This situation is called thermally neutral and can be imagined as a situation with reasonably high wind and overcast sky, compare Figure 14.2. The arguments behind this scaling approach were originally introduced by Obukhov in 1943 in a paper that has been translated later on (Obukhov, 1971). The rule is that if any local turbulence statistics is non-dimensionalized with the parameters in Eq. 14.50, then the resulting non-dimensional groups will be functions of z/L only. Below this is illustrated on Eqs. 14.46 and 14.47 where the nondimensional functions are traditionally denoted ϕ and where the terms on the right-hand side appear in the same orders as in the original equations. Eq. 14.46 is multiplied by $\kappa z/u_*^3$ to give:

$$0 = \phi_m(z/L) - z/L - z/L\frac{\partial}{\partial \frac{z}{L}}\phi_D\left(\frac{z}{L}\right) - \phi_\epsilon(z/L) \qquad (14.51)$$

Correspondingly, the two equations in Eq. 14.47 are multiplied by $\kappa z/(u_*\theta_*^2)$ and $\kappa z/(u_*q_*^2)$, respectively, to yield:

$$\begin{aligned} 0 &= 2\phi_h(z/L) \quad -z/L\frac{\partial}{\partial \frac{z}{L}}\phi_{D\theta}\left(\tfrac{z}{L}\right) \quad -2\phi_{N\theta}(z/L) \\ 0 &= 2\phi_q(z/L) \quad -z/L\frac{\partial}{\partial \frac{z}{L}}\phi_{Dq}\left(\tfrac{z}{L}\right) \quad -2\phi_{Nq}(z/L) \end{aligned} \qquad (14.52)$$

The von Kármán constant, κ, is defined such that $\phi_m(z/L)$, the non-dimensional profile function, equals 1 for $z/L = 0$. ϕ_m, ϕ_h and ϕ_q are defined as

$$\phi_m(z/L) = \frac{\kappa z}{u_*}\frac{\partial \bar{u}}{\partial z}\ , \ \phi_h\left(\frac{z}{L}\right) = \frac{\kappa z}{\theta_*}\frac{\partial \bar{\theta}}{\partial z}\ , \ \phi_q\left(\frac{z}{L}\right) = \frac{\kappa z}{q_*}\frac{\partial \bar{q}}{\partial z} \qquad (14.53)$$

The von Kármán constant must be experimentally determined and is found to be between 0.33 and 0.43 in different experiments. The other similarity functions defined in Eqs. 14.52 and 14.53 are the divergence functions, ϕ_D and the dissipation functions, ϕ_ϵ, ϕ_N.

Equation Eq. 14.52 illustrates how the buoyancy forces will produce turbulence for unstable situations, when L is negative, and destroy turbulence for stable conditions when L is positive.

The hypothesis about the surface boundary layer have been experimentally studied through a number of field experiments, starting with the wellknown Kansas 1968 experiment, from which the measuring set-up is shown in Figure 14.6.

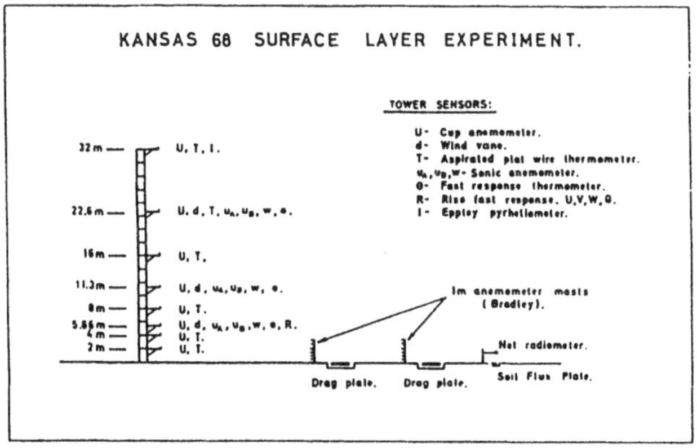

Figure 14.6: *Schematic presentation of the core instrumentation during the Kansas 1968 field experiment (Busch et al., 1979).*

Although first of its kind, the Kansas experiment is probably also the most complete surface-layer experiment carried out until today. In the context of the equations discussed above, the important missing parameters in the measuring program were humidity and pressure. As seen from the discussion above, the surface layer hyphoteses predict characteristics of the vertical variation of the mean values, fluxes, and their relation. Therefore the experiment was set up to measure mean values and turbulence at a number of heights with the turbulence (the most complicated) at three heights, the minimum number to measure the height variation of the divergence terms in the equations above.

From the Kansas experiment and its successors information has come on the behaviour of the different similarity funtions, some consensus both on the behaviour of the surface layer and on the limitation to its applicability, and some controversy that are still not properly resolved. Although controversy about the functions still exist, good qualitative agreement has been found between the different experiments (Businger et al., 1972; Wyngaard and Coté, 1972; Dyer, 1974; Högström, 1990). In Figure 14.7 is illustrated the results from such experiments.

The surface layer is only a small part of the total boundary layer. Its height is about 0.1 times the boundary layer height. Both because of this and because the surface layer studies showed that even here the parameters reflecting the total boundary layer were important for some quantities, the number of studies of all of the boundary layers increased since the early seventies.

The development in the description of simple boundary layers is best conducted seperately for the unstable, the neutral, and the stable boundary layer, with the stability being defined as in connection with the Eq. 14.49 through Eq. 14.52, and referring to Figure 14.2 for a qualitative picture.

As indicated by the name, the neutral boundary layer is controlled by mechanically produced turbulence, and the wind up through the boundary layer is described by Eq. 14.44, the height dependent balance between pressure force, Coriolis force, and the turbulent stress. The buoyancy term is unimportant. In its classical form the neutral boundary layer can be be very elegantly described by a similarity description, adding only the surface roughness to u_*, because the boundary layer height is considered to be given by the surface scale u_* and Coriolis parameter, see Tennekes (1973, 1982)

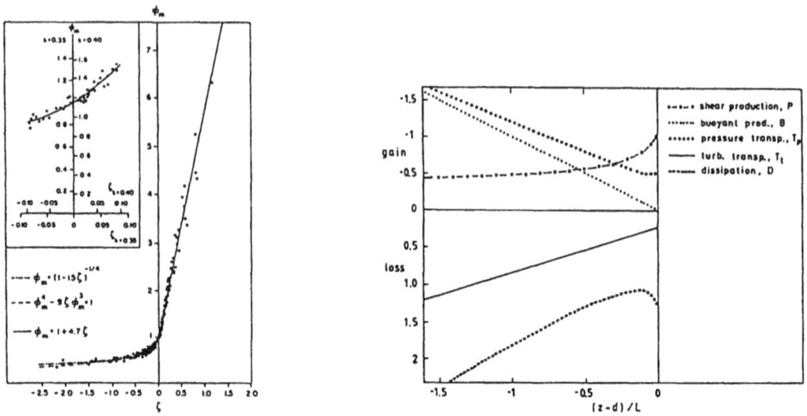

Figure 14.7: *Results from surface layer experiments. (a) shows the behaviour of ϕ_m versus $\xi = z/L$ from Businger et al. 1971). (b) shows the relative importance of the different terms in Eq. 14.51 according to measurements by Högström (1990) and plotted versus $(z-d)/L$, where d is the displacement height to be defined later. In this Figure ϕ_D has been expanded into its turbulence T_t and its pressure T_p component.*

and Tennekes and Lumley (1982). In practice, however, also the neutral boundary layer height very often is controlled by the height of the first elevated inversion, at least at mid latitude. It must be emphasized that although the neutral boundary layer carries rather little sensible heat flux, it can be responsible for large water vapour fluxes, since the neutral boundary layer is often associated with large wind speed and thereby high turbulence mixing.

Unstable boundary layers are characterized by a strong positive heat flux and often a well developed capping inversion. They are fairly conveniently described by the strong vertical fluxes because the vertical mixing is of dominating importance. At the top of the unstable surface layer the production of turbulence from the mean shear becomes unimportant compared to the buoyancy production, compare Figure 14.2(a) and Eq. 14.51. This leads to the idea that the dominating velocity scale for the turbulence in the middle of the unstable boundary layer must be constructed from the surface heat flux and the the height of the boundary layer that constitute an upper limit for the size of the eddies carrying the fluxes. In the literature this height is called either z_I or h for unstable conditions. These ideas were supported by field, laboratory, and numerical experiments (Kaimal et al., 1976; Deardorff, 1970), and led to successful efforts to describe the turbulence in large parts of the unstable boundary layer by the following scales:

$$\begin{aligned} w_* = \left(z_I\, \overline{w'\theta'_{v0}}\, g/\bar{\theta}\right)^{1/3} &= u_*(-z_I/\kappa L)^{1/3} \\ T_* = -\overline{w'\theta'_0}/w_* \,,\; Q_* &= -\overline{q'w'_0}/w_* \\ h \text{ or } z_I & \end{aligned} \tag{14.54}$$

The scales in Eq. 14.55 are usually denoted the mixed layer scales. Note that the fluxes used have to be evaluated at the surface, just as for the Monin-Obukhov scales. The mixed layer scales can be used

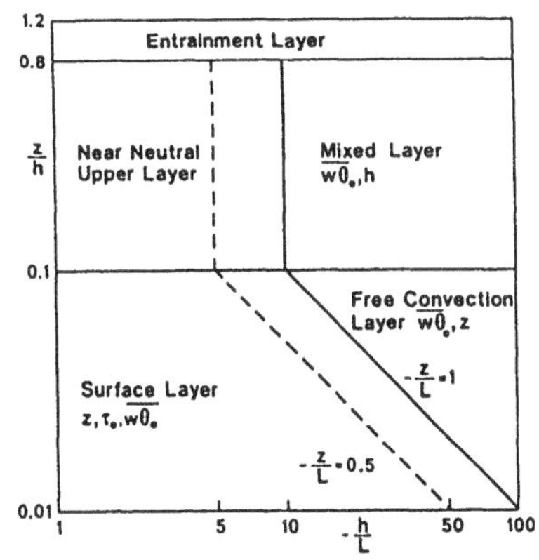

Figure 14.8: *The different scaling regions in the unstable boundary layer according to to Holtslag and Nieuwstadt (1986).*

to scale the equation of motion, just as the Monin-Obukhov scales were used in Eq. 14.51 through Eq. 14.53. The similarity functions are often called ψ and are functions of (z/h) corresponding to ϕ in the surface layer, and as an aesthetic advantage the mixed layer scales do not involve an emperical constant like the von Kármán constant in the surface layer scales. A systematic description of the turbulence equations in the mixed layer is given in Stull (1988).

In between the mixed layer and the surface layer is a matching layer where both the surface layer and the mixed layer scalings are supposed to work. This is called the matching layer or the free convection layer. It is a layer where the velocity shear is unimportant as in the mixed layer, but where also the height of the boundary layer is unimportant as in the surface layer. Based on these considerations, a number of rules can be derived for the behaviour of the turbulence statistics in the stronly unstable surface layer (Wyngaard and Coté 1971). Given the proliferation of scaling regimes in the boundary layer, several authors have tried to present systematic descriptions as to when and where the different scaling regimes apply (Nicholls and Readings, 1979; Olesen et al., 1984; Holtslag and Nieuwstadt, 1986). Figure 14.8 shows the organization of the scaling regimes in the unstable boundary layer from the latter authors.

Apart from the scaling regions discussed above, Figure 14.8 shows two new regions. The near-neutral upper layer that cannot easily be simplified. This is because here all scales retain their importance. The entrainment layer around the top of the boundary layer, through which the boundary layer interacts with and entrains air from the free atmosphere above (Zilitinkevich, 1991).

Studies have shown that in many ways the top of the boundary layer appears as a surface layer turned upside down, e.g. the scale of turbulence decreases when one approaches the entrainment zone, much in the same way as it decreases when approachindg the surface. This led Wyngaard and Brost (1984) to suggest that through the unstable boundary layer the vertical diffusion of passive scalars can be computed as a superposition of top-down and bottom-up components driven by the scalar fluxes at the bottom of the boundary layer and at the mixed layer top, respectively. For example the vertical concentration gradients and fluxes of x can be represented by:

$$\begin{aligned} \bar{x} &= \bar{x}_b + \bar{x}_t \\ \frac{\partial \bar{x}}{\partial z} &= \frac{\partial \bar{x}_b}{\partial z} + \frac{\partial \bar{x}_t}{\partial z} \\ \overline{x'w'} &= \overline{x'w'_h} \cdot z/h + (1 - z/h)\overline{x'w'_s} \end{aligned} \tag{14.55}$$

The gradients of $\bar{x}_b$ and $\bar{x}_t$ each scale with their version of the mixed layer scaling:

$$\frac{\partial \bar{x}_b}{\partial z} = \frac{-\overline{w'x'_s}}{w_* h} g_b\left(\frac{z}{h}\right), \quad \frac{\partial \bar{x}_t}{\partial z} = -\frac{\overline{w'x'_h}}{w_* h} g_t\left(\frac{z}{h}\right) \tag{14.56}$$

where the two functions must be constructed taking into account the similarities and differences between the turbulence processes at the surface and at the top of the boundary layer, see Wyngaard and Brost (1984).

Recall the discussion in section 14.3 about use of K-diffusivity and local closure in the unstable boundary layer when the vertical gradient was zero. It is seen that this problem has been avoided with this superposition, because each of the gradient $\partial \bar{x}_b/\partial z$ and $\partial x_t/\partial z$ are non-zero. The approach has been used with some succes, exactly in situations where the gradient was zero (Fairall, 1987).

Finally, we turn to the stable planetary boundary layer. It is characterized by negative heat flux, often low velocity, and reduced turbulence activity, both due to buoyancy damping and low velocity. As opposed to the unstable boundary layer, real stable boundary layers are not easy to model, because the vertical fluxes are small, and other processes, like radiation balances and advection, therefore important. In terms of Eq. 14.51 we see that as stabilty increases, z/L will eventually become so large that the loss terms in the equation will overcome the production term, ϕ_m, and the turbulence will cease. This argument can be further illustrated by the behaviour of the Richardson numbers. The flux Richardson number is built as the ratio between the two first terms on the right hand-side of Eq. 14.46

$$Rf = \frac{g}{\bar{\theta}_v} \frac{\overline{w'\theta'_v}}{\overline{u'w'}\,\frac{\partial \bar{u}}{\partial z}} \tag{14.57}$$

From Eq. 14.46 is seen that turbulence is unlikely to exist when Rf becomes of the order of or exceeds one. Since Rf contains a mixture of turbulence and mean value gradient terms it was earlier fairly difficult to estimate experimentally this quantity. Therefore, gradient Richardson number, Ri, was formulated:

$$Ri = \frac{g}{\bar{\theta}_v} \frac{\partial \bar{\theta}_v}{\partial z} \Big/ \left(\frac{\partial \bar{u}}{\partial z}\right)^2 \tag{14.58}$$

The relation between the two Richardson numbers is obvious, but as seen Ri can be determined from measurements of profiles of mean values only. In both theoretical and experimental studies Ri has been found to have a critical value, Ri_c, above which continuous turbulence cannot exist. The value

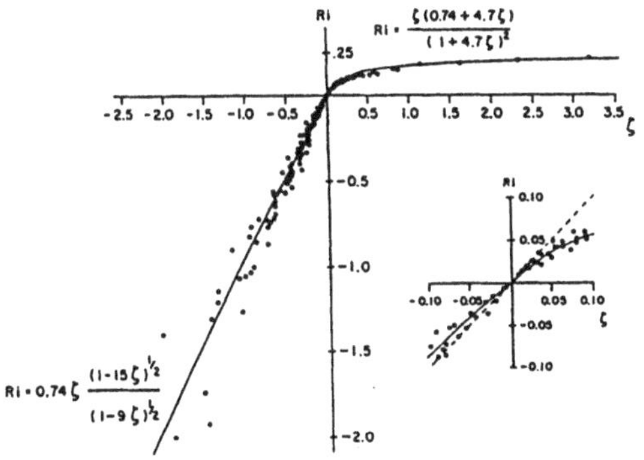

Figure 14.9: *The relation between the Richardson number and $\xi = z/L$ from (Businger et al., 1971).*

for Ri_c is generally found to be between 0.15 and 0.3 in most studies. From a theoretical point of view the value will depend on both how one relates Ri to Rf and on how the other two terms in Eq. 14.46 are estimated. From the experimental studies we present the relation between Ri and z/L from the Kansas experiment where the data seem to show a critical value around 0.2

As the stability increases, the surface layer scaling laws take a special form, called the z-less. This can be illustrated from the behaviour of the ϕ_m function reported in Figure 14.7(a). On the stable side it is seen to vary linearly with z/L, say as 1+ 5 z/L (Dyer, 1974). From the definition of ϕ_m the consequence of this form is that for large z/L, $\bar{u}/z$ becomes independent of z. Many of the other local turbulence characteristics show this behaviour, hence the name (Wyngaard, 1973).

These characteristics of the stable boundary layer led Nieuwstadt (1984) to suggest a similarity model for the stable boundary layer based on the assumption that $Rf = Ri = Ri_c$ and that the turbulence would scale with the local fluxes, not the surface values of the fluxes as in the the Monin-Obukhov and mixed layer similarity but the local values. This necessitated estimates of the height variation of stress and heat flux relative to the surface values, such that the flow in the entire boundary layer could be computed. The local scaling parameters are normally related to the surface values from expressions like:

$$
\begin{aligned}
\tau^{1/2}/u_* &= (1 - z/h)^{\alpha_1}\ ,\ \overline{w'\theta'}/\overline{w'\theta'_0} = (1 - z/h)^{\alpha_2} \\
\Lambda/L &= (1 - z/h)^{\alpha_3} \\
\Lambda &= \frac{\theta_j}{g\kappa}\tau^{3/2}/\overline{w'\theta'}
\end{aligned}
\tag{14.59}
$$

As seen Λ is a Monin-Obukhov length determined from the local fluxes. As $\Lambda \to L$ for $z \to 0$, the local scaling and the surface layer scaling overlap for small z. The α-values can be estimated from measurements or closure hypotheses. Although complete consensus is missing, they are all of the order one (Caughey, 1982).

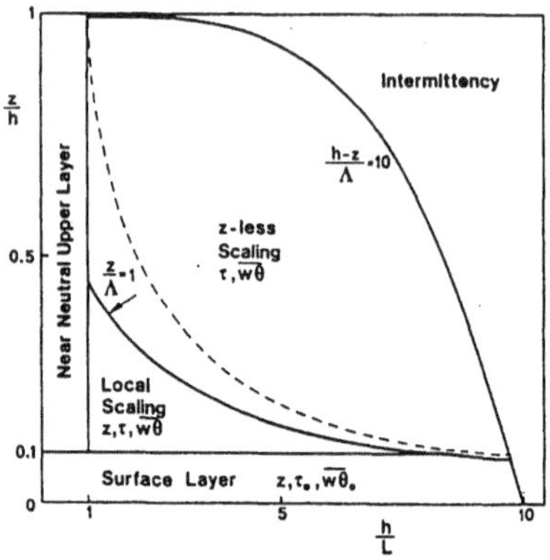

Figure 14.10: *Scaling regimes in the stable boundary layer according to Holtslag and Nieuwstadt (1986) based on the scaling parameters discussed in this section. The dashed line is given by $z/L = 1$.*

Corresponding to Figure 14.8 for the unstable boundary layer, we finally present a figure of the different scaling regions for the stable boundary layer in Figure 14.10, also taken from Holtslag and Nieuwstadt (1986).

Figure 14.10 shows a region denoted *intermittency*. In this parameter region the Richardson number is close to or larger than a crtical value for which reason continuous turbulence cannot exist. The fluxes in this region are believed to be carried by intermittet turbulence events, either induced by shear instability or internal gravity waves.

In the next sections we shall discuss the upper and lower boundary condition for the boundary layers. Here we just note that use of the z/h parameter in Figures 14.8 and 14.10 camouflages the large variability of the boundary layer height for the different situations with typical values of $h = 1000\ m$ for the unstable to neutral cases, and $h = 100\ m$ for stable cases and, as said, with huge variations.

14.5 Conditions at the surface

The similarity expressions in Eq. 14.53 can be integrated to the profiles of wind speed, temperature, and humidity:

$$\begin{aligned}\bar{u}(z) &= u_*/\kappa\left(\ln\left(\frac{z-d}{z_0}\right)-\psi_m\left(\frac{z-d}{L}\right)\right)\\ \bar{\theta}(z)-\theta_0 &= \theta_*/\kappa\left(\ln\left(\frac{z-d}{z_{0T}}\right)-\psi_h\left(\frac{z-d}{L}\right)\right)\\ \bar{q}(z)-q_0 &= q_*/\kappa\left(\ln\left(\frac{z-d}{z_{0q}}\right)-\psi_q\left(\frac{z-d}{L}\right)\right)\end{aligned} \tag{14.60}$$

where the d is a zero plane height called the displacement height, the ψ-functions are related to the ϕ-functions through:

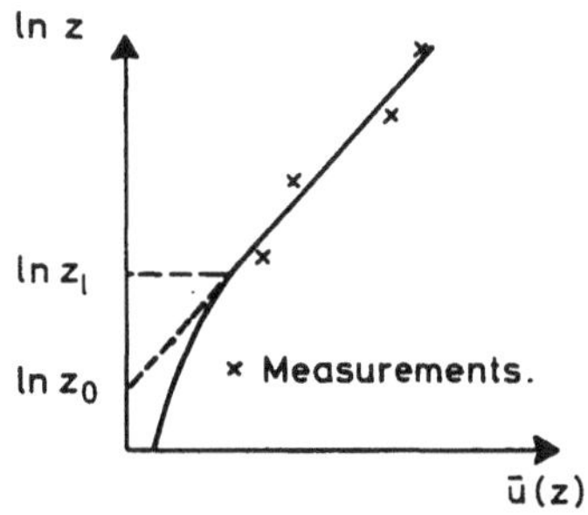

Figure 14.11: *Extrapolation of the wind profile to zero wind speed to obtain* z_0. *The lower height of validity for the neutral logarithmic law is denoted* z_l.

$$\psi\left(\frac{z-d}{L}\right)=\int_0^{(z-d)/L}(1-\phi(x))\,\frac{dx}{x} \tag{14.61}$$

All of the functions in Eq. 14.61 approach zero for $(z-d)/L \to 0$ so that Eq. 14.61 shows the profile approach to the logarithmic form for neutral conditions. Note that the z/L formulation of stability means that a boundary layer always tends towards neutral close to the surface, and that a boundary layer that appears neutral at the surface can turn more and more non-neutral as z and thereby z/L grow. Here, the different *roughness* lengths, z_0, z_0T and z_0q, are introduced formally as the heights where the surface layer forms attain the true surface values. As seen this gives a simple way of estimating z_0 as illustrated in Figure 14.11 where the measured wind profile is extrapolated down to the height z_0 where the wind speed is zero. It is worth pointing out that this method does not work so easily for the temperature and humidity profiles because surface values of these parameters are not so well-known as those for the wind speed where it is zero.

In Eq. 14.61 we have introduced the displacement height as well. If the the zero level is obviously well defined, d is of no use. But for many studies of very rough terrain like woods and built-up areas, the zero level is not well defined and it is an advantage to have d as a semi-free parameter when fitting profiles. Theoretically, the displacement height can be defined as the level at which the mean drag on the surface appears to act, and formulas for d can be worked out for specified surfaces (Jackson, 1981).

The lower height of validity for the turbulence surface layer profiles, z_l, is usually determined empericaly as $10-30$ times z_0 or for canopies $3-6$ times the height of the vegetation (Garrat, 1978; Raupach, 1979).

The method in Figure 14.11 is often not available, and z_0 must be otherwise estimated, usually based on physical considerations on, e.g. what the surface actually does to the flow, combined with the available data on the roughnesses from different measurements. As we get close to the surface we leave the surface layer as defined in section 4, and the turbulence reduces too much to carry all the momentum nessecary to brake the flow to zero velocity at the ground. Equation Eq. 14.32 shows that the braking must be done by the molecular friction or by the pressure term due to form drag on the individual obstacles constituting the surface, i.e. the roughness elements. For surfaces where the roughness elements appear as individual distinct elements, the formula of Lettau (1969):

$$z_0 = 0.5HS/A \tag{14.62}$$

Here a roughness element is characterized by its height, H and the cross wind area, S. The density of roughness elements are described by the average area available to each element, A. The equation gives reasonable estimates of z_0 when A is much larger than S, but tends to overestimate z_0 when A is of the order of S. This is because the roughness elements also contribute to a displacement height not included in the equation (Troen and Petersen, 1989). For water surfaces one will often use the Charnock (1955) relation:

$$z_0 = cu_*^2/g \tag{14.63}$$

where c is a *constant* of the order of 0.012-0.02. Equation 14.63, however, is found to describe the roughness over sand surfaces as well.

Extensive work has been done for crop covered fields, to relate d and z_0 to the vegetation characteristics and to describe the flow and flux conditions above and within the canopy (Thom, 1971; Brutsaert, 1975; Brutsaert, 1982). The relations between vegetation height, displacement length, and roughness are often presented as:

$$z_0 = \lambda(H - d) \tag{14.64}$$

with d of the order 2/3 H and λ of the order of 1/3.

In the literature many of the roughness studies finally end up in figures that could be called the accumulated conventional wisdom method for determining the roughness for a given terrain. In Figure 14.12 we present one of the latest of such figures.

Roughness estimates from figures like Figure 14.12 have been found to work quite well when used to estimate the local wind speeds for the purpose of wind energy production (Troen and Petersen, 1989). Although the figure does indicate the existence of seasonal variation of z_0, it is worth emphazising that for vegetated surfaces the roughness must be expected to show some variation with the vegetation growth cycle. This is illustrated in Figure 14.13 that shows the average seasonal variation in z_0 from three nearby fields in Denmark.

Sofar we have discussed mostly the aerodynamic roughness. Similar to z_0, the scalar roughnesses, z_{0T} and z_{0q}, are associated again with a logarithmic profile and the fluxes for a given vertical mean gradients. Hence, the magnitude of the different z_0's is controlled by transport mechanisms very close to the surface where molecular processes are important. Since the processes driving the scalar fluxes lack the mechanism of form drag around the roughness elements, that is available for the transfer of momentum, the scalar z_0's are generally less than z_0.

As an example, consider the two-layer heat flux model in Figure 14.14.

The Figure shows an upper turbulent surface layer (I) and a lower interfacial layer (II) where the heat flux is carried by molecular transport. Equating the flux in the two layers:

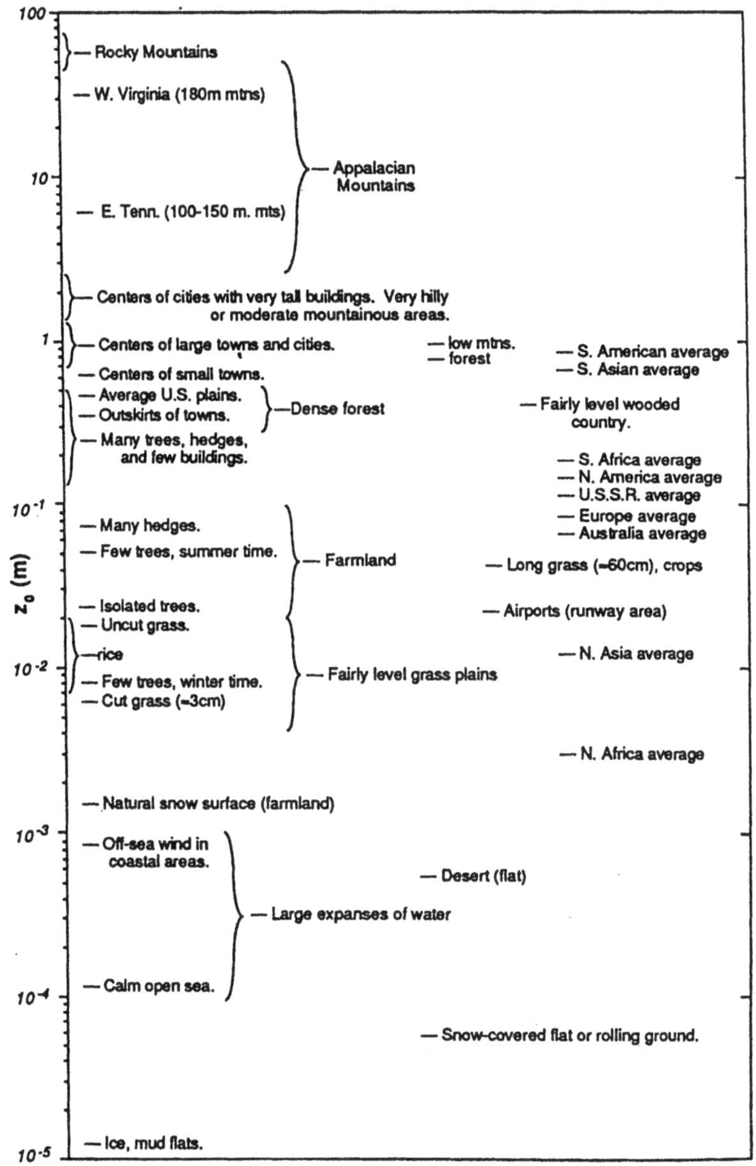

Figure 14.12: *The aerodynamic roughness length in different terrain types (Stull, 1988) integrating many earlier figures.*

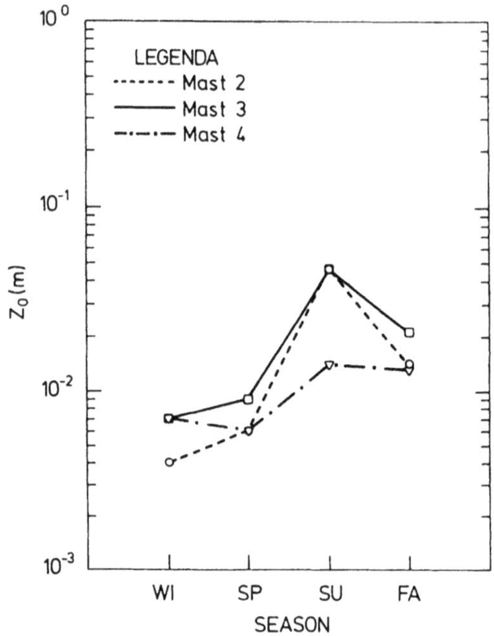

Figure 14.13: *Average seasonal variation of z_0 from two years of data at three nearby fields in Denmark (Sempreviva et al., 1989).*

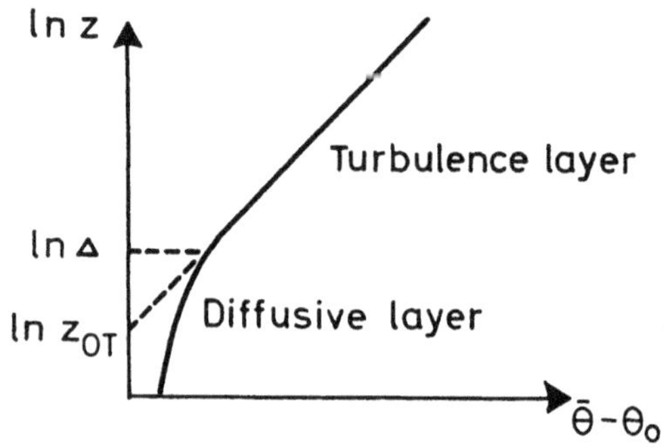

Figure 14.14: *A two-layer heat flux model to relate z_{0T} to z_0 (Jensen, 1991). The height Δ is the intersection height between the two layers.*

$$(I) - H = \kappa u_* \left(\bar{\theta_\Delta} - \theta_0\right) / \ln\left(\Delta / z_{0T}\right)$$
$$(II) - H = k_T \left(\theta_\Delta - \theta_0\right) / \Delta \tag{14.65}$$

which leads to $\ln(z_0/z_{0T}) = \kappa X$, with X given by $X = \Delta \frac{\kappa}{k_T} u_* - \ln(\Delta/z_0)$. X is usually presented as a function of the roughness Reynolds number, $Re_0 = u_* z_0/\nu$ and the Prandl number, $Pr = \nu/k_T$ for temperature, or the Scmidt number for humidity, $Sc = \nu/k_q$.

Many formulations are proposed in literature. We present the following from Brusaert (1982):

$$X(Re_0, Pr/Sc) = 7.3 Re_0^{1/4} Sc^{1/2} \tag{14.66}$$

where Pr is used for z_{0T} and Sc for z_{0q}. For a fairly wide range of atmospheric conditions this yield that the scalar roughnesses are about 0.1 z_0. The formulation used here is not applicable for fibrious surfaces, but also here the scalar roughnesses are found to be about 0.1 z_0. It would be nice to finish this section by specifying how to estimate θ_0 and q_0. in Eq. 14.61. From a modelling point of view this will obviously take estimates of the heat and moisture flux both above the surface and below in the vegetation and the soil. Also direct measurements are difficult, among others reasons because the surface in the sense of Eq. 14.61 is not necessarily very well physically defined. In spite of this, the surface radiation temperature for different surfaces has been used to test the z_{0T} expressions discussed above. In principle the surface parameters can be estimated from mesured profiles of $\bar{\theta}(z)$ and $\bar{q}(z)$ by extrapolating these down to z_{0T} and z_{0q}, if one believes in the methods for determining these parameters and the displacement height.

14.6 The upper boundary

In the three former sections the characteristics of the boundary layer flows and fluxes have been considered in terms of scales and relations defined within the boundary layer itself. One of the objectives is to describe fluxes through the boundary layer for given conditions at the surface and in the free atmosphere. Therefore we need to relate the boundary layer scales to the conditions in the free atmosphere above the boundary layer.

As a reasonable starting point we choose Eq. 14.44 because this equation contains the geostrophic wind, i.e. the wind above the boundary layer. For ideal neutral boundary layers these equation can be shown to lead to (Tennekes, 1982):

$$\begin{aligned} \kappa U_g/u_* &= (\ln(h/z_0) - A), \qquad \kappa V_g/u_* = -B, \\ h &= u_*/f_c, \qquad A \text{ and } B \text{ constants} \end{aligned} \tag{14.67}$$

It has been found from data that $A \sim 2$, while $B \sim 5$.

Equation Eq. 14.67 is called the *Resistance Law* for the neutral boundary layer because it describes the friction measured as u_*, exerted from the surface on the free stream flow. Substantial theoretical and experimental efforts have gone into generalizing Eq. 14.67, both to more general conditions and to scalar fluxes as well (Zilitinkevich, 1972 and 1975; Clarke and Hess, 1974; Melgarejo and Deardorff, 1974; Hasse, 1976; Araya, 1978):

$$\begin{aligned} \kappa U_h/u_* &= \ln(h/z_0) - A(\mu), \qquad \kappa V_h/u_* = -B(\mu) \\ \theta_h - \theta_0 &= \theta_*(\ln(h/z_{0T}) - C(\mu)) \\ \mu &= u_*/f_c L \text{ or } u = h/L \end{aligned} \tag{14.68}$$

In principle, the corresponding equation from humidity can easily be constructed from the temperature equation.

The stability parameter μ is the ratio between the boundary layer height and the Monin-Obukhov length.

Often the velocity resistance law is seen formulated in terms of the geostrophic wind speed and the angle between the surface wind and the geostrophic wind, $G^2 = U_h^2 + V_h^2$. From Eq. 14.69 the relation for G becomes:

$$\kappa G/u_* = ((\ln(h/z_0) - A(\mu))^2 + B^2(\mu))^{1/2} \tag{14.69}$$

In Figure 14.15 the stabilty functions $A(\mu)$ and $B(\mu)$ are presented. The scatter is seen to be substantial and characteristic for these types of plots. The scatter is not only in the data, also the model varies considerably among authors, some prefer μ based on the actual boundary layer height, some that it is based on u_*/f_c. Some authors prefer to use the surface geostrophic wind while others use the actual wind for $z = h$. Also several methods have been suggested on how to include the effects of instationarity, advection, baroclinity, etc. (Zilitinkevich, 1975; Hasse, 1976; Araya, 1978).

Not neglecting the uncertainty and scatter associated with the resistance laws, the neutral versions have been found to work very well for the moderate to high wind relevant for wind energy purposes (Troen and Petersen, 1989). Here data from 120 meteorological stations in Northwestern Europe were predicted from each other using Eq. 14.67 as the description of the wind above the boundary layer.

As discussed above, the initial estimate for the height of the boundary layer was formulated as:

$$h \sim c u_*/f_c \tag{14.70}$$

where c is normally choosen between 0.1 and $c = \kappa$. This expression does describe the data, although not overwhelmingly good. Also neutral boundary layer height is controlled, not only by u_*/f_c, but also by the strength and position of the lowest elevated inversion.

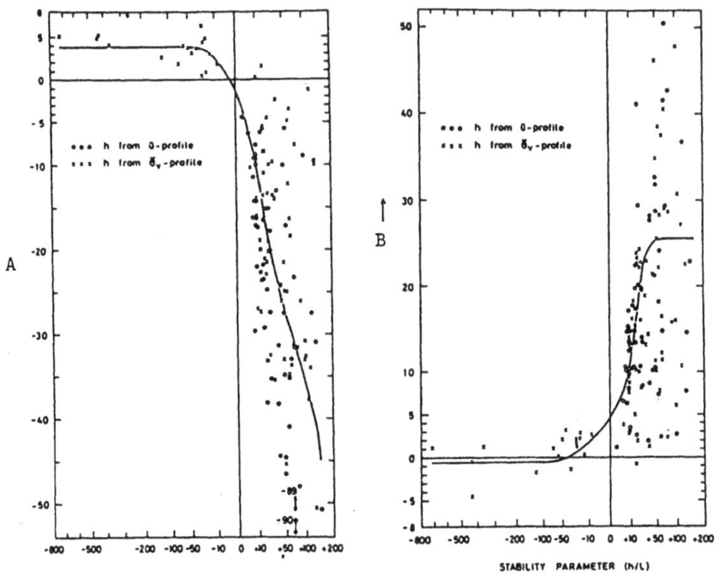

Figure 14.15: *The two stability funtions $A(\mu)$ and $B(\mu)$ as functions of h/L according to Melgarejo and Deardorff (1974).*

Such an expression can be found from the oceanographic modelling of the *inverse* problem, the depth of the mixed layer (Pollard et al., 1973)

$$h = 2^{3/4}\left(fracu_*^2 f_c N\right)^{1/2} \tag{14.71}$$

where N is the Brunt-Vaisala frequency, here used to describe the temperature stratification just outside the boundary layer.

$$N = \left(\frac{g}{\bar{\theta}}\frac{\partial\bar{\theta}}{\partial z}\right)^{1/2} \tag{14.72}$$

As a typical value of N in the free atmosphere is about 0.025 s^{-1}, Eq. 14.71 implies a value of $c \sim 0.15$ in Eq. 14.70 at midlatitude with $f_c \sim 1.5 \cdot 10^{-4} s^{-1}$.

Formulas for the height of the stable boundary layer have been worked out as well (Zilitinkevich, 1972) with:

$$h = C(u_* L/f_c)^{1/2} \tag{14.73}$$

with C around 0.7 (Caughey, 1982). The formulation simply suggests that the boundary layer height is the geometric mean between the neutral heigh, u_*/f_c and the Monin-Obukhov length.

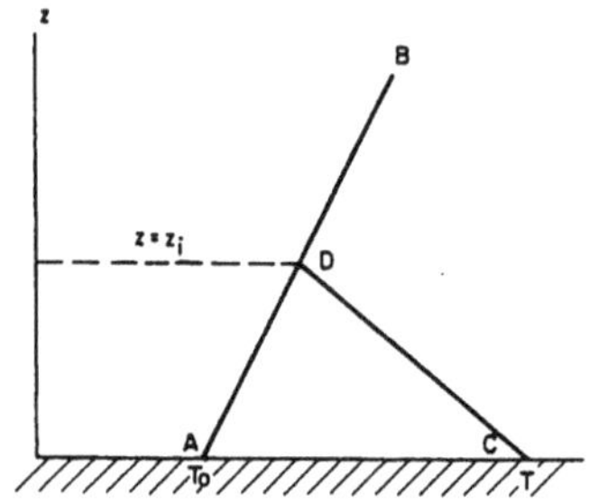

Figure 14.16: *Illustration of how the boundary layer height, $z_I = h$, grows when at sunrise the atmosphere is heated from belowe. The AB-curve represents the temperature profile at sunrise, CDB the profile after surface heating and entraining from aloft has warmed the lower layer (Panofsky and Dutton, 1984).*

For the unstable boundary layer it became obvious that formulas as the above could not be made to function. Therefore, quite soon it was proposed to determine the unstable h from a prognostic equation (Deardorff, 1972 Zilitinkevich, 1975). Consider Figure 14.16.

In Figure 14.16 we denote the temperature gradient at sunrise, line AB, as $-\gamma_n$, the gradient within the heated layer as $-\gamma_d$. γ is noramlly denoted the lapse rate. The night time gradient is likely to correspond to some type of stable situation, while the day time gradient will be close to adiabatic. Geometry then yields:

$$\gamma_d - \gamma_n = (T - T_0)/h \qquad (14.74)$$

The conservation of heat yields, including both the heat flux from the surface and from the top:

$$\overline{w'\theta'_s} - \overline{w'\theta'_h} = (T - T_0)/h \qquad (14.75)$$

Combining these two equations and differentiating, we obtain:

$$h\frac{\partial h}{\partial t} = \overline{w'\theta'_s}(1 + A)/(\gamma_d - \gamma_n) \qquad (14.76)$$

where we have included the heat delivered to the boundary layer from the free atmosphere as a fraction, A, of the surface heat flux, usually taken as approximately 20 per cent. The equations are solved to yield:

$$h(t) = (2\int_0^t \overline{w'\theta'_s}(1 + A)dt/(\gamma_d - \gamma_n))^{1/2} \qquad (14.77)$$

The rate equation for the unstable boundary layer gives quite satisfactory results. It has been much further developed and yielded a very satisfactory comparison with data on the growth of the unstable bounday layer, e.g. Gryning and Batchvarova (1990).

The development within the field is concentrated around improved description of the entrainment at the top of the boundary layer and the associated flux. The parameterization involves the structure of the entrainment zone that is as wide or wider than the boundary layer. The entrainment zone is physically described as a region where plumes and billows raise through a stable background air, with different types of turbulence and waves being generated on the interfaces (Tennekes and Driedonks, 1980; Zillitinkevich, 1991). The entrainment is often described in terms of an entrainment velocity ,w_e, defined by:

$$\overline{w'\theta'_h} = w_e \Delta\theta(EZ) \tag{14.78}$$

where $\Delta\theta(EZ)$ is the temperature gradient across the entrainment zone that is between the free atmosphere and the boundary layer. A general equation for the boundary layer height is often formulated as:

$$\partial h/\partial t = -u_j \partial h/\partial x_j + w_s + w_e + w_{ls} \tag{14.79}$$

where also advection of boundary layers with generated heights other than the local. w_s and w_e are the rise induced by the local fluxes as given by Eq. 14.76, while w_{ls} reflects larger scale movement of the atmosphere as, e.g. subsidence. Note that Eq. 14.79 allows for stationary unstable boundary layer heights.

At least for scalars the entrainment constitutes an alternative to the resistance laws to relate the conditions within the boundary layer to the conditions outside, compare Eq. 14.78 where the conditions in the free atmosphere enter through $\Delta\theta(EZ)$. For any scalar x, we can of course generate an equation like Eq. 14.78. Also note that the top-down part of the bottom-up/top-down description of the unstable boundary layer in Eqs. 14.56 and 14.56 includes the entrainment.

Finally, it should be mentioned that the success of the rate equations for the height of the unstable boundary layer has let to similar but less successful efforts for the stable boundary layer. This is due mainly to the absence of strong dominating verical fluxes here.

14.7 Demands to measurements

The plans for a measuring campaign in the atmospheric boundary layer can be organized in many ways. Below we shall try to organize the relevant points, as:

1. temporal and spatial scales that should be resolved
2. statistical demands to the data
3. instrumentation technology, calibration problems, and flow distortion.

For a simple homogeneous boundary layer average flow changes takes place only in the vertical dimension. If one can use several measuring heights, the levels should be more close at the bottom than heigher up due to the form of the profile functions as discussed in sections 4 and 5. If possible one should try to measure fluxes directly from turbulence measurements. Fluxes can be estimated from profile measurements, as can be seen from Eq. 14.61. However, both measuring uncertainties and the uncertainty about the value of the von Kármán *constant* point towards a direct measuring of the turbulence fluxes.

In the surface layer, it follows from Eq. 14.61 and the discussion about the behaviour of the ϕ-functions that profiles for a fairly large stability interval around neutral can be written as:

$$x(z) = a + b\ln(z) + cz \tag{14.80}$$

where the , b, and c coefficients contain all profile parameters other than z. If measurements of Eq. 14.80 are conducted in the two heights, z_1 and z_2, the measured gradient will correspond to $\partial\bar{x}/\partial z$ at the height $z = (z_2 - z_1)/\ln(z_2/z_1)$ as can be seen by comparing the two expressions. If flux-profile relations are measured, this height is therefore the best height for the flux instrumentation.

Next we turn to the temporal and spatial scale in the fluctuations. As mentionned in section 2, they are best described in terms of spectra. Here we want to emphazise the aspect that:

$$\begin{aligned} \overline{x'^2} &= \int_{-\infty}^{\infty} S_1(k_1)dk_1 &= \int_{-\infty}^{\infty} S(\omega)d\omega \\ \overline{u'_j x'} &= \int_{-\infty}^{\infty} Co_1(k_1)dk_1 &= \int_{-\infty}^{\infty} Co(\omega)d\omega \end{aligned} \tag{14.81}$$

which states that the contribution to the variances and covariances from different wave-number or frequency scales are described by the spectra and the co-spectra. As discussed in section 2 the relation between the wave number and the frequency is given as a good approximation by Taylor's hyphotesis, equation Eq. 14.10. Note that the co-spectra between a velocity signal and another signal describes the turbulence flux in the direction of the velocity component as discussed in section 3.

In section 2 is shown how the inertial subrange of the spectra scales with the relevant dissipations. Since the dissipations can be described as universal nondimensional functions of the relevant parameters in the different scaling regimes, this leads to the idea to describe the spectra in terms of the scaling laws considered in section 4. As an example we take the velocity spectrum in Eq. 14.8. We now substitute the Monin-Obukhov similarity function $\phi_\epsilon(z/L) = \kappa z\epsilon/u_*^3$ from Eq. 14.51 to obtain:

$$kS_u(k) = a_u\epsilon^{2/3}k^{-2/3} = a_u u_*^2 \phi_\epsilon(z/L)(kz)^{-2/3} \tag{14.82}$$

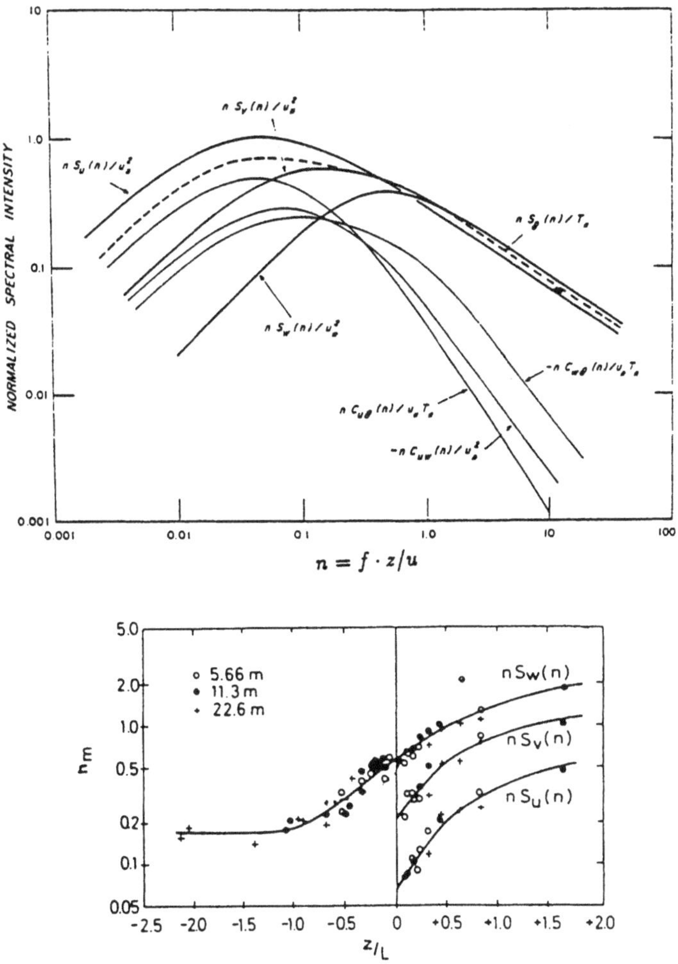

Figure 14.17: *Normalized surface-layer spectra for neutral conditions together with the variation of the peak frequency n_m with z/L for the vind spectra (Kaimal et al., 1972).*

Comparing with Eq. 14.10 we expect that after proper normalization, at least the inertial subrange forms of the spectra and co-spectra can be written as universal functions of z/L and a normalized frequency, n:

$$n = fz/u = k_1 z/2\pi = z/\lambda \tag{14.83}$$

where λ is the wave length corresponding to k_1.(NB. In micrometeorological literature, there has been a confusing change in notation. Earlier f was denoted n and vice versa).

One of the results of the Kansas experiment was that the relevant spectra could indeed be universally described over most of the frequency range according to Eqs. 14.82 and 14.83. From Kaimal et al. (1972) we present the important surface layer spectra.

Since all instruments or combinations of instruments can be characterized by a spatial and frequency resolution, the spectra in Figure 14.17 can be used to estimate the amount of variability that remains with a certain averaging time and how large a fraction of a given (co)variance a given sensor system can resolve. For example if one uses an averaging time, T, and a time resolutuion, t, one can resolve not the whole of $\overline{u_j'x'}$ in Eq. 14.81 but only $\overline{u_j'x_m'}$ given by:

$$\overline{u_j'x_m'} = 2 \int_{2\pi/T}^{2\pi/t} \mathrm{co}(\omega)d\omega, \tag{14.84}$$

where we have used that co-spectra are even functions of frequency.

Since the publication of this Figure 14.17, Kaimal et al., (1976); Højstrup (1982) have documented that also the spectra in the mixed layer can be scaled with the relevant scales here to give universal forms. Furthermore, Sorbjan (1986) has shown that also when the local scaling apply, the spectra adapt to this form as well. In the unstable surface layer it has been shown that the eddies in the mixed layer give a footprint in the surface layer spectra, especially for horizontal velocity components and temperature. The result of this is that the peak of these spectra do not scale with z/L but rather with z_I (Højstrup, 1982). This was actually the reason for the absence of n_m plots for these spectra in Figure 14.17. The z/L formulation could not be brought to describe the data. Finally, Højstrup et al. (1990) have shown that also for neutral conditions the low frequency component of the velocity spectra scale with the boundary layer height, not only with the measuring height as implied by the surface layer scaling. Finally, Olesen et al. (1984); Larsen et al. (1985, 1990) have demonstrated the importance of the gravity waves in the spectra for stable conditions. These yield spectral components that scale as:

$$kS(k) \sim N^2k^{-2} \tag{14.85}$$

where N is the Brunt-Vaisala frequency given by Eq. 14.72.

In Figure 14.18, we present examples of spectra of the horizontal velocity components in the surface layer, including the z_I-part for unstable conditions and the gravity wave part for stable conditions.

An important issue for all experimental work is the statistical confidence of the data and its relation to the choice of averaging time (or length if spatial data are available). We shall here refer to the discussion in Wyngaard (1973). Consider the variance of x around x_T where:

$$x_T(t) = \frac{1}{T} \int_{t-T/2}^{t+T/2} x(t+t')dt' \tag{14.86}$$

We can now compute the variance, δ_T^2 , as:

$$\delta_T^2 = \overline{(x(t) - x_T(t))^2} = 2\overline{x'^2}/T \int_0^T (1 - t'/T)\rho_x(t')dt'$$

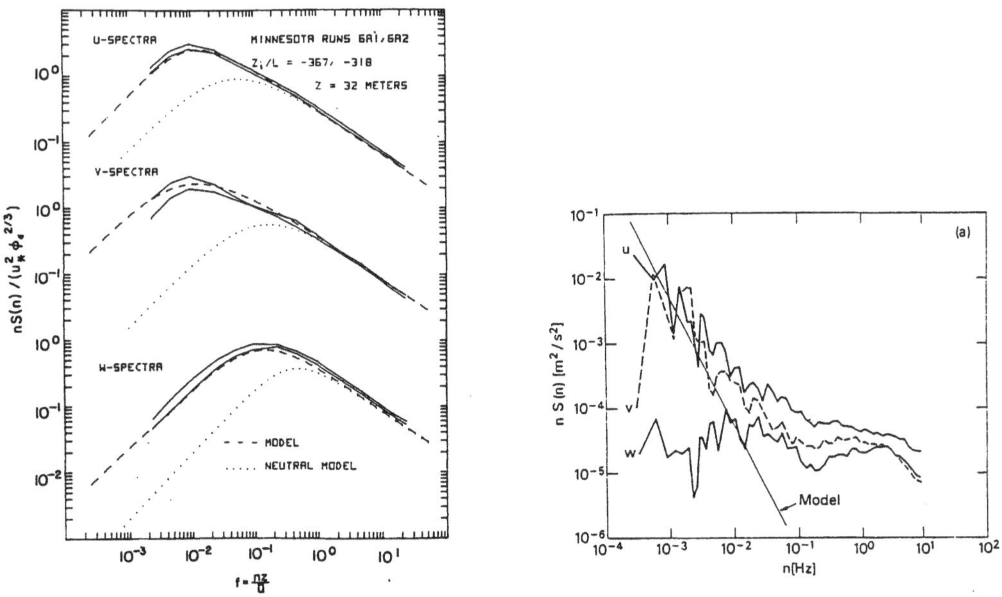

Figure 14.18: *Surface layer spectra for (a) unstable and (b) stable conditions, showing the influence of the mixed layer scaled eddies and the gravity waves, respectively (Højstrup, 1982; Larsen et al., 1985).*

where $\rho_x(t') = \overline{x(t)x(t+t')}/\overline{x'^2}$. We now assume that T is large enough for the integral to become equal to the integral of the correlation function.

With the approximation cited above, we can write as:

$$\delta_T^2 = 2\overline{x'^2}\tau/T \tag{14.87}$$

If the accuracy to be achieved is written as $a^2 = \delta_T^2/\bar{x}^2$, we can finally reverse Eq. 14.87 to give the averaging time, T_a, to achieve the relative accuracy of a^2

$$T_a = 2\frac{\tau_x}{a^2}\frac{\overline{x'^2}}{\bar{x}^2} \tag{14.88}$$

In Eq. 14.88 the integral scale, τ_x, that can be estimated by the peak frequency for the spectra presented in figures 7.1 and 7.2 will vary roughly between $z/\bar{u}$ and $h/\bar{u}$ for unstable conditions and somewhere between $z/\bar{u}$ and $L/\bar{u}$ for stable conditions, varying from signal to signal. To estimate the mean speed for example, equation Eq. 14.88 takes the form:

$$T_{a,u} = \frac{2\tau_u}{a^2}\frac{\overline{u'^2}}{\bar{u}^2} \tag{14.89}$$

For the surface layer we take $\tau_u = z/\bar{u}$ and $\overline{u'^2}/\bar{u}^2$ the turbulence intensity as 0.05. We then find that at $z = 5$ m we can achieve for $u = 5$ m/s a 1 per cent accuracy for 15 min averages.

To estimate the necessary averaging time for the variances and covariances, one must substitute these quantities for x to obtain e.g.:

$$\overline{u'^2} : \overline{x'^2}/\bar{x}^2 = \overline{(u'^2 - \overline{u'^2})^2}/(\overline{u'^2})^2 = \overline{u'^4}/(\overline{u'^2})^2 - 1$$

Correspondingly for a covariance:

$$\overline{w'y'} : \bar{x'}^2/\bar{x}^2 = \overline{w'y'^2}/(\overline{w'y'})^2 - 1$$

For the variances the result is seen to be the flatness factor. Since even order moments of turbulence are not too far from Gaussian, the flatness factor has a value of about 3. For the variances we therefore find :

$$T_{a,x^2} = 4\,\frac{\tau_x}{a^2} \tag{14.90}$$

By comparing with Eq. 14.89 this is seen to be of the order $\frac{\bar{u}^2}{u'^2}$ = 20-100 times larger than averaging time necessary to obtain the average wind speed for the same accuracy a provided the integral scale is the same for the two processes which is a reasonable assumption. It therefore seems inescapable that estimating the standard deviations with 1 per cent is out of question in the real world where the stationarity of the turbulence over periods more than a few hours is questionable. Based on data for the different moments obtained from the Kansas experiments, Wyngaard (1973) furthermore concludes that for the same accuracy the necessary averaging time for the covariances is between 3 and 20 times the averaging time for the variances. This makes it necessary to reduce the demands to the accuracy for the covariances even further than for the variances.

Furthermore, one can study the demands to averaging time for moments of higher order than the variances and covariances discussed above. One finds that it becomes almost impossible to determine these higher-order moments in the atmosphere because the averaging time must be limited to time periods substantially less than the diurnal cycle to avoid the problems with instationarity.

Finally, we shall consider the instrumentation problems. Overall the well-known instruments of today are both accurate and reliable. Most of the problems encountered, therefore, are not fundamental measuring technological problems, but are more associated with that the experimentator forget to assure himself about certain uncertainties. For velocity measurements calibration and flow distortion remain important problems. Especially for anemometers with empirical calibrations like cup anemometers, careful periodic calibrations must be recommended. Also flow distortion has been found to be important for velocity measurements and several models for flow distortion have now been developed. Flow distortion originates from the instruments themselves or from the platforms, and even very slender booms have been shown to give rise to surprisingly large flow distortion. Generally two different ways of handling flow distortion effects are used. One method consists in comprehensive wind tunnel calibrations, that ideally allow to compute a *time* velocity vector for each measured value of the same vector. This method is e.g. illustrated in Mortensen et al. (1987).

The other method involves estimating the effects on mean values and turbulence statistics, using more or less sophisticated models for the physics of the flow distortion. These models are subsequently reversed to yield *time* mean values and turbulence statistics from the measured values. The simplest *model* is here just tp rotate the coordinate system along the measured mean flow (Dyer, 1981, 1982, Wyngaard, 1982). More refined descriptions have been developed by Wyngaard (1981) for situations, where the distorting body is mauch smaller than the scale of turbulence, as is mostly the case for over land measurements in the surface layer. For scalar fluxes the flow distortion is less important generally than for velocity (Wyngaard, 1988). For temperature sensors, the dominating sneaky problem is the radiation errors, meaning that the temperature of the sensor is influenced by the radiation balance with the surroundings. Most temperature sensors need calibration, so the need to calibrate should not be a surprise, but the radiation errors might even for a well screened sensor influence the measurements systematically through some unsuspected heat bridge in the system without showing dramatic and easily detectable errors. Humidity sensors are a larger problem because both for mean value and turbulence measurements the humidity sensors are more complicated and/or less accurate than the temperature and velocity sensors. Here is really a need for sensor development. Finally should be mentioned the contamination problem with salt, a problem that of course is largest over the ocean or in the coastal region. Due to the hygroscopic qualities of salt, salt contamination can be totally destructive to the data quality of both temperature and humidity measurements. There is presently no solution to this problem aside from repeated cleaning of the sensors.

14.8 Uncertainties and unknowns for the simple PBLs

Many of the uncertainties for the simple planetary boundary layers are associated with the assumptions behind the descriptions not being sufficiently fulfilled in the real world. The models describing the total boundary layer are accepted to be more uncertain than the surface layer descriptions, and we shall defer much of the discussion about the total boundary layers to the next section that concerns inhomogenous situations.

If one looks closer on the assumptions for a surface layer, one sees that it demands that $z_0 << z << h$ (Tennekes, 1973 and 1982). As is obvious, these constrains are not always fulfilled. Therefore, one should not be too surprised to find deviations from the pure surface layer theory in the measurements. Close to very rough surfaces, as over cities and forests, the deviations from the simple surface layer ϕ functions are generally found (Garratt, 1978; Raupach, 1979). Also, if the unequality above is not well fulfilled one should expect to see processes scaling with z_0 or h show up in the surface layer. In section 7 we have already seen that h becomes an important scale for describing the spectra in what is otherwise considered the surface layer. This fact of course must have consequences for other quantities as well. Indeed, analyses of surface layer data including the pressure part of the flux divergence terms of Eq. 14.51 conclude (Elliot, 1972; McBean and Elliot, 1975) that the flux divergence term s does not scale with z/L alone as assumed in Eqs. 14.51 and 14.52. This means that one should not be surprised that some authors find some differences for the terms in these equations.

In the surface layer formalism, the von Kárman constant is an universal constant because in the analysis it cannot depend on anything. If we relax the strict surface layer formulation, then it may depend on parameters like z_0 and h. Indeed, in recent measurements variations variation of κ with the roughness Reynolds number, see Fig. 14.19.

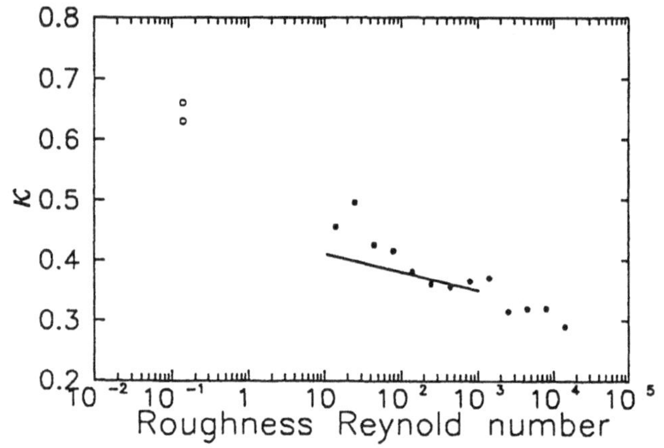

Figure 14.19: *Recent data on the variation of the von Karman constant with roughness Reynolds number, according to Oncley et al. (1990), —, Mortensen et al. (1987), •, and Gryning (1993), △. The values shown are averages over extensive data sets*

For a large fraction of the stable boundary layer situations one encounters in practice thaat the Richardson number is larger than the critical Ri, simultaneously with that vertical transport still take place through intermittently occuring turbulence, see section 4. Various ad hoc formulations have been proposed e.g. by modellers that have noticed that if they use standard flux parameterization schemes, their stable boundary layers collapse much faster and more often than the data show (Estournel and Guedalia, 1987). However, a consistent theory still has to be formulated.

Overlapping the above, the stable boundary layer is known to include different kinds of wave activities that are not systematically described, but which have at least been shown to contribute to the velocity spectra in a describable systematic way. Also the stable boundary layer is the boundary layer where the h/z_0 ratio in general is smallest because h tends to be small. Therefore the concept of surface layer might get into difficulties here before into n any other conditions. Undoubtedly, the future will see progress within this area, so that last often occurring boundary layer phenomena can be included in the list of processes, we understand.

14.9 Horizontally inhomogeneous boundary layers

After having discussed the stationary and horizontally homogeneous boundary layers, that are easy to define and almost non-existing, we turn towards the horizontally inhomogeneous boundary layers, that are everywhere, but almost impossible to specify because they come in so many versions.

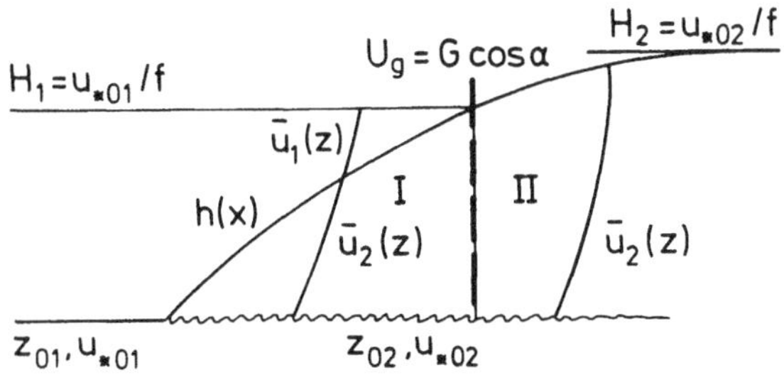

Figure 14.20: *Growth of an internal boundary layer from a smooth to rough transition under neutral conditions. Eventually the internal boundary layer height, $h(x)$ grows to establish a new boundary layer (Sempreviva et al., 1990).*

Normally one distinguishes between the effects of changes in terrain elevation and changes in surface characteristics like z_0, z_{0T} and z_{0q}, q_0 and θ_0. This is because the former influences the flow through the pressure term, while changes in surface characteristics diffuse up into the boundary layer by the turbulent diffusion.

Basically two different methods exist for handling inhomogeneous surface characteristics, when trying to estimate the flux between the atmosphere and an inhomogeneous area. In one method the inhomogeneous surface is broken down into a number of destinct subareas, that are each considered homogeneous, and with step changes in the surface characteristics in between. In the other method one estimates area averages of the surface characteristics $z_0, z_{0T}, z_{0q}, \theta_0$ and q_0 by a suitable average across the inhomogeneous terrain in question.

Considering the first approach first, we concentrate for starters on a step change in roughness, and start with neutral stability. A simple but accurate way of estimating the effect of this is to assume that the information about the surface change propagate upwards with u_*, leading to the following expression for the growth of an internal boundary layer, $h(x)$ with fetch, x, over the new surface:

$$dh/dx = au_*/u(h) \tag{14.91}$$

where a is a constant of order one. Eventually h will grow untill it fill out the boundary layer, and a new boundary layer is established. The idea is presented in Figure 14.20.

Integration of Eq. 14.91 yields:

$$cx/z_0 - 1 = (\ln(h/z_0) - 1)h/z_0 \tag{14.92}$$

where c is another constant of order unity. Equation 14.92 is seen to describe a growth of the internal boundary layer that is slightly slower that x to the power of one. To get the change in surface stress one matches the upstream and downstream velocity profiles in the $h(x)$ to obtain:

$$u_{*2}/u_{*1} = \ln(h(x)/z_{01})/\ln(h(x)/z_{02}) = 1 + \frac{\ln(z_{02}/z_{01})}{\ln h/z_{02}} \tag{14.93}$$

This equation is found to describe data very well and to works even better than second- order closure models (Larsen et al., 1982). It is seen to describe a variation with x that is strongest in the beginning when h is small, followed by a very slow change when h has become large. From Eq. 14.93 the ratio between the wind speeds are found to be:

$$\bar{u}_2(z)/\bar{u}_1(z) = \frac{u_{*2}}{u_{*1}} \ln(z/z_{02})/\ln(z/z_{01} \tag{14.94}$$

which is seen to vary less than the u_*-ratio because the two z_0's enter oppositely in Eq. 14.94 than in Eq. 14.93. Sempreviva et al. (1990) have modified the profile formulations in Eq. 14.94 and extended the formulation to include the effect of a final boundary layer height. Jensen et al., (1984) have extended it to non-neutral cases showing that the growth rate of the internal boundary varies between $x^{1/2}$ and $x^{3/2}$ times, dependent on stability conditions and height of $h(x)$. Note that the equation for the growth of an unstable boundary layer Eq. 14.75 can be used for the growth of an internal boundary layer, interpreting t as $x/\bar{u}$. The second-order modeling of Rao (1974) and Rao et al. (1974, 1974a) argely confirms the content of the above equations. The studies of Sempreviva et al. (1990) seem to indicate that it takes the surface wind speed a fetch of about 10 km before it has fully adapted to the new surface. The stability extension of the model Eq. 14.91 through Eq. 14.93 by Jensen et al. is simplistic in the sense that the same stability is assumed for both areas. There have never been tested a model system of this type that includes changes in all the surface characteristics, $z_0, z_{0T}, z_{0q}, \theta_0$ and q_0, θ_* and q_* simultaneously. The difficulties shall not be underestimated, although the roughnesses can probably be related, as described in section 14.6. Also some of the difficulties are probably not associated with the model concepts, but more associated with keeping track on the many and different subsurfaces with each their characteristics.

The direct area averaging of the surface characteristics of an inhomogeneous area sounds simpler than the approach cited above. Taylor (1987) suggests as a practical formula to average the z_0 values:

$$\ln(z_0a) = \int_A \ln(z_0(x))dx \tag{14.95}$$

where the x-integration is carried out over the area in question to obtain the effective z_{0e}. To use this type of averaging of surface characteristics, it is important to know its meaning. We can write the area averaged profile equations, from Wood and Mason (1991),corresponding to Eq. 14.61 with d neglected for simplicity:

$$\begin{aligned} < u(z) > &= \frac{(< u_*^2 >)^{1/2}}{\kappa}\left(\ln(z/z_{oe}) - \psi_m(\frac{z}{L_e})\right) \\ < \theta(z) > - < \theta_0 > &= \frac{< -\overline{w'\theta'} >}{\kappa(< u_*^2 >)^{1/2}}\left(\ln(\frac{z}{z_{0Te}}) - \psi_n(\frac{z}{L_e})\right) \end{aligned} \tag{14.96}$$

where we use <> as area average and try to relate the area averaged surface fluxes to the area averaged mean values,with the zoe value being determined such that Eq. 14.96 is true. L_e is the corresponding Monin-Obukhov length. A concept used much in these considerations is the blending height, l, a height above which the mean values has been blended enough not to vary across the inhomogenous area. The stress on the other hand is supposed still to vary. This difference between the mean values and the stress, was seen as well in connection with the step change model Eqs. 14.93 and 14.94. Using the blending height concept one can now derive corrections to Eq. 14.95. Different authors use different blending heights. Wieringa (1976) and Claussen (1991) prefer a height related to the height of the roughness element, while Mason (1988) and Wood and Mason (1991) prefer a height related to the horizontal scale for the roughness changes, and the same upward diffusion rate, use in the step change model. Mason (1988) and Wood and Mason (1991) derives estimates of z_{0e} and z_{0Te} to find that z_0e tend to be larger than z_{0a}, while z_{0Te} tend to be smaller than z_{0Ta}. However it must be emphazised that the z_{0e} values thus estimated are quite strongly dependent on which of the parameters in Eq. 14.96 one considers measured, and which one wants to estimate. They have thus become somewhat removed from the physical definition of a roughness length, and become more like a sub-grid scale process outputting some area averaged parameters, when others are inserted in the formal profile equations Eq. 14.96. The fact that the atmosphere feels a surface change through a diffusion process as depicted by Eq. 14.93 can be used when interpreting real data into a framework of the homogeneous models, considered in section 4 through 7. The simplistic way of phrasing the models of internal boundary layers (IBL) after a step change is to say that the IBL growth as between x/10 and x/200 dependent on which arameter one uses to characterise the IBL. This can be turned two ways. If one measures from a mast at height z, it means that surface changes closer than 10z, can not be seen in the data. If on the other hand, one wants to use the resistance laws in section 6 with a characteristic roughness together with the wind at z = 1km, it means that this roughness must be characteristic for an area with a width between 10 and 200 km. Next we turn to the effects of changing surface elevation. The recent advances here start with a study by Jackson and Hunt (1975) who studied the turbulent flow over a two dimensional low ridge. The result are illustrated on Figure 14.21.

The result of Jackson and Hunt (1975) and the later research (e.g. Zeman and Jensen, 1987) was that the pressure field was perturbed out to a distance of L, and the inner height, l, was the level with the maximum speed perturbation. Below l the stress achieved equilibrium with the new shear and speed ℓ was related to L through an equation similar to Eq. 14.93: $\ell \ln(l/z_0) \sim L$.

Later the results of Jackson and Hunt (1975) was combined with a linearized flow model, where the scale L was replaced by the the wave length of the Fourier decomposition (Troen and Petersen, 1989). From Eq. 14.32 the linearized equations of motion for neutral stationary flow perturbations can be written:

$$U_{0j}\frac{\partial u_i}{\partial x_j} = \frac{1}{\rho}\frac{\partial p}{\partial x_i} + \frac{\partial}{\partial z}\tau \qquad (14.97)$$

where subscript 0 indicates background flow, and where only the important terms have been retained. The kinematic boundary equations at the surface $z = H(x, y)$ are given by:

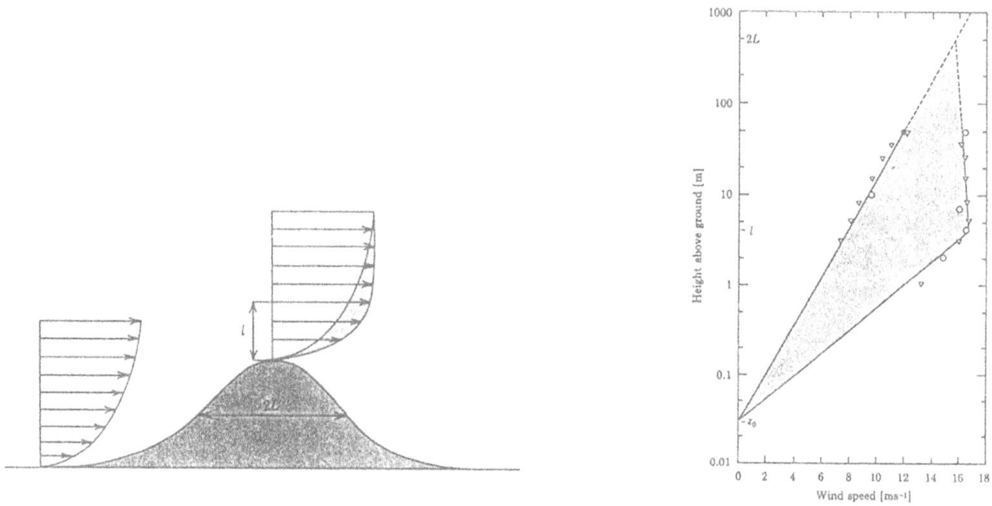

Figure 14.21: *a) Characteristic of the flow over a two dimensional low ridge according to Jackson and Hunt (1975), taken from Troen and Petersen (1989). The scale of the hill, L, and the socalled inner scale, ℓ, is shown. b) Experimental data from Jensen et al. (1984).*

$$w = \mathbf{u_0} \cdot \Delta H(x, y) \tag{14.98}$$

The equations Eqs. 14.97 and 14.98 are well suited for solution through the two dimensional Fourier transform in the coordinates horizontal coordinates. To connect to the Jackson-Hunt theory the hill scale, L , of Figure 14.21 is related to wavenumber, i.e. $L \sim |k|^{-1}$. With this relation the Jackson-Hunt model can now be used to close the vertical parts of the equation. The association of the scale L with a reciprocal wave number is neat, but it does not necessary work. Fortunately extensive test of the models again available data and models show it to work remakably well (Walmsley et al,1990). In their program for evaluating the wind potential of a given site Troen and Petersen (1989) managed to combine a roughness change model of the step change type Eq. 14.91 through Eq. 14.94 with an orographic model, just described, and a sheltering model, not described here. Aside from meteorological data from a neighbouring station, the users input to the program is three maps, one showing the location and types of shelters, one showing the terrain elevations, and one showing the distribution of roughnesses, We illustrate here the last with a roughness map.

The idea of using Fourier transforms of terrain elevation has by Belcher et al (1990) been extended to the roughness change study, Fourier transforming $\ln(z_{01}/z_{02})$, compare Eq. 14.93, from roughness maps like Figure 14.22. This is combined with an analysis of the flow perturbation adapter from the hill studies by Jackson and Hunt (1975). An attractive aspect of this Fourier transform approach is that it could be extended to three dimensions fairly easily, thereby faciliating computation of area averaged flow fields and fluxes much more than the more rigidly two dimensional formulations in Eq. 14.91 through Eq. 14.94.

It is instructive to compare the maps necessary for the wind energy evaluation with the corresponding needs if the goal was to compute the fluxes of momentum, heat and moisture from the same area.

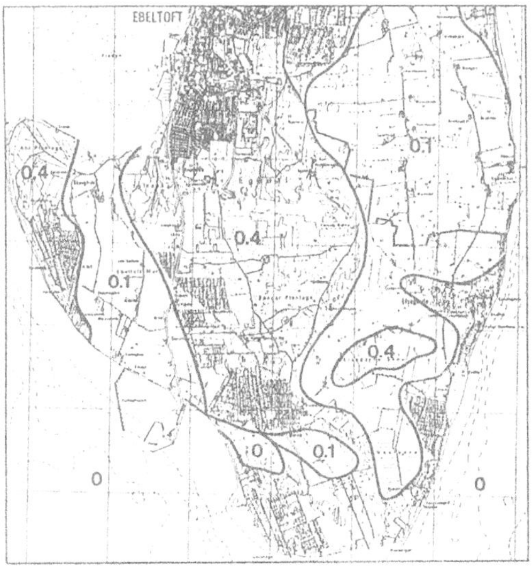

Figure 14.22: *The figure illustrate an input map with roughnesses to the analysis program for wind energy potential(Mortensen et al,1992)*

Here we would also need the orography map, and the roughness map. Probably we would not need the map of sheltering effects. Additionally we would need maps over θ_0 and q_0, z_{0T}, and z_{0q} could probably be related to z_0, as discussed in section 5. The major model modification would be, that while the wind energy model works satisfactorily as a neutral model, because the main interest is on reasonable high winds,then a model describing the fluxes would necessarily have to be be able to describe non neutral flows as well. The model should be able to handle quite extreme thermal conditions. Extremes in thermally controlled situations are found ind the arctic both associated with the polynia research and in the marginal ice zone, where the temperature difference between the water and the ice surfaces can be tens of degrees (Andreas, 1980, Claussen, 1991).

However, also in more temperate climates, large changes in surface temperature and humidity as well the associated fluxes take place on the border between land and water, so also here non neutral models become essential. In the stable boundary layer, a description is complicated by that even moderate stability and moderate terrain variation can set up thermally controlled local flow (Mahrt and Larsen, 1990). Of these effects the non-homogeneous stable boundary layers are probably most difficult to handle, because of the weak vertical fluxes. This, on the other hand may be less important, when the overall objective of the PBL study as here is to estimate the vertical fluxes. The strongly thermally driven internal boundary layers offer little problems, precisely because of the strong vertical fluxes. They can be described the same way as a growing strongly unstable boundary layer, substituting $x/\bar{u}$ for t.

The discussion above has been concentrated on the response of the boundary layer to inhomogeneous surface conditions. As a note of caution it should therefore be pointed out that inhomogeneity in the boundary layer might be due to other reasons as well. E.g. slowly developing synoptic systems and cloud systems will influence inhomogeneity that originally is independent of the surface conditions, although the surface might be influenced by for example clouds changing the radiation conditions at the surface.

Finally we should mention the studies spectra in inhomogeneous terrain, basically the low frequency part of the spectra increase for complex terrain, a fact that as least qualitatively can be associated with the contributions of the terrain variations to a larger scale roughness (Panofsky et al. 1982). In studies of spectrum response to step changes in roughness and heatflux Højstrup (1981) and Højstrup et al. (1982) describes how the inertial subrange of the spectra adjust to balance with the local stress as in Eq. 14.82, while the lower frequency parts have a larger memory, and more slowly relaxes towards the new boundary layer conditions.

14.10 Discussion and conclusion

In conclusion to this presentation, I should like to point out, that while I have discussed many aspects of the planetary boundary layer, I have not presented any recommendations for which type of description to use to describe the different processes. This is partly to save place in an already long paper, but also because I believe, that given the uncertainties on the the different descriptions, the reader will have make the choise between different and competing methods, parameters etc him/her self. In a qualitative sense most of the acknowleged methods and parameter choises are equaly good, and to prodceed deeper, one will have to dive into the litterature anyhow. I have therefore only tried to refer to papers and books, where the different expressions are summarized, while here concentrating on a general description on the current knowledge about the prosesses and parameters of importance. Finally, I hope that I have conveyed the impression of the importance of the simple ideas in turbulence closures, namely all closure is closure based on K-diffusivity and the smaller the scale of closure the more general the results.

Acknowledgement.

I wish to acknowledge inspiration and discussions with students and teachers during the meeting of the Study Institute in Glücksburg. Also thanks are due to my colleagues at Risø for support, notably Niels Otto Jensen who patiently participated in many clarifying discussions. Last but not least I wish to thank Birthe Skrumsager for her efforts with shaping up the output of my wordprocessor in all too little time.

14.11 References

Andreas EL (1980) Estimation of heat and mass fluxes over arctic leads. *Mon Wea Rev* 108: 2057-2063.

Araya SPS (1973) Contribution of form drag on pressure ridges to the air stress on arctic ice. *J Geophys Res* 78: 7092-7099.

Araya SPS (1978) Comparative effects of stability, baroclinity and the scale-height ratio on drag laws for the atmospheric boundary layer. *J Atmos Sci* 35: 40-46.

Arpaci VS, Larsen PS (1984) Convection heat transfer. *Prentice-Hall* 512pp.

Belcher SE, Xu DP, Hunt JCR (1990) The response of a turbulent boundary layer to arbitrarily distributed two-dimensional roughness changes. *Quart J Roy Met Soc* 116: 611-635.

Berkowicz R, Prahm LP (1979) Generalization of K-theory for turbulent diffusion. Part1: Spectral turbulent diffusivity concept. *J Appl Meteor* 18: 266-272.

Brown RA (1980) Longitudinal instabilities and secondary flows in the planetary boundary layer: a review. *Rev Geophys Space Phys* 18: 683-697.

Brutsaert WH (1975) Comments on surface roughness parameters and the height of dense vegetation. *J Met Soc Japan* 53: 96-97.

Brutsaert WH (1982) Exchange processes at the earth atmosphere interface. In: Engineering Meteorology (Ed. E. Plate) *Elsevier* 319-369.

Busch NE (1973) On the mechanics of atmospheric turbulence. In: Workshop on Micrometeorology. (Ed. D.A. Haugen). *Am Met Soc* Boston, Mass., 1-28.

Busch NE (1975) Fluxes in the surface boundary layer over the sea. In: Modelling and prediction of the upper layers of the ocean *NATO ASI* Ed.: E.B. Kraus 72-91.

Busch NE, Larsen SE, Thomson DW (1979) Data analysis of atmospheric measurements. *Proc Dyn Flow Conf 1978* (Eds E.G.K. Kovasznay et al.) Dantec, Skovlunde, Denmark, 887-908.

Businger JA, Wyngaard JC, Izumi Y, Bradley EF (1971) Flux-profile relationships at the atmospheric surface layer. *J Atmos Sci* 28: 191-189.

Businger JA, Araya SPS (1974) Height of the mixed layer in the stably stratified planetary boundary layer. *J Atmos Sci* 18: 181-189.

Businger JA (1982) Equations and concepts.In: Atmospheric Turbulence and Air Pollution Modelling. Eds.: F.T.M. Nieuwstadt and H. van Dop. *Reidel* Dordrecht, Holland. 1-36.

Caughey SJ (1982) Observed characteristics of the atmospheric boundary layer. In: Atmospheric Turbulence and Air Pollution Modelling (Edss. F.T.M. Nieuwstadt and H. van Dop) *Reidel Holland* 107-158.

Chapman S, Cowling TG (1970) The mathematical theory of non-uniform gases. *Cambridge University Press* Cambridge, 422 pp.

Charnock H (1955) Wind stress on a water surface. *Quart J Roy Met Soc* 81: 639-640.

Clarke RH, Hess GD (1973) On the appropriate scaling for velocity and temperature in the planetary boundary layer. *J Atmos Sci* 30: 1346-1353.

Claussen M (1991) Local advection processes in the surface layer of the marginal ice zone boundary layer. *Bound Lay Met* 30: 327-341.

Courtney M, Troen I (1990) Wind speed spectrum from one year of continuous 8 *Hz* measurements. *Proc Ninth Symp Turbulence and Diffusion* (AMS, Boston, USA), 301-304.

Deardorff JW (1970) Convective velocity and temperature scales for the unstable planetary boundary layer and for Rayleigh convection. *J Atmos Sci* 27: 1211-1213.

Deardorff JW (1972) Parameterization of the planetary boundary layer for use in general circulation models. *Mon Wea Rev* 100: 93-100.

Delage Y (1988) A parameterization of the stable atmospheric boundary layer. *Bound Lay Met* 43: 365-381.

Dyer AJ (1974) A review of flux-profile relationships. *Bound Lay Met* 7: 363-372.

Dyer AJ (1981) Flow distortion by supporting structures. *Bound Lay Met* 20: 243-251.

Dyer AJ (1982) Reply (to comment by J.C. Wyngaard). *Bound Lay Met* 22: 267-268.

Ekman VW (1905) On the influence of the earth rotation on ocean currents. *Arkiv, Math Astron O Fysik* 2: 11.

Elliott JA (1972) Microscale pressure fluctuations measured within the lower atmospheric boundary layer. *J Fluid Mech* 53: 351-383.

Estournel C, Guedalia D (1987) A new parameterization of eddy diffusivities for nocturnal boundary layer modelling. *Bound Lay Met* 39: 191-302.

Fairall CW (1987) A top-down and bottom-up diffusion model of CT_2 and Cq_2 in the entraining convective boundary layer. *J Atmos Sci* 6: 1010-1017.

Garratt JR (1978) Flux profile relations above tall vegetation. *Quart J Roy Met Soc* 104: 199-211.

Gryning SE, Batchvarova E (1990) Analytical model for the growth of the coastal internal boundary layer during onshore flow. *Quart J Roy Met Soc* 116: 187-203.

Gryning SE (1993) Meteorological measurements over Kattegat. Marine Research Program 90 Report. National Agency of Environmental Protection in Denmark, Copenhagen, Denmark.

Hasse L (1976) A resistance-law hyphothesis for the nonstationary advective planetary boundary layer. *Bound Lay Met* 10: 393-407.

Hill RJ (1989) Implications of Mounin-Obukhov similarity theory for scalar quantities. *J Atmos Sci* 46: 2236-2244.

Holtslag AAM, Nieuwstadt FTM (1986) Scaling in the atmospheric boundary layer. *Bound Lay Meteorol* 36: 201-209.

Högström U 1990: Analysis of turbulence structure in the surface layer with a modified similarity formulation for near neutral conditions. *J Atmos Sci* 47: 1949-1972.

Højstrup J (1981) A simple model for adjustment of velocity spectra in unstable conditions downstream an abrupt change in surface roughness and heat flux. *Bound Lay Met* 21: 341-356.

Højstrup J (1982) Velocity spectra in the unstable planetary boundary layer. *J Atmos Sci* 39: 2239-2248.

Højstrup J,. Larsen SE, Jensen NO (1982) Results from an experimental investigation of a step change in surface heat flux. Proceedings of First International conference on Meteorology andd Air-Sea Interaction of the Coastal Zone. *Amer Met Soc* Boston, US, 28-30.

Højstrup J, Larsen SE, Madsen PH (1990) Power spectra of horizontal wind components in the neutral atmospheric surface boundary layer. In: *Ninth Symp turbulence and diffusion* (AMS, Boston, USA), 305-308.

Jackson PS, Hunt JRC (1975) Turbulent wind flow over a low hill. *Quart J Roy Met Soc* 101: 929-955.

Jackson PS (1981) On the displacement height in the logarithmic velocity profile. *J Fluid Mech* 111: 15-25.

Jensen NO, Petersen EL, Troen I (1984) Extrapolation of mean wind statistics with special regard to wind energy application. WCP-86. *WMO*

Jensen NO (1991) Comments on surface parameterization. In: WMO Planetary Boundary Layer Model Evaluation Workshop, August, 1989, Reading UK. Ed.: J.C. Wyngaard, *WMO/TD-No. 378* 40-42.

Kaimal JC, Wyngaard JC, Izumi Y, Coté OR (1972) Spectral characteristics of surface-layer turbulence. *Quart J Roy Met Soc* 1972: 563-589.

Kaimal JC, Wyngaard JC, Haugen DA, Coté OR, Caughey SJ, Readings CJ (1976) Turbulence structure in the convective boundary layer. *J Atmos Sci* 33: 2152-2169.

Larsen SE, Hedegaard K, Troen I (1982) Change of terrain roughness problem extended to mesoscale fetches. First International conference on meteorology and air/sea interaction. *Am Met Soc* Boston, USA. 8-13.

Larsen SE, Olesen HR, Højstrup J (1985) Parameterization of the low frequency part of spectra of horizontal velocity components in the stable surface boundary layer. In: Turbulence and Diffusion in Stable Environments. (Ed. J.C.R. Hunt), *Clarendon* Oxford, UK, 181-204.

Larsen SE, Courtney M, Mahrt L (1990) Low frequency behaviour of horizontal velocity spectra in stable surface layers. In: Ninth Symposium on turbulence and diffusion, *Am Met Soc* Boston, USA. 401-404.

Lettau H (1969) Note on aerodynamic roughness-parameter estimation on the basis of roughness-element distribution. *J Appl Met* 8: 828-832.

Lumley JL, Panofsky HA (1964) The structure of atmospheric turbulence. *Interscience* New York, 239 pp.

Mahrt L (1985) The vertical structure and turbulence in the very stable boundary layer. *J Atmos Sci* 42: 2333-2349.

Mahrt L 1986: On the shallow motion approximations. *J Atmos Sci* 43: 1036-1044.

Mahrt L, Larsen S (1990) Relation of slope winds to ambient flow over gentle terrain. *Bound Lay Met* 53: 93-102.

Mason PJ (1988) The formation of areally-averaged roughness lengths. *Quart J Roy Met Soc* 114: 399-420.

McBean GA, Elliott JA (1975) The vertical transport of kinetic energy by turbulence and pressure in the boundary layer. *J Atmos Sci* 32: 753-766.

Melgargo JW, Deardorff JW (1974) Stability functions for the boundary-layer resistance laws based on observed boundary layer heights. *J Atmos Sci* 31: 1324-1333.

Mizuno T, Panofsky HA (1974) The validity of Taylor's hypothesis in the atmospheric surface layer. *Bound Lay Met* 9: 375-380

Mortensen NG, Larsen SE, Troen I, Mikkelsen T (1987) Two-years-worth of turbulence data recorded by a sonic-anemometer-based data aquisition system. Sixth Symposium on Meteorology and Instrumentation. *Am Met Soc* Boston, USA, 393-396.

Mortensen NG, Landberg L, Troen I, Petersen EL (1992) Wind atlas analysis and application program(WAsP). User's Guide. ISBN 87-550-1789-4. *Risø National Laboratory* Roskilde, Denmark, 130pp.

Nicholls S, Readings GJ (1979) Aircraft observations of the structure of the lower boundary layer over the sea. *Quart J Roy Met Soc* 105: 785-802

Nieuwstadt FTM (1984) The structure of the stable nocturnal boundary layer. *J Atmos Sci* 41: 2202-2216.

Obukhov AM (1971) Turbulence in an atmosphere with non-uniform temperature. *Bound Lay Met* 2: 7-29.

Olsen HR, Larsen SE, Højstrup J (1984) Modelling velocity spectra in the lower part of the planetary boundary layer. *Bound Lay Met* 29: 285-312.

Oncley SP, Businger JA, Friehe CA, LaRue JC, Itsweire EC, Chang SS (1990) Surface layer profiles and turbulence measurements over uniform land under neutral conditions. In: *Ninth Symposium on Turbulence and Diffusion* (AMS, Boston, USA), 235-240.

Panofsky HA, Larko D, Lipschutz RC, Stone G, Bradley FF, Bowen AJ, Højstrup J (1982) Spectra of velocity components over complex terrain. *Quart J Roy Met Soc* 108: 215-230.

Panofsky HA, Dutton JA (1983) Atmospheric Turbulence. *John Wiley*

Pollard RT, Rhines PB, THOMSON RORY (1972) The deepening of the wind mixed layer. *Geophys Fluid Dynam* 4: 297-304.

Powell DC, Elderkin CE (1974) An investigation of the application of Taylor's hypothesis to the atmospheric boundary layer turbulence. *J Atmos Sci* 31: 990-1002.

Rao KS, Wyngaard JC, Coté OR (1974) The structure of two-dimensional internal boundary layers over a sudden change of surface roughness. *J Atmos Sci* 31: 738-746.

Rao KS, Wyngaard JC, Coté OR (1974a) Local advection of momentum, heat, and moisture in micrometeorology. *Bound Lay Met* 7: 331-348.

Raupach MR (1979) Anomalies in flux-gradient relationships over forest. *Bound Lay Met* 16: 467-486.

Sempreviva AM, Larsen SE, Mortensen NG, Troen I (1989) Response of neutral boundary layers to changes of roughness. *Bound Lay Met* 50: 205-225.

Sorbjan Z (1986) Local similarity of spectral and cospectral characteristics in the stable continuous boundary layer. *Bound Lay Met* 35: 257-275.

Stull RB (1988) An introduction to boundary layer meteorology. *Kluwer Academic Publishers* 666pp.

Taylor PA (1987) Comments and further analysis on the effective roughness lenth for use in numerical three-dimensional models. *Bound Lay Met* 39: 402-419.

Tennekes H (1973) The logarithmic wind profile. *J Atmos Sci* 30: 234-238.

Tennekes H, Driedonks AGM (1980) Basic entrainment equations for the atmospheric boundary layer. *Bound Lay Met* 20: 515-531.

Tennekes H (1982) Similarity relations, scaling laws and spectral dynamics. In: Atmospheric Turbulence and Air Pollution Modelling. Eds. F.T.M. Nieuwstadt and H. van Dop. *Reidel* Holland. 37-68.

Tennekes H, Lumley JL (1982) A first course in turbulence. *MIT Press* Cambridge MA, 300 pp.

Thom AS (1971) Momentum absorption by vegetation. *Quart J Roy Met Soc* 97: 414-428.

Troen I, Mikkelsen T, Larsen SE (1980) Note on spectral diffusivity theory. *J Appl Meteor* 19: 609-615.

Troen I, Petersen EL (1989) European Wind Atlas. *Risø National Laboratory* Roskilde. Denmark.

Walmsley JL, Troen I, Lalas DP, Mason PJ (1990) Surface layer flow in complex terrain: Comparison of models and full-scale observations. *Bound Lay Met* 52: 259-289.

Wieringa J (1976) An objective exposure correction method for average wind speeds measured at a sheltered location. *Quart J Roy Met Soc* 102: 867-889.

Wood N, Mason P (1991) The influence of static stability on the effective roughness length for momentum and heat transfer. *Quart J Roy Met Soc* 117: 1025-1056.

Wyngaard JC, Coté OR (1971) Cospectral similarity in the atmospheric surface layer. *Quart J Roy Met Soc* 98: 590-603.

Wyngaard JC, Coté OR (1972) The budgets of turbulent kinetic energy andd temperature variance in the atmospheric surface layer. *J Atmos Sci* 28: 190-201.

Wyngaard JC (1973) On surface-layer turbulence. In: Workshop on Micrometeorology (Ed. D.A. Haugen), *Am Met Soc* Boston, USA, 102-149.

Wyngaard JC, Clifford SF (1977) Taylor's hypothesis and the high-frequency turbulence spectra. *J Atmos Sci* 34: 922-929.

Wyngaard JC (1981) The effects of probe-induced flow distortion in atmospheric turbulence measurements. *J Appl Met* 20: 784-794.

Wyngaard JC (1982a) Boundary-layer modelling. In: Atmospheric Turbulence and Air Pollution Modelling, Eds. F.T.M. Nieuwstadt and H. van Dop. *Reidel Publ* Dordrecht, Holland. 69-106.

Wyngaard JC (1982b) Comments on "Flow distortion by supporting structures" by AC Dyer. *Bound Lay Met* 22: 263-265.

Wyngaard JC, BrostRA (1984) Top-down and bottom-up diffusion of a scalar in the convective boundary layer. *J Atmos Sci* 41: 102-112.

Wyngaard JC (1988) The effects of probe induced flow distortion on atmospheric turbulence measurements: Extension to scalars. *J Atmos Sci* 45: 3400-3412.

Zeman O, Jensen NO (1987) Modification of turbulence characteristica in flow over hills. *Quart J Roy Met Soc* 113: 55-80.

Zilitinkevich SS (1972) On the determination of the height of the Ekman boundary layer. *Bound Lay Met* 3: 141-145.

Zilitinkevich SS (1975) Resistance laws and prediction equations for the depth of the planetary boundary layer. *J Atmos Sci* 32: 741-752.

Zilitinkevich SS (1991) Turbulent penetrative convection. *Avebury Technical* Aldershot, UK, 179 pp.

Chapter 15

THE TERRESTRIAL HYDROLOGICAL CYCLE

J.N.M. Stricker, C.P. Kim, R.A. Feddes, J.C. van Dam, P. Droogers and G.H. de Rooij
Department of Water Resources
Wageningen Agricultural University
6709 PA Wageningen
The Netherlands

15.1 Introduction

The global hydrological cycle describes the transport and occurrence of water in all three phases and its transformations from one phase to the other, on a global scale. It directly influences the cycles of other compounds, the energy cycle, the geomorphological shape of the earth as well as the global circulation of atmosphere and oceans, and by that weather and climate. Historically, different parts of the hydrological cycle have been studied by scientists from different disciplines, e.g. atmospheric water by meteorologists, oceanic water by oceanographers and terrestrial water by hydrologists. This paper will focus on the study of water present at the continent and the mutual interaction of surface and sub-surface processes with the atmosphere.

To understand hydrological processes, it is necessary to distinguish different scales. Dooge (1989) suggested division into nine scales, as shown in Table 15.1. At the micro-scales a proper understanding of the physical processes, based on the laws of conservation of mass and momentum, is present and is partly demonstrated in Section 15.2. The solutions obtained are in general non-linear. At the meso-scales and especially at the macro-scales, where spatial variability causes increasing complexity of hydrologic processes, one cannot beforehand assume a large aggregate to obey the same differential equations as a small aggregate at the micro-scales. For climate modelling, the spatial scales are comparable with those for hydrologic processes at the macro-scale. At this scale hydrologists experience difficulties in parameterization. Besides the upscaling of hydrological processes from the micro-scale up to the meso-scale and macro-scale, there is the problem of coupling the terrestrial hydrologic processes with the atmosphere. This is discussed in Section 15.3. Fig. 15.1 shows in a graphical form

NATO ASI Series, Vol. I 5
Energy and Water Cycles in the Climate System
Edited by E. Raschke and D. Jacob

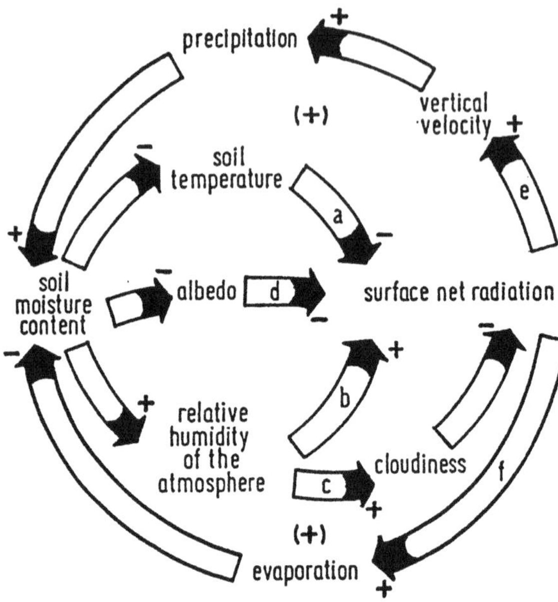

Figure 15.1: *Feedback mechanisms of sub-surface, surface and atmospheric processes in case of no vegetation (Rowntree, 1984).*

an incomplete set of interactions and feedbacks of sub-surface, surface and atmospheric processes. Variables are linked by arrows as indicated. A polarity is assigned to each arrow. A minus polarity indicates opposite effects (decrease-increase or vice versa), a plus-polarity stands for effects in the same direction. E.g. an increase in soil moisture decreases soil temperature, which in turn increases surface net radiation and consequently evaporation. By incorporating feedback processes the change in e.g. soil moisture content θ due to an initial change of $\Delta\theta$, can be written as:

$$\theta_{final} - \theta_{initial} = \Delta\theta + \Delta\Theta_{feedback} = f\Delta\theta \tag{15.1}$$

Referring to Fig. 15.1, the feedback-factor f depends on all the paths that can be made and is highly variable. Inclusion of vegetation increases the complexity of the feedbacks even more. Another point that hampers a smooth coupling is the difference in time scales of the processes in Fig. 15.1. In contrast with spatial scales, time scales of atmospheric processes are relatively small by hydrological standards. Thus coupling of terrestrial hydrologic processes to the atmospheric sub-system is essentially solving the problems caused by the different spatial and time scales and providing an accurate description of the feedbacks and interactions between the sub-surface, the surface and the atmosphere.

In modelling the terrestrial hydrologic processes, one can distinguish between different approaches: the physical, the parametric and the lumped approach. Each of those are discussed in more detail in Section 15.4. In order to construct, calibrate and verify the models described, observations are necessary. Because of the strong need for spatially representative data in hydrologic modelling

CLASS	SYSTEM	TYPICAL LENGTH SCALE [m]
	planetary	10^7
macro	continental	10^6
	large catchment	10^5
	catchment	10^4
meso	sub-catchment	10^3
	field	10^2
	pedon	10^{-1}
micro	continuum point	10^{-5}
	molecular cluster	10^{-8}

Table 15.1: *Spatial scales in hydrology (modified after Dooge,1989)*

current possibilities of satellite remote sensing are summarized in Section 15.5. Large-scale field experiments are well recognized activities in the WCRP. In Section 15.6 they are shortly discussed and the importance and opportunities for the hydrological community in participating are stressed. Some conclusions and recommendations can be found in Section 15.7.

15.2 Theory of water transport in the soil-plant system

15.2.1 Soil water balance

The one-dimensional water balance over the first few meters of the vertical soil column accounts for the incoming and outgoing fluxes. The change in water storage ΔW over the entire column for a given period of time Δt can be written as the difference of inflow, i.e. net precipitation (gross precipitation P minus interception I), net upward flow through the bottom Q_g, minus outflow, i.e. surface runoff Q_s, soil evaporation E_s and canopy transpiration E_t.

$$\Delta W = \int_{\Delta t} [(P - I) + Q_g + Q_s - E_s - E_t] dt \tag{15.2}$$

15.2.2 Main flow equation of the liquid phase

Redistribution of water inside the soil column generally be expressed as:

$$\frac{\partial \theta}{\partial t} = -\frac{\partial q_l}{\partial z} - S \tag{15.3}$$

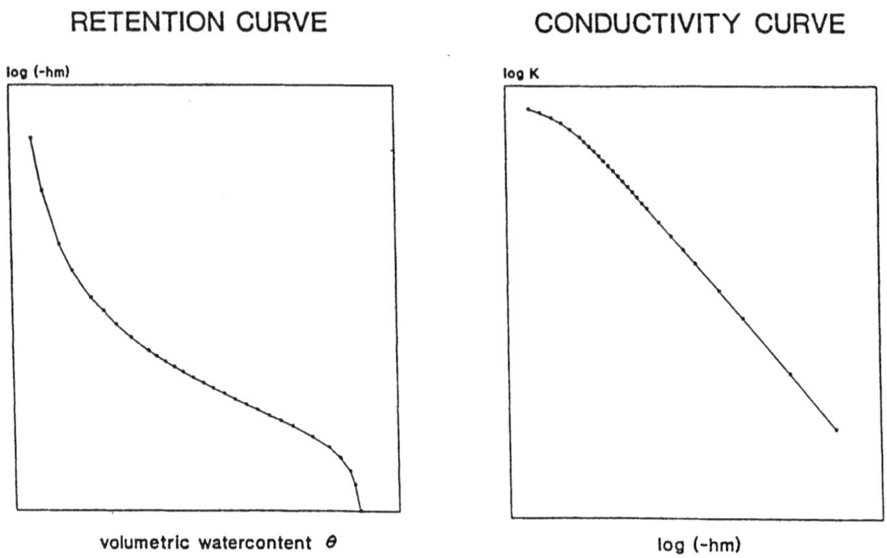

Figure 15.2: *General form of the soil hydraulic functions $K(h_m)$ and $\theta(h_m)$.*

where θ is the volumetric water content $[m^3m^{-3}]$, z the vertical coordinate, positive upwards, q_l the volumetric flux $[m^3m^{-2}s^{-1}]$ for water flow in the liquid phase and S the sink of soil water (e.g. water extraction by roots) per unit bulk volume of soil per unit of time $[m^3m^{-3}s^{-1}]$. If the changes in the system are only due to interactions between the soil matrix and the liquid water, the vertical one-dimensional Darcian flow equation in terms of heads can be written as:

$$q_l = -K(h_m)\frac{\partial(h_m + z)}{\partial z} \tag{15.4}$$

where $K(h_m)$ is the unsaturated hydraulic conductivity $[ms^{-1}]$ as a function of h_m, the matric pressure head $[m]$. The term $h_m + z$ represents the total energy potential (here expressed as head) driving the flow. For unsaturated flow the matric pressure head has a negative values due to capillary forces. In Fig. 15.2 the general form of $K(h_m)$ and $\theta(h_m)$ are depicted. Combination of Eqs. 15.3 and 15.4 and introduction of the differential soil water capacity $C(h_m) = \partial\theta/\partial h_m$ results in the following partial differential equation:

$$\frac{\partial h_m}{\partial t} = \frac{1}{C(h_m)}\frac{\partial}{\partial z}\left[K(h_m)\left(\frac{\partial h_m}{\partial z} + 1\right)\right] - \frac{S(h_m)}{C(h_m)} \tag{15.5}$$

Eq. 15.5 is known as Richards' equation and is valid for saturated and partially saturated flow. The use of pressure head h_m instead of moisture content θ as the dependent variable has the advantage of being applicable in layered soils, because h_m is continuous at the interface (Feddes et. al., 1978) and in multi-phase conditions (Menenti, 1984). Although generally the main transport of water in the soil is governed by the water potential gradient (Darcy's law, equation 15.4), other transport mechanisms may play a minor, under very dry soil conditions sometimes a major role. Mentioned are thermally induced water flow in the liquid and vapour phase and convective vapour flux (Menenti, 1984, Ten Berge, 1986). For simplicity we restrict ourselves to Darcian flow only.

15.2.3 Liquid water uptake by roots

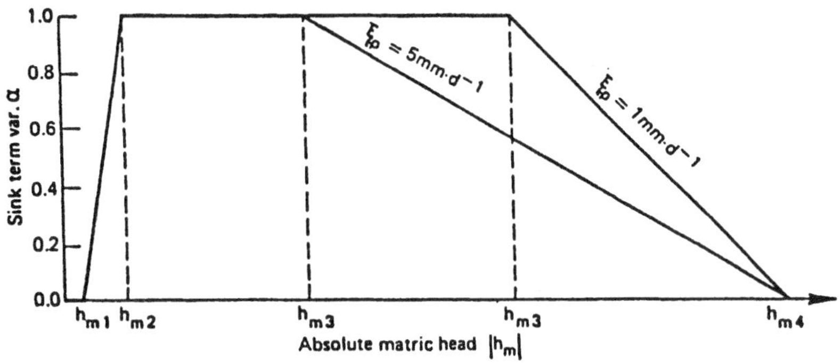

Figure 15.3: *Dimensionless sink term variable α as a function of the absolute value of the soil-water matric heat h_m (Feddes et al., 1978).*

Feddes (1981) presented an overview of possible expressions of the sink term as a function of the matric pressure head. As it seems to be impossible and unpractical to look for a complete physical description of water extraction by roots, $S(h_m)$ can be semi-empirically described as (Feddes et al, 1978):

$$S(h_m) = \alpha(h_m)S_{\max} \tag{15.6}$$

where $\alpha(h_m)$ is a dimensionless prescribed function of pressure head and $S_{\max}$ is the maximal possible water extraction by roots $[s^{-1}]$. Assuming a homogeneous root distribution over the rooting zone, $S_{\max}$ can be defined according to:

$$S_{\max} = \frac{E_{tp}}{|z_r|} \tag{15.7}$$

where E_{tp} is the potential transpiration rate $[m^3m^{-2}s^{-1}]$ and $|z_r|$ is the depth of the root zone $[m]$. Under non-optimal conditions, i.e. either too dry or too wet, $S_{\max}$ is reduced by means of the $\alpha(h_m)$-function (see Fig. 15.3). Water uptake below $|h_{m1}|$ (oxygen deficiency) and above $|h_{m4}|$ (wilting point) is set equal to zero. Between $|h_{m2}|$ and $|h_{m3}|$ (reduction point) water uptake is maximal. Between $|h_{m2}|$ and $|h_{m3}|$ a linear variation and between $|h_{m3}|$ and $|h_{m4}|$ a linear or hyperbolic variation is assumed. The value of $|h_{m3}|$ is dependent on the atmospheric demand and thus varies with E_{tp}.

15.2.4 Modelling water dynamics in the unsaturated zone of a vegetated or partly vegetated soil

A number of mechanistic simulation models are available, representing flow in the upper meters of the soil. The SWATRE-model (Feddes et al. 1978, Belmans et al. 1983) serves as an example here. This model simulates numerically transient one-dimensional water flow through the unsaturated/saturated zone, including water extraction by roots, on the basis of the partial differential flow Eq. 15.5. The evaluation of all water balance terms is performed at each time step.

The way how to treat the initial upper and lower boundary conditions has been extensively described (Feddes et al., 1988). While as a forcing function the potential evapotranspiration rate E_p is supposed to depend only on vegetation and atmospheric conditions, the evaporative flux q through the soil surface is limited by the ability of the soil matrix to transport water. Similarly, if the rainfall intensity minus the rate of interception exceeds the infiltration capacity of the soil, part of the water runs off, since the actual flux through the top layer is limited by moisture conditions in the soil. Consequently, the exact boundary conditions at the soil surface cannot be estimated a priori and solutions must be found by setting an upper bound on the absolute flux:

$$|q(K, h_m)| \leq |q_p| \tag{15.8}$$

$$h_m^{\text{lim}} \leq h_m \leq 0 \tag{15.9}$$

where $q(K, h_m)$ is the actual flux through the soil-air interface (soil evaporation flux or infiltration flux), q_p is the known potential surface flux (i.e. atmosphere controlled potential soil evaporation E_{sp} or soil controlled potential infiltration flux) and h_m^{lim} is the minimum allowed negative pressure head at the soil surface. The value h_m^{lim} can be determined by assuming thermodynamic equilibrium between soil water and atmospheric vapour at the soil surface:

$$h_m^{\text{lim}} = \frac{RT}{Mg}\ln(r) \tag{15.10}$$

where R is the universal gas constant ($8.314\,J\,mol^{-1}K^{-1}$, T is absolute temperature $[K]$), M is the molecular weight of water $[Kg\,mol^{-1}]$, g is the acceleration due to gravity $[ms^{-2}]$ and r is the relative air humidity $[-]$. In the SWATRE-model the potential evapotranspiration E_p needs to be partitioned hierarchically over required flux boundary conditions of potential soil evaporation E_{sp} and potential plant transpiration E_{tp}. Using the well known combination equation of Monteith (1965) that characterizes a vegetation as being one big leaf, E_p is described by:

$$LE_p = \frac{S_a(R_n - G_0) + \rho_a c_p(e_{sat} - e_a)r_{av}^{-1}}{S_a + \gamma(1 + r_s r_{av}^{-1})} \tag{15.11}$$

where L is the heat of vaporization $[Jkg^{-1}]$, S_a is slope of saturated vapour pressure curve $[hPaK^{-1}]$, R_n is net radiation flux $[Wm^{-2}]$, G_0 is soil heat flux $[Wm^{-2}]$, ρ_a is air density $[kg\,m^{-3}]$, c_p is specific heat at constant pressure $[Jkg^{-1}K^{-1}]$, e_{sat} is saturated air vapour pressure $[hPa]$, e_a is actual vapour pressure $[hPa]$, r_{av} is aerodynamic resistance to vapour transfer $[s\,m^{-1}]$, γ is psychrometric constant $[hPaK^{-1}]$ and r_s is minimum canopy resistance $[s\,m^{-1}]$. For further details on equation 15.11 we refer to Shuttleworth (Chapter 13). The potential soil evaporation flux E_{sp} is expressed as a function of the leaf area index L_{AI}, according to Ritchie (1972):

$$E_{sp} = e^{-\epsilon L_{AI}} E_p \tag{15.12}$$

with ϵ an extinction coefficient depending on the vegetation type. The potential plant transpiration flux E_{tp} is then obtained as:

$$E_{tp} = E_p - E_{sp} \tag{15.13}$$

Applying Eqs. 15.6 and 15.7, the actual transpiration can be calculated. Actual soil evaporation is approximated as the calculated Darcy flux through the soil-atmosphere interface, but with an upper boundary value of E_{sp}. At the lower boundary one can define three different types of conditions:

- Dirichlet conditions: the pressure head is specified (e.g. zero at the phreatic surface);
- Neumann condition: the flux is specified (e.g. free drainage or zero flux boundary);
- Cauchy condition: the flux is a function of a dependent variable. This type of boundary condition is used when the unsaturated flow model is combined with a model for regional groundwater flow or when the effects of surface water management have to be simulated.

The soil system is divided into a number of compartments of variable heights. The soil profile can be split up into several layers (containing one or more compartments) having different hydraulic properties (i.e. $\theta(h_m)$ and $K(h_m)$ relations), see Fig. 15.2. Eq. 15.5 is solved by a finite difference scheme which is implicit and applies an explicit linearization of the hydraulic conductivity and the soil water capacity. Introducing initial conditions (i.e. water content or pressure head distribution with depth) and the boundary conditions at the top and the bottom, the system of equations is solved for all compartments and for each (variable) time step.

15.3 Coupling the land-surface to the atmosphere

As stated previously, the land-surface processes and the atmospheric processes interact. In the most simplified way this interaction can be regarded as the reaction of both sub-systems on each others output. The atmosphere supplies precipitation and radiative energy to the land-surface that in return provides moisture and converted energy to the atmosphere as latent and sensible heat. The following equations expresses the fluxes that make up the terrestrial water and energy balances in terms of the main governing state variables:

$$P(\theta,\Omega) = E_a(E_p,\theta,V,S_H,z_{ov}) + Q_s(P,\theta,V,S_H,\Omega) + Q_g(\theta,S_H) + \frac{\Delta(W+C_s)}{\Delta t} \tag{15.14}$$

$$R_n(\alpha,\theta,\epsilon_s,c) = H(T_s,T_a,z_{oh}) + LE_a(E_p,\theta,V,S_H,z_{ov}) + G(\theta,S_T) \tag{15.15}$$

where W is soil moisture storage $[m]$, C_s canopy storage $[m]$, P precipitation rate $[m\,s^{-1}]$, Ω representing influence of topography, E_a latent heat flux $[m\,s^{-1}]$ $(= E_s + E_t)$, E_p potential evapotranspiration rate $[m\,s^{-1}]$, V vegetation type and density $[-]$, S_H soil hydraulic properties, z_{ov} aerodynamic roughness length for latent heat $[m]$, Q_s surface runoff $[m\,s^{-1}]$, Q_g drainage $[m\,s^{-1}]$, R_n net radiation $[Wm^{-2}]$, α the albedo $[-]$, ϵ_s emissivity $[-]$, c cloud cover $[-]$, H sensible heat flux $[Wm^{-2}]$, T_s surface temperature $[K]$, T_a air temperature $[K]$, z_{oh} aerodynamic roughness-length for sensible heat $[m]$, L heat of vaporization $[Jkg^{-1}]$, G the soil heat flux $[Wm^{-2}]$ and S_T the soil thermal properties.

Through the latent heat flux and the precipitation the hydrologic and the atmospheric sub-systems are linked. However, these two sub-systems normally apply these two terms on different temporal and spatial scales. The atmospheric sub-system requires the surface energy balance to be specified in the order of minutes, whereas in the hydrologic sub-system knowledge of daily values generally may suffice. The lateral scale at which spatial variability plays a role in the hydrologic sub-system is much smaller than in the atmospheric sub-system. Thus, although linking of the two sub-systems looks trivial, having precipitation and latent heat flux in common, it is complex because of the problem of matching scales. In numerical climate modelling, spatial grid scales are typically in the order of 10^5 m and time steps are in the order of ten minutes. This requires the description of hydrologic fluxes to be adapted to these scales. In the next chapter regionalization of hydrologic processes is further discussed.

Notify in the water and energy balance equations 15.14 and 15.15 the critical role of soil moisture (see also Fig. 15.1). Proper soil moisture accounting in the soil and near the soil surface therefore is essential in developing parameterizations.

15.4 Regionalization

In Section 15.2 the terrestrial hydrological balance was described in one dimension and in the classical way by formulating the constituting processes. Solutions are then obtained by combining the individual processes. This approach might also be applicable in describing large areas. In that case the area is assumed to be divided in a number of small, independent sub-areas. The behaviour of the total area is then obtained by combination of the behaviour of the separate sub-areas. Referring to Fig. 15.4 this approach is mainly applicable to mechanisms, having relatively low degrees of complexity and randomness. In statistical sense a very large number of objects is considered. For a sufficient degree of randomness (see again Fig. 15.4), this approach will yield prediction of the average behaviour (Dooge, 1986). To a certain extent, basic description of hydrologic processes is possible as has been shown in Section 15.2, but the high degree of spatial variability, present at all scales larger than the pedon ($10^{-1} - 10^2\,m$), poses serious problems of parameter specification. Hydrologic processes can therefore be best located in the intermediate region of Fig. 15.4. In this

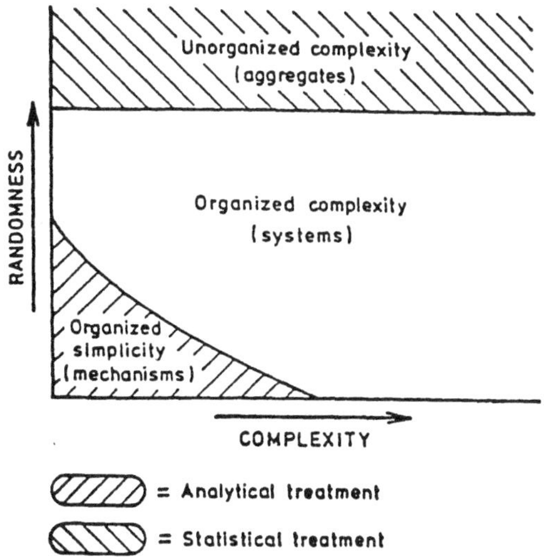

Figure 15.4: *Mechanisms, systems and aggregates (Weinberg, 1975).*

chapter different methods are discussed that apply to this intermediate region. Their features range in between the analytical and statistical approach.

15.4.1 Deterministic physical approach

When trying to model the terrestrial hydrological balance, it seems straightforward to develop numerical codes for all the components and link them together. Such a model does not need extensive calibration as the parameters may be measured directly or may be derived from literature. Also the physically based model can be used as a reference in judging the performance of more abstract, conceptual models.

An example of a physically based, distributed model on a typical catchment scale ($10^4 m$) is the quasi three-dimensional SHE (Système Hydrologique Européen) model (Abbott et al., 1986a,b). The SHE model (see Fig. 15.5) is physically based in the sense that the surface and subsurface transport of water is modelled by finite difference representations of the partial differential equations for mass and momentum. Some empirical equations are used from independent experimental research. Spatial distribution of the catchment parameters, rainfall input and hydrologic response is achieved in the horizontal through representation of the catchment by an orthogonal network of grid points, including the network of open channels, and in the vertical by a column of horizontal layers at each grid square. Each of the primary components of the land phase of the hydrological cycle is modelled - snowmelt, canopy interception, evapotranspiration, overland and channel flow and unsaturated and saturated sub-surface flow. It is assumed that flow in the unsaturated zone is essentially vertical and in the saturated zone essentially horizontal. The result is a model structure in which independent, one- dimensional, unsaturated flow columns of variable depths link an overlying two-dimensional overland flow component with an underlying two-dimensional saturated flow component. In principle

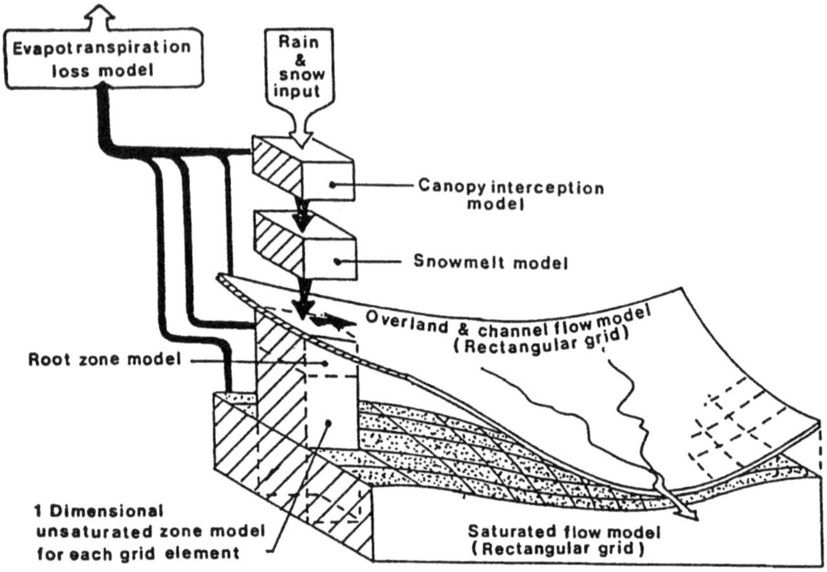

Figure 15.5: *Schematic representation of the SHE-model (Abbott et al., 1986b).*

the parameter values should not need to be calibrated, since they are supposed to be based on measurements. However, in practice calibration is likely to be required within certain limits. One good reason for this is that measured parameter values are often obtained at the point scale and may not be representative at the grid scale to which the parameters are applied. Applications of SHE are described by Bathurst (1986a) and Lumadjeng (1989). Bathurst (1986b) applied the SHE model to investigate the sensitivity of simulation results to the different parameters.

15.4.2 Inverse modelling approach

A recent possibility to simulate efficiently the vertical water movement of a large area is to use effective soil hydraulic properties in which the effects of spatial variability of the hydraulic functions are incorporated. A single run of a one-dimensional model based on Richards' equation (15.5 without sink term) would then suffice to simulate the hydrological behaviour of the entire area. If the soil hydraulic functions $\theta(h_m)$ and $K(h_m)$ are presented in a parametric form and areal soil water flow data are available, the inverse modelling technique (see e.g. Bard, 1974) may be employed to obtain the effective parameter values. Inverse modelling is a technique in which the parameters of a mathematical model are optimized to yield output that is as close as possible to a reference (in this case the areal soil water flow data). This is opposite to the practice in conventional modelling, where the parameter values are given and the model output is calculated.

A first numerical experiment was set up to test the method. The sub-surface hydrological behaviour of a bare area under wetting and drying conditions was generated by running a modified version of the one-dimensional model SWATRE (Wesseling et al., 1989) on 32 homogeneous columns of equal length. Each column had different soil hydraulic functions, taken from field measurements. The

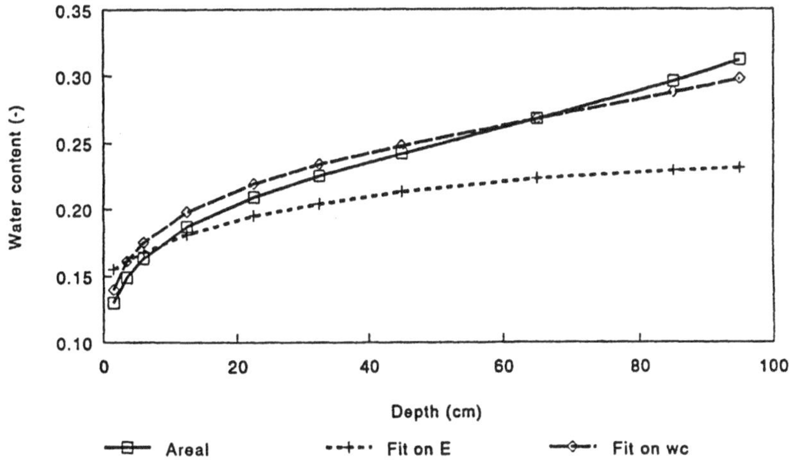

Figure 15.6: *Moisture profiles of the validation problem (at day 30).*

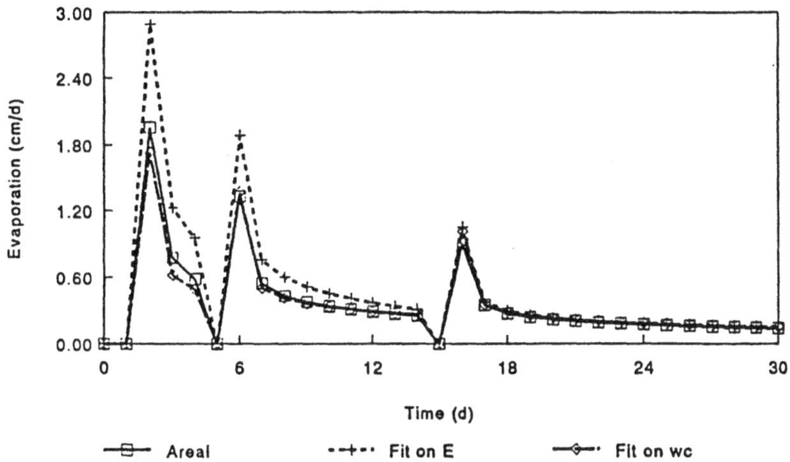

Figure 15.7: *Evaporation of the validation problem.*

inverse model SFIT (Kool and Parker, 1987) optimized effective soil hydraulic function by fitting on moisture profiles and evaporation, averaged over the thirty two columns. These effective parameters were then used to simulate the average behaviour of the columns for validation. The fit on the moisture profiles performed quite well as can be seen in Fig. 15.6 and 15.7. On the other hand, evaporation data only were insufficient to obtain a reliable set of effective parameters. For both fits, the confidence intervals of the optimized parameters of the $\theta(h_m)$- and $K(h_m)$-functions were very large.

However, some points should be kept in mind when following this approach. Richards' equation is able to describe the subtle variations in soil water behaviour. The soil is characterized by its hydraulic functions, which are incorporated in the equation. This inverse approach applies the one-dimensional version of Richards' equation far beyond the horizontal scale for which it was originally intended.

A more practical problem when applying the method is the establishment of a data set from which the inverse model can estimate the effective parameter set, especially since data regarding only the flux through the upper boundary are not sufficient as has been shown. Furthermore, it remains doubtful whether the effective hydraulic parameters are valid for circumstances that strongly differ from the conditions of the optimized problem. Further careful validation of the approach is therefore necessary.

15.4.3 Stochastic dynamic approach

As stated before at scales larger than the pedon scale, natural soil formations exhibit spatial variability in hydraulic properties and also in vegetation cover and topography. At the scales typical in climate modelling ($10^5 m$), it is practically impossible to represent these spatial features in a quasi three-dimensional model (refer to section 15.4.1). This very reason has led to the adoption of stochastic dynamic one-dimensional models in modelling land surface-atmosphere interactions. Given the stochastic properties of input parameters or variables, the stochastic dynamic method evaluates the stochastic properties of output generated by a deterministic flow model.

In general the main difference between various stochastic methods lies in the possibility to express the statistical properties of output variables in terms of statistical properties of input variables. By making assumptions about the stochastic processes, e.g. second order stationarity it is possible to find solutions of the stochastic differential equations that are representative for the system (De Marsily, 1986). The Monte Carlo-technique requires less strict assumptions. Knowing the probability distribution functions of the input parameters, realisations can be generated. Each single realisation yields corresponding values of output. Given a large number of realisations, it is possible to determine moments and distribution functions of the output. However, it is a numerical method that requires relatively much computing time.

An attractive way of introducing spatial variability in soil hydraulic functions is the concept of functional normalization (Tillotson, 1984). By this method the hydraulic functions at different points are related to a reference curve through single scaling factors. The different scaling factors and the reference curves for both functions are mathematically generated by least square methods. The scaling factors may be described by a probability density function. As an example of this method, results are shown in Fig. 15.8. Knowing the reference curve and the statistical distribution of the scaling factors, the Monte Carlo-technique can be used for generating soil hydraulic properties. Another approach is perturbation analysis (Mantoglou et al., 1987). Herein, local hydraulic properties are represented by a mean component and a second order stationary stochastic fluctuation. The stochastic properties of the output are then related to the stochastic properties of the fluctuations. Ünlü et al. (1990) presented a comparison of the last two methods. Climate models normally produce a grid-averaged precipitation rate. By distributing different precipitation intensities over the grid-area, a more realistic pattern can be achieved. This approach is followed by Entekhabi et al. (1989) who derived closed form analytical solutions for runoff, infiltration and evapotranspiration rates based on convolving two independent spatial distributions of soil moisture and precipitation intensity with a response function that represents the physical equations of the processes. Another example of stochastic dynamic modelling is by Famiglietti and Wood (1991), using a modified version

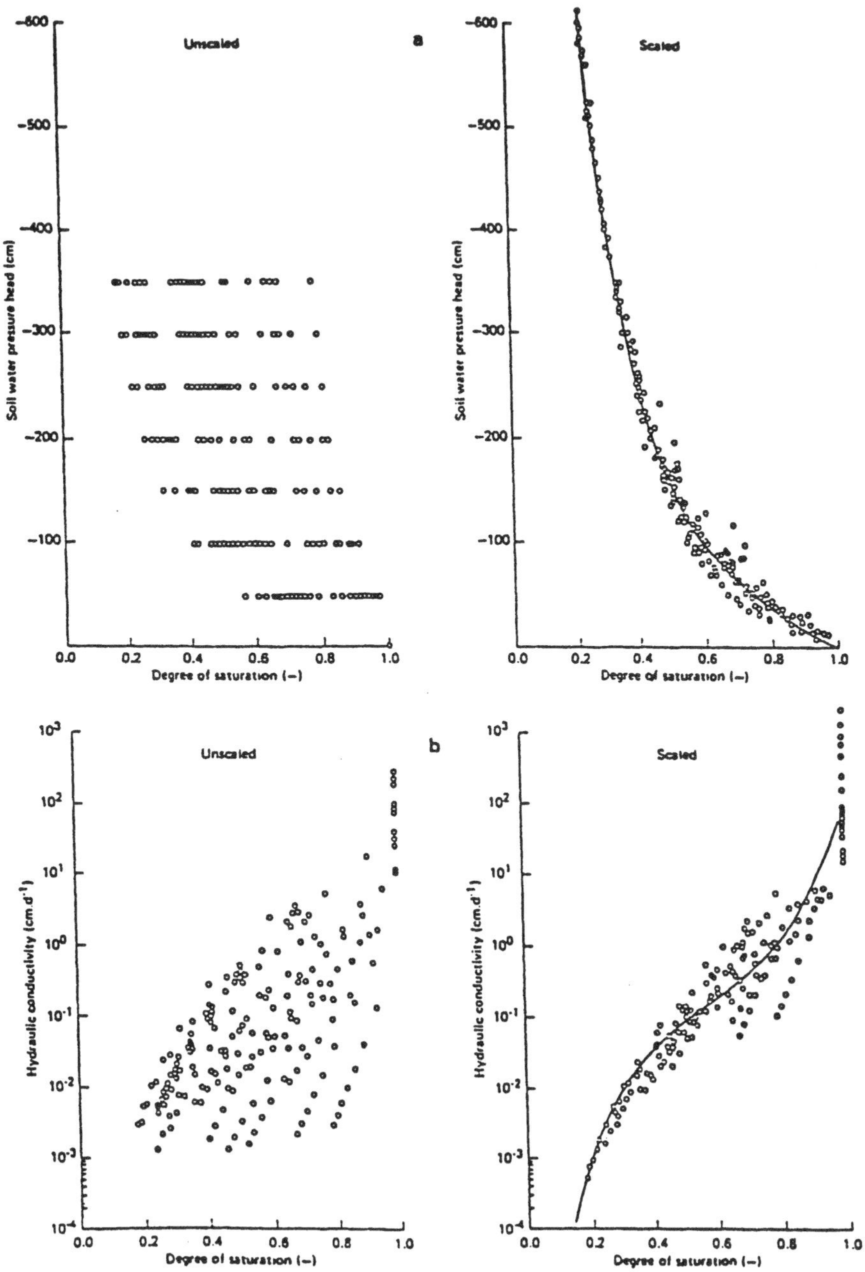

Figure 15.8: *Unscaled and scaled water retention data (a) and hydraulic conductivity data(b). An example.*

of the TOPMODEL (Beven et al., 1984). They introduce topographic and soil hydraulic variability through a probability density function of a, so called, topography-soil index. Also, a probability density function of precipitation, similar to Entekhabi et al. (1989) serves as input.

15.4.4 Parameterized approach

Following Smagorinsky (1982) and Brutsaert (1986), a parameterized model expresses the integrated effects of processes at scales smaller than the scale of interest in terms of variables that can be determined at that scale. This implies that parameterization is scale dependent. In this section, we discuss parameterized models that simplify the basic descriptions of hydrologic processes. The proper choice of the degree of parameterization depends on the problem under consideration and the available and measurable data (Brutsaert, 1986). Unbalance between these two poses either problems of parameter specification or causes oversimplification. Some examples are discussed below.

Bucket model

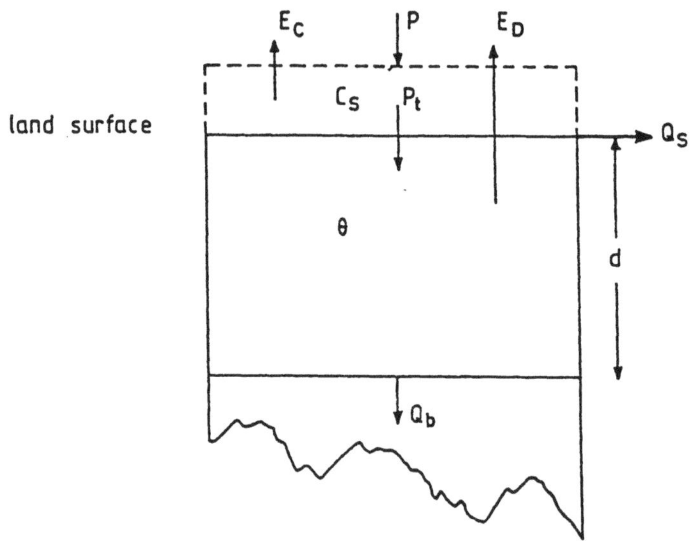

Figure 15.9: *Schematic representation of the bucket-model.*

The bucket-type soil moisture model described here is part of a one dimensional atmospheric model (Warrilow, 1986). The simple model describes the soil as a single rootzone layer underlying an interception layer due to the canopy, if present (Fig. 15.9). For each of these layers the water balance is calculated for each timestep. There is no flow-component.

During rainfall first the interception layer is filled of which the available storage depends on the ratio between the actual canopy storage C_s and the canopy capacity, C_s^{max}. Througfall either enters the rootzone or is partly discharged by surface runoff, depending on infiltration capacity, rainfall intensity and the type of rainfall, convective or advective. Referring to Fig. 15.9, the evapotranspiration is split up in evaporation of intercepted water from the wet canopy layer, E_c and transpiration from the dry canopy layer, including evaporation from the ground surface, E_t;

$$E_c = \alpha_v \left(\frac{C_s}{C_s^{\max}}\right) E_p(r_s = 0) \tag{15.16}$$

$$E_t = \left[1 - \alpha_v \left(\frac{C_s}{C_s^{\max}}\right)\right] \beta E_p(r_s \neq 0) \tag{15.17}$$

where α_v is the percentage of vegetated area $[-]$, E_p is the calculated potential evapotranspiration $[m\,s^{-1}]$, r_s is the surface resistance to transpiration $[s\,m^{-1}]$ and β a parameter indicating the evapotranspiration efficiency as a function of a critical moisture content.

In Eq. 15.16 the value of r_s is set to zero in the E_p expression of Eq. 15.11, indicating that water at the canopy can freely evaporate into the atmosphere. Critical moisture content is defined as the moisture content below which evapotranspiration is controlled by soil moisture. The rootzone can also loose water through the bottom layer by gravitational drainage, Q_b,

$$Q_b = K_s \left(\frac{\theta - \theta_{wp}}{\theta_s - \theta_{wp}}\right)^c \tag{15.18}$$

where K_s is the soil hydraulic conductivity at saturation $[m\,s^{-1}]$, θ_{wp} is the moisture content at wilting point, θ_s is the saturated water content and c a dimensionless constant, depending on soil type.

Deardorff model

The Deardorff model (Deardorff, 1977; Noilhan and Planton 1989) is made up of a canopy layer and a soil layer, including an embedded thin surface layer (see Fig. 15.10). The canopy layer may be considered as an interception bucket that can be filled by precipitation and emptied by evaporation. Precipitation in excess of the maximum canopy storage reaches the surface of the soil layer as throughfall. The soil layer gains water by throughfall precipitation and loses water by runoff (in case of saturation either of the surface or the total layer), soil evaporation and plant transpiration. Soil evaporation is assumed to occur in the surface layer only. There is no drainage through the bottom. The water-balance equations of the upper and total soil layer are:

$$\frac{\partial \theta_g}{\partial t} = \frac{C_1}{\rho_w d_g}(P_g - E_g) - \frac{C_2}{\tau}(\theta_g - \theta_2) \tag{15.19}$$

$$\frac{\partial \theta_2}{\partial t} = \frac{1}{\rho_w d_2}(P_g - E - g - E_t) \tag{15.20}$$

where C_1 and C_2 are constants depending on soil type, τ is a timescale and the other symbols are explained in Fig. 15.10. The first term on the right hand side of Eq. 15.19 can be regarded as a

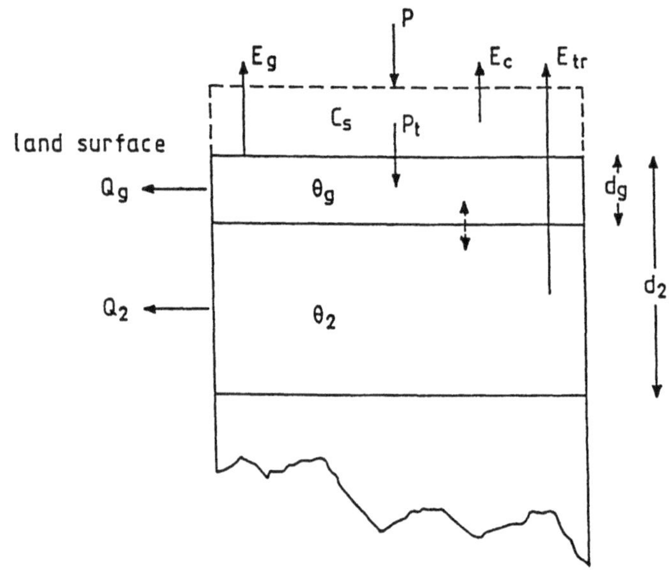

Figure 15.10: *Schematic representation of the 'Deardorff-model'.*

force term, representing the influence of surface atmospheric fluxes. The second term can then be considered as a restore term, that characterizes the diffusivity of water in the soil.

15.4.5 Lumped model approach

In terrestrial hydrology the catchment or basin acts as the natural entity to be studied. The main hydrological feature of a catchment is that the surplus of precipitated water will leave the area at a single point, the basin-outlet. This areally integrated discharge is a result of a variety of interconnected hydrologic processes, which are active at different areal and temporal scales and act in a non-linear way. It is generally accepted among hydrologists that a fully physical description of all the relevant hydrologic processes, acting at a catchment-scale, is impossible. This fact is due to the large variations in topography, land use and soil characteristics over relatively small distances, the complicated physical processes involved and to the coarse knowledge of the input, the rainfall process.

Knowing this, several more robust approaches of modelling catchment behaviour are available, the choice of which depends on our goals, data availability and the degree of parameterization we consider as being acceptable in our conceptualization of the total rainfall-runoff process (Brutsaert, 1986). Rather than describing the internal features the catchment is considered as a spatially lumped system that transfers temporally variable input into output. Modelling the system means a two step process. The first step consists of constructing a simple model that accounts for 'losses' (interception, surface storage, soil moisture replenishment) and converts measured rainfall into 'excess' or dischargeable rainfall. The second step concentrates on dynamic modelling of excess rainfall into discharge, that reaches the basin-outlet and can easily be measured. Hydrologists have always spent much attention to the second step for flood design-purposes in operational hydrology.

A typical and classical example of a lumped approach in hydrology is the Unit Hydrograph (UH)-model. It only converts excess rainfall into discharge. From a systems point of view the model represents a linear, time-invariant system (Dooge, 1973) for which the principle of superposition is valid. In other words, the catchment will always react in an identical way to a unit volume of excess rainfall, equally distributed over the basin and supplied over a short enough period in order not to exceed the characteristic reaction time of the basin (Raudkivi, 1979). Mathematically the UH-model can be written as:

$$Q(t) = \int_{0}^{t} u(t-\tau)i(\tau)d\tau \tag{15.21}$$

The process is visualized and the symbols are explained in Fig. 15.11.

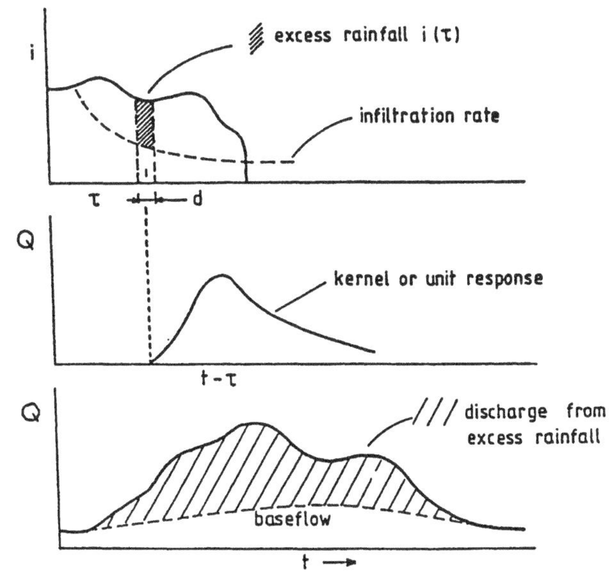

Figure 15.11: *Schematic representation of the Unit Hydrograph-model.*

Several conceptual linear models have been proposed to describe the unit response function for surface-runoff (Dooge, 1973; Clarke, 1973) as well as for groundwater discharge (Kraijenhoff van de Leur, 1958). Nowadays the UH-method is widely in use for operational water management purposes.

Generally, each model to describe the unit response function contains one or more unknown parameters which has or have to be optimized by using time series of input and output from a particular catchment. Rodriquez-Iturbe and Valdès (1979) introduced a probabilistic model-concept to derive the Instantaneous Unit Hydrograph (IUH and instantaneous because of the basins reaction to a unit pulse, Dirac function, of excess rainfall) and linked the model parameters to classical terrain characteristics of the basin as formulated already by Horton (1945). They called this the Geomorphologic IUH.

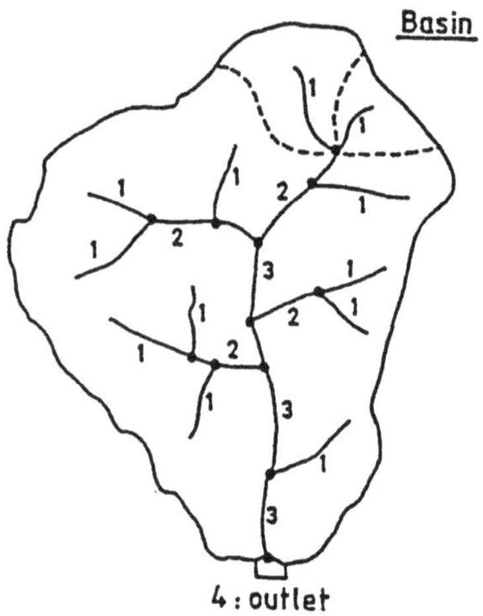

N_i = number of streamchannels of order i

A_i = average area of sub-basins of order i

L_i = average length of channels of order i

Bifurcation ratio, $R_B = N_i/(N_i+1)$

Streamlength ration, $R_L = L_i/(L_i+1)$

Area ratio, $R_A = A_i/(A_i+1)$

Figure 15.12: *Ordering of streamflow-network: An example.*

The importance of the GIUH-approach lies herein that it can be derived to a large extent from catchment features without using recorded input-output. For that, the catchment has to be internally ranked by its streamflow network and linked to Horton's numbers being the bifurcation ratio, the streamlength ratio and the stream areas ratio (see Fig. 15.12). In principle it should be possible to quantify these numbers by remote sensing techniques. Although the GIUH-approach is still not widely used, a good example of application is given by Corradini et al. (1986) for the Upper Tiber basin. Excess rainfall was estimated from measured rainfall by applying Philip's (1969) infiltration equation as a loss-term.

In general, the lumped model approach is feasible for catchments of different order of magnitude and the choice of the maximum scale depends on the typical, areal, rainfall variability and the measure of heterogeneity of general field characteristics like soils, types of vegetation and topography. A weakness of the approach lies in the first step, as mentioned. Dynamic processes like interception, infiltration and storage changes determine how rainfall intensities are converted into excess rainfall intensities. In contemporary conceptual modelling this parts of the rainfall-runoff process get much more attention.

In fact, appropriate modelling of the first step on a catchment scale opens the possibility of applying the (GI)UH-method for all seasons, not only for high flood periods. Good examples are the TOPMODEL (Wood et al., 1991) and the 'Wageningen' model (Warmerdam et al., 1982). And this in turn will open the possibility to use this type of models as land surface codes in atmospheric mesoscale models. The attractiveness over bucket-type models or other single column models can be expressed in terms of a better representation of topographic features and the dynamic behaviour of the runoff process at catchment scale, especially in hilly and mountainous areas, and in terms of model validation by comparing measured outflow from a catchment with calculated outflow.

15.5 Remote sensing

Except for the discharge term all traditional methods of observation to determine hydrological variables provide more or less local values. Examples are precipitation measurements by rain-gauges, micro-meteorological measurements (including the radiation budget) for determination of actual and potential evapotranspiration, soil moisture changes by neutron probe measurements and soil hydraulic properties by in-situ field methods or laboratory methods. This situation may be held responsible to a large extent for the poor calibration and validation of most of the hydrologic models in use, especially those developed for regional purposes.

What has remote sensing to offer to the hydrological community to improve the methods for observation? In principle, remote sensing (electro-magnetic radiance) can offer indirect information about the state of the surfaces over large areas and, in case of satellites, for regular time intervals. This ability to yield spatially integrated observations is very attractive for regional and large scale hydrological monitoring and model validation and looks most promising for use in global programs like GEWEX. However, the information gathered by remote sensing has always to be converted into the desired information by one or more algorithms. The conversion can either be achieved by empirical statistical means (regression) or by more theoretical means based on known physical laws. Inevitably, both approaches need validation by local, ground-based data. Thus, what causes poor validation of hydrological models equally creates problems of validation in remote sensing applications. Nevertheless remote sensing by satellite-mounted sensors may be considered as the only way for monitoring large scale processes.

Recently some excellent reviews have been presented by Schmugge & Becker (1991) and Choudhury (1991a,b) about the state-of-the-art of remote sensing research in hydrology.

From the viewpoint of terrestrial hydrology monitoring of regional and large scale water balances is of primary interest, preferably applied to catchment areas. Of second interest and in line with the first is the further development of parametric models that represent the dynamic behaviour of the terrestrial hydrologic processes at the selected scale.

Based on information sampled from literature an attempt is made to summarize in Tab. 15.2 all the variables, which are relevant for hydrology and which can or can not be detected by remote sensing sofar. Additional columns of information are added to the table.

It has to be notified that evapotranspiration can only be derived from remote sensing through monitoring of the surface energy balance with E_T as the remaining unknown quantity. Furthermore, additional information is needed from local meteorological stations.

Although a number of variables are detectable it has to be realized that the overall accuracy of any derived quantity is quite low in many cases and that still much research efforts (of which a synergetic approach is the new trend) have to be employed to reach a satisfactory situation for operational use.

Variable	Detectable	Spectral region	Main interfering factors	Literature
Rainfall	yes	visible/thermal IR.	only convective rainfall	Shih (1989)
Snowcover	yes	microwave	snow wetness grain size	Rott & Aschbacher (1989) Choudhury (1991a)
Surface soil & moisture	yes	microwave	vegetation optical roughness	Choudhury (1991a)
Subsurface s.m.	no	-	-	-
Discharge	no	-	-	-
Evapotranspiration	yes	visible/	atmospheric attenuation	Carlson (1991)
(including surface		near infrared/		Choudhury (1991b)
energy balance)		thermal infrared	aerodyn. surface roughness	Sugita (1990)
			plant cover %	V.d. Griend et al.(1991)
			near surface temp.	Menenti (1984)
			cloudiness	Bastiaanssen (1991)
Interception	no	-	-	-
Vegetation char.	yes	visible and near IR/ microwave	plant cover % vegetation type	Choudhury (1991b)
Topography	yes	visible	vegetation	Running et al.(1989)
Channel network config.	yes	visible/near IR	-	Menenti et al.(1989)
Soil hydraulic properties	no	-	-	-

Table 15.2: *Possibilities in Remote Sensing*

15.6 Observational studies and terrestrial hydrology

Over the last six years several large scale land surface/atmosphere field ex periments have been carried out successfully. Full descriptions of the objectives and programs have been published ('HAPEX-MOBILHY' by André et al., 1990; 'FIFE' by Sellers et al., 1990) or will soon be published ('EFEDA-Spain' by Bolle et al., 1992). The next experiment in row, 'HAPEX-Niger', has been planned for autumn 1992.

These field experiments cover in the first place intensive periods of micro-meteorological measurements in order to determine surface fluxes, of observations in the atmospheric boundary layer by different type of sensors (e.g. balloons, lidar, sodar) and include incidental airborne observations of fluxes and surface characteristics by aircraft- or satellite-mounted sensors. Hydrologic activities were and will be part of the experimental plans. They concentrate on monitoring of the soil moisture storage and soil moisture redistribution in the upper few meters of the soil (Goutorbe et al. 1989) and on characterizing soil properties in the experimental area.

One of the main objectives to undertake these experiments is to improve and calibrate one-dimensional Soil-Vegetation-Atmosphere Transfer Schemes (SVATS) used in meso- and large-scale atmospheric models. SVATS are basically doing the same as mechanistic one-dimensional unsaturated flow models, namely keeping the water balance and regulating external fluxes. They differ, however, in two aspects. SVATS are interactive with the atmosphere, where unsaturated flow models are forced by the atmosphere and secondly, SVATS pretend, unproved, to represent a horizontal scale far exceeding the accepted scales of mechanistic flow models.

Why deserve these field experiments much more input from hydrologists (and soil physicists) than they get until now? Several arguments can be given. First of all, if these experiments aim at improving SVATS then it can be equally true for (often weakly validated) mechanistic models provided that the additional needed data are collected (e.g. runoff, drainage, soil hydraulic properties, streamflow discharge). If mechanistic models can be well validated and spatial variability can be (stochastically) incorporated, these models may serve as calibration and validation tools for more simple SVATS. Secondly, hydrologists are familiar with experiments and observations at a catchment scale and over a longer period. However, direct determination of actual evapotranspiration is normally not included in their measuring programs and thus forms the missing link in closing the water balance by its separate terms. Large-scale field experiments, in contrast, stands out by their micro-meteorological and boundary layer activities to determine regional E_T.

Much information would be gained if hydrologists would select and equip one or more catchments inside the larger experimental area and the meteorologists would partly continue their efforts for a longer period (say a couple of years) in monitoring atmospheric fluxes. This would open excellent conditions for long-term validation on consistent behaviour of a variety of hydrological models and SVATS and of remote sensing applications at the meso-scale.

A third argument for strengthening the hydrological participation relates to widespread activities in modelling the hydrological consequences of climate change. We know from feedback-studies with climate models the importance of terrestrial hydrology, as modelled in SVATS, on the continental weather and climate. See for an excellent review the article of Avissar and Verstraete (1990). Outcomes of simulation studies of climate change by General Circulation Models and, more regionally, by embedded meso-scale models (Giorgi et al., 1991) are expressed, among others, in terms of shifts in precipitation, temperature, air humidity and net radiation regime. These outcomes are taken as forcing conditions in a regional hydrologic model other than the SVATS in order to simulate the hydrological consequences of climate change in terms of shifts in actual and potential evapotranspiration, in catchment-discharge and groundwater recharge.

In fact this procedure is only correct as far as the selected hydrologic model and the SVATS may be supposed to yield nearly identical outputs for the same area and under identical forcing. Otherwise the hydrological consequences will be biased by the procedure. This point has not yet been given attention by hydrologists. Further thoughts on the role of hydrology in climate studies and global scale activities like GEWEX and EOS are clearly expressed in recent articles of Shuttleworth (1991) and Eagleson (1991). Especially, the GEWEX-Continental- Scale International Project planned in the Mississippi basin may be considered as a testcase for intensive hydrological participation.

15.7 Some conclusions and recommendations

Hydrologic research has undergone dramatic changes in the field of topics to be studied over the last twenty or thirty years. It evolved from a typical design-aided science into a theoretical and applied science of all facets of water- and vapour transport under natural conditions. The familiar hydrologic time scale is a day or longer and the typical space scale is a basin or a field. However, present environmental problems are manifest at a global scale and this fact forces hydrology to observe and model hydrologic processes and balances over a much wider range of scales in space and time than traditionally.

Observations in terrestrial hydrology have mainly a local character. These local values are subject to non-deterministic, spatial variability. Extending the spatial scale in hydrologic modelling means to find the scale-adjusted parameterization and according parameter specification. This is still an unsolved problem.

Hydrologists should become more actively involved in multi-disciplinary large scale field experiments. Concerted action offers excellent opportunities for improving the calibration and validation of hydrologic models.

Remote sensing as a spatially integrating method of observation has not yet reached an operational level in hydrology. Radiative signals are the result of emission, reflectance and scattering of several surface characteristics and, by that, they can not be interpreted unambiguously. New directions in remote sensing research, like e.g. a synergetic approach, seem to be needed for further progress.

Hydrological models and SVATS have much in common. There is a need for several reasons to intercompare and validate both classes of models for identical areas and data sets. One consequence may be that a General Circulation Model or atmospheric meso-scale model needs different SVATS for different landscapes.

15.8 References

Abbott MB, Bathurst JC, Cunge JA, O'Connell PE, Rasmussen J (1986a) An introduction to the European Hydrological System - Systéme Hydrologique Européen, "SHE", 1: History and philosophy of a physically-based, distributed modelling system. *J Hydrol* 87: 45-59

Abbott MB, Bathurst JC, Cunge JA, O'Connell PE, Rasmussen J (1986b) An introduction to the European Hydrological System - Systéme Hydrologique Européen, "SHE", 2: Structure of a physically-based, distributed modelling system. *J Hydrol* 87: 61-77

André JC, Schmugge TJ, Perrier A, Goutorbe JP (1990) HAPEX-Mobilhy: Results from a large scale field experiment. In: O'Connell PE (ed) Recent advances in hydrological modelling. *NATO* 13-20

Avissar R, Verstraete MM (1990) The representation of continental surface processes in atmospheric models.*Rev Geophys* 28(1): 35-52

Bard Y (1974) Nonlinear parameter estimation. *Academic Press Inc* London

Bastiaansen WGM (1991) Derivation of areal soil physical data from satellite measurements. In: Kienitz, G et al. (eds) Hydrological interactions between atmosphere, soil and vegetation *IAHS publ* no. 204: 95-105

Bathurst JC (1986a) Physically-based distributed modelling of an upland catchment using the Système Hydrologique Européen. *J Hydrol* 87: 79-102

Bathurst JC (1986b) Sensitivity analysis of the Système Hydrologique Européen for an upland catchment. *J Hydrol* 87: 103-123

Belmans C, Wesseling JG, Feddes RA (1983) Simulation model of the water balance of a cropped soil providing different types of boundary conditions: SWATRE. *J Hydrol* 63: 271-286

Beven KJ, Kirkby MJ, Schofield N, Tagg AF (1984) Testing a physically-based flood forecasting model (TOPMODEL), for three U.K. catchments. *J Hydrol* 69: 119- 143

Bolle HJ et al. (1992) EFEDA-Spain: Description of a field experiment and some first results. *Ann Geophys* (in preparation)

Brutsaert W (1986) Catchment-Scale Evaporation and the Atmospheric Boundary Layer. *Wat Res Res* 22(9): 39S-45S

Carlson TN (1991) Recent advances in modelling the infrared temperature of vegetation canopies. In: Schmugge TJ, André JC (eds) Land Surface Evaporation, Measurements and Parameterization. *Springer Verlag* New York, 349-358

Choudhury BJ (1991a) Passive microwave remote sensing contribution to hydrological variables. *Surv Geophys* 12: 63-84

Choudhury BJ (1991b) Multispectral satellite data in the context of land surface heat balance. *Rev Geophys* 29(2), p 217-236

Clarke RT (1973) Mathematical models in hydrology. *FAO* Irrigation and Drainage Paper no. 19

Corradini C, Melone F, Ubertini L, Singh VP (1986) Geomorphic approach to synthesis of direct runoff hydrograph from the Upper Tiber River Basin, Italy. In: Gupta VK et al. (eds) Scale problem[illegible] drology; Runoff generation and basin response. *Reidel Publ Comp* Dordtrecht, 57-80

[illegible] (1977) A parameterization of ground surface moisture content for use in atmospheric [illegible]. *J Appl Meteor* 16: 1182-1185

De Marsily G (1986) Quantitative Hydrogeology. *Academic Press Inc* San Diego, California

Dooge JCI (1973) Linear theory of hydrologic systems. *Agr Res* Serie USDA, Techn. Bull. no. 1468

Dooge JCI (1986) Looking for Hydrologic Law. *Wat Res Res* 22(9): 46S-58S

Dooge JCI (1989) The Role of the Hydrological Cycle in Climate. In: Berger et al. (eds) Climate and Geo-Sciences. *Kluwer*, Dordtrecht, 355-366

Eagleson PS (1991) Hydrologic science: A distinct Geoscience. *Rev Geophys* 29(2): 237-248

Entekhabi D, Eagleson PS (1989) Land surface parameterization for atmospheric General Circulation Models including subgrid scale spatial variability. *J Clim* 2: 816-831

Famiglietti JS, Wood EF (1991) Evapotranspiration and runoff from large land areas: Land surface hydrology for atmospheric general circulation models. In: Wood EF (ed) Land Surface - Atmosphere interactions for climate modeling. *Kluwer Acad Publ* Dordtrecht, 179-204

Feddes RA (1981) Water use models for assessing root zone modification. In: Modifying the plant root environment.*Monograph 4* ASAE, St. Joseph, 347-390

Feddes RA, Kabat P, Van Bakel PJT, Halbertsma J (1988) Modelling soil water dynamics in the unsaturated zone, state of art. *J Hydrol* 100 (special issue) 69-111

Feddes RA, Kowalik PJ, Zaradny H (1978) Simulation model of the water balance of a cropped soil. *Simulation Monograph* PUDOC, Wageningen

Giorgi F, Mearns LO (1991) Approaches to the simulation of regional climate change: A review. *Rev Geophys* 28(2): 191-216

Goutorbe JP, Noilhan J, Valancogne C, Cuenca RH (1989) Soil moisture variation during HAPEX-Mobilhy. *Ann Geophys* 7(4): 415-426

Horton RE (1945) Erosional development of streams and their drainage basins; a hydrophysical approach to quantitative morphology. *Geol Soc Am Bull* 56: 275- 370

Kool JB, Parker JC (1987) Estimating soil hydraulic properties from transient flow experiments: SFIT user's guide. Dept. of Soil and Environmental sciences, *Virginia Polytechnic Institute and State University* Blacksburg, Virginia

Kraijenhoff van de Leur DA (1958) A study of non-steady groundwater flow with special reference to a reservoir coefficient. *Ingenieur* 70 (19): 87-94

Lumadjeng HS (1989) Modelling the hydrological response to human activities. In: Flow regimes from experimental and network data (FREND). *Institute of Hydrology* Wallingford, UK

Mantoglou A, Gelhar LW (1987) Large-scale transient unsaturated flow. *Wat Res Res* 23(1): 37-46

Menenti M (1984) Physical aspects and determination of evaporation in deserts applying remote sensing techniques. Ph.D. Thesis and Report 10 (special issue), *The Winand Staring Centre* Wageningen

Menenti M, Visser TNM, Morabito JA, Drovandi A (1989) Appraisal of irrigation performance with satellite data and georeferences information. In: Rydzewsky JR, Ward K (eds) Irrigation Theory and Practice, *Pentech Press* London, 785-801

Monteith JL (1965) Evaporation and environment. In: The state and movement of water in living organisms. *19th Symp Soc Exp Biol* 205-234

Noilhan J, Planton S (1989) A simple parameterization of land surface processes for meteorological models. *Mon Wea Rev* 117: 536-549

Philip JR (1969) Theory of infiltration. In: *Advances of Hydroscience* vol. 5. 216- 296

Raudkivi AJ (1979) Hydrology, an advanced introduction to hydrological processes and modelling. *Pergamon Press*

Ritchie JT (1972) A model for predicting evaporation from a crop with incomplete cover. *Wat Res Res* 8(5): 1204-1213

Rodriquez-Iturbe J, Valdés JB (1979) The geomorphologic structure of hydrologic response. *Wat Res Res* 15(6): 1409-1420

Rott H, Aschbacher J (1989) On the use of satellite microwave radiometers for large-scale hydrology. In: Rango A (ed) Remote sensing and large-scale global processes. *IAHS Publ* no. 186: 21-31

Rowntree PR (1984) Review of General Circulation Models as a basis for predicting the effects of vegetation change on climate. In: Reynolds ERC, Thompson (eds) Forests, Climate and Hydrology: regional impacts. *United Nations University* 1988, Oxford, 162-196

Running SW, Nemani R, Peterson DL, Band LE, Potts DF, Preice LL, Spanner MA (1989) Mapping regional forest evapotranspiration and photosynthesis by coupling satellite data with eco-system simulation. *Ecology* 70: 1090-1111

Schmugge TJ, Becker F (1991) Remote sensing observations for the monitoring of land surface fluxes and water budgets. In: Schmugge TJ, André JC (eds) Land Surface Evaporation, Measurement and Parameterization. *Springer Verlag* 337- 347.

Sellers P et al. (1990) First ISLSCP field experiment: Experiment, execution and preliminary analysis. In: O'Conell PE (ed) Recent advances in hydrological modelling. *NATO* 49-58.

Shih S (1989) Potential application of satellite data for rainfall estimation. In: Rango A (ed) Remote sensing and large-scale global processes. *IAHS Publ* no. 186: 97-104

Shuttleworth WJ (1991) The role of hydrology in global science. In: Kienitz, G et al. (eds) Hydrological interactions between atmosphere, soil and vegetation *IAHS Publ* no. 204: 361-375

Smagorinsky J (1982) Large-scale climate modelling and small-scale physical processes. In: Eagleson PS (ed) Land surface processes in atmospheric general circulation models. *Cambridge University Press* Cambridge

Sugita M, Brutsaert W (1990) Regional surface fluxes from remotely sensed skin temperature and lower boundary layer measurements. *Wat Res Res* 26(12): 2937-2944

Ten Berge H (1986) Heat and water transfer at the bare soil surface. Ph.D. Thesis. *Agricultural University Wageningen* 214

Tillotson PM, Nielsen DR (1984) Scale factors in soil science. *Soil Sci Soc Am J* 48(5): 953-959

Ünlü K, Nielsen DR, Biggar JW (1990) Stochastic analysis of unsaturated flow: One-dimensional Monte Carlo simulations and comparisons with spectral perturbation analysis and field observation. *Wat Res Res* 26(9): 2207-2218

Van de Griend A, Owe M, Groen M, Stoll MP (1991) Measurement and spatial variation of TIR-surface emissivity in a savanna environment. *Wat Res Res* 27(3): 371-379

Warrilow DA (1986) Indications of the sensitivity of European climate to land use variations using a one-dimensional model. In: *Proc ISLSCP Conference* Rome, Italy, 2-6 December 1985

Weinberg GM (1975) An introduction to general systems thinking. *Wiley- Interscience* New York

Wood EF, Sivapalan M, Beven KJ (1990) Similarity and scale in catchment storm response. *Rev Geophys* 28(1): 1-18

Warmerdam PMM, Stricker JNM (1982) Estimation of the water balance in the Hupselse Beek basin over a period of three years and a first effort to simulate the rainfall-runoff process for a complete year. In: *Proc of the Int Symp on Hydrological Research Basins and their use in Water Resources Planning* Bern 1982, 379-388

Wesseling JG, Kabat P, Van Den Broek BJ, Feddes RA (1989) Instructions for input. SWACROP. *The Winand Staring Centre* Wageningen

Appendix A

DATABASES FOR GEWEX RESEARCH

Jeff Dozier
Laboratory for Atmospheres
NASA-Goddard Space Flight Center
Greenbelt, MD 20771
USA

A.1 Introduction

Sequoia 2000 is a project to design a next-generation information system for accessing, archiving, distributing, managing, and visualizing data for global change research. Funded by the Digital Equipment Corporation, it has investigators from computer science and Earth science departments on five campuses of the University of California.

A.2 Motivation and objectives for Sequoia 2000

Global change researchers, particularly users of satellite remote sensing data, are seriously impeded by inadequate information systems. Users of remote sensing data need substantial knowledge about electromagnetic properties of the surface and atmosphere, and an ability to decipher arcane data formats. Data-intensive research in global change is also impeded by the immaturity of algorithms for estimating geophysical and biological variables from measurements of electromagnetic radiation from aircraft or spacecraft. Furthermore, agencies responsible for data focused too much on data acquisition, rather than storage, management, and distribution, hence the phrase *write-only archive* describes the status of many terabytes of historical satellite data (Dutton, 1990). The study of global change has suffered because many of its brightest, most innovative practitioners have avoided the analysis of data and the use of data to verify and calibrate their models.

NATO ASI Series, Vol. I 5
Energy and Water Cycles in the Climate System
Edited by E. Raschke and D. Jacob

Hence the major objective of Sequoia 2000 is to develop an improved data and *information* system for global change researchers that fosters synergistic interactions between observations and models and enables and encourages interdisciplinary research.

A.3 Earth Observing System (EOS)

NASA's part of Global Change Research Program is the Earth Observing System, a long-term project extending through 2013 (Dozier, 1990). Precursor missions began with the launch of the Upper Atmosphere Research Satellite (UARS) in 1991. The core of the EOS program consists of 6 satellites with about 20 sensors launched between 1998–2002, with instrument teams to develop scientific products (Moore et al., 1991). The volume of raw data and scientific products will exceed 1 terabyte per day by 2000. There are 29 interdisciplinary investigations—one Sequoia co-PI[1] is also an interdisciplinary PI—and 570 named co-investigators, including one Sequoia co-I (NASA, 1991). The total user population of EOSDIS is estimated at 10,000.

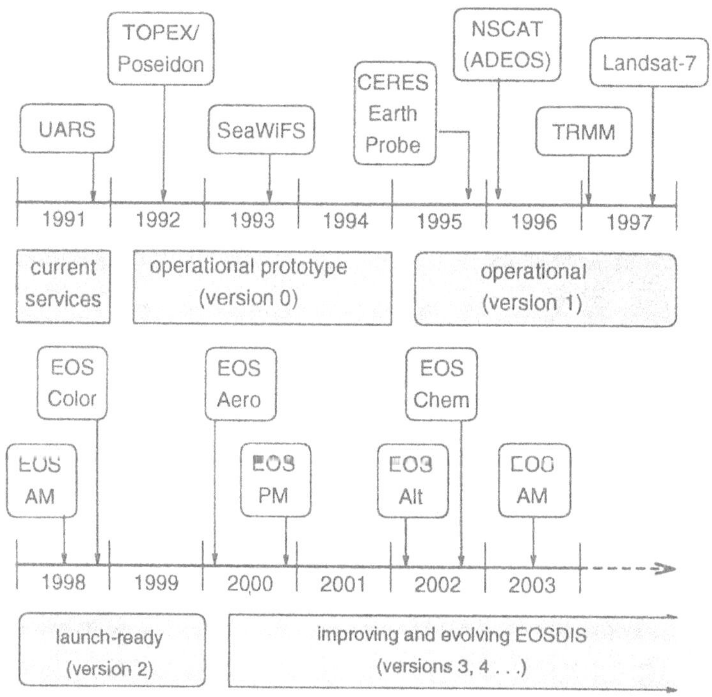

Figure A.1: *Schedule of U.S. Satellite Launches and EOSDIS Versions*

Sequoia 2000 is a research project linking computer science and global change research (Stonebraker, 1992). The EOS Data and Information System (EOSDIS) includes operational system and next-generation prototypes, but because of their requirement to deliver soon an operational system, the

[1]PI – Principle Investigator

prototypes are conservative. EOSDIS Version 0, providing access to existing operational satellite data, is scheduled for availability to users in 1994 (Dozier and Ramapriyan, 1994). Features to be incorporated include:

- interoperability between EOS archive centers;
- science processing library;
- ingestion of Pathfinder data sets (scientific products developed from current operational sensors).

Prototypes developed in Sequoia 2000, described in Section A.4, will be available for incorporation into EOSDIS.

A.4 Issues in information systems for global change

A.4.1 Data base management

Current data base management systems are too primitive for global change research, hence they have not been adopted by the science community to the extent, for example, that modern numerical methods are incorporated in scientific models. Existing data base management software is typically designed for industrial applications with a huge number of short records, but where the records are characters or numbers. Queries on mature systems are usually optimized.

However, the paucity of data types and the absence of geographic queries reduce the usefulness of common data base management systems. Earth system scientists want to query and analyze:

- raster data (e.g. satellite images, digital elevation grids)
- vectors
  - polygons (e.g. drainage basin boundaries)
  - directed graphs (e.g. river networks)
  - simple vectors (e.g. profiles of atmospheric humidity)
- point data (e.g. river discharge records)
- text data (e.g. algorithm descriptions, processing histories)

The challenge faced by Sequoia 2000 is to develop or modify a data base management system that satisfies needs for remote sensing, geographic information systems, diverse data types, and operations. These require different query indexing and access methods.

A.4.2 Physical data organization

Scientific problems faced by global change scientists require fast network access to huge amounts of distributed, heterogeneous data. The volumes are large — stored satellite data, derived products, and model output will add up to 500 terabytes/year by 2000 — and the objects themselves are large. For example a single Landsat Thematic Mapper image is 300 megabytes.

The challenge faced by Sequoia 2000 is to develop:

- a file system and database management system that understand distributed, tertiary memory;
- a data base management system that efficiently indexes and accesses large objects (including clever abstracts);
- a system that optimizes performance both for users' models and users' queries.

A.4.3 Sample queries

Some sample queries illustrate the problems we face:

Raster queries

- "Select all day-time AVHRR images for the Southern Sierra Nevada between October 1991 and June 1992 and sort chronologically."
- "Select
  - band 1 $\geq$ shadow threshold (anything less is not snow)
  - band 5 $\leq$ cloud threshold (gets rid of clouds)
  - $\frac{\text{band 4} - \text{band 5}}{\text{band 4} + \text{band 5}}$ [snow grain-size index]

  for Landsat Thematic Mapper image covering Bubbs Creek area nearest to April 1, 1992."

Directed-graph queries

- "Identify all reaches of the San Joaquin River above Friant Dam."

Polygon queries

- "Show land use and land cover within 500 km of Ürümqi."

Point queries

- "Show all snow courses in the Kern River basin and tabulate their snow accumulation in April 1992."

Joins between types

- "Show ocean phytoplankton concentration underneath the ozone hole in an ice-free area for dates nearest October 15, 1991."
- "Show all snow courses in the Kern River basin having more than 500 *mm* of snow water equivalence in April 1992."

A.4.4 Development of applications

Science software must interface with the data base management system. Model codes are created by Sequoia scientists. They require:

- interface between algorithms and data base management system;
- libraries for common functions;
- easy import of user-supplied algorithms.

The challenge is to provide a system whereby ideas are more easily translated into code that can access the data. Access to data should be provided through the data base management system, instead of through files. The DBMS should be designed for easy "recipe management" (e.g. boxes and arrows paradigm).

A.4.5 Remote visualization

We remotely visualize large data sets stored on a multi-terabyte storage system. We render data on local workstations so that we can: browse through many compressed images; manage, use, and manipulate large data sets through a data base management system; and use "computational steering" to guide models.

Challenges for the visualization system are:

- it must handle large objects (single images of several hundred megabytes);
- it must exploit the capabilities of a distributed, heterogeneous computing environment, including workstations, vector and parallel supercomputers, and networks;

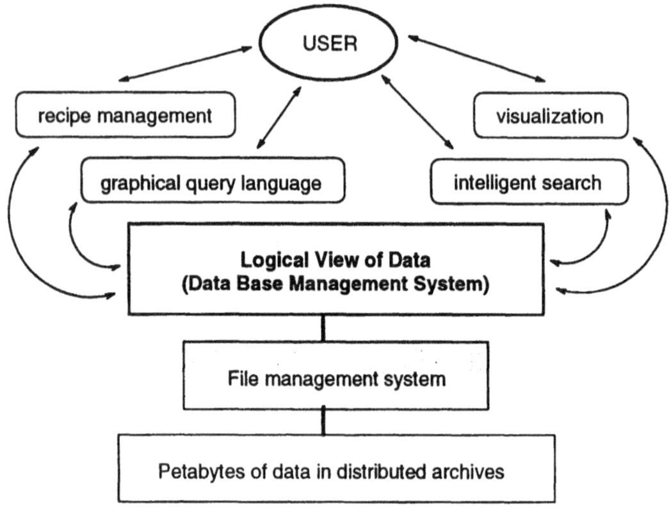

Figure A.2: *"DBMS-centric" View of Scientists' Interaction with Data*

- it must provide interactive I/O with models, so that changes in model output, caused by interactively changing parameters, are immediately displayed;
- the data base must be queried using graphs, maps, and images (as well as text).

To support these functions, we couple the visualization packages (we support IDL, AVS, and Khoros) to the data base management system (Postgres). We provide interfaces to a wide variety of data sets, and output to a spectrum of hard-copy devices. Also coupled to the data base system are a graphical query language and a recipe manager.

A.4.6 Networks

Current applications in Earth System science bring networks to their knees. The success of a system to provide access to many terabytes of data in distributed archives hinges on networks. Sequoia 2000 researchers will be connected by a 45 megabit/sec network.

The challenges faced by the network component are to develop new protocols that address needs for particular applications, e.g. guaranteed delivery for reserved periods. Moreover, the operating system must not be the bottleneck; fast networks are expensive, and we want to make sure we use the band-width.

A.4.7 Data base administration

Scientists need to trust the functions for

- integrity (e.g. time variation of calibration coefficients);
- concurrency and triggers (processing of scientific products depends on availability of input data);
- version control and tracing (algorithms change).
- distribution, replication, and backup.

The challenge is that the system must provide "data lineage," a reproducible audit trail for every item. It must also provide information about algorithms and functions, as well as data objects, and must prevent invalid operations on data.

A.4.8 Evolution

As Sequoia 2000 develops, it must not build in obsolesence. Requirements are driven not only by changing computing technology (processing, storage, networks, software), but also by changing science (knowledge and algorithms) and changing scientists' behavior and familiarity with computing.

Sequoia 2000 (and EOSDIS to a much larger clientele) provides operational services, yet must allow for evolution. It must incorporate advances in computing technology and not lock into obselesence. Information management must allow for insertion of next-generation DBMS. Implementation of parallel processing hardware might be constrained by difficulty of porting existing software. Both near-term and long-term goals must be accommodated.

Acknowledgements

This research was sponsored by the University of California and the Digital Equipment Corporation under Digital's flagship research project Sequoia 2000. Other government and industrial partners include the California Department of Water Resources, the U.S. Geological Survey, ESL, Hewlett-Packard, Hughes Aircraft Company, MCI, Metrum, Research Systems, and Science Applications International. The work on EOS by J. Dozier is supported by the National Aeronautics and Space Administration.

A.5 References

Committee on Earth and Environmental Sciences (1991) Our Changing Planet: The FY 1992 U.S. Global Change Research Program. *Office of Science and Technology Policy*, Washington, D.C.

Dozier J (1990) Looking ahead to EOS: the Earth Observing System. *Computers in Physics*, 4: 248–259.

Dozier J and Ramapriyan HK (1991) Planning for the EOS Data and Information System (EOSDIS). *In: Corell, R W and Anderson, P A, editors, Global Environmental Change*, NATO ASI Series I: Global Environmental Change, Vol. 1, pp. 155–180, *Springer-Verlag*, Berlin.

Dutton JA (1990) The EOS Data and Information System: concepts for design. *IEEE Transactions on Geoscience and Remote Sensing*, 27: 109–116.

Houghton JT Jenkins GJ and Ephraums JE, editors (1990) Climate Change: The IPCC Scientific Assessment. *Cambridge University Press*, Cambridge.

Moore B, Dozier J, Abbott MR, Butler DM, Schimel D, and Schoeberl MR (1991) The restructured Earth Observing System: instrument recommendations. *Eos, Transactions, American Geophysical Union*, 72: 505, 510, 516.

NASA (1991) EOS Reference Handbook. *NASA Goddard Space Flight Center*, Greenbelt, MD.

Stonebraker M (1992) An overview of the Sequoia 2000 project. *In: Proceedings 1992 COMPCON Conference*, San Francisco.

US Committee for an International Geosphere-Biosphere Program (1986) Global Change in the Geosphere-Biosphere: Initial Priorities for an IGBP. *National Academy Press*, Washington, D.C.

Appendix B

Summary of Contributions by Participants of the NATO-ASI on Energy and Water Cycles in the Climate System

Markus Quante
GKSS Research Center
D-2054 Geesthacht
Germany

Jennifer Francis
University of Washington
Seattle, WA 98105
USA

Alev Temel
Bogazigi University
Istanbul, 80815
Turkey

B.1 Introduction

The Advanced Study Institute on **Energy and Water Cycles in the Climate System** attempted to bring together young scientists who already work or start to work in this active and important area of climate research. All participants were invited to prepare posters which document their own current research work. Many students accepted the invitation and 30 poster contributions were exhibited during the two weeks at Glücksburg, covering many subjects closely related to GEWEX. Lively discussions occurred around the poster area during coffee and lunch breaks, demonstrating the great interest in the displayed results. In order to encourage these discussions special poster sessions were arranged, during which the authors had the opportunity to give an oral presentation of their work in front of the plenum and to answer questions of general interest.

The 30 posters were reviewed and grouped under five major topics, according to their contents and methods that were applied. During five late afternoon sessions, chaired by some of the lecturers, the posters were presented in a short talk followed by questions and discussions. The organization of the sessions followed closely the topics addressed in the primary lectures. The principle authors and the

NATO ASI Series, Vol. I 5
Energy and Water Cycles in the Climate System
Edited by E. Raschke and D. Jacob

titles of their posters and presentations are listed in Tables B.1 – B.5. The five selected topics were titled:

- general circulation and energetics,
- clouds and radiation,
- ocean-atmosphere interaction,
- soil and vegetation,
- mesoscale phenomena.

Overall it can be said that in addition to enlighting lectures, the poster and oral presentations by participants contributed substantially to the scientific success of this ASI on **Energy and Water Cycles in the Climate System**. They also facilitated many contacts between research groups. Scientific and social exchange was one of the major goals of the meeting.

In order to document the poster sessions we briefly summarize the presentations in the following paragraphs. More detailed information can be obtained from the individual authors.

B.2 General Circulation and Energetics

Several studies related to this broad subject were presented in student posters. P. Siegmund investigates the generation of available potential energy in the atmosphere by comparing various components of the energy cycle computed from numerical weather prediction model output to quantities retrieved from satellite measurements. Another important and poorly known component of the energy cycle is the exchange of water between the surface and the atmosphere. Using a global set of radiosonde profiles, D. Gaffen studies the annual and long-term variations in tropospheric water vapor. She found that precipitable water is a function of relative humidity (rather than the absolute water content) in the tropics and that large amplitude variations in the relative humidity occur in the tropics and mid-latitudes. Passive satellite-borne microwave data from the instruments SMMR and SSMI are used by M. Schrader to calculate latent heat fluxes over the ocean. His results will in future be validated with measurements obtained from field experiments in the North Atlantic. The water cycle over continents, particularly the causes of droughts and wet cycles, is under investigation by K. Brubaker.

The contribution of *recycled moisture* (moisture that falls on land, re-evaporates, and precipitates again) to the total precipitation appears to be important to the development of dry and wet epochs in certain regions. Another aspect of precipitation is the focus of work by C. Cosgrove. She analyses radar data to study temporal and spatial rainfall patterns over Florida, in an effort to better understand extreme precipitation events, to provide validation data for the Tropical Rainfall Monitoring Mission (TRMM), and to improve operational forecasts.

The increasing interest on the role of troposphere-stratosphere-exchange processes in climate and ozone research motivated S. Schultka to numerically study the penetration of air parcels through the tropopause. She uses a 1-dimensional linear parcel model and for comparison a 3-dimensional non-hydrostatic mesoscale model (MESOSCOP) to simulate the overshooting turret as a rising air parcel to investigate dynamical aspects of the phenomenon. The results show the complex 3-dimensional structure of the overshooting and mixing processes, which call for the use of an adequate dynamical model for this type of studies. Finally in this section A. Swamy investigates the deposition of pollution products by precipitation, particularly in context of the Indian Monsoon. His results show that the southwest monsoon brings large concentrations of atmospheric ions into India.

P. SIEGMUND, Royal Netherlands Meteorological Institute, De Bilt, Netherlands:
The generation of available potential energy in the atmosphere.

D. GAFFEN, NOAA, Silver Spring, U.S.A.:
The annual cycle of tropospheric water vapor.

M. SCHRADER, Institut fr Meereskunde, Kiel, Germany:
Estimation of evaporation by satellites with microwaves.

K.L. BRUBAKER, D. ENTEKHABI, P.S. EAGLESON, MIT, Cambridge, U.S.A.:
Recycling of continental precipitation estimated from atmospheric data.

C.M. COSGROVE, University of Virginia, Charlotteville, U.S.A.:
Temporal and spatial distribution of rainfall.

S. SCHULTKA, T. HAUF, DLR, Oberpfaffenhofen, Germany:
Three dimensional simulation of an air parcel penetrating into the stratosphere.

A. NARAYAMA SWAMY, Andhra University, Visakhapatnam, India:
Atmospheric deposition from bulk precipitation at Visakhapatam, India.

Table B.1: *Poster contributions to* **General Circulation and Energetics.**

B.3 Clouds and Radiation

Clouds are an important modulator of the radiation budget of our planet and therefore directly affect the global energy cycle. Clouds are also an important component in the water cycle, they are responsible for vertical and horizontal transport of water and the development of precipitation in the atmosphere. Many of the ASI lectures and posters addressed clouds and radiation in either satellite or modelling studies or measurement programs. M. Rieland presented results of a major satellite project, the Earth Radiation Budget Experiment (ERBE). His special interest is the influence of clouds on the earth radiation budget. Using data for the period from February 1985 to January 1987 to evaluate the cloud forcing, he showed regional distributions, annual variation and hemispheric averages. One of the major results was the dominance of the albedo effect of the present cloud distribution.

Data products from another satellite project, the International Satellite Cloud Climatology Project (ISCCP), are used by C. Poetzsch-Heffter to calculate the influence of clouds on the radiation at the top of the atmosphere. Thirty-five cloud types are distinguished and the relative contribution of each class to the greenhouse effect of all clouds is calculated. High level, optically thick clouds showed the biggest contribution to the longwave cloud forcing. The zonal cloud albedo was estimated in a study introduced by P. Post. She took the opportunity to compare 2 total cloud amount data sets, which were available for the same time and spatial averaging scale, and looked for regional differences. At lower latitudes she found an annual cycle of the cloud albedo whereas this quantity was nearly constant for midlatitudes, a result she attributed to cloud microphysical processes and the redistribution of water between cloud layers.

In order to address the question, of whether bulk formulae yield sufficiently accurate long time series of net longwave radiation at sea from routine (ship log) data, F. Bignami compared such data sets with direct measurements gathered during ar esearch cruising in the Mediterranean Sea. The comparisons of the measured data with results of present bulk formulae showed differences of up to 40 %. In the future, new formulae will be developed for applications in regional climate studies, e.g. for the Mediterranean Sea area. In polar regions for example the role of sea ice in global weather and climate is not well known, primarily because of the lack of conventional data in high latitudes. J. Francis is addressing part of this problem by employing retrievals from the TOVS system to estimate air/ice radiative and turbulent energy budgets of the sea surface.

Three posters addressed the ice phase of water in the atmosphere. Using ray tracing techniques A. Macke calculated the scattering phase function of arbitrarily shaped ice particles. Particle shapes are taken both from ice crystal classifications and in situ observations during the International Cirrus Experiment (ICE). The spectral variability was discussed extensively. The comparison of different particle types shows that backscattering is a sensitive indicator for ice crystal type identification. M. Quante analyses aircraft measurements of microphysical and dynamical quantities in cirrus clouds during ICE, concentrating on the spatial variability. The observations show that cirrus is highly variable on the meso- and microscale. It often appears multilayered and is embedded in a complex flow field. During the same field campaign measurements also could be made in a contrail segment; some results were presented by F. Albers. In this encountered contrail many more small particles were measured compared to cirrus segments in the vicinity. The level of turbulence was about 2 orders of magnitude higher than in the surrounding clear air. Contrail studies are important with regard to the projected doubling of air traffic up to the year 2005.

In order to better understand radiative properties of polluted cloud particles, F. Müller investigates the collision aerosol scavenging. E.g. he hopes to reveal the dependence of arerosol uptake, multiphase chemistry, and radiative properties of collector particles on the spectral distribution of aerosols and collector particles. R. Boime demonstrated how ozone concentrations can be retrieved from satellite measurements (AVHRR) using the visible absorption band of O_3. The technique can only be used over Antarctica, where ozone is the only important absorber in the relevant channel. An advantage of using AVHRR data for ozone concentration determinations compared to other techniques is the relatively long time series of this data, which extends from 1974 to present.

M. RIELAND, S. EWALD, C. STANDFUSS, University of Hamburg, Hamburg, Germany:
The influence of clouds on the earth's radiation budget.

C. POETZSCH-HEFFTER, University of Kiel, Kiel, Germany:
Effects of clouds on the OLR calculated with the ISCCP cloud products.

P. POST, Tartu University, Tartu, Estonia:
An estimate of zonal cloud albedo.

F. BIGNAMI, R. SANTOLERI, M.E. SCHIANO, Instituto di Fisica dell'Atmosfera, Rome, Italy:
Net longwave radiation in the western Mediterranean Sea.

J. FRANCIS, University of Washington, Seattle, U.S.A.:
Surface energy budget in polar regions studied from space.

A. MACKE, GKSS Research Center, Geesthacht, Germany:
Scattering properties of atmospheric ice crystals.

M. QUANTE, P.R.A. BROWN *, P. UDELHOFEN, F. ALBERS, GKSS Research Center, Geesthacht, Germany, * Meteorological Research Flight, Farnborough, UK:
Spatial variability in cirrus: observations from ICE 1989.

F. ALBERS, M. QUANTE, E. RASCHKE, GKSS Research Center, Geesthacht, Germany:
Aircraft measurements in high altitude contrails during ICE 1989.

F. MÜLLER, Heinrich-Hertz Institut, Berlin, Germany:
Collision aerosol scavenging by different collector types.

R. BOIME, University of Washington, Seattle, U.S.A.:
Mapping Antarctic ozone from visible channel data.

Table B.2: *Poster contributions to* **Clouds and Radiation**.

B.4 Ocean-Atmosphere Interaction

A significant portion of this ASI was devoted to better understanding the role of oceans in climate, including topics in air-sea interaction and modelling techniques. Several posters were presented in this section that also address these issues.

L. Nicolini, for example, is studying the sensitivity of the ocean thermohaline circulation to pertubations in the climate. In particular he is interested in the response of Antarctic bottom water formation processes to changes in the surface energy budget. A related study was presented by A. Coward and S. Thompson, in which a group from the UK is developing a high-resolution model of the Southern Ocean. They hope to better understand the roles of eddies, topography and sea-ice in governing ocean circulation.

S. Peng has undertaken an investigation in the opposite side of the world: a correlation study of North Atlantic sea-surface temperatures (SSTs) and output from two major rivers in the former USSR that flow into the Arctic Ocean. She found that cold (warm) winter SSTs in the North Atlantic are correlated with large (small) runoff volume during the following summer. Also on a regional scale, A. Raabe is studying the mesoscale circulation near a coast line, and the behaviour of the atmospheric boundary layer near this surface-discontinuity. He is particularly interested in the Baltic Sea coast, where a regional study is planned as a part of GEWEX. Another aspect of this coastal zone is studied by T. Wolf. Using a wave model, he is investigating the effect of current patterns and wave activity on sediment transport.

L. NICOLINI, University of Pisa, Pisa, Italy:
The impact on climate of changes in the thermohaline circulation.

S. PENG, L. MYSAK, McGill University, Montreal, Canada:
A teleconnection study of interannual sea surface temperature fluctuations in the Northern North Atlantic and precipitation and runoff over western USSR.

A. RAABE, H.J. SCHÖNFELD, University of Leipzig, Leipzig, Germany:
Mesoscale model of the North Sea - Baltic sea saltwater inflow and the air-sea interaction near a coastline.

A. COWARD, S. THOMPSON, Institute of Oceanographic Science, Wormley, UK:
F.R.A.M. (Fine Resolution Antarctic Model).

T. WOLF, University of Leipzig, Leipzig, Germany:
Sea wave modelling in the Baltic Sea.

Table B.3: *Poster contributions to* **Ocean-Atmosphere Interaction**.

B.5 Soil and Vegetation

Recent GCM studies suggest that biology plays a crucial role in climate modulation and that present parametrization schemes are crude and highly variable between climate models. Four posters illustrated on-going studies to model or measure energy fluxes from various types of vegetation. Beginning with low agricultural plant covers, M.G. Inclan discussed a 20-layer land-atmosphere model (CUPID) that was used to compute moisture and heat fluxes from four different crops. She found large differences in fluxes from each of these vegetation types. In the future she plans to enhance the model to include a forest canopy.

F. Boseveld is developing a model of a mid-latitude coniferous forest. Ultimately he plans to investigate the effect of pollution on the vitality of a forest. Results from two observational studies were presented by J.I. Pitman. The first, conducted in the dry tropical forest of Thailand, was designed to investigate the correlation between stomatal conductance in the canopy and atmospheric forcing. Strong correlations were found between conductance and irradiance, temperature, humidity and soil water potential. The second study focuses on a plant species with an annual growth cycle (bracken fern). Results show that biomass production is proportional to the photosynthetic active radiation, thus leaf area and plant height are predictable from climate data.

Finally a study of marine vegatation was presented by S. Podewski. Using data obtained from drifting buoys that measure temperature and fluorescence through the euphotic zone, as well as the flux of organic matter into the deep ocean, the organic carbon cycle or *biological pump* in the ocean is studied.

M.G. INCLAN, DLR, Oberpfaffenhofen, Germany:
A multilayer vegetation model for estimating energy fluxes.

F. BOSEVELD, W. BOUTEN, F. NOPPERT, E.G. STEINGRÖVER, A. TIKTAK, Royal Netherlands Meteorological Institute, De Bilt, Netherlands:
Evapo-transpiration of a Douglas fir forest.

J.I. PITMAN, University of London, London, UK:
Stomata behavior of a tropical dry evergreen forest in Thailand.

R. PEINERT, S. PODEWSKI ET AL., University of Kiel, Kiel, Germany:
Short-term variability of chlorophyll distribution in relation to its vertical flux.

Table B.4: *Poster contributions to* **Soil and Vegetation**.

B.6 Mesoscale Phenomena

B. Geerts reports on observed thermodynamical and cloud microphysical parameters in a frontal rainband during the GALE experiment. He introduces a technique to combine a thermodynamical retrieval method from dual-Doppler radar data with a detailed bulk-parameterized microphysical model of cloud processes to derive estimates of the thermodynamical and microphysical structure of the rain band. An extensive experimental investigation of the development of a mixed layer in an urban area was described by S. Nuñez. The results allow the identification of mechanisms controlling the evolution of the daytime urban boundary layer and provide a data base for the validation and improvement of models used for urban air pollution studies.

Applying a non-hydrostatic mesoscale model (GESIMA), D. Jacob simulated a sea-breeze circulation over northern Germany. The applicability of the model and its cloud module to studying this phenomenon was demonstrated by a good agreement of the numerical results with observed data. The strong influence of convective cloud formation on the dynamics of the sea breeze circulation is one of the main results. S. Reyes reported on on-going studies related to atmospheric processes influencing the water reserve of Northern Mexico. The work is motivated by the rapidly increasing population on one hand, and the large variety of regional climates in that region on the other. The studies have an interdiciplinary character and combine many aspects of meteorology and ecology.

B. GEERTS, MacQuarie University, Melbourne, Australia:
The retrieval of thermodynamic and cloud microphysical variables from the airflow as observed by dual-Doppler radars in a narrow cold-frontal rainband (NCFR).

S. NUÑEZ, B. ARTINANO, CIEMAT, Madrid, Spain:
Experimental study on the mixed layer in an urban area.

W. KOCH, D. JACOB, D. EPPEL, GKSS Research Center, Geesthacht, Germany:
Land-See-Windzirkulation im norddeutschen Küstenbereich.

S. REYES, Group of Meteorology CICESE, San Isidro, U.S.A.:
Atmospheric studies over northern Mexico.

Table B.5: *Poster contributions to* **Mesoscale Phenomena**.

Index

GPSR Compliance
The European Union's (EU) General Product Safety Regulation (GPSR) is a set of rules that requires consumer products to be safe and our obligations to ensure this.

If you have any concerns about our products, you can contact us on

ProductSafety@springernature.com

In case Publisher is established outside the EU, the EU authorized representative is:

Springer Nature Customer Service Center GmbH
Europaplatz 3
69115 Heidelberg, Germany

www.ingramcontent.com/pod-product-compliance
Lightning Source LLC
Chambersburg PA
CBHW061152240926
51846CB00006B/996

9783642769597